W. Künzel A. Jensen (Eds.)

The Endocrine Control of the Fetus

Physiologic and Pathophysiologic Aspects

With 152 Figures and 18 Tables

Springer-Verlag Berlin Heidelberg New York
London Paris Tokyo

Professor Dr. Wolfgang Künzel
Universitäts-Frauenklinik
Klinikstraße 32
6300 Gießen, FRG

Professor Dr. Arne Jensen
Universitäts-Frauenklinik
Klinikstraße 32
6300 Gießen, FRG

ISBN-13: 978-3-642-72977-5   e-ISBN-13: 978-3-642-72975-1
DOI: 10.1007/978-3-642-72975-1

Library of Congress Cataloging-in-Publication Data
The Endocrine control of the fetus : physiologic and pathophysiologic aspects / W. Künzel,
A. Jensen (eds.) p. cm. Includes bibliographies and indexes.

1. Fetus--Physiology   2. Endocrine glands.   3. Fetus--Diseases--Endocrine
aspects.   4. Endocrinology, Comparative.   I. Künzel, W. (Wolfgang), 1936-   . II. Jensen,
A. (Arne), 1951- . [DNLM: 1. Endocrine Glands--physiology.   2. Fetal
Development.   3. Fetal Diseases--physiopathology.   4. Fetus--physiology.   WQ210.5 E558]
RG616.E53   1988   612'.647--dc19   88-18332

© Springer-Verlag Berlin Heidelberg 1988
Softcover reprint of the hardcover 1st edition 1988

Typesetting: Appl, Wemding
2123/3145-543210

Dedicated to K.-H. Wulf
on the occasion of his 60th anniversary

Preface

From the implantation of the fertilized ovum into the endometrium to the delivery of the baby, the growth of the fetus is embedded in a series of physiological interactions between the mother and the fetus aimed to guarantee normal fetal development.

Immunoregulatory mechanisms are the basic response in the mother in accepting the embryo as a transplant in a foreign environment. The transport of nutrients and oxygen to the fetus and the clearance of waste products from the fetus are achieved by an increase of blood flow to the pregnant uterus and by a rise of the fetal and placental circulation. This interaction relies upon the endocrine response of mother and fetus in the regulation of fetal, maternal and placental circulation.

This book originated from a conference held in Schloß Rauischholzhausen (the conference center of the Justus-Liebig-Universität Gießen, West Germany) from July 29th to August 1st, 1987. The meeting aimed to combine both aspects of fetal development: the physiological adaptations throughout pregnancy and the endocrine aspects involved. This book is organized to provide thoroughgoing information on the endocrine control of circulation with its humoral and neuronal mechanisms; it reviews fetal growth and fetal lung development, and gives an insight into the endocrine control of thermoregulation. Special chapters are devoted to the fetal carbohydrate metabolism and the mechanisms of parturition. Findings from both animal and human studies are integrated to give a comprehensive view of each topic and to provide mutual stimulation for experimental and clinical investigators.

I wish to thank the pharmaceutical corporation Nourypharma GmbH of Oberschleißheim, West Germany, for their generous support of this conference. The organizational responsibilities fell primarily on Prof. Dr. A. Jensen and Dr. Dr. Kirschbaum, with assistance from Karin Jakobi and Edith Bausch.

I particularly wish to thank Springer-Verlag Heidelberg for their courtesy and generosity in the process of seeing the manuscript through to press.

Gießen, May 1988   W. Künzel

Table of Contents

Endocrine Control of the Fetus – Basic Mechanisms

Endocrine Control of the Fetal Growth

Endocrine Aspects of Fetal Behaviour

Endocrine Control of Thermoregulation

Endocrine Control of the Fetal Carbohydrate Metabolism

Mechanisms of Parturition

Index of Authors

The page numbers indicate the article to which each author has contributed.

Growth Hormone and Fetal Growth: Historical Perspective*

G. D. THORBURN[1], C. A. BROWNE, A. W. HEY, S. MESIANO and I. R. YOUNG

In this paper, we will be describing the development of our knowledge of the roles of pituitary growth hormone in the regulation of fetal growth in the sheep. The approach will be essentially historical, but since the role of history is to illuminate the present and to prepare us for the future, we beg no forgiveness for devoting a part of this review to a discussion of the most recent findings and hypotheses in this area.

Growth Hormone in the Fetus

While growth hormone (GH) had been identified in the pituitary of the fetal lamb as early as 50 days of gestation (Stokes and Bode 1968), there was little known about GH in fetal plasma. The plasma concentrations of GH in newborn lambs had been measured and found to be frequently very high (Bassett and Alexander 1970). It was soon realized that GH concentrations decreased during the first 24 h after birth, reaching concentrations characteristic of the adult sheep (Wallace and Bassett 1970).

We decided to collect blood samples from fetal lambs to see if the GH concentrations in fetal plasma were high. For this purpose, a technique was developed for the catheterization of the carotid artery and jugular vein of fetal lambs. As these catheters remained patent for long periods, blood samples could be collected and test substances could be administered into the conscious ewe and her fetus. To our knowledge these were the first chronically catheterized preparations in which the catheters were inserted into fetal blood vessels rather than umbilical blood vessels.

We found that GH concentrations in the plasma of fetal lambs were ten or more times those of ewes during the last third of pregnancy (Bassett et al. 1970). There was a steady rise in fetal plasma GH from 40–50 ng/ml at 100–110 days of gestation to a maximum value of 110–120 ng/ml between 130–140 days of gestation and then a rapid decrease around the time of birth. These findings have been confirmed since in several different studies (Gluckman et al. 1979; Mesiano et al. 1987).

* This work was supported by a program grant awarded to Professor G. D. Thorburn by the National Health and Medical Research Council of Australia.
[1] Department of Physiology, Monash University, Clayton, Victoria, Australia 3168

The Endocrine Control of the Fetus
Ed. by W. Künzel and A. Jensen
© Springer-Verlag Berlin Heidelberg 1988

To establish the source of the GH in the fetal circulation, we ablated the pituitary using the transphenoidal approach described by Ferguson (1951) for use in adult sheep. Following *acute* hypophysectomy of the fetus, GH disappeared from the fetal plasma, showing that there was no significant placental source of GH, and that circulating GH was produced by the pituitary.

The high GH concentrations in the fetus, and their dramatic decrease after parturition, raised questions about the secretion and metabolism of GH before and after birth. One suggestion was that the high GH concentrations could be explained by the inability of the immature fetal kidneys to remove GH from the circulation as efficiently as in the adult sheep (Wallace et al. 1972). The change in GH concentrations at birth could then have been due to a change in renal function at birth. In our earlier study, the rate of disappearance of both exogenous GH and endogenous GH following hypophysectomy suggested that the rate of removal of GH from fetal plasma was slower than that in lambs or adults (Wallace and Bassett 1970). However, in a subsequent study using radiolabelled GH (Wallace et al. 1973), we found that the metabolic clearance rates per kilogram body weight of fetuses and lambs were not significantly different. The production rate for the fetuses (0.5 mg kg^{-1} day^{-1}) was 10–12 times higher than for the lambs. It was concluded from these experiments that the high concentrations of GH in the plasma of fetal lambs were *not* due to a low metabolic clearance rate, and that the change in GH levels at birth was *not* due to a change in clearance rate. We proposed that GH was secreted maximally by the fetal pituitary under the influence of a hypothalamic releasing factor, and that, following birth, the secretion rate falls due to the inhibition of this hypothalamic stimulus.

It is of interest that these observations have been recently confirmed by infusing hypophysectomized lamb fetuses with a preparation of GH which was judged to be 90% pure by radioimmunoassay (Stevens and Alexander 1986). Infusion at the calculated production rate (0.5 mg kg^{-1} day^{-1}) resulted in plasma GH concentrations which were about one-third of normal values. Indeed, they found that an infusion rate of 1.5 mg kg^{-1} day^{-1} was needed to give plasma concentrations of the right order. By contrast, Parkes and Hill (1985) found that the infusion of 400 μg/day of GH (a National Institutes of Health preparation) into hypophysectomized fetuses produced normal GH concentrations.

To examine the influence of the fetal hypothalamus on GH secretion, we examined the effect of sectioning the pituitary stalk. This led to a rapid decrease in circulating concentrations of GH. This experiment confirmed that the high concentrations of GH in the plasma of fetal lambs were not caused by a low metabolic clearance rate, but by a high rate of GH secretion, which in turn resulted from active stimulation of the fetal pituitary by the hypothalamus (Wallace et al. 1973).

We had some reservations about conclusions drawn from the stalk-section fetuses. The technique of Liggins and Kennedy (1968) had been used, in which a spatula was run along the base of the skull to section the pituitary stalk. At autopsy, the pituitaries were found to have a large central area of necrosis, presumably caused by damage to the arterial circulation to the hypophysis. The pituitary infarct may have been responsible for part of the decrease in GH concentration. However, recently in our laboratory, Ross Young and Marion Silver have sectioned by pituitary stalk under direct vision (I. R. Young and M. Silver 1986, personal communication) using a technique devised by Jim Cummins and Iain Clarke (Clarke and Cummins 1982) for adult sheep and in which the blood supply to the anterior pituitary is preserved. In these hypothalamic-pituitary disconnected fetuses, the plasma GH fell rapidly, confirming the earlier data and reaffirming the role of the hypothalamus.

We had raised the question of whether GH secretion by the fetal pituitary was actively regulated or whether there was uncontrolled GH secretion before the maturation of the hypothalamic control system (Bassett et al. 1970). To investigate this, we infused glucose into fetal lambs at rates sufficient to cause marked hyperglycaemia (Bassett et al. 1970). Plasma glucose concentrations in the fetal lamb are very low and it seemed possible that hypoglycaemia might be acting as a powerful stimulus for GH release, thus explaining the high GH concentrations. However, the glucose infusion failed to alter GH concentrations significantly in the fetus, whereas a similar challenge in the newborn lamb suppressed GH concentrations and in older lambs caused a significant increase. These results suggested an immaturity of the hypothalamic regulatory mechanisms in the fetus and that the high fetal GH levels represented a lack of inhibitory control. We found that the β agonist, isoprenaline, caused a rapid initial decrease in plasma GH concentrations in the fetus, followed by an increase in GH concentrations above the pre-infusion values, suggesting a time-dependent inhibition of an active secretory process (Bassett et al. 1970). Since earlier work had indicated that adrenergic mechanisms appeared to be involved in GH regulation (Blackard and Heidingsfelder 1968), we suggested that inhibitory adrenergic effects on the pathways regulating GH secretion, either in the hypothalamus or in the pituitary, may be less active in the immature fetus. An increase in adrenergic activity at birth (Bassett and Alexander 1970) was proposed as an explanation of the decrease in GH concentrations at birth (Bassett et al. 1970).

These results in the sheep fetus were consistent with the observations in both the human (Kaplan and Grumbach 1967) and monkey (Mintz et al. 1969), in which it had been found that fetal GH concentrations were also high. Kaplan et al. (1972) also correlated the temporal changes in GH during fetal life and around birth to a maturation of other neural functions, particularly electroencephalographic and motor behaviour.

Growth Hormone and Fetal Growth

In reviewing the earlier studies on small laboratory animals, where fetal decapitation was used as a means of hypophysectomy, Jost (1966) concluded that the fetal pituitary played a minimal role in the regulation of body-weight gain. In contrast, Heggestad and Wells (1965) observed a 20% decrease in the body weight of fetal rats following decapitaiton and, unlike other workers, examined skeletal growth and found that it was retarded. Moreover, these workers administered GH to the decapitated fetuses and found that normal body weight was restored. A similar reduction in fetal body weight (25%–33%) was observed by Honnebier and Swaab (1973) after aspiration of the fetal rat brain. In these experiments the fetal pituitary was left behind, although no assessment of its function was made. The studies of Beam (1968), Jack and Milner (1975) and Hill et al. (1979) failed to show any effect of fetal decapitation on growth of the fetal rat. A note of caution should be raised in the use of body weight as the sole index of body growth, since a decrease in body weight can be masked by oedema. For instance, Stryker and Dziuk (1975) failed to find any decrease in body weight after decapitation of fetal pigs, but observed that the fetuses were oedematous and contained 5%–10% less protein.

The fetal lamb pituitary clearly secreted large amounts of GH, but we had not addressed the question of whether GH was needed for the growth of the fetus. The pioneering studies of Liggins and Kennedy (1968), while not specifically addressing the role of GH, did examine the effects of fetal hypophysectomy on the growth of the lamb fetus. It is now well known that although these experiments were mainly directed at the role of the pituitary in the initiation of labour (Liggins et al. 1967), they provided important information on the role of fetal pituitary in the growth of the fetus. In these experiments, the uterus was not opened, but an electrocoagulation probe was passed through the uterine wall, through the fetal skull anteriorly and into the sella turcica. No fetal catheters were inserted. In subsequent experiments a self-retaining catheter was inserted into the peritoneal cavity through the uterine wall for the infusion of test substances.

Compared to the technique used in the studies in small laboratory animals, Liggin's technique was a major step forward, since the fetal pituitary alone was removed. However, in many instances electrocoagulation failed to ablate the pituitary completely (Liggins and Kennedy 1968; Kendall et al. 1977; Parkes and Hill 1985). Some studies failed to include any functional assessment of the completeness of hypophysectomy, such as measurement of fetal GH and prolactin concentrations. In addition, in one study (Parkes and Hill 1985), the electrocoagulation apparently interrupted the blood supply to the brain, leading to brain degeneration in 10 out of 13 fetuses. Clearly, in most cases, some hypothalamic damage must have been caused by the surgical approach.

Recently, we have used the trans-sphenoidal approach so that the fetal pituitary can be totally removed under direct vision, and the completeness of the hypophysectomy has been assessed by the measurement of GH and prolactin concentrations in response to thyrotrophin releasing hormone (TRH). The results of these experiments will be discussed later.

Liggins and Kennedy (1968) reported that hypophysectomy of the sheep fetus resulted in marked reduction in skeletal maturation, hypoplasia of the adrenal cortices, thyroid glands and testes, and disturbances in the metabolism of carbohydrate and fat. Although this paper clearly demonstrated a reduction in skeletal growth and especially skeletal maturation, the interpretation was complicated for a number of reasons.

While these experiments clearly demonstrated the importance of fetal pituitary hormones, they failed to delineate the role of GH specifically. The delayed ossification of the epiphyseal centres in their hypophysectomized fetuses clearly demonstrated a lack of thyroid hormone. The subsequent experiments of Hopkins and Thorburn (1974) showed growth retardation and delayed skeletal maturation in fetuses thyroidectomized between 80 and 100 days of gestation. These changes were reversed by administration of thyroid hormones. Clearly replacement therapy with thyroid hormone is needed to assess the effect of GH deficiency per se on fetal growth when using the hypophysectomy approach.

Liggins and Kennedy (1968) also reported that there was an accumulation of subcutaneous fat and this was later confirmed by others (Barnes et al. 1977; Stevens and Alexander 1986). Stevens and Alexander (1984) showed that the deposition of subcutaneous fat in hypophysectomized fetuses was prevented by the administration of GH but not adrenocorticotrophic hormone or triiodothyronine. Thus, it appears that GH is lipolytic in the fetal sheep, as it is in the adult sheep (Bassett and Wallace 1966; Wallace et al. 1970). This is of especial interest since it has been shown (Gluckman et al. 1983) that there is a lack of GH receptors in the fetal liver, and, as a result, there has been a tacit assumption that all fetal tissues lack GH receptors and that the high GH concentrations serve no physiological role in the sheep fetus.

Our recent data (Mesiano et al. 1987) show that following hypophysectomy on about day 115, the limbs of hypophysectomized fetuses were considerably shorter at 147 days of gestation than sham-operated controls. However, considerable growth had occurred after the hypophysectomy; the increase in fore- and hind-limb length over the 115–147 days period was 70% of the corresponding value for the shams. When some of the hypophysectomized fetuses were examined at 163 days of gestation (16 days longer than normal term), no further growth was observed in the appendicular skeleton, whereas there was a significant increase in body weight (Mesiano et al. 1987). These data indicated that skeletal and soft tissue growth may be separately controlled. This data is similar to the earlier studies in the rat. Walker and his co-workers (1950) showed that, following hypophysectomy, growth was only slightly decreased in animals less than 10 days of age, but that thereafter the rate of growth progressively decreased and ceased completely (in terms of body weight) 28 days after birth, demonstrating the gradual increase in GH dependence of postnatal growth. Similarly the Snell-Smith mouse, which does not secrete GH or thyroid-stimulating hormone (TSH) due to a recessive gene defect, grows at a near-normal rate during the 1st week of postnal life, but thereafter the growth rate gradually declines, finally stopping 1 month after birth (Smeets and van Buul-Offers 1983). These studies did not address the potential difference in the regulation of skeletal and soft tissue growth.

It seems, then, that a gradual increase in GH dependence appears during late gestation in the sheep and in the early postnatal period in the rat. Gluckman et al. (1983) failed to demonstrate any somatogenic receptors in the fetal liver, but these receptors appeared in the immediate postnatal period and the high binding capacity observed in adults was soon reached. A similar increase in GH binding sites in the rat liver as a function of age was observed by Maes et al. (1983). Data are not yet available concerning the appearance of GH binding sites in the epiphyseal plates, but the data on the growth of the hypophysectomized lamb would suggest that they may appear during late gestation in the sheep fetus.

Gluckman (1984) suggested that a similar lack of receptors for GH in the hypothalamus of the fetal lamb may inhibit the negative feedback loop and result in the high circulating GH concentration in the fetal circulation. Indeed, the appearance of GH receptors in the hypothalamus during late gestation may account in part for the rapid decrease in GH levels starting about 72 h before birth. The cause of the increase in GH receptors in any tissues is as yet unknown.

It is apparent that the fetal lamb can still grown, albeit more slowly, in the absence of the pituitary (and therefore without GH, TSH and prolactin). We have assumed that lack of GH accounts for the growth retardation we observed following hypophysectomy but prolactin may still be important. The ovine fetal liver contains lactogenic receptors but their function has not been elucidated. There is evidence in the rat that prolactin can stimulate somatomedin production. Clearly it is important in the hypophysectomized ovine fetus to institute replacement therapy with both GH and/or prolactin.

It is of interest that Stevens and Alexander (1986) failed to find any decrease in body weight following hypophysectomy of the lamb fetus, whereas they found a significant body weight increase in the hypophysectomized lambs given GH replacement. However, they did not measure skeletal growth.

The above discussion might suggest, because of the lack of GH receptors in some sites in the fetal lamb, that GH has no role in fetal growth. In the sheep fetus, hypophysectomy results in increased deposits of brown fat and the appearance of a large persistent depot of subcutaneous white adipose tissue which was absent from the intact fetus except around 100–120 days of gestation (Liggins and Kennedy 1968; Barnes et al. 1977; Alexander 1978). Stevens and Alexander (1986) found that treatment of the hypophysectomized fetus with GH resulted in the disappearance of the depot of subcutaneous fat. They concluded that GH itself has a major role in the control of lipid metabolism in the fetus. Moreover, they proposed that the high GH levels in the fetus counterbalance the strong lipogenic actions of insulin in the fetus, so that fetal nutrients are channelled into protein rather than fat synthesis. This study highlights the danger of using body weight as the sole criterion of growth, when a change in body composition, such as obesity or oedema, can conceal changes in skeletal growth and muscle mass.

Growth Hormone and Somatomedins in Fetal Lamb Growth

In 1957, Salmon and Daughaday first suggested that the growth promoting action of GH was mediated by some other hormone that was induced by GH, and subsequently the term 'somatomedins' was coined (Daughaday et al. 1972). Since that time, it has been shown that the somatomedins are the insulin-like growth factors (IGFs), peptide mitogens that are structurally homologous to proinsulin (Klapper et al. 1983). Two major IGFs, IGF-I (also known as somatomedin C) and IGF-II, have been characterized. In humans, these IGFs have been purified from plasma and their amino acid sequence has been determined (Rinderknecht und Humble 1978) and the complementary DNAs encoding precursor forms of these peptides have been isolated (Jansen et al. 1983; Bell et al. 1984; Rotwein 1986). Recently, ovine IGF-I and IGF-II have been purified in our laboratory by Allan Hey and Chris Browne. However, the fact that the ovine IGFs have not been available for investigators has not hampered studies unduly, as there is a large degree of parallel cross-reactivity in the commonly used assay systems between the human IGFs and the ovine IGFs (C. A. Browne, A. W. Hey and S. Mesiano, unpublished observations 1987).

The initial observations (Table 1) on the regulation of IGFs in the circulation of fetal lamb were reported by Falconer et al. (1979), who used a porcine costal cartilage bioassay, Brinsmead and Liggins (1979), who used a similar bioassay and a multiplication stimulating activity (MSA) binding assay (MSA is the rat homologue of human IGF-II), and by Gluckman et al. (1979), who used a radioreceptor assay (RRA). It seems likely that the porcine costal cartilage bioassay is mainly measuring IGF-I-like activity, whereas the MSA RRA is clearly measuring IGF-II. It is now known that IGFs are bound to binding proteins in plasma, but there is some question whether they are biologically active in the bound form. Indeed, it has now been suggested that the binding proteins can directly enhance the activity of the IGFs (Clemmons et al. 1987). Since plasma is active in the porcine cartilage bioassay (Falconer et al. 1979) and fetal calf serum stimulates proteoglycan synthesis in bovine articular cartilage (McQuillen et al. 1986), one must suggest that protein-bound

Table 1. Effect of hypophysectomy on somatomedins in the fetal lamb

Reference	Assay		Extraction method	Effect
Falconer et al. 1979	SLA	Bioassay	None	*Lowered*
Brinsmead and Liggins 1979	MSA	RRA	None	No effect
	SLA	Bioassay	None	No effect
Parkes and Hill 1985	SLA	Bioassay	None	No effect
Gluckman and Butler 1985	IGF$_1$	RIA	Acid–ethanol	No effect
	IGF$_2$	RRA	Acid–ethanol	No effect
Mesiano et al. 1987	IGF$_1$	RIA	Acid–ethanol	No effect
	IGF$_2$	RRA	Acid–gel chrom	No effect
S. Mesiano, I. R. Young,	IGF$_1$	RIA	Acid–gel chrom	*Lowered*
C. A. Browne and	IGF$_2$	RRA	Acid–gel chrom	No effect
G. D. Thorburn,				
unpublished data, 1987				

SLA, somatomedin-like activity; MLA, multiplication stimulating activity; RRA, radioreceptor assay; RIA, radioimmunoassay

IGF is active. While it is not clear whether it is necessary for the IGF-I to be released from the binding protein for it to exert its action, McQuillen et al. (1986) found that a monoclonal antibody to IGF-I blocked the stimulatory action of fetal calf serum on bovine cartilage. The antibody could combine with IGF-I as it is released from the binding protein or while it is still attached to the binding protein. These results also suggest that the active factor stimulating sulphation in fetal calf serum is IGF-I. McQuillen et al. (1986) also showed that when plasma was chromatographed under neutral conditions, the sulphation factor activity in the plasma was in the high molecular weight fraction, suggesting that the bound form of IGF was biologically active.

While the binding protein does not appear to interfere in bioassays, it clearly does in RIA or RRAs. Normally the IGFs are separated from their binding proteins by acid ethanol extraction (Daughaday et al. 1980) or acid gel chromatography (Horner et al. 1978). However, in the studies of Brinsmead and Liggins (1979) and Gluckman et al. (1979), no extraction techniques were used prior to the receptor assays.

Falconer et al. (1979) showed that their "somatomedin-like activity' (SLA) was approximately the same in fetal and maternal plasma, and that fetal hypophysectomy, nephrectomy and carunclectomy all reduced the fetal plasma SLA concentration by about 50%. By way of contrast. Brinsmead and Liggins (1979) reported that with their SLA bioassay they failed to detect any reduction in SLA in the plasma of hypophysectomized fetuses compared to controls, but their animal numbers were small. Later, Parkes and Hill (1985) also reported that fetal hypophysecomty failed to affect plasma SLA, although they showed a significant decrease in SLA in hypophysectomized fetuses that survived for 10 days or longer.

In 1983, Gluckman and Butler reported individual plasma concentrations for IGF-I and IGF-II throughout gestation, the perinatal and the postnatal periods in the sheep. They were also careful to use acid-ethanol extraction of the samples prior to assay so that, on the face of it, valid determinations of ovine fetal IGF-I and IGF-II plasma concentrations could be achieved. They reported that IGF-II concentrations were higher in the fetus than in the adult, whereas for IGF-I the converse was true, and that the switchover from relatively high IGF-II to relatively high IGF-I occurred around the time of birth. Subsequently, they used similar assays to investigate the effect of fetal decapitation and of lesioning median eminence on fetal IGF concentrations, and found that neither treatment had any effect (Gluckman and Butler 1985). They concluded that GH "is not essential for the maintenance of circulating concentrations of insulin-like growth factor-I or II."

Recently we have reinvestigated these same problems, this time performing the hypophysectomy via a trans-sphenoidal route (Mesiano et al. 1987). We found that there was no pituitary dependence of fetal IGF-I or IGF-II plasma concentrations, but we did find that fetal hypophysectomy did affect fetal growth (see above). Again, IGF-II concentrations exceeded IGF-I concentrations in fetal plasma. We now have been forced to modify our views on the pituitary regulation of IGFs.

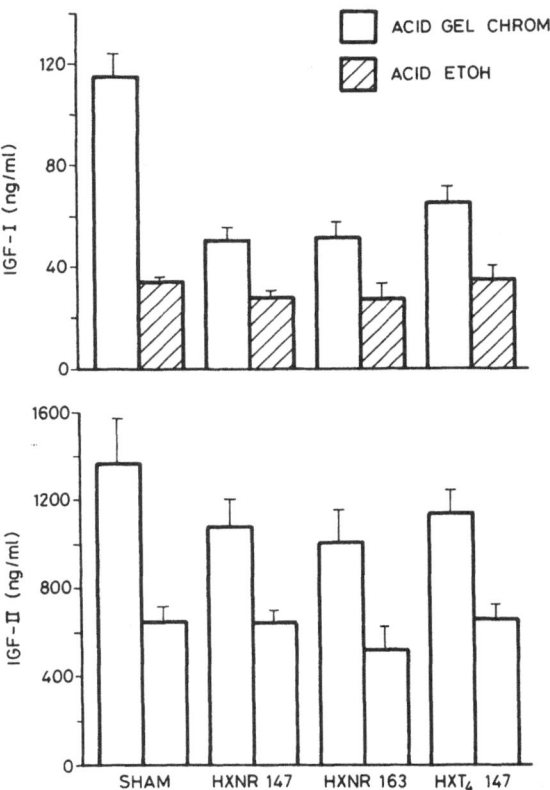

Fig. 1. Effect of hypophysectomy on plasma IGF-I and IGF-II concentrations in the late-gestation fetal lamb. Samples were obtained from hypophysectomized fetuses operated at day 115 and sampled prior to normal term at day 147 *(HXNR 147)*, hypophysectomized fetuses left in utero for a further 16 days beyond normal term *(HXNR 163)*, hypophysectomized fetuses which were given daily thyroxine replacement therapy (100 µg T_4/24 h) from the time of operation until day 147 *(HXT_4 147)* and sham-operated controls *(SHAM)*. The upper panel shows IGF-I concentrations assayed after acid gel chromatography *(open bars)* or acid ethanol extraction *(hatched bars)* using a human IGF-I radioimmunoassay (RRA) in both cases. The lower panel shows IGF-II concentrations determined by radioreceptor assay after acid gel chromatography. The *lefthand bars* in the *lower panel* show the IGF-II values obtained in a homologous RRA using ovine placental membranes and ovine IGF-II standard and tracer. The *righthand bars* are IGF-II values obtained using a heterologous RRA using rat placental membranes and human IGF-II as standard and tracer

Our recent studies (Mesiano et al. 1987) clearly indicated that the ovine fetal pituitary, probably mediated by the action of GH, plays a significant role in the skeletal growth of the fetus during the last third of gestation. In the post-weaning lamb, however, hypophysectomy resulted in a complete cessation of skeletal growth and a 50% reduction in plasma-IGF-I concentrations (Young et al. 1986). In this study, recombinant hGH, when administered to the hypophysectomized lamb, caused a resumption of the normal growth rate. We have recently repeated our IGF assays on fetal plasma, using acid gel chromatography rather than acid–ethanol extraction for the fetal IGF-I determinations. Following hypophysectomy, there was now a marked decrease in the circulating IGF-I concentrations in plasma (Fig. 1). In the fetus the concentrations of IGF-I decreased from a mean of 110 ng/ml to 50 ng/ml after hypophysectomy. Thyroid hormone replacement partially restored the IGF-I concentration. Hypophysectomy did not alter the plasma concentrations of IGF-II in either the lamb or the fetus. These results indicated that the pituitary (probably GH) influence on growth may be mediated via IGF-I both in the lamb and the fetus.

It is worthwhile to issue a word of warning here. When these samples were first assayed, acid–ethanol extraction was used and a decrease in IGF-I concentrations in the fetus was not observed after hypophysectomy (Mesiano et al. 1987) (Fig. 1). For various reasons, we suspected that acid–ethanol extraction did not completely remove the IGF binding proteins. This problem was overcome by using acid gel chromatography prior to assay, which we had used on the advice of Ray Hintz for IGF-II. The use of acid–ethanol extraction gave us false results and led us to draw the wrong conclusions (Mesiano et al. 1987). Other workers have also used acid-ethanol extraction when measuring IGF-I and IGF-II concentrations in sheep plasma (Gluckman and Butler 1983, 1985) and we believe that their results should be viewed with some reservation. While we are on the question of assays, there is also a problem with cross-reaction of IGF-II in the IGF-I radioimmunoassay because of the very high levels of IGF-II in fetal plasma. The 50 ng/ml IGF-I measured in fetal plasma after hypophysectomy may be partly an artefact due to cross-reactivity of IGF-II in the IGF-I assay. We are at present attempting to determine whether any IGF-I is present after fetal hypophysectomy, by separating the IGF-I and IGF-II prior to assay. If there is any remaining IGF-I after hypophysectomy, this residual IGF-I may be adequate to maintain fetal growth in the absence of GH. Clearly this is not the case in the postnatal lamb, since hypophysectomy prevents further skeletal growth.

There have been several suggestions that placental lactogen (PL) may play a role in the regulation of IGF concentrations in the fetus. In the rat, oPL but not oGH stimulated rat embryo fibroblasts (Adams et al. 1983) to produce IGF-II, whereas oGH and oPL stimulated the production of IGF-I in fibroblasts from 25-day-old rats but not in those from fetal rats.

Plasma PL concentrations are high in the fetal lamb and they fall in late gestation in a manner that parallels the fall in IGF-II (Lowe et al. 1984). The finding that the GH receptor is not detectable in the ovine fetal liver had added further weight to these arguments, especially since the ovine fetal liver possessed a lactogenic receptor (Gluckman et al. 1983) and an oPL receptor (Chan et al. 1978). The role of the IGFs as circulatory mediators of GH action - that is, as classical endocrine factors - still remains in some doubt. Recently, several investigators have raised the possibility of alternate autocrine or paracrine mechanisms. These are now considered in the next few pages.

Paracrine and Autocrine Regulation of Somatomedins in the Fetus

Although originally the liver was thought to be the primary site of IGF synthesis, evidence from a number of studies now suggests that the IGFs are synthesized in many tissues. This has led to the hypothesis that the IGFs may act locally to elicit their biological response in a paracrine or autocrine manner. In the fetus, this is supported by the findings that the IGFs are synthesized in many fetal tissues, that they exert mitogenic actions on cultured cells from fetal tissues and that specific IGF receptors exist in many fetal tissues. For instance, D'Ercole et al. (1980) showed that immunoreactive IGFs are secreted by cultured explants of a number of embryonic mouse organs, and IGF receptors have been shown to be present in fetal tissues in the human (Sara et al. 1983), rodent (Kaplowitz et al. 1982; Hill et al. 1983), and the sheep (Owens et al. 1980, 1985).

Until recently, it has been generally agreed that IGF-I is responsible for most of the growth effects ascribed previously to GH (Daughaday 1983). However, this hypothesis is now a subject of some controversy. Green and his co-workers (Morikawa et al. 1982) noted that GH specifically promoted the differentiation of cloned lines of preadipose 3T3 cells into adipose cells. This was due to a direct action of GH on the cells. IGF-I, which is regarded as the obligatory mediator of the hormone in the promotion of growth, did not promote differentiation. The findings seemed to rule out the classical somatomedin hypothesis as enunciated, at least in the case of adipose tissue. In 1985 Green et al. (1985) proposed the 'dual effector' theory based on the concept that the growth of tissues occurs in two steps; (1) differentiated cells are formed from their precursors (stem cells) and (2) the number of young differentiated cells is increased through limited multiplication (clonal expansion). This theory states that both stages are stimulated by GH, the first directly by GH and the second indirectly through its mediator, IGF-I. In a recent study, Zezulak and Green (1986) found that GH, but not IGF-I, directly promoted the differentiation of preadipocytes to adipocytes. Adipocytes newly differentiated from precursor cells in response to GH were shown to be far more sensitive to the mitogenic effects of IGF-I than the precursor cells. Again IGF-I had a weak mitogenic effect on preadipose cells, but it produced no differentiation. It only had a strong mitogenic effect on preadipose cells when administered together with GH. This effect was seen mainly on differentiating cells. They suggested that the action of IGF-I is

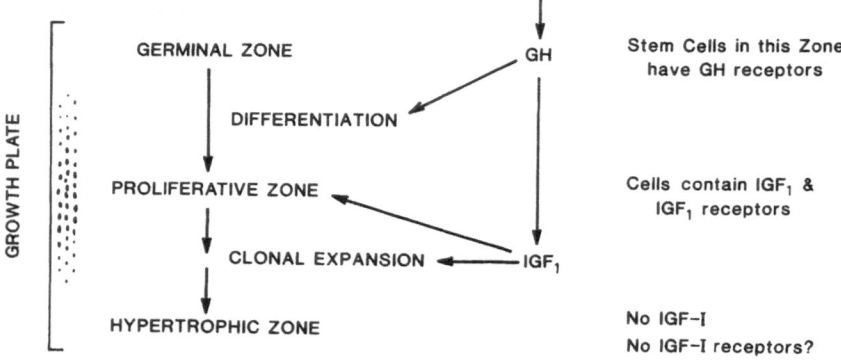

HYPOTHESIS:

GH works by binding directly to the growth plate where it can
stimulate local production of IGF-I

Fig. 2. Hypothetical scheme for the hormonal regulation of bone growth, after the ideas of Isaksson and co-workers (Nilsson et al. 1986)

to multiply selectively young differentiated cells (clonal expansion), and the preferred target cells for IGF-I action are created by the direct action of GH.

Recently, Isaksson and his colleagues (Nilsson et al. 1986; Isgaard et al. 1986) have shown that the local injection of GH into the epiphyseal growth plate of the proximal tibia of hypophysectomized rats stimulates unilateral longitudinal bone growth. This observation has been confirmed by Russell and Spencer (1985). Chondrocytes isolated from the rabbit ear and epiphyseal growth plate have specific binding sites for GH, and GH can stimulate DNA and proteoglycan synthesis in cultured chondrocytes from cartilage of rabbit and rat rib growth plate (Madsen et al. 1983). These results were considered to be inconsistent with an effect of GH mediated through circulating IGF-I. In the growth plate, chondrocytes are arranged in a columnar pattern reflecting the stage of maturation. Cells from the germinal zone, the stem cell area of the growth plate, differentiate and enter the proliferative zone where the cells undergo limited clonal expansion during the process of longitudinal bone growth (Fig. 2).

Using immunocytochemical methods, Nilsson et al. (1986) localized IGF-I immunoreactivity to cells of the proliferative zone of the proximal tibial growth plate of normal rats, whereas the cells of the germinal and hypertrophic zones stained only weakly. In hypophysectomized rats, the number of cells with detectable IGF-I immunoreactivity were markedly reduced. When hypophysectomized rats were treated with GH, either systemically or directly in the growth plate, the number of IGF-immunoreactive cells in the proliferative zone increased. Nilsson et al. (1986) concluded that the number of IGF-I-containing cells was directly regulated by GH and that IGF-I was produced in proliferating chondrocytes in the growth plate.

They further suggested that IGF-I had a specific role in the clonal expansion of differentiated chondrocytes. It would seem that GH stimulates the differentiation of the prechondrocyte, and that genes that code for the growth factors and their receptors are expressed locally within the growth plate. Activation of these genes results in the increased local production of IGF-I and its receptor, which in turn promotes the clonal expansion of the chondrocytes through paracrine and autocrine mechanisms.

Before we go on to consider the relevance of these findings to the fetal lamb, we would like to consider some recent work on the cellular localization of IGF mRNA in the human fetus. Han et al. (1987) have shown, using in situ hybridization histochemistry, that IGF-I and IGF-II mRNAs are found in connective tissue or cells of mesodermal origin in the human fetus. In the liver, the mRNAs were found in the perisinusoidal cells (i.e. mesenchymal cells such as endothelial and reticuloendothelial cells and fibroblasts), whereas the hepatocytes showed no detectable IGF mRNA. These results would indicate that the connective tissue framework of an organ can synthesize IGFs to act on the adjacent ectodermally-derived cells, which in turn should possess IGF receptors (a paracrine mechanism). The mesenchymal cells may also possess IGF receptors and be acted upon by their own IGFs (an autocrine mechanism).

The above observations would seem inconsistent with some of the findings of Nilsson et al. (1986) on the rat cartilage and their own earlier finding in human fetal tissue, where they found that IGF immunoreactivity was not always localized to mesenchymal cells or connective tissues but was instead localized to a wide variety of cells, such as proliferating chondrocytes, renal tubule cells and myoblasts. This problem has not yet been resolved but one consideration is that the immunoreactive IGF found in those cells may represent an accumulation of IGF, possibly via receptor binding (Fig. 3). The failure to detect IGF mRNAs in non-mesenchymal cells could reflect either a low level of IGF gene expression or a short half-life of IGF mRNA in those cells. In the costal cartilage of human fetal tissues, mRNA was found in the perichondrium and fibrous sheath, but not in the chondrocytes. In general, in the human fetus, the abundance of the IGF-II mRNA was greater in all tissues than that of IGF-I, a somewhat surprising finding in the case of the cartilage. One important corollary of the above findings would be that GH should act on mesenchymal cells, if one of its perceived roles is to switch on IGF-I synthesis, and that GH receptors should be present on mesenchymal cells (see Fig. 3).

Further evidence suggesting that the direct growth-promoting effect of GH on cartilage is mediated by the local production of somatomedin has been presented by Schlechter et al. (1986). They infused GH and human IGF-I into the arterial supply of one limb of hypophysectomized rats and stimulated growth of the epiphyseal plate of the infused limb. Rabbit antiserum to human IGF-I, but not normal rabbit serum, completely abolished the growth promotion inducible by the rat GH. This study did not exclude the possibility that circulating IGFs may also play a role, but it indicated that local somatomedin production may be more important for promoting growth in vivo than was previously realized. These studies support the primary

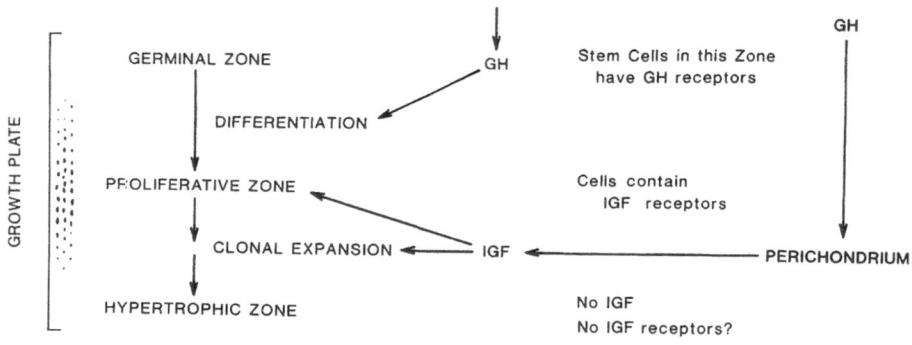

HYPOTHESIS:

GH acts directly on the Germinal Zone to cause differentiation and on the
Perichondrium to stimulate IGF production. The IGFs then promote proliferation.
The Hypertrophic Zone is then unresponsive to both GH and IGF.

Fig. 3. Hypothetical scheme for the hormonal regulation of bone growth. Direct actions of growth hormone may be locally mediated by IGFs via a paracrine mechanism. (Modified from Nilsson et al. 1986 after consideration of Han et al. 1987)

component of the somatomedin theory, that is, that IGFs mediate the growth-promoting action of GH in vivo but they suggest that local production is more important. They also suggest that a peripheral site for GH action exists, but they do not address the previous question of whether GH has a local action which is independent of IGFs.

Conclusion

In the last 20 years, considerable progress has been made in elucidating the role of the fetal pituitary in the growth of the fetal lamb. Whereas hypophysectomy of the post-weaning lamb caused a cessation of skeletal growth, in the fetus it only results in a slowing of growth of the appendicular skeleton. This finding agrees with the earlier bioassay data (Falconer et al. 1979), where a 50% decrease in somatomedin-like activity was observed after hypophysectomy of the fetal lamb, and our recent unpublished data (S. Mesiano, I. R. Young, C. A. Browne and G. D. Thorburn, 1987), where we observed a similar decrease in IGF-I, but not IGF-II, levels after hypophysectomy. These results suggest that in the lamb, GH/IGF-I is solely responsible for skeletal growth, whereas in the fetus, it accounts for about 30% during late gestation and more in the post-term hypophysectomized fetus. Our data would be consistent with a gradual evoluation of pituitary (GH)/IGF-I dependence with gestational age.

It is not clear as yet what mechanism is responsible for the majority of skeletal growth in the fetus. It is possible that a placental growth factor (oPL) acts on receptors in the growth plate to stimulate the synthesis of IGFs and the differentiation of the prechondrocytes. Whether these receptors are oPL or GH-like is not known, nor is it known whether any local oPL action is mediated by IGF-I or IGF-II. Clearly, there are still many fascinating questions which need to be elucidated.

Acknowledgements. The authors wish to acknowledge the assistance of Mrs. Judi Herschell and Mr. Frank Groeneveld in the preparation of this manuscript.

References

Adams SO, Nissley P, Greenstein LA, Yang YWH, Rechler MM (1983) Synthesis of multiplication-stimulating activity (rat insulin-like growth factor II) by rat embryo fibroblasts. Endocrinoloy 112: 979–987

Alexander G (1978) Quantitative development of adipose tissue in foetal sheep. Aust J Biol Sci 31: 489–503

Barnes RJ, Comline RS, Silver M (1977) The effect of bilateral adrenalectomy or hypophysectomy of the foetal lamb in utero. J Physiol (Lond) 264: 429–447

Bassett JM, Alexander G (1970) Insulin, growth hormone adn corticosteroids in neonatal lambs: normal concentrations and effect of cold. Biol Neonate 17: 112–115

Bassett JM, Wallace ALC (1966) Short term effects of ovine growth hormone on plasma glucose, free fatty acids and ketones in sheep. Metabolism 15: 933–944

Bassett JM, Thorburn GD, Wallace ALC (1970) The plasma growth hormone concentrations of the foetal lamb. J Endocrinol 48: 251–263

Bearn JG (1968) The thymus and the pituitary-adrenal axis in anencepaly. A correlation between experimental foetal endocrinology and human pathological observations. Br J Exp Pathol 49: 136–144

Bell GI, Merryweather JP, Sanchez-Pescador R, Stempien MM, Priestley L, Scott J, Rall LB (1984) Sequence of a cDNA clone encoding human preproinsulin-like growth factor II. Nature 310: 775–777

Blackard WG, Heidingsfelder SA (1986) Adrenergic receptor control mechanism for growth hormone secretion. J Clin Invest 47: 1407–1414

Brinsmead MW, Liggins GC (1979) Serum somatomedin activity after hypophysectomy and during parturition in fetal lambs. Endocrinology 105: 297–305

Chan JSD, Robertson HA, Friesen HG (1978) Distribution of binding sites for ovine placental lactogen in the sheep. Endocrinology 120: 632–640

Clarke IJ, Cummins JT (1982) The temporal relationship between gonadotropin releasing hormone and luteinizing hormone secretion in ovarectomized ewes. Endocrinology 111: 1737–1739

Clemmons DR, Elgin RG, Busby WH, McCusker RH (1987) Insulin-like growth factor binding proteins are biologically active and modulate the cellular growth response to somatomedin-C. Proc Endocrinol Soc, 69th meeting, Indianapolis June 1987, abstract no 414

Daughaday WH (1983) The somatomedin hypothesis: origins and recent developments. In: Spencer EM (ed) Insulin-like growth factors/somatomedins. Basic chemistry, biology and clinical importance. de Gruyter, Berlin, pp 3–11

Daughaday WH, Hall K, Raben MS, Salmon WD, van der Brande JL, Van Wyk JJ (1972) Somatomedin: proposed designation for sulphation factor. Nature 235: 107

Daughaday WH, Mariz IK, Blethen SL (1980) Inhibition of access of bound somatomedin to membrane receptor and immunobinding sites: a comparison of radioreceptor and radioimmunoassay of somatomedin in native and acid-ethanol-extracted serum. J Clin Endocrinol Metab 51: 781–788

D'Ercole AJ, Applewhite GT, Underwood LE (1980) Evidence that somatomedin is synthesized by multiple tissues in the fetus. Dev Biol 75: 315–328

Falconer J, Forbes JM, Hart IC, Robinson JS, Thorburn GD (1979) Somatomedin-like activity in the plasma after foetal hypophysectomy or nephrectomy and in experimental intra-uterine growth retardation in sheep. J Endocrinol 83: 119–127

Ferguson KA (1951) The effect of hypophysectomy on wool growth. J Endocrinol 7: I xi (abstract)

Gluckman PD (1984) Functional maturation of the neuroendocrine system axis in the perinatal period: studies of the somatotropic axis in the ovine fetus. J Dev Physiol 6: 301–312

Gluckman PD, Butler JH (1983) Parturition-related changes in insulin-like growth factors I and II in the perinatal lamb. J Endocrinol 99: 223–232

Gluckman PD, Butler JH (1985) Circulating insulin-like growth factor-I and -II concentrations are not dependent on pituitary influences in the midgestation fetal sheep. J Dev Physiol 7: 405–409

Gluckman PD, Mueller PL, Kaplan SL, Rudolph AM, Grumbach MM (1979) Hormone ontogeny of the ovine fetus I. Circulating growth hormone in mid- and late gestation. Endocrinology 104: 162–168

Gluckman PD, Uthne K, Styne DM, Kaplan SL, Rudolph AM, Grumbach MM (1979) Hormone ontogeny of the ovine fetus IV. Serum somatomedin activity in the fetal and neonatal lamb and pregnant ewe: correlation with maternal and fetal growth hormone, prolactin, and chorionic somatomammotrophin. Pediatr Res 14: 194–196

Gluckman PD, Butler JH, Elliot TB (1983) The ontogeny of somatotropic binding sites in ovine hepatic membranes. Endocrinology 112: 1607–1612

Green H, Morikawa M, Nixon T (1985) A dual effector theory of growth-hormone action. Differentiation 29: 195–198

Han VKM, D'Ercole AJ, Lund PK (1987) Cellular localization of somatomedin (insulin-like growth factor) messenger RNA in the huma fetus. Science 236: 193–197

Heggestad CB, Wells LJ (1965) Experiments on the contribution of somatotrophin to prenatal growth in the rat. Acta Anat (Basel) 60: 348–361

Hill DJ, Davidson P, Milner RDG (1979) Retention of plasma somatomedin activity in the foetal rabbit following decapitation in utero. J Endocrinol 81: 93–102

Hill DJ, Fekete M, Milner RDG, De Prins F, Van Assche A (1983) Reduced plasma somatomedin activity during experimental growth retardation in the fetal and neonatal rat. In: Spencer EM (ed) Insulin-like growth factors/somatomedins. Basic chemistry, biology and clinical importance. de Gruyter, Berlin, pp 345–352

Honnebier WJ, Swaab DF (1973) The influence of anencephaly upon intrauterine growth of fetus and placenta and upon gestation length. J Obstet Gynaecol Br Commonw 80: 577–588

Hopkins PC, Thorburn GD (1974) The effect of foetal thyroidectomy on the development of the ovine foetus. J Endocrinol 54: 55–66

Horner JM, Liu F, Hintz RL (1978) Comparison of [^{125}I]somatomedin A and [^{125}I]somatomedin C radioreceptor assays for somatomedin peptide content in whole and acid-chromatographed plasma. J Clin Endocrinol Metab 47: 1287–1295

Isaksson OGP, Jansson JO, Gause IAM (1982) Growth hormone stimualtes longitudinal bone growth directly. Science 216: 1237–1239

Isgaard J, Nilsson A, Lindahl A, Jansson JO, Isaksson OGP (1986) Effect of local administration of GH and IGF-I on longitudinal bone growth in rats. Am J Physiol 250: E367–E372

Jack PMB, Milner RDG (1975) Effect of decapitation and ACTH on somatic development of the rabbit fetus. Biol Neonate 26: 195–204

Jansen M, van Schaik FMA, Ricker AT, Bullock B, Woods DE, Gabbay KH, Nussbaum AL, Sessenbach JS, van den Brande JL (1983) Sequence of cDNA encoding human insulin-like growth factor I precursor. Nature 306: 609–611

Jost A (1966) Anterior pituitary function in foetal life. In: Harris GW, Donavan BT (eds) The pituitary gland, vol 2. Butterworth, London, pp 299-323

Kaplan SL, Grumbach MM (1967) Growth hormone secretion in the human fetus and in anencephaly. Int congr ser no 142, Abstract no 98. Excerpta Medica Foundation, Amsterdam

Kaplan SL, Grumbach MM, Shepherd TH (1972) The ontogenesis of human fetal hormones. I. Growth hormone and insulin. J Clin Invest 51: 3080-3093

Kendall JZ, Challis JRG, Hart IC, Jones CT, Mitchell MD, Ritchie JWK, Robinson JS, Thorburn GD (1977) Steroid and prostaglandin concentrations in the plasma of pregnant ewes during infusion of adrenocorticotrophins or dexamethasone to intact or hypophysectomized foetuses. J Endocrinol 75: 59-71

Klapowitz PB, D'Ercole AJ, Underwood LE (1982) Stimulation of embryonic mouse limb bud mesenchymal cell growth by peptide growth factors. J Cell Physiol 112: 353-359

Klapper DG, Svoboda ME, van Wyk JJ (1983) Sequence analysis of somatomedin-C: conformation of identity with insulin-like growth factor I. Endocrinology 112: 2215-2217

Liggins GC, Kennedy PC (1968) Effects of electrocoagulation of the foetal lamb hypophysis on growth and development. J Endocrinol 40: 371-381

Liggins GC, Kennedy PC, Holm LW (1967) Failure of initiation of parturition after electrocoagulation of the pituitary of the fetal lamb. Am J Obstet Gynecol 98: 1080-1086

Lowe KC, Jansen CAM, Gluckman PD, Nathanielz PW (1984) Comparison of changes in ovine plasma chorionic somatomammotropin concentrations in the fetus and mother before spontaneous vaginal delivery at term and adrenocorticotropin-induced premature delivery. Am J Obstet Gynecol 150: 524-527

Madsen K, Friburg U, Roos P, Eden S, Isaksson O (1983) Growth hormone stimulates the proliferation of cultured chondrocytes from rabbit ear and rat rib growth cartilage. Nature 304: 545-547

Maes M, De Hertogh R, Watrin-Granger P, Ketelslegers JM (1983) Ontogeny of liver somatotropic and lactogenic binding sites in male and female rats. Endocrinology 113: 1325-1332

McQuillen DJ, Handely CJ, Campbell MA, Bolis S, Milway VE, Herington AC (1986) Stimulation of proteoglycan biosynthesis by serum and insulin-like growth factor-I in cultures bovine articular cartilage. Biochem J 240: 423-430

Mesiano S, Young IR, Baxter RC, Hintz RL, Browne CA, Thorburn GD (1987) Effect of hypophysectomy with and without thyroxine replacement on growth and circulating concentrations of insulin-like growth factors I and II in the fetal lamb. Endocrinology 120: 1821-1830

Mintz DH, Chez RA, Horger EO (1969) Fetal insulin and growth hormone metabolism in the subhuman primate. J Clin Invest 48: 176-186

Morikawa M, Nixon T, Green H (1982) Growth hormone and the adipose conversion of 3T3 cells. Cell 29: 783-789

Nilsson A, Isgaard J, Lindahl A, Dahlström A, Skottner A, Isaksson OGP (1986) Regulation by growth hormone of the number of chondrocytes containing IGF-I in rat growth plate. Science 233: 571-574

Owens PC, Brinsmead MW, Waters MJ, Thorburn GD (1980) Ontogenic changes in multiplication-stimulating activity binding to tissues and serum somatomedin-like receptor activity in the ovine fetus. Biochem Biophys Res Commun 96: 1812-1820

Owens PC, Waters MJ, Thorburn GD, Brinsmead MW (1985) Insulin-like growth factor receptor in fetal lamb liver: characterization and developmental changes. Endocrinology 117: 982-990

Parkes MJ, Hill DJ (1985) Lack of growth hormone-dependent somatomedins or growth retardation in hypophysectomized fetal lambs. J Endocrinol 104: 193-199

Rinderknecht E, Humble RE (1978) Primary structure of human insulin-like growth factor II. FEBS Lett 89: 283-286

Rotwein P (1986) Two insulin-like growth factor I messenger RNAs are expressed in human liver. Proc Natl. Acad Sci USA 83: 77-81

Russell SM, Spencer EM (1985) Local injections of human or rat growth hormone or of purified somatomedin C stimulate unilateral tibial growth in hypophysectomized rats. Endocrinology 116: 2563-2567

Salmon WD, Daughaday WH (1957) A hormonally controlled serum factor which stimulates $^{35}SO_4$ incorporation by cartilage. J Lab Clin Med 49: 825-836

Sara VR, Hall K, Misaki M, Frykland L, Christensen N, Wetterberg I (1983) Ontogenesis of somatomedin and insulin receptors in the human fetus. J Clin Invest 71: 1084-1094

Schlechter NL, Russell SM, Greenberg S, Spencer EM, Nicoll CS (1986) A direct growth effect of growth hormone in rat hindlimb shown by arterial infusion. Am J Physiol 250: E231-E235

Smeets T, van Buul-Offers S (1983) A morphological study of the development of the tibial proximal epiphysis and growth plate of normal and dwarf Snell mice. Growth 47: 145-159

Stevens D, Alexander G (1986) Lipid deposition after hypophysectomy and growth hormone treatment in the sheep fetus. J Dev Physiol 8: 139-145

Stokes H, Boda JM (1968) Immunofluorescent localization of growth hormone and prolactin in the adenohypophysis of fetal sheep. Endocrinology 83: 1362-1366

Stryker JL, Dziuk PJ (1975) Effect of fetal decapitation on fetal development, parturition and lactation in pigs. J Anim Sci 40: 282-287

Walker DG, Simpson ME, Asling CW, Evans HM (1950) Growth and differentiation in the rat following hypophysectomy at 6 days of age. Anat Record 106: 539-554

Wallace ALC, Bassett JM (1970) Plasma growth hormone concentrations in sheep measured by radioimmunoassay. J Endocrinol 47: 21-36

Wallace ALC, Stacy BD, Thorburn GD (1970) The effect of kidney ligation on the release of plasma free fatty acids following the injection of growth hormone. J Endocrinol 48: 297-298

Wallace ALC, Stacy BD, Thorburn GD (1972) The fate of radioiodinated sheep-growth hormone in intact and nephrectomized sheep. Pflugers Arch 331: 25-37

Wallace ALC, Stacy BD, Thorburn GD (1973) Regulation of growth hormone secretion in the ovine fetus. J Endocrinol 58: 89-95

Young IR, Mesiano S, Browne CA, Thorburn GD (1986) Growth hormone and testosterone stimulated growth in hypophysectomized lambs. Proc Endocrinol Soc Aust, vol 29, abstract no 144

Zezulak KM, Green H (1986) The generation of insulin-like growth factor-I-sensitive cells by growth hormone action. Science 233: 551-553

Endocrine Control of Circulation I – Humoral Mechanisms

Endocrine Control of the Fetal Circulation

A. M. RUDOLPH[1]

Introduction

The ability of circulating hormones to affect the fetal circulation was cited by Barcroft (1946), who showed that large doses of epinephrine increased arterial pressure in the fetal lamb. Subsequently, studies were directed to assessing the ability of the circulation to respond to infusion of various hormones, with little regard to the amounts given or to plasma concentrations achieved. Only recently, with the development of techniques to measure plasma concentrations of many hormones, has the role of endocrine systems in controlling fetal circulation under resting conditions and in response to stress begun to be defined. Several issues need to be addressed. These include: whether the fetus is capable of producing the hormone, what concentrations are achieved at rest, whether secretory responses to appropriate stimulation occur, and comparison of plasma concentrations with adult values. Furthermore, the ability of the heart and blood vessels to respond to various concentrations of hormones as well as the magnitude and direction of response have to be defined. In addition, it has become apparent that hormones may be important in the maturation of vascular structures during fetal development, in a manner similar to that shown for lung maturation (Ballard et al. 1977; Mescher et al. 1975). In this review, the role of hormones in normal regulation of the circulation, their responses to stress, and their influence on circulatory responses to stress will be presented briefly, while the effects of hormones on maturation of the circulation, particularly the ductus arteriosus and the heart, will be emphasized.

Circulatory Responses to Hormones

Catecholamines

To separate the effects of catecholamines on the circulation from those of sympathetic nervous influences has been difficult. This has been accomplished by producing a chemical sympathectomy by administration of 6-hydroxydopamine to lamb

---

[1] Cardiovascular Research Institute, Box 0544, HSE 1403, University of California San Francisco, San Francisco, CA 94143, USA

The Endocrine Control of the Fetus
Ed. by W. Künzel and A. Jensen
© Springer-Verlag Berlin Heidelberg 1988

fetuses (Iwamoto et al. 1983; Lewis et al. 1984); this abolishes sympathetic nervous responses but not catecholamine release, although the latter response may be blunted. The effects of catecholamines on the circulation after chemical sympathectomy are discussed by Iwamoto et al. (this volume).

Most studies on sympathoadrenal influences on the circulation have centered on the effects of α- and β-adrenergic blockers both at rest and during stress responses.

Catecholamines have been detected in the adrenal medulla of fetal sheep by 80 days' gestation and their concentrations increase with gestational age; innervation of the medulla, however, occurs much later, at about 120 days (Comline and Silver 1961). Hypoxemia produces marked increases in plasma concentrations of epinephrine and norepinephrine in fetal lambs (Jones and Robinson 1975; Cohen et al. 1982). The response is partly the result of direct effects of hypoxemia on the adrenal medulla and partly due to reflex stimulation, the latter being more important during later gestation (Comline and Silver 1961).

The combined role of sympathetic nerves and catecholamines on the fetal circulation at rest has been studied in fetal lambs by administering selective β- and α-adrenergic receptor blockers at different periods of gestation. β-adrenoreceptor blockade with propranolol demonstrated a 7%–10% decrease of heart rate by 90 days' gestation or even earlier, with a somewhat greater fall of about 12%–13% near term in fetal lambs (Vapaavouri et al. 1973). Similar responses were observed by Walker et al. (1978). However, propranolol produced no significant effects on cardiac output or regional blood flows (Court et al. 1984). α-Adrenoreceptor blockade with phentolamine or phenoxybenzamine resulted in a small fall in arterial blood pressure in fetal sheep. Before 100 days' gestation, only a 5%–6% drop in mean pressure resulted, but in older fetuses a drop of about 10% occurred (Vapaavouri et al. 1973). Thus there is some resting sympathoadrenal influence on the fetal circulation, but it does not appear to have an important role in circulatory regulation.

α-Adrenergic and β-adrenergic mechanisms are of much greater importance, however, during fetal responses to stress. During acute hypoxemia, α-adrenoreceptor blockade significantly reduced arterial pressure, with a resultant fall in blood flow to the umbilical–placental circulation. In addition, the increases in blood flow to the myocardium, brain, and adrenal gland usually achieved during hypoxemia were not achieved (Reuss et al. 1982). β-Adrenoreceptor blockade resulted in a much greater fall in heart rate and in cardiac output than usually occurs during hypoxemia in fetal lambs, indicating the important role of β-adrenoreceptor stimulation during acute stress (Cohen et al. 1982; Court et al. 1984). The circulatory responsiveness to β-adrenergic receptor stimulation has been demonstrated in fetal lambs at an early period of gestation. Barrett et al. (1972) showed that β-adrenoreceptor stimulation by infusion of isoproterenol increased heart rate by 60 days' gestation and also resulted in a fall in blood pressure. Recent studies by Picardo and Rudolph (unpublished observations) in fetal lambs at 120–125 days' gestation, in which isoproterenol was infused at 0.2 μg/kg^{-1}/min^{-1}, a dose that produces maximal increase in fetal heart rate, showed a rise of heart rate from about 175 to about

260 per min. Fetal mean arterial blood pressure fell from 46 to 39 mm Hg. Combined ventricular output did not change significantly, but blood flow to the myocardium more than doubled, flow to the lungs increased threefold, and flow to the adrenal also doubled. A dramatic five- to sixfold increase in blood flow to brown fat also was noted. Blood flow to the kidneys fell significantly to about 75% of control values. There was a tendency for blood flow to the peripheral circulation to increase, suggesting that peripheral vascular resistance fell, but the changes were not statistically significant in this study. However, the studies do indicate that the fetal circulatory response to β-adrenoceptor stimulation is similar to that postnatally and is evident quite early in gestational development.

Renin–Angiotensin System

This hormonal system is developed quite early during gestation. In the fetal lamb, the juxtaglomerular cells in the kidney have been demonstrated by 90 days' gestation, and renin and angiotensin have been detected in fetal blood (Smith et al. 1974). Renin secretion is very sensitive to reduction in blood volume in the fetus, and small amounts of hemorrhage produced an increase in plasma renin and angiotension II concentrations (Broughton-Pipkin et al. 1974). Saralasin produced a modest reduction in arterial blood pressure. Although combined ventricular output did not change, blood flow to the fetal body increased, and umbilical–placental flow fell. Thus, angiotensin II exerts a mild tonic vasoconstriction on the peripheral circulation in the fetal body, which maintains arterial pressure, and because the umbilical–placental blood flow is dependent on perfusion pressure, angiotensin II helps to maintain umbilical blood flow.

When angiotensin II was infused into fetal lambs in amounts that achieve plasma concentrations similar to those that occur as a result of a rapid reduction of blood volume by 15%–20%, arterial pressure was increased, and heart rate initially fell, then increased. Combined ventricular output increased modestly, but umbilical blood flow did not change. Renal blood flow fell, whereas pulmonary and myocardial blood flows increased (Iwamoto and Rudolph 1981a). The renin–angiotensin system is most important in maintaining cardiovascular function in the fetal response to blood loss, as demonstrated by the administration of saralasin acetate to lambs subjected to hemorrhage (Iwamoto and Rudolph 1981b). The response to hemorrhage in the fetal lamb includes a fall in heart rate and blood pressure and a decrease in combined ventricular output. When the animals were subjected to the same amount and rate of blood loss during infusion of saralasin, there was a much greater decrease in heart rate, blood pressure, and combined ventricular output, and umbilical blood flow fell profoundly.

Neurohypophyseal Hormone (Vasopressin)

Vasopressin is detectable in fetal lambs at a very early point in gestation. Drummond et al. (1980) showed that a vasopressin response to blood volume reduction could be demonstrated by 60 days' (0.4 of) gestation. Plasma arginine vasopressin concentrations have been shown to increase markedly in response to hypoxia in fetal lambs (Rurak 1978), and blood volume reduction by about 15% results in a marked increase in plasma vasopressin concentrations to about 50 pg/ml (Drummond et al. 1980).

Resting plasma vasopressin concentrations average about 5 pg/ml, and vasopressin does not appear to exert any effect on the circulation at rest (Kelly et al. 1983). Infusion of vasopressin to achieve average plasma concentrations of about 60 pg/ml produced systemic arterial hypertension and a modest decrease in heart rate (Iwamoto et al. 1979). The bradycardia was not entirely related to the baroreflex response to hypertension because it was not completely abolished by atropine administration. Combined ventricular output did not increase in response to vasopressin infusion, but the proportion of the cardiac output distributed to the fetal body decreased, whilst the umbilical-placental blood flow increased. The role of vasopressin in cardiovascular response to hemorrhage was examined by Kelly et al. (1983), who measured the arterial pressure response before and after administration of a vasopressin antagonist. In the presence of the antagonist, the pressure fall was somewhat greater and was also more prolonged following a rapid hemorrhage.

Hormonal Influences in Circulatory Maturation

Ductus Arteriosus

During fetal life, the ductus arteriosus is a large channel that diverts blood ejected by the right ventricle away from the lungs to the descending aorta. Postnatally, the ductus closes functionally within minutes in rabbits (Hornblad 1969) and hours in human infants (Rudolph et al. 1961). It was proposed by Kennedy and Clark (1942) that closure of the ductus arteriosus after birth is related to the constrictor effects of oxygen on the smooth muscle of the ductus. Exposure of isolated perfused ductus from fetal lambs of different gestational ages to various oxygen concentrations showed that the more mature the lamb, the greater the degree of constriction (McMurphy et al. 1972). This relative insensitivity of the immature ductus arteriosus to oxygen-induced constriction was suggested as an explanation for the delayed closure of the ductus which had been reported in premature infants (Danilowicz et al. 1966). This difference in the response of the immature ductus is not related to a difference in the ability of the ductus smooth muscle to contract effectively, because in response to other constrictor agents, such as potassium or acetylcholine, developed tension is similar in isolated ductus rings obtained from mature and immature fetal lambs (Clyman 1980).

Earlier, it had generally been held that the ductus arteriosus is passively held open during fetal life by the high intraluminal pressure, but the potential role of prostaglandins in maintaining ductus arteriosus patency in the fetus was raised by the observations of Coceani and Olley (1973), that prostaglandins could relax the isolated ductus, and of Sharpe et al. (1974), that prostaglandin synthetase inhibitor could close the ductus arteriosus in utero in fetal rats. The importance of prostaglandins in maintaining ductus arteriosus relaxation was shown in hemodynamic studies in fetal lambs in utero, in which inhibition of prostaglandin synthesis with acetylsalicylic acid caused ductus arteriosus constriction, which could be reversed by infusion of prostaglandin E_1 (Heymann and Rudolph 1976). These studies did not specify whether the effect of the cyclooxygenase inhibitors in preventing prostaglandin synthesis was exerted locally on the ductus or on circulating prostaglandins, and also which arachidonic acid derivative was involved. It has been shown that the ductus arteriosus produces PGE_2, $PGF_{2\alpha}$, and PGI_2 (Pace-Asciak and Rangaraj 1977; Fink and Powell 1985). Although PGI_2, a potent peripheral vasodilator, is produced in much greater quantities than PGE_2, it is very ineffective in producing relaxation of the ductus arteriosus. Clyman (1980) and Coceani et al. (1978) showed that the ability of PGI_2 to dilate the isolated ductus arteriosus of fetal lambs was two to three orders of magnitude less than that of PGE_2. The findings led to the conclusion that the amounts of both PGI_2 and PGE_2 produced locally by the ductus arteriosus are probably inadequate to maintain ductus arteriosus relaxation, and that the relaxation is most likely maintained by circulating PGE_2.

Administration of glucocorticoids to mothers in preterm labor, in an attempt to produce lung maturation in the fetus by stimulating surfactant production, has resulted in a reduced incidence of clinically significant patent ductus arteriosus (Clyman et al. 1981 a). This is probably the result of a direct effect of the glucocorticoid on the ductus itself, rather than due to its role in maturing the lung with resultant alteration of pulmonary circulation. Intravenous infusion of hydrocortisone into premature fetal lambs achieved plasma cortisol concentrations similar to levels present in fetal lambs just before normal birth. After 48 h infusions, the circulatory hemodynamics were examined (Clyman et al. 1981 c) after the lambs were delivered. Hydrocortisone-treated animals showed a much higher ductus vascular resistance than control animals, indicating a greater degree of ductus arteriosus constriction. However, there was no difference in circulating prostaglandin E_2 concentrations, leading to the conclusion that the effect of glucocorticoid was to alter the responsiveness of the ductus, making it less sensitive to the relaxant effects of PGE_2.

In studies with ductus arteriosus rings derived from lambs at different gestational ages, Clyman (1980) showed that on exposure to both oxygen and indomethacin, although the maximal tension developed was similar at all ages, a progressive increase in the proportion of the constriction produced by oxygen and a decrease in that produced by indomethacin occurred with advancing age. Ductus arteriosus rings obtained from immature lambs pretreated with glucocorticoid showed a lesser component of contraction on exposure to indomethacin than the ductus rings obtained from control lambs at a similar age, thus behaving in a manner similar to ductus rings obtained from mature fetuses.

On the basis of these findings, the concept has been generated that the frequency of patent ductus arteriosus after birth in preterm infants is related to greater responsiveness to circulating prostaglandins. An additional factor of importance is that plasma clearance and pulmonary metabolism of PGE_2 are reduced in preterm as compared to full-term fetal lambs after birth (Clyman et al. 1981b). In full-term lambs, by about 2 h after birth, circulating concentrations of prostaglandin are below the threshold which will cause the ductus arteriosus to relax. However, in lambs delivered prematurely, circulating PGE_2 concentrations are higher after birth, whether or not they have been pretreated with glucocorticoid. Therefore, the combination of the higher circulating PGE_2 concentration and the greater sensitivity of the immature ductus to PGE_2 account for the higher incidence of postnatal persistent patency of the ductus.

The mechanism by which glucocorticoid decreases the relaxant response of the ductus arteriosus to PGE_2 is yet to be determined.

Circulation

During fetal life, the combined ventricular output of the heart is about 450 ml min^{-1} kg^{-1} body weight in the lamb (Rudolph and Heymann 1970). Of this, about 300 ml min^{-1} kg^{-1} is ejected by the right, and 150 ml min^{-1} kg^{-1} by the left ventricle. After birth, the cardiac output, or output of each ventricle, is about 300–425 ml min^{-1} kg^{-1}, with a dramatic increase of left ventricular output by two- to threefold (Klopfenstein and Rudolph 1978). Heart rate also increases soon after birth in lambs. On the basis of studies in which rapid infusion of volume into fetal lambs resulted in an increase in right and left atrial pressures, but caused little increase in left or right ventricular output, it was suggested that the fetal heart is operating near the top of its function curve (Gilbert 1980; Thornburg and Morton 1986). Yet the newborn lamb, in spite of the high resting cardiac output, is capable of increasing output substantially in response to volume loading (Klopfenstein and Rudolph 1978). This led to the idea that some mechanism affected the circulation in the perinatal period to improve myocardial performance. Because marked changes in hormonal milieu occur at this time, we considered the possible role of the marked increase in catecholamine concentration (Padbury et al. 1981) and of the circulating thyroid hormone concentrations (Fisher et al. 1977) that occur immediately after birth. It did not appear likely that sympathoadrenal responses were entirely responsible for the increase in left ventricular output and heart rate because administration of propranolol, a β-adrenoreceptor blocker, caused only a 12% fall in heart rate and a 15% fall in cardiac output in 5- to 7-day-old lambs (Klopfenstein and Rudolph 1978).

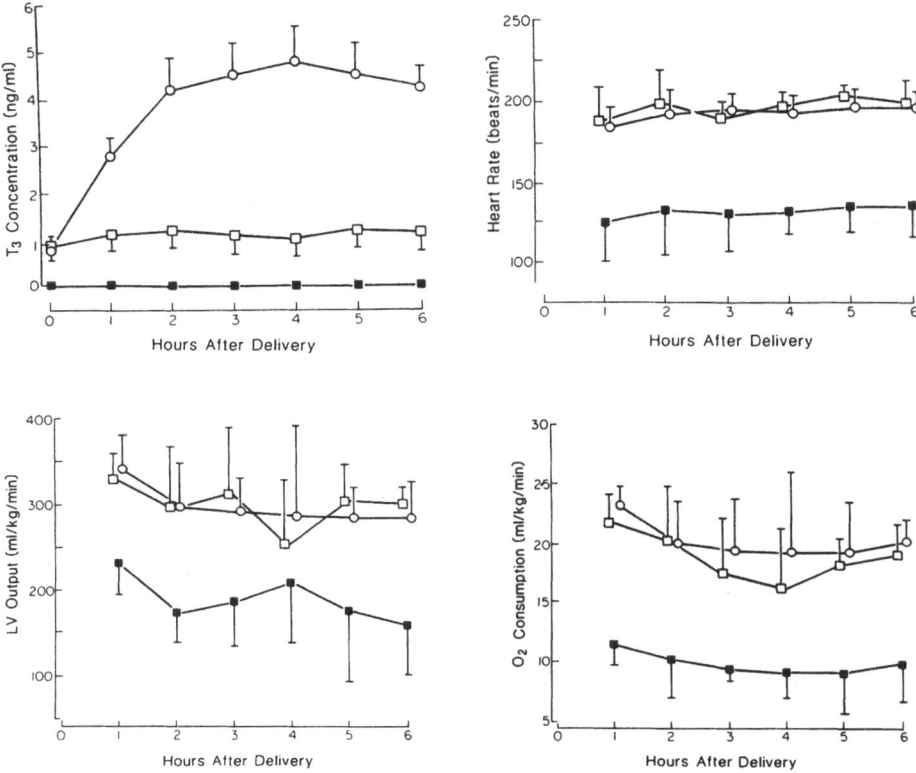

Fig. 1. Changes at various points of time after delivery are shown for plasma triiodothyronine (T₃) concentrations, heart rate, left ventricular output, and oxygen consumption in three groups of lambs. ○ : Control group; ■ : lambs in which thyroidectomy was performed about 2 weeks prior to delivery; □ : lambs in which thyroidectomy was performed immediately prior to delivery. T₃ concentrations at 0 time are those measured in fetal blood just prior to delivery. Data are mean ± SD

The possible role of thyroid hormone was appealing because it had been shown in adult animals that thyroid increases myocardial contractility (Buccino et al. 1967; Taylor et al. 1969). However, the effect of thyroid administration was not immediate; yet the rise in plasma triiodothyronine (T₃) levels occurred within 30 min after birth, and the cardiovascular responses also were rapid. To assess the role of thyroid hormone, we studied three groups of fetal lambs (Breall et al. 1984). Surgery was performed on all lambs in each group at 128–129 days' gestation. Catheters were placed in the left atrium directly at thoracotomy, into an internal thoracic artery and advanced to the aorta, and into a hindlimb artery and vein. In one group, the thyroid gland was removed completely at the time of fetal surgery. All animals were allowed to progress to term (about 142 days' gestation) and were delivered by cesarian section. In the second group, prior to permitting the lamb to breathe and prior to separating the umbilical cord, the thyroid gland was removed under local anesthesia. The third group of lambs served as a nonthyroidectomized control.

The results are depicted in Fig. 1. It is apparent that in the control group, left ventricular output, heart rate, and total body oxygen consumption increased to normal neonatal levels, which were sustained for several hours. Also, plasma T_3 concentrations showed the expected postnatal increase. In the animals thyroidectomized during fetal life, there was no detectable T_3, and left ventricular output, heart rate, and oxygen consumption remained at normal fetal levels for the 6 h of measurement. However, in the animals thyroidectomized immediately prior to delivery, plasma T_3 concentrations did not increase above fetal levels after birth, yet the heart rate, cardiac output, and oxygen consumption responses were identical to those seen in the control lambs.

We concluded that the postnatal increase in plasma T_3 concentrations is not responsible for the circulatory changes after birth, but that prenatal thyroid action is essential. The mechanisms by which the thyroid permits these responses have not been determined. It has been suggested that thyroid hormone increases myocardial contractility through an effect on cardiac myosin adenosine triphosphatase activity (Morkin 1979). Also, Fink and Morton (1977) indicated that increased contractility could be explained in part by appearance of a new myosin with greater actin-activated adenosine triphosphatase activity. Several studies have shown that excess thyroid hormones increase β-adrenoreceptor activity (Whitsett et al. 1982a) and also that myocardial β-adrenoreceptors are markedly reduced in developing rat pups made hypothyroid in utero (Whitsett et al. 1982b). We have found that numbers of myocardial β-adrenoreceptors are reduced by 50%, and there is a reduction in isoproterenol-stimulated adenylate cyclase in lamb hearts derived after birth when thyroidectomy was performed at about 130 days' gestation (E. Birk, K. Jacobs, J. Roberts, A. M. Rudolph, preliminary observations). Further studies are currently in progress to define the responsiveness of the heart to β-adrenergic stimulation in fetal lambs following thyroidectomy and in lambs made hypothyroidic.

There is increasing evidence that prenatal hormonal activity is important in maturation of many biochemical and physiological functions. The circulation is also dependent on fetal hormonal activity for appropriate postnatal adaptation.

References

Ballard PL, Benson BJ, Brehier A (1977) Glucocorticoid effects in the fetal lung. Am Rev Respir Dis 115: 29–36

Barcroft J (1946) Researches on prenatal life. Blackwell, Oxford

Barrett CT, Heymann MA, Rudolph AM (1972) Alpha and beta adrenergic receptor activity in fetal sheep. Am J Obstet Gynecol 112: 1114–1121

Breall JA, Rudolph AM, Heymann MA (1984) Role of thyroid hormone in postnatal circulatory and metabolic adjustments. J Clin Invest 73: 1418–1424

Broughton-Pipkin F, Kirkpatrick SML, Lumbers ER, Mott JC (1974) Renin and angiotensin-like levels in foetal, newborn, and adult sheep. J Physiol (Lond) 241: 575-588

Buccino RA, Spann JF Jr, Pool PE, Sonnenblick EH, Braunwald E (1967) Influence of thyroid state on the intrinsic contractile state and energy stores of the myocardium. J Clin Invest 46: 1669-1682

Clyman RI (1980) Ontogeny of the ductus arteriosus response to prostaglandins and inhibitors of their synthesis. Semin Perinatol 4: 115-124

Clyman RI, Ballard PL, Sniderman S, Ballard RA, Roth R, Heymann MA, Granberg JP (1981a) Prenatal administration of betamethasone for prevention of patent ductus arteriosus. J Pediatr 98: 123-126

Clyman RI, Mauray F, Heymann MA, Roman C (1981b) Effect of gestational age on pulmonary metabolism of prostaglandin E_1 and E_2. Prostaglandins 21: 505-513

Clyman RI, Mauray F, Roman C, Heymann MA, Ballard PL, Rudolph AM, Payne B (1981c) Effects of antenatal glucocorticoid administration on the ductus arteriosus of preterm lambs. Am J Physiol 241: H415-H420

Coceani F, Olley PM (1973) The response of the ductus arteriosus to prostaglandins. Can J Physiol Pharmacol 51: 220-225

Coceani F, Bodach E, White E, Bishai I, Olley PM (1978) Prostaglandin I_2 is less relaxant than prostaglandin E_2 on the lamb ductus arteriosus. Prostaglandins 15: 551-556

Cohen WR, Piasecki GJ, Jackson BT (1982) Plasma catecholamines during hypoxemia in fetal lambs. Am J Physiol 243: R520-R525

Cohn HE, Piasecki GJ, Jackson BT (1982) The effect of β-adrenergic stimulation on fetal cardiovascular function during hypoxemia. Am J Obstet Gynecol 114: 810-816

Comline RS, Silver M (1961) The release of adrenaline and noradrenaline from the adrenal glands of the foetal sheep. J Physiol (Lond) 156: 424-444

Court DJ, Parer JT, Block BSB, Llanos AJ (1984) Effects of beta-adrenergic blockade on blood flow distribution during hypoxemia in fetal sheep. J Dev Physiol 6: 349-358

Danilowicz D, Rudolph AM, Hoffman JIE (1966) Delayed closure of ductus arteriosus in premature infants. Pediatrics 37: 74-78

Drummond WH, Rudolph AM, Keil LC, Gluckman PC, MacDonald AA, Heymann MA (1980) Arginine vasopressin and prolactin after hemorrhage in the fetal lamb. Am J Physiol 238: E214-E219

Fink CD, Morkin E (1977) Evidence for a new cardiac myosin species in thyrotoxic rabbits. FEBS Lett 81: 391-394

Fink CD, Powell WS (1985) Release of prostaglandins and monohydroxy and trihydroxy metabolites of linoleic and arachidonic acids by adult and fetal aortae and ductus arteriosus. J Biol Chem 260: 7481-7488

Fisher DA, Dussault JH, Sack J, Chopra IJ (1977) Ontogenesis of hypothalamic-pituitary-thyroid function and metabolism in man, sheep, and rat. Recent Prog Horm Res 33: 59-166

Gilbert RD (1980) Control of fetal cardiac output during changes in blood volume. Am J Physiol 238: H80-H86

Heymann MA, Rudolph AM (1976) Effects of acetylsalicylic acid on the ductus arteriosus and circulation in fetal lambs in utero. Circ Res 38: 418-422

Hornblad PY (1969) Experimental studies on closure of the ductus arteriosus utilizing whole-body freezing. Acta Pediatr Scand [Suppl] 190: 6-21

Iwamoto HS, Rudolph AM (1979) Effects of endogenous angiotensin II on the fetal circulation. J Dev Physiol 1: 283-293

Iwamoto HS, Rudolph AM (1981a) Effects of angiotensin II on the blood flow and its distribution in fetal lambs. Circ Res 48: 183-189

Iwamoto HS, Rudolph AM (1981b) Role of the renin-angiotensin system in the fetal response to hemorrhage. Am J Physiol 240: H848-H854

Iwamoto HS, Rudolph AM, Keil LC, Heymann MA (1979) Hemodynamic responses of the sheep fetus to vasopressin infusion. Circ Res 44: 430-436

Iwamoto HS, Rudolph AM, Mirkin BL, Keil LC (1983) Circulatory and humoral responses of sympathectomized fetal sheep to hypoxemia. Am J Physiol 245: H767–H772

Jones CT, Robinson RO (1975) Plasma catecholamines in foetal and adult sheep. J Endocrinol 73: 11–20

Kelly RT, Rose JC, Meiss PJ, Hargrave BY, Morris M (1983) Vasopressin is important for restoring cardiovascular homeostasis in fetal lambs subjected to hemorrhage. Am J Obstet Gynecol 146: 807–812

Kennedy JA, Clark SL (1942) Observations on the physiological reactions of the ductus arteriosus. Am J Physiol 136: 140–147

Klopfenstein HS, Rudolph AM (1978) Postnatal chagnes in the circulation and responses to volume loading in the sheep. Circ Res 423: 839–845

Lewis AB, Wolf WJ, Sischo W (1984) Fetal cardiovascular and catecholamine responses to hypoxemia after chemical sympathectomy. Pediatr Res 18: 318–322

McMurphy DM, Heymann MA, Rudolph AM, Melmon KL (1972) Developmental changes in constriction of the ductus arteriosus: responses to oxygen and vasoactive substances in the isolated ductus arteriosus of the fetal lamb. Pediatr Res 6: 231–238

Mescher EJ, Platzker ACG, Ballard PL, Kitterman JA, Clements JA, Tooley WH (1975) Ontogeny of tracheal fluid, pulmonary surfactant, and plasma corticoids in the fetal lamb. J Appl Physiol 39: 1017–1021

Morkin E (1979) Stimulation of cardiac myosin adenosine triphosphate in thyrotoxicosis. Circ Res 44: 1–7

Pace-Asciak CR, Ragnaraj G (1977) The 6 keto-prostaglandin $F_{1\alpha}$ pathway in the lamb ductus arteriosus. Biochim Biophys Acta 486: 583–585

Padbury JF, Diakomanolis ES, Hobel CJ, Perelman A, Fisher DA (1981) Neonatal adaptation: sympatho-adrenal response to cord cutting. Pediatr Res 15: 1483–1487

Reuss ML, Parer JT, Harris JL, Krueger TR (1982) Hemodynamic effects of alpha-adrenergic blockade during hypoxia in fetal sheep. Am J Obstet Gynecol 142: 410–415

Rudolph AM, Heymann MA (1970) Circulatory changes during growth in the fetal lamb. Circ Res 26: 289–299

Rudolph AM, Drorbraugh JE, Auld PAM, Rudolph AJ, Nadas AS, Smith CA, Hubbell JP (1961) Studies on the circulation in the neonatal period. The circulation in the respiratory distress syndrome. Pediatrics 27: 551–566

Rurak DW (1978) Plasma vasopressin levels during hypoxemia and the cardiovascular effects of exogenous vasopressin in fetal and adult sheep. J Physiol (Lond) 277: 341–357

Sharpe GL, Thalme B, Larsson KS (1974) Studies on closure of the ductus arteriosus. XI. Ductal closure in utero by a prostaglandin synthetase inhibitor. Prostaglandins 8: 363–368

Smith FG Jr, Lupu AN, Barajas L, Bauer R, Bashore RA (1974) The renin-angiotensin system in the fetal lamb. Pediatr Res 8: 611–620

Taylor RR, Covell JW, Ross JJ (1969) Influence of the thyroid state on left ventricular tension-velocity relations in the intact sedated dog. J Clin Invest 48: 775–784

Thornburg KL, Morton MJ (1986) Filling and arterial pressures as determinants of left ventricular stroke volume in fetal lambs. Am J Physiol 251: H961–H968

Vapaavouri EK, Shinebourne EA, Williams RL, Heymann MA, Rudolph AM (1973) Development of cardiovascular responses to autonomic blockade in intact fetal and neonatal lambs. Biol Neonate 22: 177–188

Walker AM, Cannata J, Dowling MH, Ritchie B, Maloney JE (1978) Sympathetic and parasympathetic control of heart rate in unanesthetized fetal and newborn lambs. Biol Neonate 33: 135–143

Whitsett JA, Noguchi A, Moore JJ (1982a) Developmental aspects of alpha and beta-adrenergic receptors. Semin Perinatol 6: 125–141

Whitsett JA, Pollinger J, Matz S (1982b) Beta-adrenergic receptors and catecholamine sensitive adenylate cyclase in developing rat ventricular myocardium: effect of thyroid status. Pediatr Res 16: 463–469

Control of the Pulmonary Circulation in the Perinatal Period

M. A. HEYMANN[1]

Introduction

In the fetus, gas exchange occurs in the placenta and pulmonary blood flow requirements are therefore low. In the undisturbed near-term fetal lamb (the species in which accurate measurements have been made), pulmonary blood flow is 90–110 ml min^{-1} 100 g^{-1} wet lung tissue or 30–40 ml min^{-1} kg^{-1} fetal body weight. This represents about 8% of total fetal cardiac output, which near term is normally about 450 ml min^{-1} kg^{-1} fetal body weight (Rudolph and Heymann 1970; Heymann et al. 1973). The low pulmonary blood flow at this stage of gestation is maintained by a high pulmonary vascular resistance. Shortly after birth, with the initiation of pulmonary ventilation, pulmonary vascular resistance and pulmonary arterial pressure fall rapidly. There is an associated eight- to ten-fold increase in pulmonary blood flow, which in lambs reaches 300–400 ml min^{-1} kg^{-1} body weight shortly after birth (Kuipers et al. 1982). In the sheep, pulmonary arterial blood pressure falls towards adult levels within several hours; however, in humans at 24 h of age, normal mean pulmonary arterial blood pressure may still be about twice normal adult levels (Moss et al. 1963). After the initial quite rapid fall in pulmonary vascular resistance and pulmonary arterial blood pressure, there is a slow progressive fall to near-adult levels by 3–6 weeks after birth (Lucas et al. 1961) related to growth of new vessels, to the involution of the large amount of medial smooth muscle found normally in the small pulmonary arteries in the fetus, and also possibly to changes in viscosity as hematocrit falls.

The exact mechanisms controlling the initial dramatic changes in pulmonary vascular resistance with the start of ventilation are not yet fully understood. Many factors produce pulmonary vasodilatation, and the perinatal changes in the pulmonary circulation very probably involve a complex interaction of many of these factors (Fig. 1).

[1] Cardiovascular Research Institute and Department of Pediatrics, Department of Physiology, and Department of Obstetrics, Gynecology and Reproductive Sciences, University of California, San Francisco, CA 94143, USA

The Endocrine Control of the Fetus
Ed. by W. Künzel and A. Jensen
© Springer-Verlag Berlin Heidelberg 1988

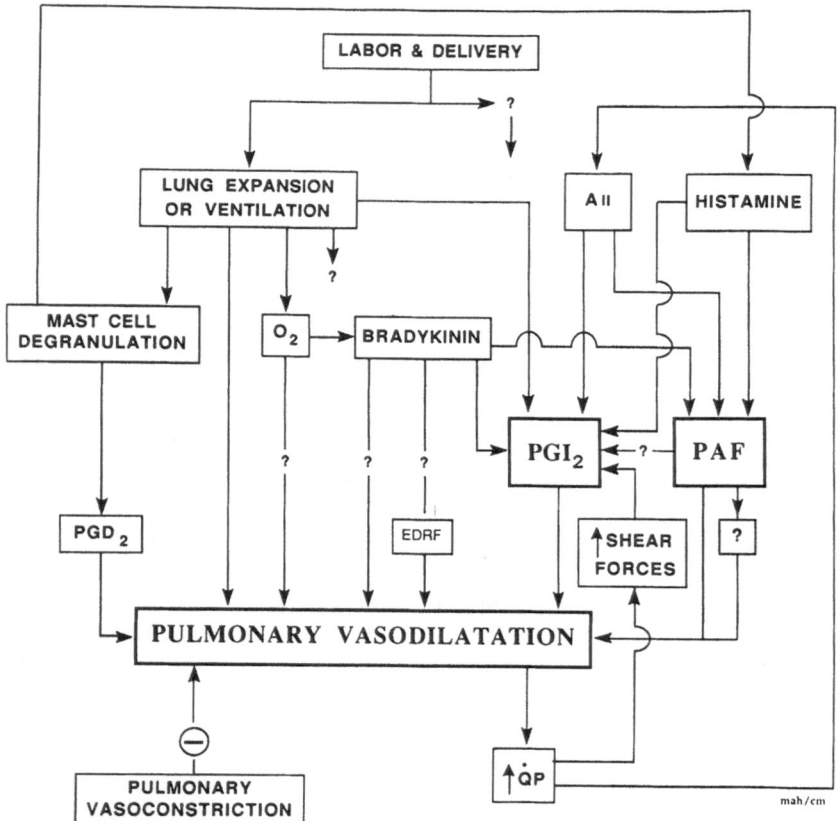

Fig. 1. Possible factors in postnatal pulmonary vasodilatation

Role of O₂

In the postnatal period, the level of O_2 to which the small pulmonary arteries (resistance vessels) are exposed most likely plays a direct role in regulating pulmonary vascular resistance and thereby pulmonary blood flow; however, perhaps even more important is the interrelationship of an increase in O_2 environment and the subsequent production of vasoactive substances by the lungs. In late gestation, the pulmonary circulation responds readily to changes in O_2 environment, and these responses may be even more pronounced after birth (Heymann et al. 1969; Lewis et al. 1976; Rudolph and Yuan 1966). Reducing PaO_2 close to fetal levels in newborns produces a marked increase in pulmonary vascular resistance and concurrent fall inn pulmonary blood flow; conversely, increasing fetal PaO_2 above normal, either by ventilating fetuses with oxygen (Cassin et al. 1964a) or by exposing fetal lambs to hyperbaric oxygenation (Heymann et al. 1969), increases pulmonary blood flow. Nevertheless, although a change in O_2 environment could directly relax or constrict pulmonary vascular smooth muscle, it also could modify the local production and release of vasoactive substances, which then could act directly or indirectly on the pulmonary vasculature.

Role of Vasoactive Substances

Bradykinin is a potent vasoactive peptide that for some time has been considered a possible factor in regulating perinatal cardiovascular changes. It produces marked pulmonary vasodilatation when infused into fetal lambs (Campbell et al. 1967) and is released from fetal lungs either by ventilation with air or during hyperbaric oxygenation without ventilation (Heymann et al. 1969). Bradykinin could induce pulmonary vasodilatation by several different mechanisms (see Fig. 1). It could act directly on the pulmonary vascular smooth muscle or, as is now considered more likely, through local stimulation of production of a second vasoactive substance, such as endothelial-dependent relaxing factor (EDRF), or more particularly by stimulating production of prostaglandin I_2 (PGI_2) or platelet activating factor (McIntyre et al. 1985; see below).

Rhythmic ventilation of fetal lungs with a gas mixture (but not a liquid) that does not alter fetal arterial PO_2 at all reduces pulmonary vascular resistance, an action possibly related to two mechanisms. The first, which plays only a minimal role, if any, is a change in surface tension factors acting at the alveolar air-liquid interface to reduce perivascular tissue pressure (Cassin et al. 1964a, b; Enhorning et al. 1966). The second mechanism probably has greater direct physiologic importance: *mechanical* distortion, stretch, or distension of the lung, leading to the release of vasoactive substances, particularly PGI_2, which in turn affect the pulmonary vasculature (see below).

Prostaglandins also are potent vasoactive substances, and their possible role in perinatal pulmonary vascular physiologic changes has been studied quite extensively. In near-term fetal animals, exogenous prostaglandin E_1 (PGE_1) and prostaglandin E_2 (PGE_2) both produce modest pulmonary vasodilatation (Cassin 1980; Cassin et al. 1979). PGI_2 is also pulmonary vasodilator (Cassin 1980; Cassin et al. 1981; Leffler and Hessler 1979) and is considerably more potent than either PGE_1 or PGE_2 (Cassin 1980). None of these vasodilating prostaglandins is truly specific for the pulmonary circulation, and all affect sytemic vascular resistance to a similar degree. However, these nonspecific systemic effects do not exclude the possibility that endogenous local prostaglandin production and rapid metabolism occur in the perinatal period, and there now is quite strong evidence that PGI_2 does play a physiologic role in postnatal pulmonary vasodilatation. Several very different mechanisms could be responsible for the production of PGI_2. Lung distortion, distension, or mechanical stimulation leads to PGI_2 production (Gryglewski 1980; Gryglewski et al. 1978), and spontaneous or mechanical ventilation of fetal lungs, even without changing O_2 environment, is associated with the net production and release into left atrial blood of the stable metabolite of PGI_2 (6-keto-$PGF_{1\alpha}$; Leffler et al. 1984a, b; Leffler et al. 1980). Circulating concentrations of this metabolite fell after several hours, further suggesting a pulmonary vasodilatory role for PGI_2 in the immediate postnatal period.

As indicated previously, bradykinin is produced by the lung in response to oxygenation (Heymann et al. 1969). Bradykinin stimulates PGI_2 production both by the intact lung (Gryglewski 1980; Leffler et al. 1984a) and by vascular endothelial cells in culture (McIntyre et al. 1985); therefore, this is a possible mechanism by which PGI_2 production could be stimulated in the transitional phase. An additional mechanism whereby PGI_2 could be produced involves angiotensin II (AII), which increases in concentration with establishment of the pulmonary circulation at birth and which also is known to stimulate PGI_2 production (Dusting 1981). Furthermore, neonatal lung is more sensitive to AII stimulation of PGI_2 production than is fetal or adult lung (Omini et al. 1983). A further mechanism for PGI_2 release involves the physical effects of increased flow and the resultant increase in shear forces, which has been shown to induce PGI_2 production in pulmonary vascular endothelial cells in culture (VanGrondelle et al. 1984). If pulmonary blood flow (QP) were to increase (e.g., in direct response to an increase in O_2 or bradykinin), then subsequent shear-induced PGI_2 production could produce further or sustained pulmonary vasodilatation.

Prostaglandin D_2 (PGD_2) also has been shown to be a pulmonary vasodilator in the perinatal period. In newborn lambs with hypoxia-induced pulmonary hypertension, PGD_2 reduces pulmonary arterial blood pressure and calculated pulmonary vascular resistance and increases pulmonary arterial blood pressure and calculated pulmonary vascular resistance and increases pulmonary as well as systemic blood flows. Beyond about 10 days of age, however, PGD_2 produces pulmonary vasoconstriction (Soifer et al. 1983). These fairly specific early perinatal pulmonary vascular effects suggest a physiologic role for PGD_2 in establishing a normal pulmonary circulation in the immediate newborn period. This hypothesis is further supported by studies in rhesus monkeys that showed a progressive increase in the number of mast cells in fetal lungs towards the end of gestation, with fewer mast cells observed in the lungs after birth (Schwartz et al. 1974). Because PGD_2 is released from mast cells, degranulation of these cells with the release of PGD_2 may be involved in regulation of postnatal pulmonary vasodilatation. Unlike the specific physiologic mechanisms that could initiate the release of PGI_2, those for perinatal release of PGD_2 are not known. It is interesting that histamine, like PGD_2, is a pulmonary vasodilator in the immediate perinatal period (Goetzman and Milstein 1980); histamine also is released when mast cells degranulate.

The active role of vasodilating prostaglandins in the immediate postnatal period has been further substantiated by studies in perfused fetal lungs using inhibition of prostaglandin synthesis (Leffler et al. 1978). When fetal lungs were ventilated, the ensuing fall in pulmonary vascular resistance had two separate phases: an initial rapid fall in the first 30–60 s after onset of ventilation, followed by a slower fall over about 15 min. This latter phase was attenuated by pretreatment with indomethacin, an inhibitor of prostaglandin synthesis. These studies indicate neither which prostaglandin is responsible nor which is more important, but do confirm the role of one or other vasodilating cyclo-oxygenase product of arachidonic acid metabolism.

An additional possible lipid mediator of perinatal pulmonary vasodilatation that should be considered is platelet activating factor (PAF). Although generally intracellular, PAF can be released in the presence of cellular injury. Recent work has shown that small doses of PAF can produce pulmonary vasodilatation (Voelkel et al. 1986), and endothelial cells in culture produce PAF when stimulated by the same factors (viz. bradykinin, histamine) that stimulate PGI_2 production (McIntyre et al. 1985).

An additional important consideration is that the fetal pulmonary circulation is actively maintained in a constricted state, and inhibition of the factors responsible for this constriction will allow for pulmonary vascular smooth muscle relaxation, a reduction in pulmonary vascular resistance, and increased pulmonary blood flow. The factors regulating fetal pulmonary vascular resistance are not fully understood, but clearly involve the normally low-O_2 environment in the fetus and also the lipoxygenase products of arachidonic acid metabolism, the leukotrienes. Administration of either a putative leukotriene receptor blocker (FPL 57231) or a putative lipoxygenase inhibitor (U60257B) to normal fetal lambs dramatically increases pulmonary blood flow (Soifer et al. 1985; LeBidois et al. 1987), strongly suggesting a role for the production of leukotrienes in active regulation of pulmonary vascular resistance in the fetus. Further studies indicate that this effect most likely does not involve the secondary, leukotriene-induced, release of thromboxane (Clozel et al. 1985).

Therefore, regulation of the fetal and immediately postnatal pulmonary circulation probably reflects a balance between factors producing active pulmonary vasoconstriction and those producing pulmonary vasodilatation. The great increase in pulmonary blood flow after birth could reflect a major shift from active pulmonary vasoconstriction to active pulmonary vasodilatation. Possibly, arachidonic acid or even phospholipid precursor metabolism shifts from lipoxygenase products (leukotrienes) in the low-O_2 environment of the fetus toward cyclo-oxygenase products (prostaglandins), either with mechanical stimulation due to lung expansion or in the higher O_2 environment after birth.

Acknowledgement. The work on which this paper is based was supported in part by the United States Public Health Service Program Project Grant HL-24056 from the National Heart, Lung, and Blood Institute.

References

Campbell AGM, Dawes GS, Fishman AP, Hyman AI (1967) Pulmonary vasoconstriction and changes in heart rate during asphyxia in immature foetal lambs. J Physiol (Lond) 192: 93-110

Cassin S (1980) Role of prostaglandins and thromboxanes in the control of the pulmonary circulation of the fetus and newborn. Semin Perinatol 4: 101-107

Cassin S, Dawes GS, Mott JC, Ross BB, Strang LB (1964a) The vascular resistance of the foetal and newly ventilated lung of the lamb. J Physiol (Lond) 171: 61-79

Cassin S, Dawes GS, Ross BB (1964b) Pulmonary blood flow and vascular resistance in immature foetal lambs. J Physiol (Lond) 171: 80-89

Cassin S, Tyler TL, Leffler C, Wallis R (1979) Pulmonary and systemic vascular responses of perinatal goats to prostaglandin E_1 and E_2. Am J Physiol 236: H828-H832

Cassin S, Winikor I, Tod M, Philips J, Frisinger S, Jordan J, Gibbs C (1981) Effects of prostacyclin on the fetal pulmonary circulation. Pediatr Pharmacol (New York) 1: 197-207

Clozel M, Clyman RI, Soifer SJ, Heymann MA (1985) Thromboxane is not responsible for the high pulmonary vascular resistance in fetal lambs. Pediatr Res 19: 1254-1257

Dusting AJ (1981) Angiotensin-induced release of a prostacyclin-like substance from the lungs. J Cardiovasc Pharmacol 3: 197-206

Enhorning G, Adams FH, Norman A (1966) Effect of lung expansion on the fetal lamb circulation. Acta Paediatr Scand 55: 441-451

Goetzman BW, Milstein JM (1980) Pulmonary vascular histamine receptors in newborn and young lambs. J Appl Physiol 49: 380-385

Gryglewski RJ (1980) The lung as a generator of prostacyclin. Ciba Found Symp 78: 147-164

Gryglewski RJ, Korbut R, Ocetkiewicz A (1978) Generation of prostacyclin by lungs in vivo and its release into the arterial circulation. Nature 273: 765-767

Heymann MA, Rudolph AM, Nies AS, Melmon KL (1969) Bradykinin production associated with oxygenation of the fetal lamb. Circ Res 25: 521-534

Heymann MA, Creasy RK, Rudolph AM (1973) Quantitation of blood flow patterns in the foetal lamb in utero. In: Comline KS, Cross KW, Dawes GS, Nathanielsz PW (eds) Proceedings of the Sir Joseph Barcroft Centenary Symposium. Foetal and Neonatal Physiology. Cambridge University Press, Cambridge, pp 129-135

Kuipers JRG, Sidi D, Heymann MA, Rudolph AM (1982) Comparison of methods of measuring cardiac output in newborn lambs. Pediatr Res 16: 594-598

LeBidois J, Soifer SJ, Clyman RI, Heymann MA (1987) Piriprost, a putative leukotriene inhibitor, increases pulmonary blood flow in fetal lambs. Pediatr Res 22: 350-354

Leffler CW, Hessler JR (1979) Pulmonary and systemic vascular effects of exogenous prostaglandin I_2 in fetal lambs. Eur J Pharmacol 54: 37-42

Leffler CW, Tyler TL, Cassin S (1978) Effect of indomethacin on pulmonary vascular response to ventilation of fetal goats. Am J Physiol 234: H346-H351

Leffler CW, Hessler JR, Terragno NA (1980) Ventilation-induced release of prostaglandin-like material from fetal lungs. Am J Physiol 238: H282-H286

Leffler CW, Hessler JR, Green RS (1984a) Mechanism of stimulation of pulmonary prostaglandin synthesis at birth. Prostaglandins 28: 877-887

Leffler CW, Hessler JR, Green RS (1984b). The onset of breathing at birth stimulates pulmonary vascular prostacyclin synthesis. Pediatr Res 18: 938-942

Lewis AB, Heymann MA, Rudolph AM (1976) Gestational changes in pulmonary vascular responses in fetal lambs in utero. Circ Res 39: 536-541

Lucas RV Jr, St Geme JW Jr, Anderson RC, Adams P Jr, Ferguson DJ (1961) Maturation of the pulmonary vascular bed: a physiologic and anatomic correlation in infants and children. Am J Dis Child 101: 467-475

McIntyre TM, Zimmerman GA, Satoh K, Prescott SM (1985) Cultured endothelial cells synthesize both platelet-activating factor and prostacyclin in response to histamine, bradykinin, and adenosine triphosphate. J Clin Invest 76: 271–280

Moss AJ, Emmanouilides G, Duffie ER Jr (1963) Closure of the ductus arteriosus in the newborn infant. Pediatrics 32: 25–30

Omini C, Vigano T, Marini A, Pasargiklian R, Fano M, Maselli MA (1983) Angiotensin II: A releaser of PGI_2 from fetal and newborn rabbit lungs. Prostaglandins 25: 901–910

Rudolph AM, Heymann MA (1970) Circulatory changes during growth in the fetal lamb. Circ Res 26: 289–299

Rudolph AM, Yuan S (1966) Response of the pulmonary vasculature to hypoxia and H^+ ion concentration changes. J Clin Invest 45: 399–411

Schwartz LW, Osburn BI, Frick OL (1974) An ontogenic study of histamine and mast cells in the fetal rhesus monkey. J Allergy Clin Immunol 56: 381–386

Soifer SJ, Morin FC III, Kaslow DC, Heymann MA (1983) The development effects of PGD_2 on the pulmonary and systemic circulations in the newborn lamb. J Dev Physiol 5: 237–250

Soifer SJ, Loitz RD, Roman C, Heymann MA (1985) Leukotriene end organ antagonists increase pulmonary blood flow in fetal lambs. Am J Physiol 249: H570–H576

VanGrondelle A, Worthen S, Ellis D, Mathias MM, Murphy RC, Murphy RJ, Strife J, Reeves JT, Voelkel NF (1984) Altering hydrodynamic variables influences PGI_2 production by isolated lungs and endothelial cells. J Appl Physiol 57: 388–395

Voelkel NF, Chang SW, Pfeffer KD, Worthen SG, McMurtry IF, Henson PM (1986) PAF antagonists: Different effects on platelets, neutrophils, guinea pig ileum and PAF-induced vasodilation in isolated rat lung. Prostaglandins 32: 359–372

Maternal and Fetal Responses to Long-term Hypoxemia in Sheep*

T. KITANAKA, R. D. GILBERT, and L. D. LONGO[1]

Abstract

To determine the maternal and fetal cardiovascular and hematological responses to long-term hypoxemia, we studied three groups of animals: (a) pregnant ewes ($n=20$) at 110 to 115 days' gestation, subjected to hypoxemia for up to 28 days; (b) pregnant ewes ($n=4$) which served as a normoxic control group, and (c) nonpregnant ewes ($n=6$) which were subjected to hypoxemia for up to 28 days. In the ewes we measured mean arterial pressure, heart rate, uterine blood flow, and uterine vascular resistance continuously for 1 h/day. Arterial PO_2, O_2 saturation, hemoglobin concentration, hematocrit, arteriovenous O_2 difference, uterine O_2 uptake, and blood volume were measured daily or biweekly. In the fetus we measured the above except the uterine blood flow and O_2 difference. Following the control measurements the ewes were placed in a chamber with an atmosphere of 12%–13% O_2 for up to 28 days.

In response to long-term hypoxemia both pregnant and nonpregnant sheep experienced only minor cardiovascular and hematologic responses. In the ewe hemoglobin and erythropoietin concentrations increased slightly. The arterial O_2 content, glucose concentration, mean arterial pressure, and cardiac output decreased slightly. In contrast, body weight, heart rate, blood volume, uterine blood flow, uterine O_2 flow, and uteroplacental O_2 uptake remained relatively constant.

The fetuses underwent somewhat more dramatic responses. In relation to controls the hemoglobin concentration, hematocrit, whole blood volume, erythrocyte mass, and erythropoietin concentration increased moderately. Perhaps surprisingly, these fetuses were able to compensate, so that at term their body weights were normal.

* This study was supported in part by United States Public Health Service grants HD 03807 and HD 22190.
[1] Division of Perinatal Biology, Departments of Physiology and Obstetrics and Gynecology, School of Medicine, Loma Linda University, Loma Linda, CA 92350, USA

The Endocrine Control of the Fetus
Ed. by W. Künzel and A. Jensen
© Springer-Verlag Berlin Heidelberg 1988

Introduction

Pregnancy is associated with a number of physiological changes in the maternal cardiovascular, respiratory, endocrine, metabolic, and immune systems which serve to ensure normal fetal development. In humans and some other species these changes include increases in maternal blood volume, cardiac output, and uterine blood flow to optimize the delivery of oxygen and substrates to the developing fetus via the placenta.

Under normal conditions the fetal arterial O_2 tension is about 25 Torr, i.e., about 25% of the adult value. These low O_2 tensions combined with a relatively steep oxyhemoglobin saturation curve make the fetus particularly vulnerable to hypoxemia (Longo 1987). Theoretical studies suggest that one of the most important factors that affects fetal O_2 levels is the maternal arterial O_2 tension (Longo et al. 1972), thus the question arises: To what extent are fetal functions altered when the maternal arterial O_2 content is abnormally low? A corollary question is: What are the maternal and fetal physiologic adaptations that allow pregnancy to progress and succeed under these circumstances?

Although numerous studies have explored maternal and fetal adaptations to acute or short-term hypoxemia, i.e., up to 4 h, there is little information on these adaptations in response to more prolonged hypoxemia, i.e., several weeks. Long-term hypoxemia, such as occurs in residents at high altitude, can pose a threat to both mother and fetus. In addition, long-term hypoxemia may be a factor in the development of the fetus that is growth-retarded, i.e., small for gestational age, in instances of multifetal pregnancy, mothers with cyanotic congenital heart disease, and pregnancy-induced hypertension or other conditions with reduced uteroplacental blood flow (Longo 1987).

At high altitude a number of investigators have reported that infant birth weight is reduced (Howard et al. 1957b; Lichty et al. 1957; McClung 1969; Kruger and Arias-Stella 1970; Moore et al. 1982; Sobrevilla et al. 1971), and that infant mortality is increased (Grahn and Kratchman 1963; Lichty et al. 1957; Mazess et al. 1965; McCullough et al. 1977). Nonetheless, not all newborn infants were small for gestational age (McCullough et al. 1977; Moore et al. 1982) and umbilical venous oxyhemoglobin saturations were not necessarily low (Howard et al. 1957b). O_2 saturation initially decreased dramatically, then over a period of 6–18 days increased to values similar to those existing at lower altitude (Makowski et al. 1968). In humans at 4200 m fetal scalp O_2 tension averaged 19 Torr, a value not significantly different from the sea level value of 22 Torr (Sobrevilla et al. 1971). Therefore, it would appear that some maternal and/or fetal mechanisms help to optimize O_2 delivery under these circumstances.

In an effort to define the roles of maternal and fetal cardiovascular, hematologic, and endocrinologic responses to long-term hypoxemia, the time course of their change, and their relative importance, we measured a number of physiologic variables in chronically catheterized sheep. In the mother these included: respiratory blood gases, plasma glucose concentration, hemoglobin concentration, hematocrit, blood volume, blood pressure, cardiac output, total uterine blood flow, uteroplacental O_2 uptake, and the concentrations of erythropoietin, the catecholamines, and cortisol. Fetal measurements included all of the above except cardiac output, uterine blood flow, and uteroplacental O_2 uptake.

Methods

Principle of Methods

Three to 5 days after pregnant ewes were received into the animal care facility, we performed surgery under general anesthesia at 105–115 gestational days. Postoperatively they were kept in a normobaric chamber in which they remained during the experiments. Five days after surgery we measured continuously for 1 h, as a control value, maternal and fetal heart rates, maternal arterial and uterine venous pressures, fetal arterial and venous pressure, and uterine blood flow. We also measured maternal and fetal respiratory blood gases, plasma glucose, hemoglobin concentrations, hematocrits, blood volumes, and maternal cardiac output, calculated uteroplacental O_2 uptake, and drew maternal and fetal blood samples for hormone determinations. Ambient O_2 concentration was lowered by flowing N_2 gas into the chamber. Daily, we measured respiratory blood gases, heart rates, vascular pressures, uterine blood flow, and uteroplacental O_2 uptake. Weekly, we collected blood for hormone determinations and measured maternal cardiac output and maternal and fetal blood volumes.

Surgery

We performed surgery in two steps, using general anesthesia [1.5% halothane (Fluothane), 30% nitrous oxide, balance O_2]. In the first phase, we placed the ewe in the right lateral position for a retroperitoneal approach to the distal branches of the aorta. We ligated the lateral and dorsal sacral arteries and placed an electromagnetic flow probe of appropriate size [about 8 mm internal diameter (ID); Dienco Instruments, Los Angeles, Calif] around the common internal iliac artery to measure total uterine blood flow, as previously described (Lotgering et al. 1983 a).

Immediately after the first phase we placed the ewe in the supine position. Through a hysterotomy incision we placed catheters [Tygon, 1.5 mm outside diameter (OD)] in a fetal pedal artery and vein (passing them into the descending aorta and inferior vena cava, respectively), and into the amniotic fluid, as previously described (Lotgering et al. 1983 b). After closing the uterus we inserted a catheter (2.2 mm OD) into the common uterine vein via a small branch. We also placed a catheter (1.5 mm OD) into the maternal pedal artery to measure maternal blood pressure and heart rate, and placed a catheter (5 mm ID) into the maternal jugular vein for later insertion of a Swan-Ganz catheter for cardiac output determination.

Postoperatively we daily flushed each catheter with saline, injected 500 mg ampicillin and 40 mg gentamicin into the amniotic fluid, and administered 2.5 ml combiotic intramuscularly to the ewe.

Three groups of animals were studied:

(a) 20 pregnant ewes subjected to hypoxemia for up to 28 days,
(b)   4 pregnant ewes which served as a control group not subjected to hypoxemia,
(c)   6 nonpregnant ewes which were subjected to hypoxemia for up to 28 days.

Among the hypoxemic pregnant ewes instrumented, 8 were studied for 7 days, 6 for 14 days, 4 for 21 days, and 2 for 28 days.

Environmental Chamber

In a specially modified room in the animal quarters the inspired O_2 concentration (FIO_2) was maintained at 12%–13% by inflowing N_2 gas. This FIO_2 resulted in an altitude equivalent of about 4270 m (14000 ft). Here the ewes remained throughout the study. To maintain the FIO_2 at 12% the N_2 flow rate was set at about 30 l/min. During experiments we continuously monitored the chamber O_2 concentration by use of an O_2 analyzer (Sensor Medics, Model OM-14, Anaheim, Calif) and recorder (Beckman, Model MMC/CET, Fullerton, Calif). Twice daily we calibrated the O_2 analyzer with a 13% O_2 gas mixture. Chamber humidity was kept almost constant (around 60%) by use of a dehumidifier (Sears, Kenmore, Coldspot Model 106, Chicago, Ill), the temperature was maintained at $20° \pm 1.0$ C, the light-dark cycle was maintained at 12 L, 12 D, and a fan circulated the air. Two ewes could be kept in the chamber simultaneously. Because the catheters and electronic cables passed from the chamber into an anteroom which housed the polygraph, flowmeter, computer, O_2 analyzer, liquid N_2 cylinder, and other equipment, we needed to enter the chamber only once a day to clean, feed, and water the ewes. Also, when we entered the chamber we temporarily increased the inflowing N_2 rate so as not to alter the FIO_2 significantly.

Blood Gases, Uteroplacental O_2 Uptake, and Glucose Concentrations

After we had recorded the maternal and fetal vascular and amniotic pressures during 1 h of uterine quiescence, we withdrew 1 ml samples simultaneously from the maternal pedal artery and uterine vein, and we analyzed the blood gases (Radiometer, ABL2, Copenhagen, Denmark), O_2 saturation and hemoglobin concentration (Radiometer, OSM Hemoximeter), and hematocrit. We calculated arterial O_2 content from O_2 saturation and hemoglobin concentration, uteroplacental O_2 delivery ($\dot{Q}_{O_2}$) from the product of uteroplacental blood flow and arterial O_2 content, and

uteroplacental O_2 uptake ($\dot{V}_{O_2}$) from the product of arteriovenous O_2 difference and
uterine blood flow. We also measured the maternal and fetal plasma glucose concentrations on the control day and twice weekly thereafter. Following the hematocrit measurements we utilized 10 µl of plasma from the hematocrit tube to determine the glucose concentration, taking the mean of three values.

Blood Volumes and Maternal Weight

We measured maternal and fetal whole blood volumes by use of ^{99}Tc-labeled erythrocytes on the control day and weekly thereafter. Following completion of the daily
determination of all variables we obtained maternal (5 ml) and fetal (2 ml) blood
samples and tagged the erythrocytes. We then injected the tagged erythrocytes
through the venous catheters into ewe and fetus, and took arterial blood samples
(maternal: 2 ml; fetal: 1 ml) 10, 20, and 30 min after injection. We quantified the
sample radioactivity in a gamma counter (Packard Instruments, Model 5912,
Downers Grove, Ill) at a 122–163 keV window (peak 143 keV), and corrected this
value for hematocrit and sample weight. From these counts and standard counts we
extrapolated to time zero and calculated blood volume, and then calculated plasma
volume from the microhematocrit (Brace 1983).

We weighed the ewes weekly following the blood volume measurements. Because they were weighed outside the chamber, a nylon bag was placed over the
ewe's head with 12%–13% O_2 (21% O_2 for the pregnant control group) flowing into
the bag during weight determination. This procedure took about 15 min.

Maternal Cardiac Output

We measured maternal cardiac output by thermodilution on the control day, every
hour for 6 h on day 0, on the 1st and 3rd day, and weekly thereafter. After complet-
ing the daily measurement of other variables we inserted a Swan-Ganz catheter
through the jugular vein, placing it in the appropriate position by monitoring the
pressure changes. We then injected 10 ml cold saline (0 °C) to calculate cardiac output by use of the computer (Instrumentation Laboratories, Model 601–602, Lexing-
ton, Mass). For each value we made seven measurements, deleting the maximum
and minimum values and taking the mean of the other five.

Uterine Blood Flow ($\dot{Q}_{ut}$), Vascular Pressures, and
Uterine Vascular Resistance (R_{ut})

We measured continuously total uterine blood flow with an auto-zeroing electromagnetic flowmeter (Dienco) with a DC amplifier and 8-channel polygraph
(Gould, Cleveland, Ohio) for 1 h daily. At the same time each day we recorded ma-

ternal and fetal heart rates, and maternal arterial (P_{ma}) and uterine venous (P_{uv}) pressures. We also measured fetal arterial and venous and amniotic fluid (P_{af}) pressures using pressure transducers and amplifiers (Gould). We calculated uterine vascular resistance [$R_{ut} = (P_{ma} - P_{uv})/\dot{Q}_{ut}$]. By use of a digital computer (IBM PC-AT) we calculated 1-min averages of all variables and stored the values on hard disk.

Erythropoietin Measurement

On the control day, 1st and 6th hour on day 0, on the 1st and 3rd day, and weekly thereafter we withdrew 4 ml blood samples from the maternal pedal artery and 2 ml from the fetal artery using a Monovette syringe containing EDTA solution. Plasma sample was stored at $-80\,°C$ for later analysis. The radioimmunoassay of erythropoietin was performed by Dr. G. Clemons. This assay's sensitivity is equivalent to 0.4 mU/ml. The inter- and intra-assay variabilities were <12 and <9, respectively (Clemons 1986).

Statistical Analysis

As noted above, we calculated 1-min mean values for all continuously measured variables and stored them on hard disk. After eliminating the data during uterine concentrations and/or contractures, we calculated 1-h averages for vascular pressures, uterine blood flow, and uterine vascular resistance. For all data we tested differences between means by two-way analysis of variance and Duncan's multiple range test.

Results

Body Weight and Glucose Concentration

During the first few days of hypoxemia both the pregnant and nonpregnant ewes lost their appetite, but then it rapidly returned to a more normal level. Nonetheless, during the experiments there was no significant change in maternal body weight in these two groups. In contrast, the pregnant control ewes gained about 5.2 kg during the 4 weeks of the study, a change experienced by similar animals in our laboratory.

In both groups of hypoxemic ewes the plasma glucose concentration reflected their initial appetite loss, decreasing from about 70.7 ± 11 mg/dl to 55.6 ± 11 mg/dl on the first day of the hypoxemia and then returning to control values by the 17th day. Nonetheless, there was no significant change in plasma glucose values in any of the three groups during the course of the study.

Respiratory Blood Gases

Figure 1 shows the blood gas changes in the three groups of ewes during the 3 weeks of the study. Within a few minutes of exposure to the lowered O_2 concentration, the maternal P_{O_2} decreased about 40% from 101.5 ± 5.1 Torr to 56.5 ± 5.1 Torr ($p < 0.001$), where it remained during the remainder of the study (Fig. 1). In the hypoxemic nonpregnant group the changes were similar. During the remainder of the study the arterial P_{O_2} tended to increase slowly in both groups. To keep these values relatively constant we periodically lowered the chamber O_2 concentration, e.g., from 13% to 12.5% on day 7 and to 12% on day 14. Although arterial P_{O_2}

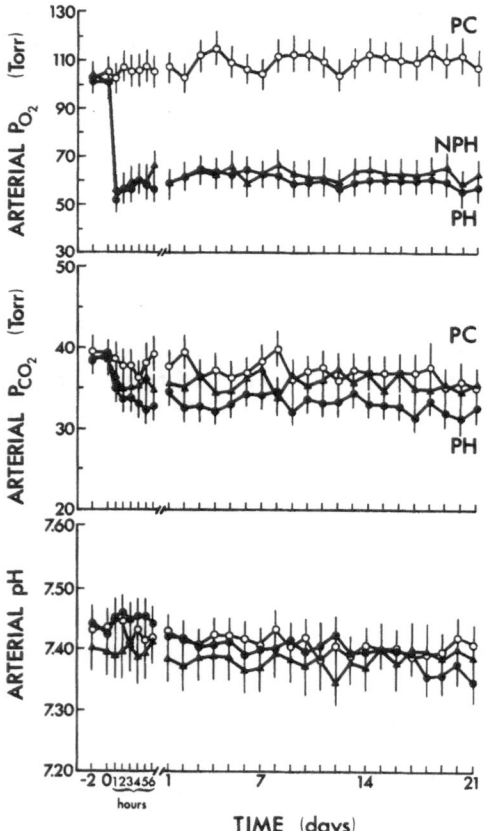

Fig. 1. *Upper panel* Arterial O_2 tension in the three groups of animals during the study period. *PH*, hypoxemic pregnant ewes; *NPH*, hypoxemic nonpregnant ewes; *PC*, control pregnant ewes not subjected to hypoxemia. Values are means $\pm 95\%$ confidence intervals of the means. *Middle panel* Arterial CO_2 tension and *lower panel* pH values in these ewes. Note that in this and subsequent figures the abscissa shows the initial period of 6 h after the onset of hypoxia, followed by the subsequent 21 days

tended to be higher in the nonpregnant than in the pregnant hypoxemic ewes (Fig. 1), these differences were not significant. Arterial P_{O_2} in the pregnant control ewes did not change significantly during the course of the study (Fig. 1).

In association with maternal hyperventilation arterial P_{CO_2} decreased from 38.3 ± 1.5 Torr to 34.6 ± 1.5 Torr ($p < 0.01$; Fig. 1) in the hypoxemic group. Although the hypoxemic nonpregnant ewes hyperventilated, their degree of hypocarbia was less extreme, and changes in the pregnant control group were not significant (Fig. 1). Following the initial respiratory adjustments arterial pH did not change significantly during the experimental period in any of the groups (Fig. 1).

In hypoxemic pregnant ewes maternal oxyhemoglobin saturation decreased about 18% from $93.2\% \pm 4.2\%$ to $76.2\% \pm 4.2\%$ on the first day, where it remained during the study (Fig. 2). Similar changes were seen in the hypoxemic nonpregnant ewes, while [HbO$_2$] in the control group remained normal (Fig. 2).

Figure 2 also shows the arterial O$_2$ content during the experimental period. Shortly after the onset of hypoxemia O$_2$ content in the pregnant ewes decreased about 12% from 11.4 ± 0.6 ml/dl to 10.1 ± 0.6 ml/dl ($p < 0.05$) and then continued to decrease gradually to 9.3 ± 0.6 ml/dl by the 10th day. Afterwards it increased slightly, but by the 21st day was still lower than control value ($p < 0.01$). The hypoxemic nonpregnant ewes also experienced a significant decrease in arterial O$_2$ content which, although rising slowly after 8–12 days, failed to return to control values (Fig. 2). Arterial O$_2$ content changes in the pregnant control group were not significant (Fig. 2).

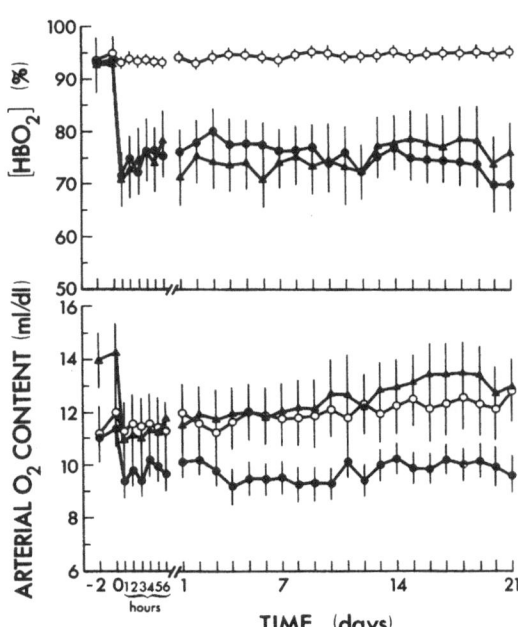

Fig. 2. *Upper panel* Arterial oxyhemoglobin saturation in the three groups of sheep. *Lower panel* Arterial O$_2$ content in the three study groups. Symbols are as in Fig. 1

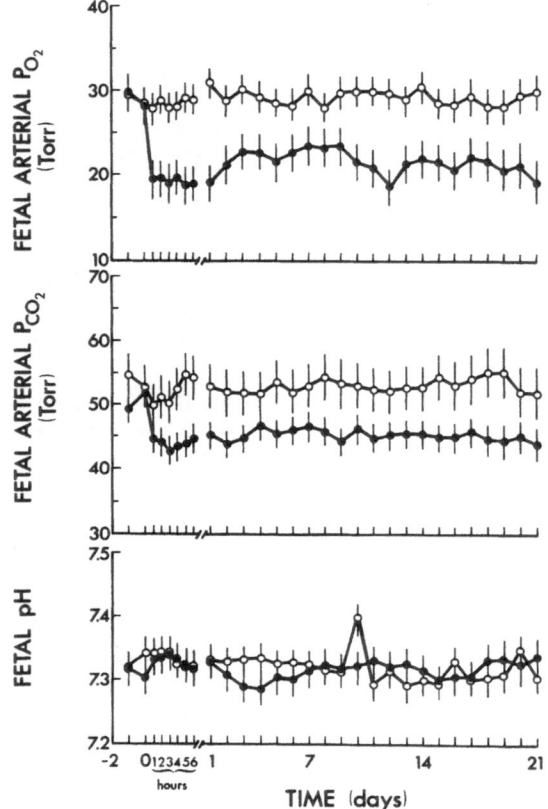

Fig. 3. *Upper panel* Arterial O_2 tensions in the hypoxemic (●) and control (○) fetuses. *Middle panel* Fetal arterial CO_2 tensions. *Lower panel* Fetal pH values

Figure 3 shows the changes in fetal arterial P_{O_2}, P_{CO_2} and pH during the 3-week study. In the hypoxemic group, following the onset of lowered maternal fractional inspiratory O_2 (FIO_2) fetal arterial P_{O_2} rapidly dropped about 36% from 29.7 ± 2.1 Torr to 19.1 ± 2.1 Torr ($p < 0.01$). During the remainder of the study the fetal P_{O_2} tended to increase slowly, as did that of the ewe. Arterial P_{O_2} values did not change significantly in the control group.

Again, in the hypoxemic fetuses arterial P_{CO_2} decreased from 49.4 ± 2.0 Torr to 45.3 ± 2.0 Torr (Fig. 3) in association with the decrease in maternal arterial P_{CO_2}. The arterial pH values did not change significantly during the experimental period in either group.

In the hypoxemic fetuses oxyhemoglobin saturation decreased about 36%, from $69.1\% \pm 6.6\%$ to $44.3\% \pm 6.6\%$, where it remained during the course of the study (Fig. 4). Control fetuses showed no significant changes.

Figure 4 also shows the changes in arterial O_2 content which in the hypoxemic group decreased about 34% from 9.2 ± 1.1 ml/dl to 6.1 ± 1.1 ml/dl. However, dur-

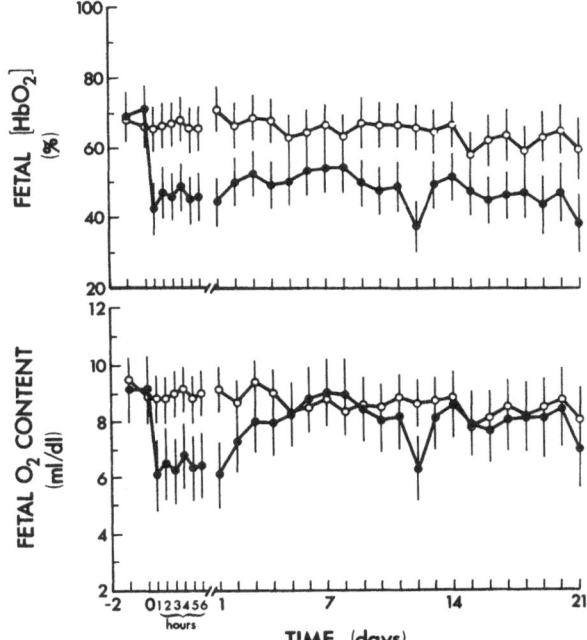

Fig. 4. *Upper panel* Arterial oxyhemoglobin saturation, and *lower panel* arterial O_2 content in the hypoxemic and control fetuses. Symbols are as in Fig. 3

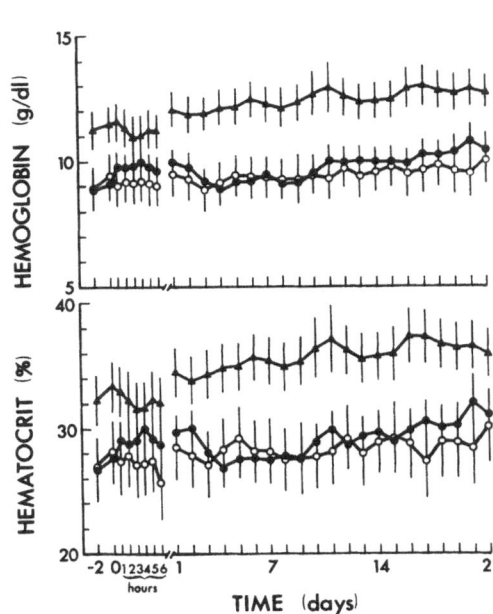

Fig. 5. *Upper panel* Ewes' hemoglobin concentration and *lower panel* hematocrit in the three study groups. Symbols are as in Fig. 1

ing the course of the study fetal arterial O_2 content increased to near control values in concert with its increase in hemoglobin concentration (Fig. 4). In the control group of these fetuses these variables showed no significant change.

Fetal arterial glucose concentrations decreased slightly from 27.1 ± 3.7 mg/dl to 23.2 ± 3.7 mg/dl (NS) in the hypoxemic animals.

Hemoglobin, Hematocrit, and Blood Volume

Figure 5 shows the changes of maternal hemoglobin concentration and hematocrit in the three study groups. In the pregnant ewes on the first day of hypoxemia, hemoglobin concentration and hematocrit increased about 12% from 8.9 ± 0.5 g/dl and $26.7 \pm 1.5\%$ to 10.0 ± 0.5 g/dl and $29.7 \pm 1.5\%$, respectively, returning to their control values by the 3rd day. From the 10th day onward these values increased significantly, so that by day 21 both variables were increased about 18% above their

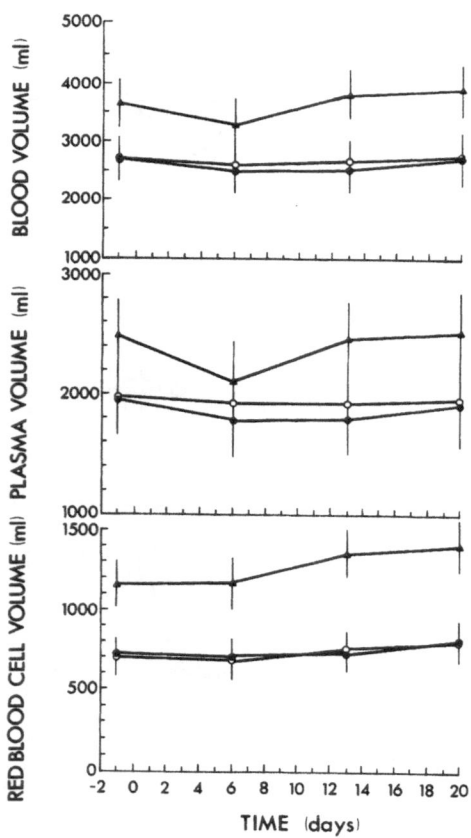

Fig. 6. Ewes' whole blood volume *(upper panel)*, plasma volume *(middle panel)*, and erythrocyte mass *(lower panel)* in the three groups of ewes during the course of the study. Symbols are as in Fig. 1

control values. Similar changes occurred in the hypoxemic nonpregnant ewes (Fig. 5), while values in the pregnant control group did not change significantly.

In addition, whole blood volume did not change significantly in any of the groups during the experiments (Fig. 6). Although in the hypoxemic pregnant ewes erythrocyte volume increased from 720 ± 44 ml to 820 ± 38 ml (and from 16.9 ± 2.4 ml/kg to 20.0 ± 3.0 ml/kg) by the 20th day, these changes were not significant. Again, in the hypoxemic, nonpregnant ewes erythrocyte mass increased from 1161 ± 145 ml to 1411 ± 159 ml (and from 20.2 ± 2.3 ml/kg to 24.1 ± 2.5 ml/kg) but these changes, like the minor changes seen in the pregnant control animals were not significant (Fig. 6).

Figure 7 shows the changes in hemoglobin concentration and hematocrit in the hypoxemic and control fetuses. In contrast to these parameters in the ewe, by day 7 of hypoxemia fetal hemoglobin concentration increased about 29% from 10.0 ± 1.0 g/dl to 12.9 ± 1.0 g/dl, rising slightly further to 14.0 ± 1.2 g/dl by day 21. The fetal hematocrit changes (Fig. 7) tended to reflect the hemoglobin increase. Again, these functions did not alter significantly in the control animals.

As shown in Fig. 8, in both hypoxemic and control fetuses whole blood volume increased during the course of the study. For the hypoxemic sheep this increase equaled about 112%, from a control value of 320 ± 90 ml to 680 ± 120 ml by day 21. In the control fetuses, whole blood volume increased 62% from 261 ± 65 ml to 424 ± 86 ml during the same period (Fig. 8). During this time in hypoxemic fetuses the erythrocyte volume increased 170% from 100 ± 40 ml to 270 ± 50 ml (Fig. 8),

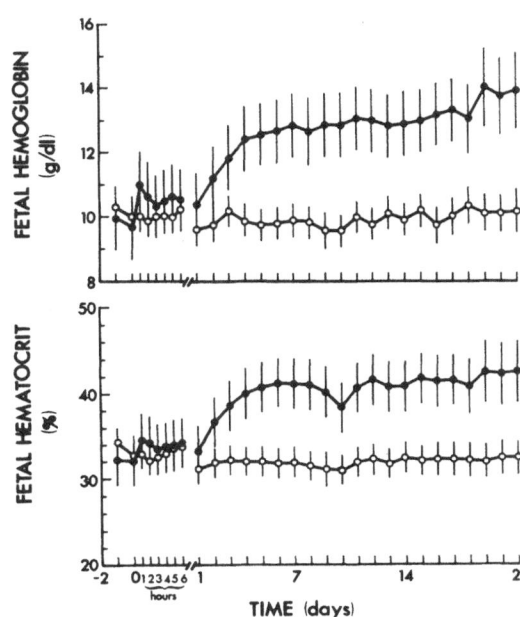

Fig. 7. *Upper panel* Fetal hemoglobin concentration and *lower panel* hematocrit in the two study groups. Symbols are as in Fig. 3

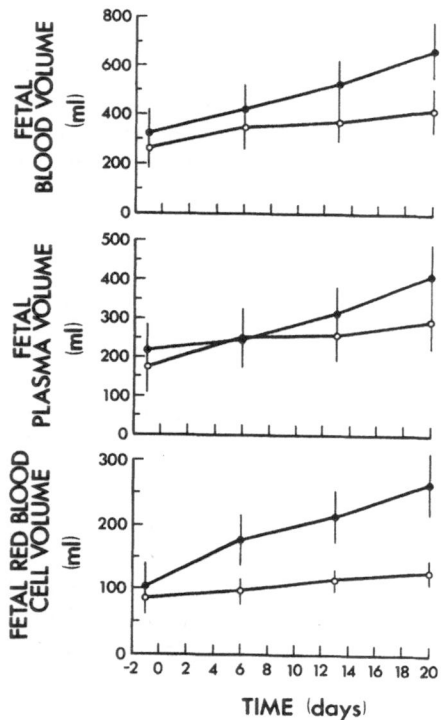

Fig. 8. Whole blood volume *(upper panel)*, plasma volume *(middle panel)*, and erythrocyte mass *(lower panel)* in hypoxemic and control fetuses during the course of the study. Symbols are as in Fig. 3

while in the controls the increase was 47% from 86 ± 16 ml to 127 ± 18 ml (Fig. 8). When calculated per estimated fetal weight, only the hypoxemic fetuses showed an increase, i.e., about 85% from 130 ml/kg to 240 ml/kg.

Blood Pressure and Heart Rate

During the 1st and 2nd hours of hypoxemia mean arterial pressure in the pregnant ewes increased about 7% from $91.9\% \pm 4.5$ mm Hg to 99.6 ± 4.9 mm Hg, later returning to control values (Fig. 9). Afterwards, arterial pressure in these ewes gradually decreased, so that by the 21st day of hypoxemia it had dropped about 12% to 81.4 ± 5.5 mm Hg (Fig. 9). Although in the hypoxemic, nonpregnant ewes mean arterial pressure was somewhat higher, it followed a similar trend (Fig. 9). In the pregnant control group arterial pressure decreased slightly during the study ($p < 0.01$).

Shortly after the onset of hypoxemia maternal heart rate increased about 18%, returning to control value by the 3rd day where it remained (Fig. 9). Except for the change during the initial hours of hypoxemia, the changes recorded did not differ from those in the pregnant control animals. Again, although heart rates tended to

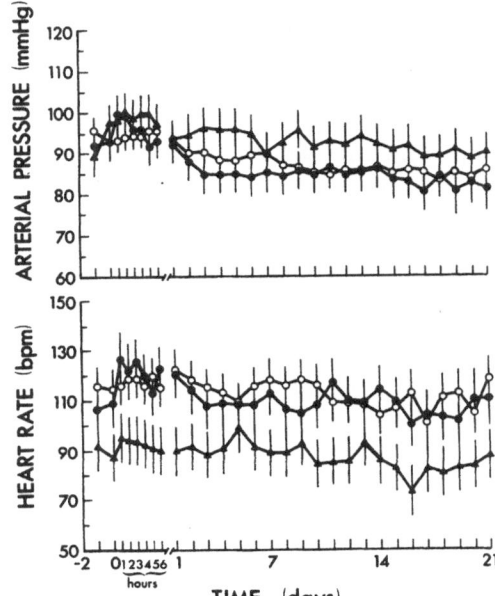

Fig. 9. *Upper panel* Ewes' arterial pressure, and *lower panel* heart rate in the three experimental groups during the study. Symbols are as in Fig. 1

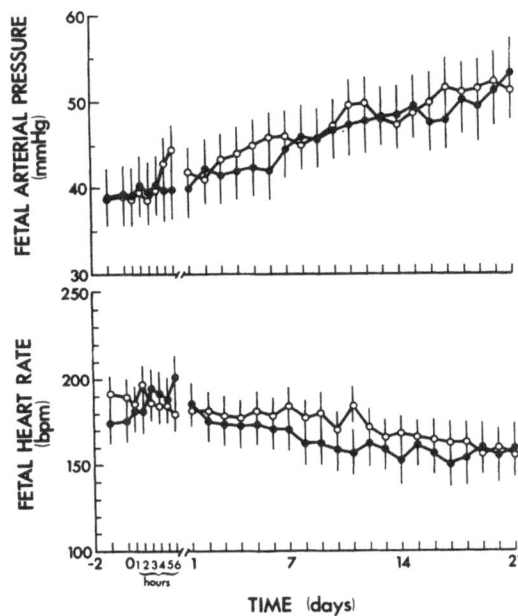

Fig. 10. *Upper panel* Fetal arterial pressure and *lower panel* fetal heart rate, in the hypoxemic and control animals during the study. Symbols are as in Fig. 3

be lower in the nonpregnant ewes, the changes after the first few hours of hypoxemia were nonsignificant (Fig. 9).

During the first few hours of hypoxemia fetal mean arterial pressure increased slightly from 38.9 ± 3.4 mm Hg to 40.4 ± 3.4 mm Hg, then returned to normal within 24 h (Fig. 10). Over the course of the 3-week hypoxemic exposure, mean fetal arterial pressure rose 38% to 53.3 ± 4.2 mm Hg at day 21 (Fig. 10). In the control fetuses, mean arterial pressure rose similarly to 51.4 ± 3.3 mm Hg during the experimental period (Fig. 10).

With the initial hypoxemic insult fetal heart rate increased 16% from 173.9 ± 11.8 bpm to 201.4 ± 11.8 bpm. After returning to normal values within 24 h, it then decreased gradually to 152.4 ± 11.8 bpm by day 14 where it remained (Fig. 10). In control fetuses the fetal heart rate showed a similar decrease during the 3-week study.

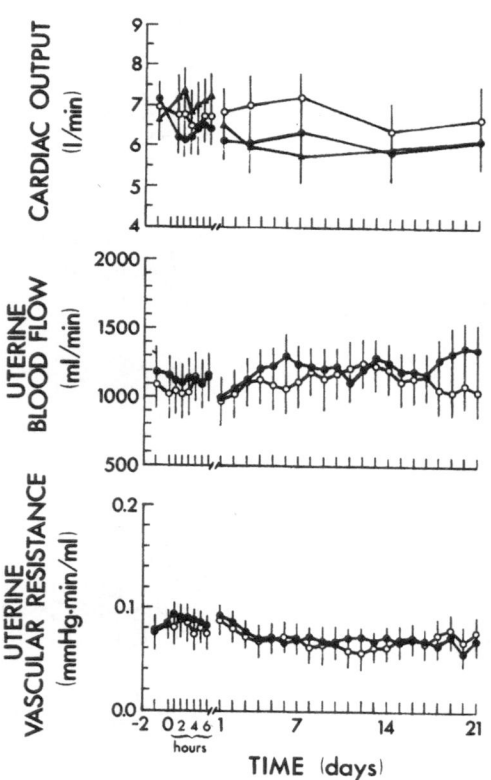

Fig. 11. Ewes' cardiac output (upper panel), uteroplacental blood flow (middle panel), and uterine vascular resistance (lower panel) during the experimental period in the three groups of ewes. Symbols are as in Fig. 1

Maternal Cardiac Output, Uterine Blood Flow, and Uterine Vascular Resistance

In hypoxemic pregnant ewes cardiac output decreased about 14% from 7.15 ± 0.46 l/min to 6.13 ± 0.46 l/min ($p < 0.05$) on the first day, where it remained during the experimental period (Fig. 11). In neither the pregnant control nor hypoxemic, nonpregnant ewes did cardiac output change significantly during the studies (Fig. 11).

As also shown in Fig. 11, within a few hours after the onset of hypoxemia uterine blood flow decreased about 15% from 1180 ± 134 ml/min to 990 ± 130 ml/min ($p < 0.001$). Thereafter it increased significantly (compared to the first day) to 1360 ± 190 ml/min by the 21st day (Fig. 11). With the onset of hypoxemia uterine vascular resistance increased during the first few hours, but thereafter it returned to normal values where it remained during the study, and did not differ from controls (Fig. 11).

Uteroplacental O_2 Delivery, Arteriovenous O_2 Difference, and Uteroplacental O_2 Uptake

Uteroplacental O_2 delivery initially decreased during the first few hours, thereafter gradually increasing, but insignificantly, in both the hypoxemic, pregnant ewes and the control group (Fig. 12). Arteriovenous O_2 differences did not change significantly either.

In addition, uteroplacental O_2 uptake did not change during the 3 weeks of hypoxemia (Fig. 12). Because we could not measure the weight of uterus and products of conception during the study, we did not caluculate O_2 uptake per unit weight of the uterus.

Hormone Concentrations

In hypoxemic, pregnant ewes plasma erythropoietin concentrations increased from a control value of 16.6 ± 24.4 IU/ml to 35.3 ± 25.9, 36.8 ± 24.4, and 35.4 ± 24.4 IU/ml on the 1st, 6th, and 24th hour of hypoxemia, respectively. However, by two-way ANOVA the changes were not significant. Afterwards it returned to control values on the 3rd and 7th day. By day 14 it increased to 83.2 ± 36.6 IU/ml, which was five times the control value. The changes in the hypoxemic nonpregnant and pregnant control ewes were also not significant.

Within 6 h of the onset of hypoxemia fetal plasma erythropoietin concentration had increased from 23 ± 51 IU/ml, and by 24 h it had reached 160 ± 51 IU/ml. By day 7 it had returned to 52 ± 69 IU/ml, remaining near that level until day 21 of hypoxemia. Again, the control fetuses showed only insignificant changes.

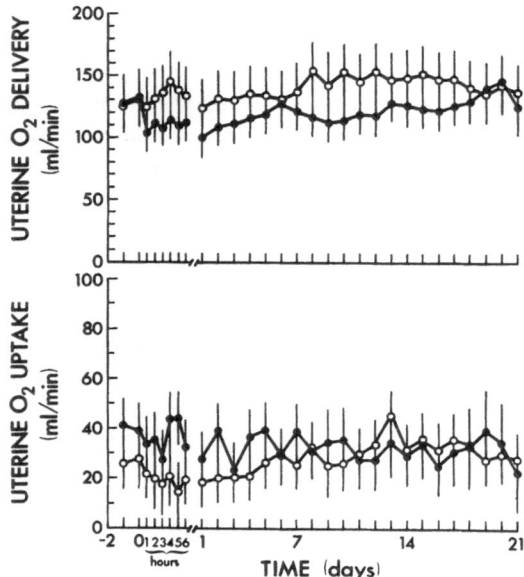

Fig. 12. *Upper panel* Uterine O_2 delivery and *lower panel* utero-placental O_2 uptake during the study. Symbols are as in Fig. 1

Fetal Breathing Movements

During the first few hours of hypoxemia the incidence of fetal breathing movements decreased from a control value of 40% of the time to 20%. However, within 24 h the incidence had returned to normal, suggesting that the mechanism which regulate this activity adapt rapidly. Details of this aspect of the work are reported elsewhere (Koos et al. 1988).

Body Weight

Our estimate of fetal weight at the time of surgery (115 days) was 1.65 ± 0.5 kg. At 138.8 ± 1.6 days gestation the hypoxemic fetuses weighed 3.3 ± 0.3 kg, while the weight of the controls was 3.5 ± 0.4 kg (NS).

Discussion

High Altitude Adaptations

Theoretical studies suggest that one of the most important factors which affects fetal O_2 levels is maternal arterial O_2 tension (Longo et al. 1972). The barometric pressure, and thus the ambient inspired O_2 tension, decreases exponentially with distance above sea level, and thus arterial O_2 tension also decreases. Between 1500 and 3000 m relatively minor compensatory adjustments serve to maintain normal tissue oxygenation. Above 3000 m (9840 ft) hypoxic effects become more apparent, particularly with the added stresses of exercise or cold exposure. In man and many mammals immediate changes include hyperventilation and increased cardiac output. Long-term effects include increases in erythrocyte mass, hemoglobin concentrations, intra-erythrocyte 2,3-diphosphoglycerate concentration, muscle myoglobin concentration, and tissue capillary density, as well as decreased cardiac output, and other compensations. At about 4500 m (14750 ft) an unacclimatized healthy individual may show signs of hypoxemia and "altitude sickness". At 5490 m (18000 ft) atmospheric pressure equals only one-half that at sea level (380 Torr), and a healthy unacclimatized individual may show signs of inadequate oxygenation. Despite the hypoxia associated with high altitude, about 16 million individuals live at elevations above 3000 m, and in the Andes a million or so permanent residents live higher than 4600 m (Metcalfe et al. 1967).

Pregnancy under normal circumstances is associated with a number of maternal adaptations of the cardiovascular, hematologic, and endocrine systems. Sojourn at high altitudes also is accompanied by a host of compensatory adjustments of these and other systems. When the pregnant individual undergoes the additional stress of high altitude exposure, questions arise regarding the extent to which the several systems adapt to allow successful fetal development, the time course of these changes, and their relative roles.

Fetal Responses to Long-term Hypoxia

Although the determinants of fetal oxygenation under normal conditions have attracted a great deal of attention, less is known about fetal adaptations to conditions which may result in long-term hypoxemia such as high altitude. Under normal conditions, fetal arterial oxygen levels are a function of a number of maternal, placental, and fetal factors. The normal fetal arterial O_2 tension equals about 25 Torr, i.e., about one-quarter of adult values. Because near-term, fetal hemoglobin concentrations are greater than or equal to maternal, and fetal blood O_2 affinity is greater than maternal, fetal arterial O_2 content equals or is greater than that of the mother despite its low O_2 tension. Nonetheless, in face of relatively low O_2 tensions and a steep oxyhemoglobin saturation curve, the fetus is particularly vulnerable to hypoxemia of whatever cause.

Long-term hypoxemia can pose a threat to the fetus, as compensatory physiologic adjustments would be required over a prolonged period, on contrast to those compensations induced by short-term hypoxemia. Theoretical studies suggest that one of the most important factors which affect fetal O_2 levels is maternal arterial O_2 tension (Longo et al. 1972), which in turn is a function of inspired O_2. The barometric pressure, and thus the ambient inspired O_2 tension, decreases exponentially with distance above sea level. In adults, relatively minor compensatory adjustments serve to maintain normal tissue oxygenation between 1500 and 3000 m, whereas above 3000 m hypoxic effects become more apparent.

Several studies have reported fetal and newborn blood gas levels at actual or simulated high altitude. For instance, at about 3050 m umbilical venous oxyhemoglobin saturation [HbO_2] in newborn infants before the onset of respiration averaged 50%, as compared with 58% at 1525 m (Howard et al. 1957a). The O_2 tensions of sheep and goats acutely exposed to a simulated altitude of 3060 m were 15 and 11 Torr in the umbilical vein and artery, respectively (Blechner et al. 1968), as compared with control values of about 26 and 20 Torr, respectively. In another study at a simulated altitude of 5490 m the PO_2 values in the umbilical vein and artery averaged 21 and 8 Torr, respectively (Kaiser et al. 1958).

In chronically catheterized sheep which had been moved from 1525 to 4300 m, the O_2 tensions and oxyhemoglobin saturations in uterine and umbilical vessels initially fell dramatically (Makowski et al. 1968). For instance, umbilical venous and arterial [HbO_2] decreased from control values of 74% and 61%, respectively, to 45% and 28% during the first few days at the higher elevation. Then, over a period of 6–18 days, [HbO_2] increased to values similar to those existing at the lower altitude (Makowski et al. 1968).

Apparently the only study of human blood gas values at high elevations is from Cerro de Pasco, Peru. Here at 4200 m fetal scalp O_2 tension averaged 19 Torr, a value only slightly lower than the sea level value of 22 Torr (Sobrevilla et al. 1971).

Taken together, these results indicate that in prolonged hypoxic hypoxemia various physiologic mechanisms operate to maintain fetal O_2 values only 10%–20% below those at sea level. A number of questions arise as to the compensatory mechanisms which serve to maintain fetal oxygenation under these hypoxic conditions, and their relative importance. Unfortunately, few specific details of this adaptive process have been described.

Some of the compensatory changes which the fetus might make during long-term hypoxemia include:

(a) an increase in erythrocyte mass or total blood volume;
(b) an increase in cardiac function and venous return, resulting in an increase in cardiac output;
(c) a redistribution of blood flow to vital organs (brain, heart, etc.);
(d) a reduction in total oxygen consumption; and
(e) increased secretion of various hormones (catecholamines, arginine vasopressin, renin, erythropoietin) which may initiate or facilitate the above changes.

Environmental Chamber

To study adaptations to high altitude, ideally one might conduct such studies in the field. However, because scientific investigation under such conditions is fraught with numerous problems, many workers have chosen the more controlled conditions of an environmental chamber. Because use of a hypobaric chamber is associated with other problems, we chose to conduct the present studies in a chamber with a lowered ambient O_2 concentration. Usually, two ewes were kept in the chamber at a time with the electronic cables and catheters exiting via an airtight seal to an anteroom which housed the analyzers and recording equipment. Although we entered the chamber daily, the time the door was open was less than 3 sec and the ambient O_2 concentration did not rise significantly. During the weekly weighing of the ewe the FIO_2 was maintained with a portable system allowing the ewe to inhale a 12%–13% O_2 mixture. Thus the long-term hypoxemic exposure was essentially continuous.

Maternal Body Weight

During the 3 weeks of the study maternal body weight failed to change significantly from the control value at 116 days' gestation. In part, this was because of the loss of appetite by the ewe, particularly during the 1st week of hypoxemia, which was reflected in the plasma glucose concentrations (see below). In the pregnant control ewes, body weight increased 12% during a similar period.

Blood Gases and Glucose

Following exposure to the lowered FIO_2 the maternal P_{O_2} and P_{CO_2} rapidly decreased to the levels at which they remained. Later during the course of the study maternal arterial P_{O_2} tended to increase gradually. Thus, to maintain it at a relatively constant level the FIO_2 had to be decreased periodically, as noted above. In sheep taken to high altitude, a similar increase in maternal arterial P_{O_2} occurred during the week or two that maternal compensatory adjustments occurred (Makowski et al. 1968). In the present studies maternal [HbO$_2$] followed the P_{O_2} changes, while arterial O_2 content tended to increase gradually after day 10 in association with both the increased P_{O_2} and increased hemoglobin concentration. P_{O_2} values and oxyhemoglobin saturations in the hypoxemic animals were similar to those values in both ewes (Metcalfe et al. 1962a, b) and humans (Hellegers et al. 1961) at about 4420 m.

The maternal plasma glucose concentration decreased rather strikingly and remained depressed despite the apparent return of maternal appetite. Fetal blood gases followed those of the ewe within a few minutes (Fig. 3). As noted in the mother, during the course of the study fetal arterial P_{O_2} tended to increase gradually, so that to maintain P_{O_2} values relatively constant the FIO_2 had to be decreased periodically. Fetal [HbO$_2$] (Fig. 4) followed the P_{O_2} changes, while arterial O$_2$ content increased gradually after day 1 in association with increased hemoglobin concentration. These changes are not unlike those reported by Makowski and colleagues (1968) in which fetal [HbO$_2$] gradually increased during 1–2 weeks of high altitude exposure.

The fetal plasma glucose concentration decreased within a few hours of the onset of hypoxemia and remained depressed throughout the study, following the maternal level.

Hemoglobin, Hematocrit, and Blood Volume

Shortly after the onset of hypoxemia the maternal hemoglobin concentration and hematocrit increased about 12% (Fig. 4), probably reflecting mobilization of erythrocytes from the spleen and other stores. However, by the 3rd day these variables had returned to control values. Not until day 10 were they again significantly higher than prehypoxic values, and thereafter they rose gradually throughout the remainder of the experiment but were never different from levels in the pregnant, nonhypoxic control animals. Although the nonpregnant ewes started at a higher hemoglobin concentration and hematocrit than the pregnant animals, during hypoxemia their overall increase was of similar magnitude (13% vs 17%, NS).

Whole blood volume did not change significantly during the course of hypoxemia in any of the three groups, whether calculated as ml or as ml/kg maternal weight. Although red cell volume in the hypoxemic, pregnant group had increased about 12% by day 20, the change was not statistically significant. The lack of a significant increase in either whole blood volume or erythrocyte mass in the hypoxemic, nonpregnant ewes is particularly surprising. Pregnant sheep at high altitude (Metcalfe et al. 1962a, b) were reported as having hemoglobin concentrations of about 14 g/dl as compared with sea-level control of about 11 g/dl. Even in ewes at simulated altitude for only 10 days the hemoglobin concentration increased to 14–16 g/dl (Kaiser et al. 1958).

Following the onset of hypoxemia fetal hemoglobin concentration and hematocrit increased, until at day 3 they were significantly higher than control values (Fig. 7). The tendency to increase continued throughout the remainder of the experiment (Fig. 7). This increase in fetal hemoglobin concentration was similar to that reported by Kaiser and coworkers (1958) for fetuses in a hypobaric pressure chamber for 10 days at a similar FIO$_2$. The fetal hematocrits in the present series averaged 40.8 vol % after 12 days' hypoxemia, a value somewhat less than the 52.7% in the fetuses of sheep reared at high altitude (Prystowsky et al. 1960).

With acute hypoxemia in human fetuses, fetal hematocrit increases (Linderkamp 1982). Fetal sheep at high altitude and guinea pig fetuses made hypoxemic throughout pregnancy had elevated hematocrits near term (Prystowsky et al. 1960; Gilbert et al. 1979). Whether this increased hematocrit indicates true polycythemia or a reduction in plasma volume is not known. Fetal sheep do not appear to have very large reservoirs of red cells (Brace 1983); however, over long periods red cell production could increase in response to erythropoietin. Erythropoietin concentration increases in the sheep fetus made hypoxemic either by having the ewe inspire a low FIO_2 for 4-6 h or by continuous glucose infusions for several days (Phillips et al. 1982); however, in that study data were presented from only four animals.

Fetal whole blood volume (ml) increased significantly from week to week during the course of hypoxemia, but it also increased in the control group, although to a lesser extent (Fig. 8). When expressed as milliliters per estimated weight, blood volume increased 85% in the hypoxemic fetuses but not in the controls. In hypoxemic fetuses erythrocyte volume increased strikingly so that by day 21 it had more than doubled (Fig. 8). Again, the value for increase in whole blood volume is not as great as reported in fetuses reared at high altitude throughout gestation (Prystowsky et al. 1960). However, this may be accounted for by the use of Evans blue in the altitude studies (Prystowsky et al. 1960), a method known to give larger blood volumes than that with labeled erythrocytes (Longo and Hardesty 1985).

Blood Pressure, Heart Rate, and Cardiac Output

During the first few hours of hypoxemia both maternal arterial pressure and heart rate increased strikingly (Fig. 9). Thereafter, mean arterial pressure decreased to 85 ± 4.5 mm Hg by day 3, where it remained until after day 21, when it further decreased to 81 ± 5.5 mm Hg. These changes were similar to those in the control ewes. Following the initial rise and fall in the hypoxemic ewes the heart rate changes were not significant.

In pregnant ewes shortly after the onset of hypoxemia, cardiac output fell about 14%, suggesting that stroke volume decreased even more, since the heart rate increased or remained normal. Thereafter, it failed to increase significantly. In contrast, in nonpregnant ewes there was a sharp increase in cardiac output during the first few hours, which then returned to control values.

During the first few hours of hypoxemia both fetal mean arterial pressure and heart rate increased strikingly (Fig. 10). After returning to the control value, arterial pressure slowly increased during the course of gestation, but no more so than that in the control group (Fig. 10).

Following its initial hypoxemic rise and fall, the fetal heart rate decreased slowly, probably reflecting the rise in arterial pressure, and was similar to that of the controls (Fig. 10).

Uterine Blood Flow, Uterine Vascular Resistance, and Uteroplacental O_2 Uptake

Following the onset of hypoxemia, uterine blood flow decreased, in part reflecting the lower cardiac output (Fig. 11). Although cardiac output remained below normal, uteroplacental blood flow returned to control values by day 3, thereafter remaining relatively constant. The finding that hypoxemia did not affect uterine blood flow is in agreement with observations in chronically catheterized sheep of insignificant uterine blood flow changes in response to hypoxia (Greiss et al. 1972; Makowski et al. 1973). Somewhat surprisingly, in view of the increased mass of the products of conception, uterine blood flow did not increase significantly in either group as gestation advanced (Fig. 11). Except for a brief rise in uterine vascular resistance in the hypoxemic ewes, the subsequent changes were also not significant in either group.

With the onset of hypoxemia uteroplacental O_2 delivery in the hypoxemic, pregnant sheep fell dramatically, reflecting the decreases in both blood flow and arterial O_2 content. However, as with uterine blood flow, by day 3 this had returned to control values. Despite some variability, uteroplacental O_2 uptake remained constant during the course of the study in both hypoxemic and control ewes.

Hormonal Changes

The control values of maternal erythropoietin concentration were similar to those reported by others (Widness et al. 1986). During the first day of hypoxemia they doubled, returning to baseline values on days 3 and 7. By days 14 and 21 they had increased fivefold. Because of the variability, these changes were not statistically significant. However, they were consistent.

The control values of fetal erythropoietin concentration were similar to those reported by others (Widness et al. 1986). During the first day of hypoxemia they increased significantly about sixfold, returning to near-baseline values by day 2. Thereafter the erythropoietin concentration increased fivefold by day 21; however, because of the variability, these changes were not statistically significant. According to Clemons increases of only 5 IU/ml above baseline are sufficient to result in an erythropoietic response.

Acute hypoxemia also leads to increases in the plasma concentrations of several hormones including the catecholamines (Jones 1980; Cohen et al. 1982), vasopressin (Rurak 1978; Lewis et al. 1982; Stark et al. 1982), adrenocorticotropic hormone and cortisol, the endorphins (Stark et al. 1982), and others. Each of these hormones could potentially mediate changes in blood volume and other cardiovascular functions, but their effects during long-term hypoxemia have not been explored.

Fetal Body Weight

Following the 3 weeks of the study, the hypoxemic fetuses appeared normal and their weights were not significantly different from the control animals'. Similar findings were noted by Jacobs and coworkers (1986) in fetuses of ewes made hypoxemic during days 120 to 140 of gestation. In contrast, these latter workers observed intrauterine growth retardation only in those fetuses that were hypoxemic from days 30 to 135 of gestation (Jacobs et al. 1986). Thus, during the last trimester hypoxemia of the magnitude to produce growth retardation must be maintained longer than 20–30 days. This appears somewhat paradoxial since it is during this period that the fetus experiences almost exponential growth. Numerous studies in humans (Grahn and Kratchman 1963; Haas 1976; Lichty et al. 1957; Moore et al. 1982) and laboratory animals (Johnson and Roofe 1965; Timiras et al. 1957) have reported lowered newborn birth weights and increased mortality (Grahn and Kratchman 1963; Lichty et al. 1957; Mazess 1965; McCullogh et al. 1977) at high altitude. However, almost without exception these studies have been in subjects at altitude throughout the entire gestation. In addition, several studies have failed to show an altitude effect on weight in sheep (Metcalfe et al. 1962b). In humans Moore et al. (1982) showed that birth weight at altitude is a function of maternal O_2 levels.

Conclusion

The present studies have sought to examine the maternal and fetal adaptations to relatively prolonged hypoxemia and the regulatory mechanisms of these adaptations. In addition, they examined the temporal changes in these responses during the course of long-term hypoxemia. In response to long-term hypoxemia both pregnant and nonpregnant sheep experience moderate cardiovascular and hematologic responses. The following maternal functions increase slightly: hemoglobin concentration and erythropoietin concentration. The following maternal functions decrease slightly: arterial O_2 content, glucose concentration, mean arterial pressure, and cardiac output. In contrast, the following maternal variables remain relatively constant: body weight, heart rate, blood volume, uterine blood flow, uterine O_2 flow, and uteroplacental O_2 uptake.

In response to long-term hypoxemia the fetus undergoes somewhat dramatic responses. By comparison to controls the following functions increased moderately: hemoglobin concentration, hematocrit, whole blood volume, erythrocyte mass, and erythropoietin concentration. Perhaps surprisingly, these fetuses were able to compensate, so that at term their body weights were normal. Obviously, it would be of interest to know to what extent cardiac function, the distribution of cardiac output, placental diffusing capacity, and other functions were altered during the course of hypoxia to account for the compensations. These questions are being pursued in current studies.

Acknowledgment. We thank Benjamin L. Siu and Alan J. Stevenson for their technical assistance.

References

Blechner JN, Cotter JR, Hinkley CM, Prystowsky H (1968) Observations on pregnancy at altitudes. II. Transplacental pressure differences of oxygen and carbon dioxide. Am J Obstet Gynecol 102: 794–805

Brace RA (1983) Blood volume and its measurement in the chronically catheterized sheep fetus. Am J Physiol 244: H487–H494

Clemons GK (1986) Comparison of radioimmunoassay and bioassay of erythropoietin. In: Zanjani ED, Tavassoli M, Ascensao JL (eds) Humoral and cellular regulation of erythropoiesis. Scientific, New York

Cohen WR, Piasecki GT, Jackson BT (1982) Plasma catecholamines during hypoxemia in fetal lambs. Am J Physiol 243: R520–R525

Gilbert RD, Cummings LA, Juchau MR, Longo LD (1979) Placental diffusing capacity and fetal development in exercising or hypoxic guinea pigs. J Appl Physiol 46: 828–834

Grahn D, Kratchman J (1963) Variation in neonatal death rate and birth weight in the United States and possible relations to environmental radiation, geology and altitude. Am J Hum Genet 15: 329–352

Greiss FC, Anderson SG, King LC (1972) Uterine vascular bed: effects of acute hypoxia. Am J Obstet Gynecol 113: 1057–1064

Haas J (1976) Prenatal and infant growth and development. In: Baker PT, Little MA (eds) Man in the Andes: a multidisciplinary study of high-altitude quechua. Dowden, Hutchison and Ross, Stroudsburg, Pa, pp 161–179

Hellegers A, Metcalfe J, Huckabee WE, Prystowsky H, Meschia G, Barron DH (1962) Alveolar P_{CO_2} and P_{O_2} in pregnant and nonpregnant women at high altitude. Am J Obstet Gynecol 83: 241–245

Howard RC, Bruns PD, Lichty JA (1957a) Studies of babies born at high altitudes. III. Arterial oxygen saturation and hematocrit values at birth. AMA J Dis Child 93: 674–678

Howard RC, Lichty JA, Bruns PD (1957b) Studies of babies born at high altitudes. II. Measurement of birth weight, body length, and head size. AMA J Dis Child 93: 670–674

Jacobs R, Falconer J, Robinson JS, Webster MED (1986) Effect of hypoxia on the initiation of secondary wool follicles in the fetus. Aust J Biol Sci 39: 79–83

Johnson D, Roofe PG (1965) Blood constituents of normal newborn rats and those exposed to low oxygen tension during gestation; weight of newborn and litter size also considered. Anat Rec 153: 303–310

Jones CT (1980) Circulating catecholamines in the fetus, their origin, actions and significance. In: Parves H, Parves S (eds) Biogenic amines and development. Elsevier/North-Holland Biomedical, New York

Kaiser IH, Cummings JN, Reynolds SRM, Marbarger JP (1958) Acclimatization response of the pregnant ewe and fetal lamb to diminished ambient pressure. J Appl Physiol 13: 171–178

Koos BJ, Kitanaka T, Matsuda K, Gilbert RD, Longo LD (1988) Fetal breathing adaptation to prolonged hypoxemia. J Dev Physiol 10: 38–40

Kruger H, Arias-Stella J (1970) The placenta and the newborn infant at high altitudes. Am J Obstet Gynecol 106: 586–591

Lewis AB, Evans WN, Sischo W (1982) Plasma catecholamine responses to hypoxemia in fetal lambs. Biol Neonate 41: 115–122

Lichty JA, Ting RY, Burns PD, Dyar E (1957) Studies of babies born at high altitude. I. Relation of altitude to birth weight. AMA J Dis Child 93: 666–670

Linderkamp O (1982) Placental transfusion: determinants and effects. Clin Perinatol 9: 559–592

Longo LD (1987) Respiratory gas exchange in the placenta. In: Fishman AP, Farhi LE, Tenney SM, Geiger SR (eds) Handbook of physiology. Sect 3, The respiratory system; vol 4, Gas exchange. American Physiological Society, Bethesda, Md, pp 351–401

Longo LD, Hardesty J (1985) Maternal blood volume: measurement, hypothesis of control, and clinical considerations. Rev Perinatol Med 5: 35–59

Longo LD, Hill EP, Power GG (1972) Theoretical analysis of factors affecting placental O_2 transfer. Am J Physiol 222: 730–739

Lotgering FK, Gilbert RD, Longo LD (1983a) Exercise responses in pregnant sheep: oxygen consumption, uterine blood flow, and blood volume. J Appl Physiol 55: 834–841

Lotgering FK, Gilbert RD, Longo LD (1983b) Exercise responses in pregnant sheep: blood gases, temperatures, and fetal cardiovascular system. J Appl Physiol 55: 842–850

Makowski EL, Battaglia FC, Meschia G, Behrman RE, Schruefer J, Seeds AE, Bruns PD (1968) Effect of maternal exposure to high altitude upon fetal oxygenation. Am J Obstet Gynecol 100: 852–861

Makowski EL, Battaglia FC, Meschia G, Behrman RE, Schruefer J, Seeds AE, Bruns PD (1973) Effect of maternal exposure to high altitude upon fetal oxygenation. Am J Obstet Gynecol 115: 624–629

Mazess RB (1965) Neonatal mortality and altitude in Peru. Am J Phys Anthrop 23: 209–214

McClung J (1969) Effects of high altitude on human birth. Observations on mothers, placentas, and the newborn in two Peruvian populations. Harvard University Press, Cambridge, Mass

McCullough RE, Reeves JT, Liljegren RL (1977) Fetal growth retardation and increased infant mortality at high altitude. Arch Environ Health 32: 36–39

Metcalfe J, Meschia G, Hellegers A, Prystowsky H, Huckabee W, Barron DH (1962a) Observations on the growth rates and organ weights of fetal sheep at altitude and sea level. Q J Exp Physiol 47: 305–313

Metcalfe J, Meschia G, Hellegers A, Prystowsky H, Huckabee W, Barron DH (1962b) Observations on the placental exchange of the respiratory gases in pregnant ewes at high altitude. Q J Exp Physiol 47: 74–92

Metcalfe J, Bartels H, Moll W (1967) Gas exchange in the pregnant uterus. Physiol Rev 47: 782–838

Moore LG, Rounds SS, Jahnigen D, Grover RF, Reeves JT (1982) Infant birth weight is related to maternal arterial oxygenation at high altitude. J Appl Physiol 52: 695–699

Moore LG, Brodeur P, Chumbe O, D'Brot J, Hofmeister S, Monge C (1986) Maternal hypoxic ventilatory response, ventilation, and infant birth weight at 4300 m. J Appl Physiol 60: 1401–1406

Phillips AF, Widness JA, Garcia JF, Raye JR, Schwartz R (1982) Erythropoietin elevation in the chronically hyperglycemic fetal lamb. Proc Soc Exper Biol Med 170: 42–47

Prystowsky H, Hellegers A, Meschia G, Metcalfe J, Huckabee W, Barron DH (1960) The blood volume of fetuses carried by ewes at high altitude. Q J Exp Physiol 45: 292–297

Rurak DW (1978) Plasma vasopressin levels during hypoxaemia and the cardiovascular effects of exogenous vasopressin in foetal and adult sheep. J Physiol (Lond) 277: 341–357

Sobrevilla LA, Cassinelli MT, Carcelen E, Malaga JM (1971) Human fetal and maternal oxygen tension and acid-base status during delivery at high altitude. Am J Obstet Gynecol 111: 1111–1118

Stark RI, Wardlaw SL, Daniel SS, Husain MK, Sanocka UM, James LS, Vande Wiele RL (1982) Vasopressin secretion induced by hypoxia in sheep: developmental changes and relationship to β-endorphin release. Am J Obstet Gynecol 143: 204–215

Timiras PS, Krum A, Pace N (1957) Body and organ weights of rats during acclimatization to an altitude of 12470 feet. Am J Physiol 191: 598–604

Widness JA, Teramo KA, Clemons GK, Garcia JF, Cavalieri RL, Piasecki GJ, Jackson BT, Susa JB, Schwartz R (1986) Temporal response of immunoreactive erythropoietin to acute hypoxemia in fetal sheep. Pediatr Res 20: 15–19

Adrenal Endocrine and Circulatory Responses to Acute Prolonged Asphyxia in Surviving and Non-Surviving Fetal Sheep near Term*

A. Jensen[1], H. Gips, M. Hohmann, and W. Künzel

Introduction

A previous study on the effects of acute prolonged asphyxia produced some interesting results on the changes in blood flow of surviving and non-surviving fetal sheep near term (Jensen et al. 1985, 1987). It revealed that fetal circulatory centralization during acute asphyxia is a rapid process in which blood flow to peripheral organs falls and that to central organs increases after 1 and 2 min asphyxia respectively. That study also demonstrated that the ability of the fetal circulation to maintain centralization throughout asphyxia is essential for survival, because circulatory *decentralization* on the nadir of asphyxia always preceded fetal death. These observations and the fact that in non-surviving fetuses adrenal blood flow failed to increase during asphyxia suggested that adrenal function and fetal survival of acute prolonged asphyxia might be interrelated. However, since non-surviving fetuses were more asphyxic than surviving fetuses, the possibility also had to be considered that severe hypoxaemia and acidaemia may have direct adverse effects on the maintenance of circulatory centralization and hence may reduce the fetus' chances of surviving asphyxia.

This study was designed to test these hypotheses and to improve the understanding of the mechanisms involved in the development of circulatory decentralization during acute prolonged asphyxia. We studied the effects of arrest of uterine blood flow for 4 min on fetal blood gases, acid-base balance, organ blood flow distribution and plasma concentrations of hormones produced by the adrenal in fetal sheep near term. To separate the effects of asphyxia on the adrenal medulla from those on the adrenal cortex, plasma concentrations of both catecholamines and corticosteroids were determined. Corticosteroids were selected so as to differentiate between effects on hormones deriving from the three different cell layers of the adrenal cortex, i.e. zona glomerulosa, zona fasciculata and zona reticularis.

* This investigation was supported by the Deutsche Forschungsgemeinschaft (Je 108/4-1).

[1] Department of Obstetrics and Gynaecology, Justus-Liebig-Universität Gießen, Klinikstr. 32, D-6300 Giessen, Federal Republic of Germany

Material and Methods

Animal Preparation

Nine fetal sheep (merino land breed) near term (term is at 147 days) weighing
4.6 ± 0.5 (SEM) kg were prepared as described in detail elsewhere (Jensen et al.
1987). Briefly, all ewes were anaesthetized by subarachnoid or epidural injection of
2-4 ml 1% (w/v) tetracaine HCl at the lower spine, and were operated on under
sterile conditions. Polyvinyl catheters were placed in a maternal iliac artery and
vein through tibial vessels. The ewe's flank was opened and a balloon catheter was
implanted extraperitoneally around the descending aorta below the renal artery in
order to arrest uterine and ovarian blood flow. Care was taken to not include any
nerves. Under local anaesthesia (xylocaine 1%, w/v), fetal catheters were inserted
via a tibial vein and artery of each hindleg and advanced so that the tips were in the
caudal vena cava and the abdominal aorta respectively. An additional catheter was
placed in the abdominal aorta via a tributary of the femoral artery in the groin. A
catheter was placed in the carotid artery through a cranial thyroid artery or through
a purse-string suture to avoid occlusion. The uterus was closed around another
catheter in the amniotic cavity after the fluid lost had been replaced by warm saline
(0.9%, w/v). Five fetal lambs received antibiotics daily, injected intravenously into
the ewe and into the amniotic cavity (4 g cefotaxime and 80 mg gentamycin). These
fetal lambs were allowed to recover for 3 days before being studied (chronic preparation). For comparison, the other four fetal lambs were studied in utero 2 h after
the operation (acute preparation).

Experimental Protocol

Nine unanaesthetized fetal sheep were studied in utero at a mean gestational age of
130 ± 2 days. Acute prolonged asphyxia was produced by arrest of uterine blood
flow for 4 min. To determine the redistribution of blood flow to fetal organs and the
placenta during the asphyxic episode at short intervals, isotope-labelled microspheres were injected into the inferior vena cava during a control period, i.e.
5-10 min before asphyxia and 1, 2, 3 and 4 min after uterine blood flow was arrested. In one fetus blood flows were measured before asphyxia and after 2, 3 and
4 min of asphyxia. Blood samples were drawn from the fetal femoral artery during
the control period and after 0, 1, 2, 3, 4, 5, 10, 30 and 60 min to measure blood
gases, acid-base balance, lactate, haemoglobin and catecholamine concentrations.
Corticosteroid concentrations were measured at control and at 5, 10, 30 and 60 min.
Particular care was taken to time all manoeuvres precisely.

Measurements

All measurements were made in unanaesthetized fetal sheep in utero in the absence of uterine contractions. Fetal heart rate and arterial blood pressure, amniotic fluid pressure and maternal arterial blood pressure were recorded continuously on a polygraph (Hellige, FRG). Complete occlusion of the ewe's descending aorta throughout the 4-min study period was checked by the fall in pressure below the balloon. Blood gas values and acid-base balance were determined by an automatic analyzer (Technicon BG I A, Bad Vilbel, FRG). Haemoglobin concentrations and oxygen saturation of haemoglobin were measured photometrically (OSM 2, Radiometer, Copenhagen, Denmark). Plasma concentrations of norepinephrine and epinephrine were measured by radioenzymatic assay, which had a detection limit below 5 pg/ml and an intra-assay variance below 20% (Da Prada and Zürcher 1976). Plasma concentrations of pregnenolone, 17α-pregnenolone, progesterone, 17α-progesterone, aldosterone, cortisol, dehydroepiandrosterone, dehydroepiandrosterone sulphate, oestrone and 17β-oestradiol were determined by radioimmunoassay (Gips 1983).

To assess the changes in total oxygen consumption of the fetus, placenta and membranes during asphyxia, the decrease (Δ) of the oxygen content (ml/dl), i.e. fractional oxygen saturation $\times$ haemoglobin concentration (g/dl) $\times$ 1.34 (dissolved oxygen is negligible), in the blood of the descending aorta was determined at 1-min intervals (e.g. ΔO_2 content/$min_1 = O_2$ content at control $- O_2$ content at 1 min asphyxia) and multiplied by the estimated fetal blood volume (dl/kg) (Creasy et al. 1970). Values are expressed as ml O_2/min per kg fetus.

Blood Flow Measurements

The isotope-labelled microspheres (^{141}Ce, ^{113}Sn, ^{103}Ru, ^{95}Nb and ^{46}Sc, 16 µm diameter, New England Nuclear), suspended in 10% (w/v) dextran containing 0.01% polyoxyethelene 80 sorbitan mono-oleate (Tween 80), were sonicated and checked for size (range 15.8 ± 1.1 to 16.5 ± 0.4, mean and SD), shape, and aggregation. The specific activity determined on 1000 microspheres was low (< 5 counts per minute and microsphere) to avoid coincidence loss during counting, and there was no significant leaching. Depending on the activity, 0.7–2.8 million microspheres per batch were flushed into the inferior vena cava through a mixing chamber over 20 s by 5 ml warm (39 °C) saline (Heymann et al. 1977) to a total of 7.8 ± 0.5 (SEM) million microspheres. Batches containing lower concentrations of microspheres were injected first. Reference blood samples were withdrawn from both a carotid and a femoral artery at a rate of 2.5 ml/min in the control period and during asphyxia over 75 s and 270 s respectively. Using a catheter in the inferior vena cava this volume of blood was simultaneously replaced by maternal blood maintained at 39 °C in a waterbath.

Because it was necessary to measure blood flow at short intervals we modified the original microsphere method (Rudolph and Heymann 1967) in that we injected batches of differently isotope-labelled microspheres sequentially during continuous withdrawal of reference blood over 4.5 min, the volume being simultaneously replaced. Thus, the methodological error due to the variance of four separate reference sampling procedures was reduced to a quarter and observations of acute blood flow changes at intervals of 1 min were possible. Furthermore, the time of withdrawal (270 s vs 360 s) and hence the volume exchanged were reduced by 25%, an advantage of increasing importance in smaller animals, and in proportion to both the number of reference samples and the rate of withdrawal. However, this technique has limitations that apply in part to the microsphere method in general. The blood flow is determined by the ratio of the number of microspheres trapped to the number collected per volume during withdrawal of the reference sample (usually 75 or 90 s in fetal lambs) when the mixing and the total number of microspheres are adequate. Hence, the calculated value reflects integral flow (unlike measurements with electromagnetic flow transducers), and biphasic blood flow changes of the same magnitude during the time of both injection and entrapment of the microspheres cannot be detected either under resting conditions or during asphyxia. In the present study this limitation was overcome by sequential injections of microspheres which permitted sequential blood flow measurements, rendering the loss of information of a single measurement less significant; thus, the trend of the changes could be observed during progressive asphyxia.

The time of circulation of microspheres varies with the cardiac output and therefore may be prolonged during asphyxia. To assure that the sampling time for reference blood was adequate, the number of microspheres still circulating 30-40 s after each injection was determined by counting the activity of centrifuge vials containing pellets of blood samples of a known volume drawn at 2, 3, 4 and 5 min. After 2 min asphyxia, 2.6% ± 1.1% of the microspheres collected in the reference sample were still circulating. After 3 and 4 min, i.e. at the nadir of asphyxia, the figure was 4.5% ± 1.3%, and after 5 min, i.e. after 1 min of recovery, 4.2% ± 1.6% residual microspheres were found. Thus, the error introduced by incomplete collection of microspheres from the last batch 60-70 s after the injection was considered negligible. A mean of 3.2% (range 2.3%-6.1%) of the total blood volume was exchanged by the withdrawal/infusion procedure, assuming a blood volume of 135 ml/kg body weight (Creasy et al. 1970). The haemoglobin concentration at 10 min (11.5 ± 0.7 g/dl) was not significantly different from the control value (12.1 ± 0.8 g/dl).

After the experiment a lethal dose of sodium pentobarbitone was given to each ewe and the lambs were perfused with 300 ml formalin (15%, w/v, saline). Fetal organs and cotyledons were weighed and placed in vials, which were filled to the same height to reduce variations in geometry. The intestines were separated from the mesentery, opened and cleared of contents. Paired organs (lungs, kidneys and adrenals) were counted separately, as were the right and left sides of the cerebrum and of the brain stem parts. No significant preferential streaming of microspheres was found. Specimens of the skin and muscle were taken from the hips and shoulders of each side; there were no differences in blood flow between the intact side and the side on which the femoral or brachial arteries were catheterized. There were also no differences in blood flow between tissues from the upper and lower body;

the results for all skin specimens from these sites have therefore been combined. Similarly, the results for all muscle specimens were combined.

The solid-state semiconductor gamma counter [Ge (Li)] used had a high energy resolution of about 3 keV and was connected to a multichannel (2048) pulse height analyzer (Nuclear Data Inc.). The results were normalized with respect to time and sample weight, punched on paper tape, transferred to a disc file and processed with a programmable computer (PDP 11/45) (Winkler et al. 1982).

Calculations

For statistical evaluation the Wilcoxon rank test and linear regression analysis were used. Logarithmic transformation was performed where necessary. The results are given as means $\pm$ SEM. The results from chronically and acutely prepared lambs were not significantly different during the control period or during asphyxia, as judged by the heart rate, arterial blood pressure, blood flow to the principal organs, acid-base balance, pCO_2, PO_2 and lactate concentrations; the results of the two groups were therefore combined. Vascular resistance was calculated by dividing arterial blood pressure (corrected for amniotic fluid pressure) by blood flow and was expressed in mm Hg/ml $\times$ min per 100 g tissue.

Results and Discussion

Blood Gases and Oxygen Consumption

Arrest of uterine blood flow for 4 min caused severe fetal asphyxia: heart rate, arterial oxygen saturation of haemoglobin, oxygen content and pH decreased, and arterial blood pressure, PCO_2 and lactate concentrations increased (Fig. 1).

The fall of the mean arterial oxygen content (from 6.4 ± 1.0 ml/dl at control to 0.7 ± 1.0 ml/dl after 4 min) was paralleled by an exponential fall of the total oxygen consumption of fetus, placenta and membranes with time to zero ml O_2/min ($r = -0.80$, $n = 36$, $2a < 0.0001$), and there was a close linear correlation between these two variables after each minute of asphyxia.

Blood Flow Changes in Peripheral and Central Organs

Within 1 min asphyxia there was a significant fall in blood flow to most of the peripheral organs, including skin, scalp and choroid plexus, to between 46% and < 1% of the control values, and to the cerebrum, whereas the blood flow to the cardiac ventricles and that to the brain stem were maintained.

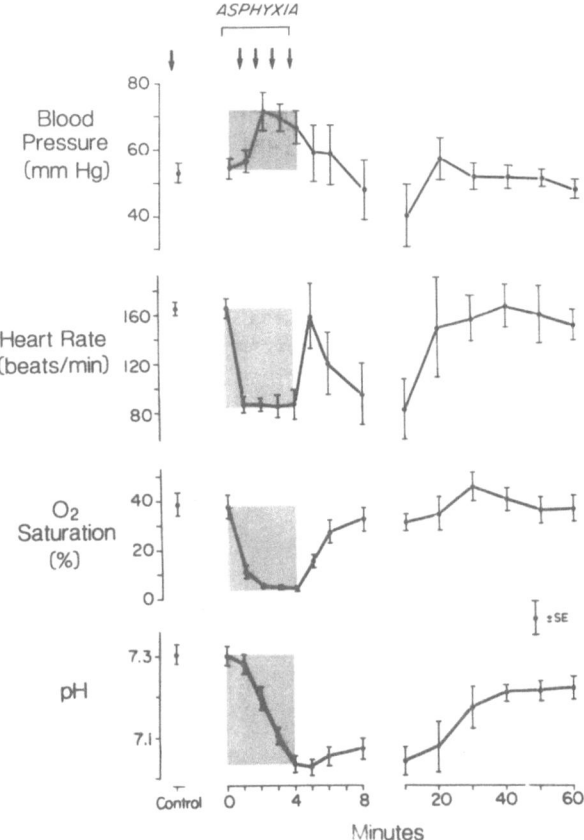

Fig. 1. Changes during acute asphyxia in nine unanaesthetized fetal sheep near term. Asphyxia was caused by arrest of uterine blood flow for 4 min. The distribution of blood flow to the fetal organs was measured during the control period (5–10 min before asphyxia) and after each minute of asphyxia (↓) by injection of isotope-labelled microspheres 16 μm in diameter. (Jensen et al. 1987)

After 2 min asphyxia, blood flow to the heart, brain stem and (in surviving fetuses only) adrenals increased; the blood flow to the total brain was unchanged. Blood flow to peripheral organs and total oxygen consumption continued to decrease. Blood flows to placenta and lungs were maintained at about 40%–50% of control throughout asphyxia.

After 4 min arrest of uterine blood flow the fetuses were severely asphyxic and had lower heart rates (87 ± 13 beats/min), but arterial blood pressure was still maintained (67 ± 5 mm Hg). There were reduced blood flows to the heart and brain, whereas blood flow to the kidneys, gastrointestinal tract, pancreas, thyroid gland, diaphragm and brown adipose tissue increased, suggesting a redistribution towards these peripheral organs at the expense of the central organs.

Differences Between Surviving and Non-surviving Fetuses

Blood Gases and Arterial Blood Pressure

Two of the five chronically prepared fetuses and three of the four acutely prepared fetuses died 10–35 min after the end of the asphyxic period caused by arrest of uterine blood flow. Before asphyxia was induced there were few differences between surviving and non-surviving fetuses; only arterial blood pressure and adrenal blood flow were higher in the non-surviving than in the surviving fetuses. However, during asphyxia there were lower mean pH, base excess and PCO_2 values, lower heart rates and smaller increases in arterial peak pressure above control values ($18\% \pm 17\%$ vs $62\% \pm 20\%$, $P < 0.05$) in the non-surviving fetuses.

Blood Flow

In the majority of individual organs there were no significant blood flow differences between non-surviving and surviving fetuses during the first 3 min asphyxia. But, unlike survivors, in non-survivors adrenal blood flow failed to increase (Fig. 2), and blood flow to cardiac ventricles, cerebrum and brain stem, particularly to the medulla, fell after 4 min asphyxia.

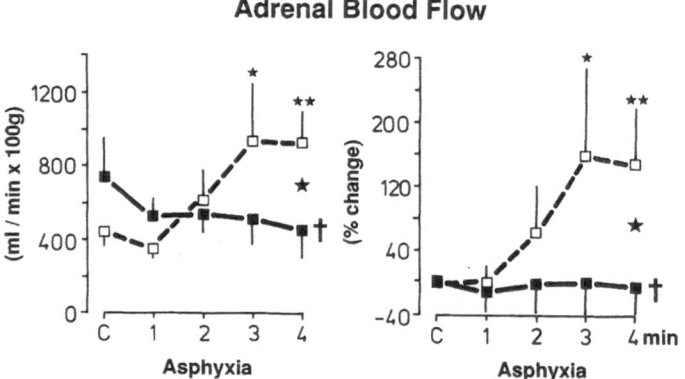

Fig. 2. Changes in adrenal blood flow as ml/min × 100 g *(left)* and as percent of control *(right)* during acute asphyxia caused by arrest of uterine blood flow for 4 min in surviving (- - □ - -) and non-surviving (— ■ —) fetal sheep near term. Note that in non-surviving fetuses adrenal blood flow fails to increase during asphyxia. * $P < 0.05$, ** $P < 0.01$; *small asterisks*, significantly different from control; *large asterisks*, significant difference between groups

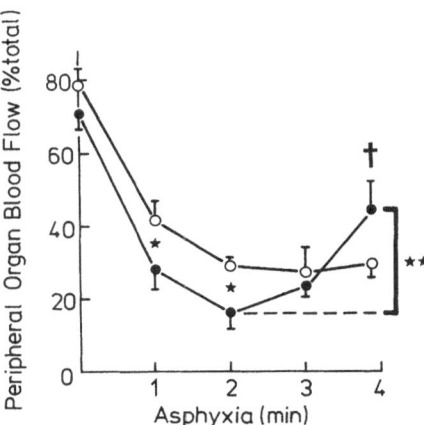

Fig. 3. Changes in blood flow to peripheral organs during acute asphyxia in four surviving (○) and five non-surviving (●) fetal sheep near term. Note that the results at 1 and 2 min asphyxia differ significantly between the groups (* $P < 0.05$). Note also that, unlike in surviving fetuses, at the nadir of asphyxia (4 min) in non-surviving fetuses, the portion of blood flow directed to peripheral organs was significantly higher than at 2 min asphyxia (** $P < 0.01$)

In non-surviving fetuses the proportion of blood flow (%) directed to peripheral organs was significantly lower after 1 and 2 min ($P < 0.05$), but higher after 4 min of asphyxia ($P < 0.06$), than in surviving fetuses (Fig. 3). In contrast to survivors, in which the reduction in peripheral blood flow was maintained at the nadir of asphyxia, there was a significant increase in peripheral blood flow after 4 min of asphyxia in the non-survivors ($P < 0.01$). These blood flow changes in non-surviving fetuses at 4 min, which were accompanied by a rise and a fall in vascular resistance (mean arterial blood pressure did not change) of central and peripheral organs respectively, are consistent with *decentralization* of the fetal circulation at the nadir of asphyxia (Fig. 3).

After 6 min recovery (i.e. at 10 min) the non-surviving fetuses were hypotensive and severely acidaemic and their heart rate was low (after transient recovery of both heart rate and oxygen saturation), whereas in the surviving fetuses arterial blood pressure, blood gases, acid-base balance and heart rate gradually returned towards normal.

Catecholamines

To address the question of whether the failure of adrenal blood flow to increase during asphyxia might have influenced survival by affecting adrenal function, plasma catecholamine concentrations of non-surviving fetuses were compared with those of surviving fetuses. During asphyxia there was a steep rise in both norepinephrine and epinephrine concentrations, but there were no significant differences between the groups (Fig. 4), indicating that plasma concentrations of these hormones, which derive from the adrenal medulla, are unrelated to fetal survival of acute asphyxia. Therefore we explored the possibility that deficient concentrations of steroids deriving from the adrenal cortex might be related to fetal survival.

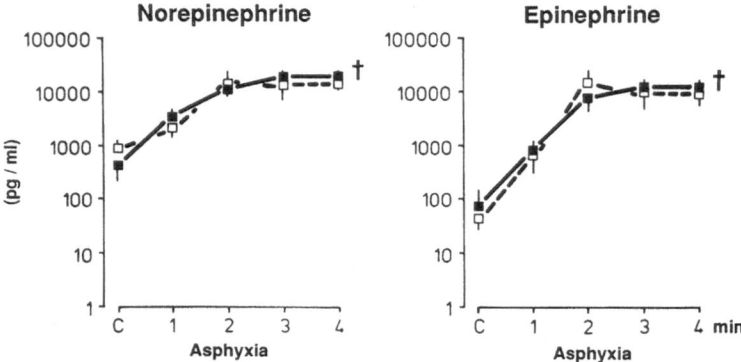

Fig. 4. Changes in plasma norepinephrine *(left)* and epinephrine *(right)* concentrations before and during acute asphyxia caused by arrest of uterine blood flow for 4 min in surviving (- - □ - -) and non-surviving (— ■ —) fetal sheep near term. Note that there are no significant differences between groups

Steroid Hormones and Asphyxia

A schematic representation of the adrenal corticosteroid biosynthesis is given in Fig. 5. In order to assess the effects of asphyxia on steroids predominantly synthesized in the zona glomerulosa, zona fasciculata and zona reticularis of the adrenal cortex, the concentrations of the principal end-products of steroid synthesis characteristic of the respective layer and their main precursors were measured.

After 4 min asphyxia plasma concentrations of pregnenolone decreased, but those of both progesterone and aldosterone (zona glomerulosa) increased (Fig. 6). Concentrations of 17α-hydroxypregnenolone, 17α-hydroxyprogesterone, cortisol (zona fasciculata) (Fig. 7) and dehydroepiandrosterone (DHEA, zona reticularis) decreased after asphyxia (Fig. 8), whereas those of dehydroepiandrosterone sulphate (DHEAS) and those of oestrone and oestradiol-17β, which are non-adrenal derivatives of DHEA and DHEAS, did not change. None of these hormone concentrations, except for those of DHEAS, which where higher in non-survivors, differed significantly between surviving and non-surviving fetuses, suggesting that plasma concentrations of the principal corticosteroids are unrelated to fetal survival of acute asphyxia.

The effects of asphyxia on enzymes of the corticosteroid biosynthesis, e.g. 17,20-desmolase and 17α-hydroxylase, was assessed by calculating the substrate-product ratios 17α-hydroxypregnenolone/DHEA and pregnenolone/17α-hydroxypregnenolone in the control period and after asphyxia (Fig. 8). The ratio 17α-hydroxypregnenolone/DHEA increased significantly, whereas that of pregnenolone/17α-hydroxypregnenolone tended to decrease with both decreasing arterial oxygen saturation of haemoglobin ($r = 0.84$; $P < 0.001$) and decreasing pH ($r = 0.78$; $P < 0.01$) (not illustrated), indicating that 17,20-desmolase, but not 17α-hydroxylase, is

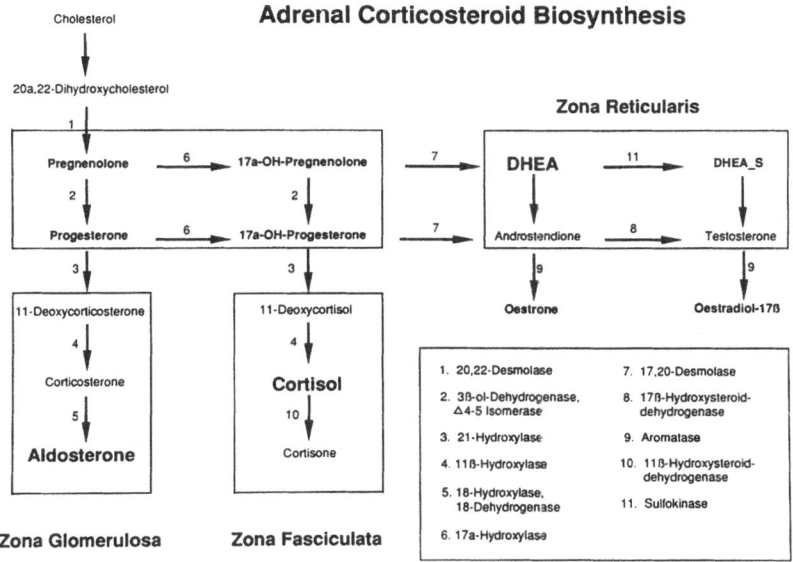

Fig. 5. Adrenal corticosteroid biosynthesis

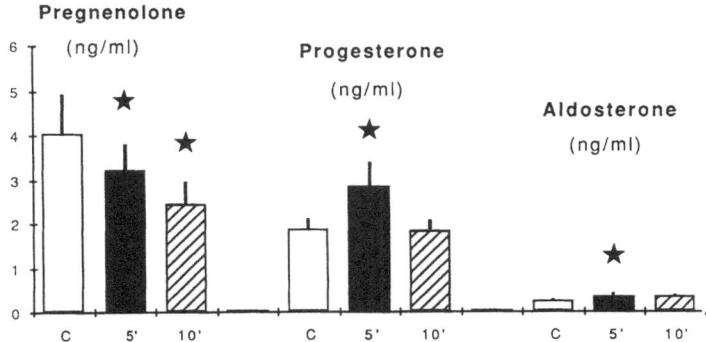

Fig. 6. Changes in plasma concentrations of adrenal steroid hormones after acute asphyxia caused by arrest of uterine blood flow for 4 min in nine unanaesthetized fetal sheep near term. Note that pregnenolone, which is the key precursor of adrenal corticosteroid formation (see Fig. 5), decreases progressively after asphyxia. Note also that progesterone and aldosterone concentrations increase in spite of decreasing pregnenolone concentrations, suggesting that progesterone, which is converted to aldosterone in the zona glomerulosa of the adrenal cortex, may in part be of placental origin. * $P < 0.05$; C, 5', 10', results at control, at 5 min (i.e. after 1 min recovery) and at 10 min (i.e. after 6 min recovery)

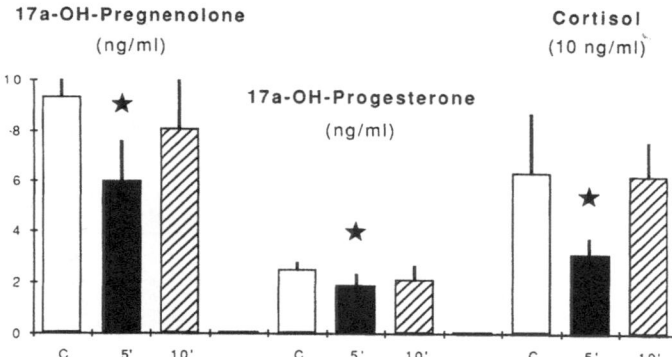

Fig. 7. Changes in plasma concentrations of adrenal steroid hormones after acute asphyxia caused by arrest of uterine blood flow for 4 min in nine unanaesthetized fetal sheep near term. Note that 17α-OH-pregnenolone, 17α-OH-progesterone and cortisol, which is formed in the zona fasciculata of the adrenal cortex, decrease transiently after acute asphyxia. * $P<0.05$; C, 5', 10', results at control, at 5 min (i.e. after 1 min recovery) and at 10 min (i.e. after 6 min recovery)

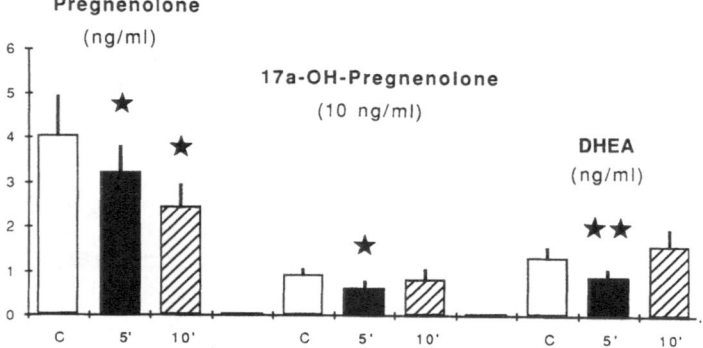

Fig. 8. Changes in plasma concentrations of adrenal steroid hormones after acute asphyxia caused by arrest of uterine blood flow for 4 min in nine unanaesthetized fetal sheep near term. Note that 17α-OH-progesterone and dehydroepiandrosterone (DHEA), which is formed in the zona reticularis of the adrenal cortex, decrease transiently after acute asphyxia. * $P<0.05$; C, 5', 10' results at control, at 5 min (i.e. after 1 min recovery) and at 10 min (i.e. after 6 min recovery)

inhibited by asphyxia. Hence, synthesis of androgens in general and that of DHEA in particular is inhibited during asphyxia. Thus, the pool of 17α-hydroxypregnenolone, which is in short supply because pregnenolone decreases during asphyxia, is relatively maintained and hence may constitute a substrate reserve for synthesis of cortisol.

In spite of these various significant changes in corticosteroid concentrations and enzyme activity in the adrenal cortex during asphyxia, no overt deficiency of adrenal medullary or cortical hormones could be detected in non-surviving as compared with surviving fetuses. In a further attempt to determine why some of the fetuses developed circulatory decentralization and died, steroid hormones were related to fetal organ blood flows before and during asphyxia.

Gastro-Intestinal Tract

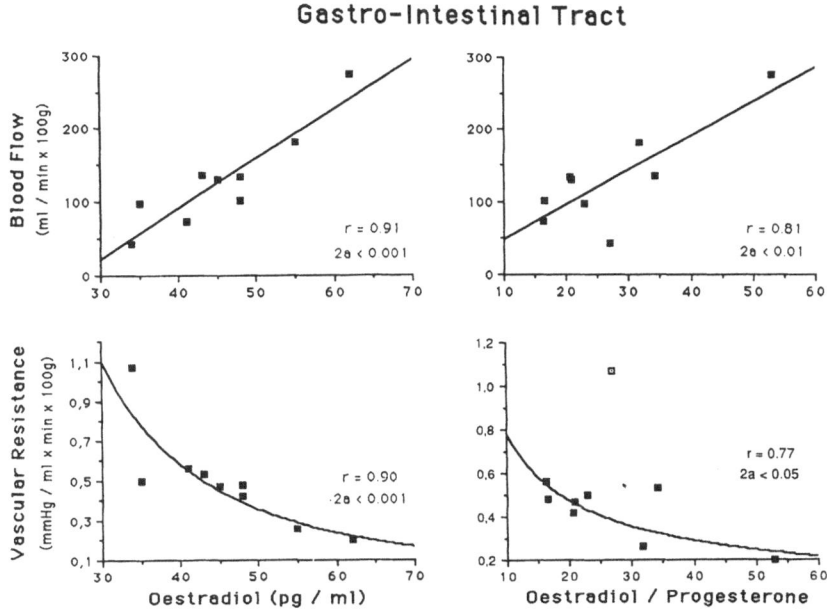

Fig. 9. Close correlations between both plasma concentrations of oestradiol-17β and the oestradiol-17β/progesterone ratio and blood flow to and vascular resistance of the gastrointestinal tract under resting conditions (control measurements) in nine unanaesthetized fetal sheep near term. Note that gastrointestinal blood flow increases and vascular resistance decreases when both oestradiol-17β concentrations and the oestradiol-17β/progesterone ratio increase. One possible explanation for this unexpected result is that these hormones may change phospholipase A_2 activity and hence may change prostaglandin concentrations locally, as suggested by Liggins (1980) (see Fig. 10)

Steroid Hormones and Blood Flow Under Resting Conditions

At control there were surprisingly close correlations between both oestradiol-17β concentrations and the ratio oestradiol-17β/progesterone and blood flow to a number of central and peripheral organs, including those of the gastrointestinal tract (Fig. 9), kidneys, brown fat and skin. Blood flow to these organs was high and vascular resistance was poor when both oestradiol-17β concentrations and the ratio oestradiol-17β/progesterone were high. These results indicate for the first time that the balance of oestradiol-17β and progesterone concentrations is an important determinant of fetal organ blood flow under resting conditions (Fig. 9). This was confirmed by multiple regression analysis of all factors that could possibly affect blood flow in the control period, including blood gases, acid-base variables, catecholamines, corticosteroids, and blood pressure. None of these factors except oestradiol-17β concentrations and the ratio oestradiol-17β/progesterone bore a significant relationship to fetal organ blood flows under resting conditions.

The mechanisms involved are not clear; however, it is likely that changes in oestradiol-17β and progesterone concentrations act locally on the vascular smooth muscle through activation or inhibition of phospholipase A_2. Thus, formation or degradation of short-acting vasoactive prostaglandins, e. g. prostacyclin and thromboxane, may change organ blood flow by changing vascular resistance (Liggins 1980) (Fig. 10).

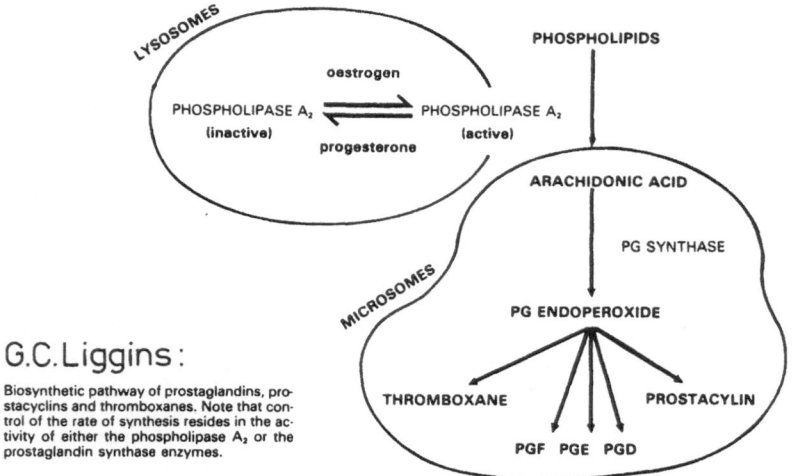

G.C.Liggins:

Biosynthetic pathway of prostaglandins, prostacyclins and thromboxanes. Note that control of the rate of synthesis resides in the activity of either the phospholipase A_2 or the prostaglandin synthase enzymes.

Fig. 10. Effects of oestrogen and progesterone on phospholipase A_2 activity and prostaglandin synthesis (Liggins 1980) (see Fig. 9)

Catecholamines and Blood Flow During Asphyxia

During the first 2 min asphyxia in a number of organs, changes in vascular resistance and those in blood flow were related to changes in plasma concentrations of norepinephrine and epinephrine, but not to corticosteroids. Particularly a decrease in the norepinephrine/epinephrine ratio correlated with decreasing vascular resistance in and increasing blood flow to parts of the brain, e.g., brain stem, as reported previously (Jensen et al. 1985).

After 3 and 4 min of asphyxia there were no correlations between catecholamines or corticosteroids and actual vascular resistance and blood flow, which could explain the loss in vascular resistance of peripheral organs preceding circulatory decentralization in the non-surviving fetuses.

Circulatory Decentralization During Severe Asphyxia

Direct effects of asphyxia on peripheral vasomotor responses were explored by correlating blood pH at 4 min of asphyxia with the change (i.e. the values at 4 min minus those at 2 min) in both vascular resistance and blood flow of the skeletal muscle at the nadir of asphyxia. There was a linear decrease in the vascular resistance of and a linear increase in the blood flow to skeletal muscle with decreasing arterial pH (Fig. 11) and increasing norepinephrine ($r=0.86$, $2a < 0.01$) and cortisol concentrations ($r=0.83$, $2a < 0.01$) (not illustrated). Only those three fetuses in which blood pH was about 7.1 were able to increase vascular resistance of skeletal

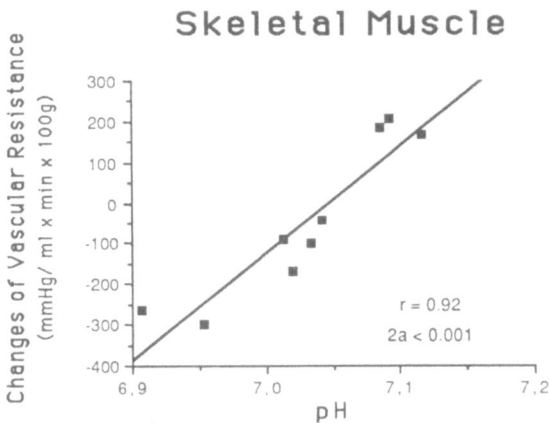

Fig. 11. Close correlation between blood pH after 4 min of asphyxia and the change of the vascular resistance in skeletal muscle, i.e. resistance at 4 min minus that at 2 min, in nine unanaesthetized fetal sheep near term. The loss of vascular resistance in skeletal muscle during severe acidaemia is largely responsible for circulatory decentralization at the nadir of asphyxia, which precedes fetal death (see Fig. 3)

muscle; in the remaining fetuses, in which blood pH values ranged from 7.05 to 6.90, vascular resistance decreased, even though norepinephrine and cortisol concentrations increased linearly. Similar correlations were found for organs of the gastrointestinal tract, suggesting that severe blood acidaemia and an inferred local acidosis in the vascular smooth muscle cell may prevent peripheral vessels from maintaining constriction in spite of high plasma concentrations of vasoconstrictive substances. Thus, it seems likely that circulatory decentralization, at the nadir of asphyxia, which precedes fetal death is caused by a loss of vascular resistance in both skeletal muscle and the gastrointestinal organs induced by local acidaemia. However, there is no firm evidence for adrenal malfunction as a primary cause of fetal death after acute prolonged asphyxia, even though adrenal blood flow did not increase in non-surviving fetuses.

Summary and Conclusions

In summary, a number of conclusions can be drawn from this study on unanaesthetized fetal sheep near term. First, arrest of uterine blood flow for 4 min causes acute asphyxia, exponential reduction in oxygen consumption to zero and circulatory centralization. Secondly, acute prolonged asphyxia changes plasma concentrations of adrenal medullary and cortical hormones, in that concentrations of epinephrine and norepinephrine (medulla) and aldosterone (cortex, zona glomerulosa) increase, but those of, for example, cortisol (zona fasciculata) and DHEA (zona reticularis) decrease. Thirdly, adrenal corticosteroid biosynthesis is redistributed during asphyxia at the expense of the androgen and in favour of the cortisol pathway. Fourthly, under resting conditions fetal organ blood flows depend largely on plasma oestradiol-17β concentrations and on the oestradiol-17β/progesterone ratio, whereas in the initial phase of acute asphyxia, in which circulatory centralization develops, catecholamines are the major determinants of blood flow. Fifthly, maintenance of circulatory centralization at the nadir of asphyxia is essential for survival of the fetus; circulatory decentralization is irreversible and hence fatal, even though arterial oxygen saturation and heart rate recover transiently after release of uterine blood flow. Sixthly, the loss of peripheral vascular resistance which causes circulatory decentralization is strongly related to acidaemia but not to deficient adrenal hormones. Thus, local vascular effects of asphyxia rather than adrenal hormonal factors cause circulatory decentralization and eventually death of the fetus.

Acknowledgements. We thank S. and J. Jelinek, P. Becker, I. Denstedt, B. Maid, R. Achenbach and O. Adam for technical assistance and Dr. R. Dietz and Dr. A. Schömig, Dept. of Pharmacology, University of Heidelberg, FRG for catecholamine analyses. We are also indebted to Dr. B. Winkler and Prof. Dr. W. Schaper, Max Planck Institute for Experimental Cardiology, Bad Nauheim, FRG for the blood flow analyses.

References

Creasy RK, Drost M, Green MV, Morris JA (1970) Determination of fetal, placental and neo-natal blood volumes in sheep. Circ Res 27: 487-497

Da Prada M, Zürcher G (1976) Simultaneous radioenzymatic determination of plasma and tis-sue adrenaline, noradrenaline and dopamine within the femtomole range. Life Sci 19: 1161-1174

Gips H (1983) Die Funktion der mütterlichen Nebennierenrinde in der Schwangerschaft und im Wochenbett. Thesis, University of Gießen

Heymann MA, Payne BD, Hoffman JIE, Rudolph AM (1977) Blood flow measurements with radionuclide-labeled particles. Prog Cardiovasc Dis 20: 55-79

Jensen A, Künzel W, Hohmann M (1985) Dynamics of fetal organ blood flow redistribution and catecholamine release during acute asphyxia. In: Jones CT, Nathanielsz PW (eds) The physiological development of the fetus and newborn. Academic, London, pp 405-410

Jensen A, Hohmann M, Künzel W (1987) Dynamic changes in organ blood flow and oxygen consumption during acute asphyxia in fetal sheep. J Devel Physiol 9: 543-559

Liggins GC (1980) Etiology of premature labor. Mead Johnson Symp Perinat Dev Med 15: 3-7

Rudolph AM, Heymann MA (1967) The circulation of the fetus in utero: methods for study-ing distribution of blood flow, cardiac output and organ blood flow. Circ Res 21: 163-184

Winkler B, Stämmler G, Schaper W (1982) Measurement of radioactive tracer microsphere blood flow with NaJ (Tl)- and Ge-well type detectors. Basic Res Cardiol 77: 292-300

Production and Inactivation of Catecholamines During Hypoxia and Recovery

O. Schwab[1], E. Kastendieck[2], R. Paulick[2], and H. Wernze[3]

It is well established that levels of fetal free catecholamines, i.e. norepinephrine (NE), epinephrine (E), and dopamine (DA) rise during hypoxia. High levels of free catecholamines cause a redistribution of fetal cardiac output favoring the brain, the heart, the adrenal glands, and the umbilical circulation (Cohn et al. 1974). Moreover, free catecholamines effect an increase in pulmonary surfactant efflux (Lawson et al. 1978) and inhibit fetal pulmonary fluid secretion (Walters and Olver 1978). Finally, free catecholamines play a part in perinatal glucose homeostasis (Sperling et al. 1984). However, little is known about the inactivation of catecholamines. Therefore, in order to investigate the behavior of catecholamines during hypoxia and recovery and to study mechanisms of catecholamine inactivation, we performed experiments with chronically instrumented sheep and measured catecholamines in the cord blood of human newborns. The aim of our study was to answer the following four questions:

1. At what degree of hypoxia do catecholamine levels rise in the sheep fetus?
2. How quickly are free catecholamines inactivated in the sheep fetus at the end of hypoxia?
3. What role does sulfoconjugation of free catecholamines play in catecholamine inactivation in human newborns?
4. What is the role of the human placenta in catecholamine inactivation?

Effect of Hypoxia on Catecholamines

We performed experiments in eight chronically instrumented sheep fetuses. A catheter was advanced into the fetal aorta for blood sampling. An inflatable occluder was placed around the maternal aorta to produce a progressive reduction of uterine blood flow. The experiments were performed 4–5 days after surgery. We determined fetal free catecholamines radioenzymatically according to the principles of

[1] Department of Pediatrics, University of Würzburg, Joseph-Schneider-Str. 2, 8700 Würzburg, FRG

[2] Department of Gynecology and Obstetrics, University of Würzburg, Joseph-Schneider-Str. 4, 8700 Würzburg, FRG

[3] Department of Internal Medicine, University of Würzburg, Joseph-Schneider-Str. 2, 8700 Würzburg, FRG

The Endocrine Control of the Fetus
Ed. by W. Künzel and A. Jensen
© Springer-Verlag Berlin Heidelberg 1988

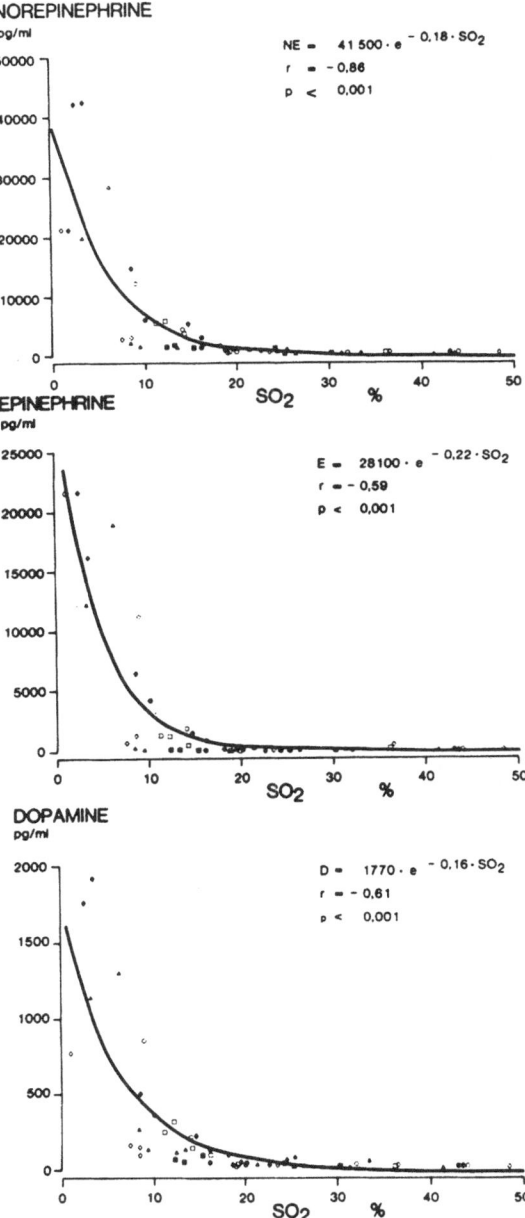

NOREPINEPHRINE
pg/ml

$$NE = 41\,500 \cdot e^{-0.18 \cdot SO_2}$$
$$r = -0.86$$
$$p < 0.001$$

EPINEPHRINE
pg/ml

$$E = 28\,100 \cdot e^{-0.22 \cdot SO_2}$$
$$r = -0.59$$
$$p < 0.001$$

DOPAMINE
pg/ml

$$D = 1770 \cdot e^{-0.16 \cdot SO_2}$$
$$r = -0.61$$
$$p < 0.001$$

Fig. 1. Inverse relationship between oxygen saturation and the free catecholamines norepinephrine, epinephrine, and dopamine. An oxygen saturation of 15%–20% seems to be a critical threshold for exponential increase of free catecholamines

Peuler and Johnson (1977) We also measured the oxygen saturation, pH, and lactate in fetal arterial blood (Paulick et al. 1987).

A progressive decrease of fetal arterial oxygen saturation was induced by gradual reduction of uterine blood flow. An inverse relationship was found between oxygen saturation and fetal free NE, E, and DA (Fig. 1). When the oxygen saturation fell below 15%–20%, an exponential increase of catecholamine concentrations occurred. With oxygen saturation values below 15%, lactate production increased by $0.1-0.3$ mmol $\times l^{-1} \times min^{-1}$ and pH decreased by $0.003-0.009 \times min^{-1}$. These values were significantly different to those given at oxygen saturation values above 15% (lactate $0.01-0.03$ mmol $\times l^{-1} \times min^{-1}$; pH $0.0003-0.0012 \times min^{-1}$). Therefore, our experiments suggest that an oxygen saturation level of 15%–20% represents a critical threshold for development of severe acidosis and shock syndrome in the sheep fetus.

Posthypoxic Inactivation of Catecholamines

In order to investigate the posthypoxic inactivation of catecholamines, we induced acute fetal hypoxia by total compression of the maternal aorta for 5 min ($n=5$; Paulick et al. 1988).

Oxygen saturation reached the zero line within 3 min and rapidly increased to initial values during recovery. The pH values decreased from 7.33 ± 0.02 to 6.94 ± 0.02 but did not reach the initial values during recovery. This is due to lactic acid, which rose from 2.3 ± 0.2 to 7.1 ± 0.3 mmol $\times l^{-1}$ and remain above 6 mmol $\times l^{-1}$ during the whole observation period. As shown in Fig. 2, peak concentration of all free catecholamines were reached within 3 min after occlusion was started. Free NE increased 60-fold, E 370-fold and DA 13-fold compared to control values. The posthypoxic elimination of catecholamines is characterized by fast inactivation during the first 10 min and a period of slower inactivation during the following 20 min. The mean-half-lives of free NE, E, and DA were 3.2 min, 4.3 min, and 2.9 min respectively during the first 10 min and 10.6 min, 6 min, and 8.3 min respectively during the 20 min following release of the occlusion. By comparison, the half-life of infused free E ranged from $0.23-0.27$ min in sheep fetuses (Jones and Robinson 1975). The observed difference may be attributed to the persistent secretion of high amounts of catecholamines into the circulation after cessation of occlusion. This results in a prolongation of the half-life measured in our experiments.

Role of Sulfoconjugation in Catecholamine Inactivation

The inactivation of catecholamines is not well understood. DA can be enzymatically transformed into NE and E by dopamine-β-hydroxylase and phenylethanolamine-N-methyltransferase, respectively. In addition we must remember that large amounts of NE and DA function as neuronal transmitters and may thus be inacti-

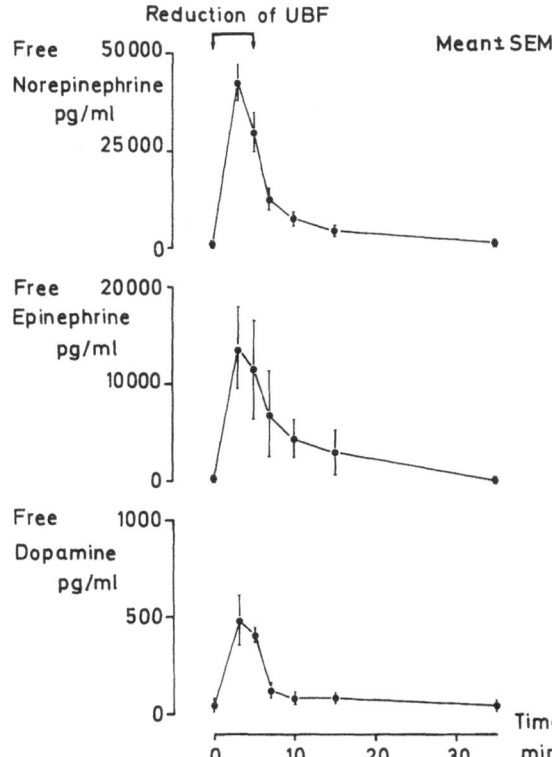

Fig. 2. Increase of free norepinephrine, epinephrine, and dopamine after a total stoppage of uterine blood flow (5 min) and subsequent decrease to initial values. The reduction in catecholamines is characterized by a fast inactivation during the first posthypoxic period and a slower inactivation afterwards

vated by neuronal reuptake. The inactivation of catecholamines by catechol-*O*-methyltransferase and monoamine oxidase predominantly occurs in the liver and kidneys (Tipton 1973). The pathway of inactivation of free catecholamines by sulfoconjugation via phenolsulfotransferase and by glucuronide conjugation via glucuronidase has raised increased interest in the last decade. It has been established that sulfoconjugation predominates in humans (Claustre et al. 1983), whereas glucuronide conjugation is predominant in rats (Wang et al. 1983) and sheep fetuses (O. Schwab, R. Paulick, E. Kastendieck, and H. Wernze, unpublished observations, 1986). In humans, the sulfoconjugation occurs mainly in red blood cells, platelets, liver, intestine, and brain (Vandongen 1984). Defects of sulfoconjugation resulting in raised levels of free catecholamines have been described in patients with essential hypertension, pheochromocytoma, and diet-induced migraine (Vandongen 1984). A further pathway of catecholamine inactivation is the accumulation of free and sulfoconjugated catecholamines in red blood cells (Alexander et al. 1984).

Data concerning sulfoconjugation in human newborns have not reported until now. Therefore, we wanted to obtain reference data of sulfoconjugated catechol-

84 O. Schwab et al.

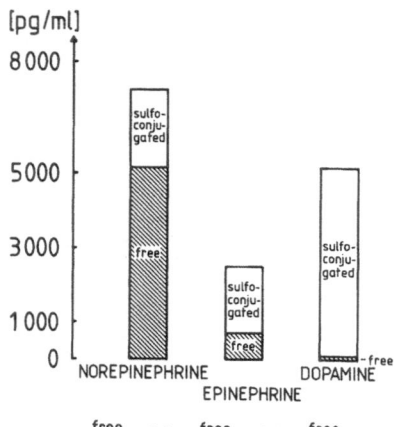

Fig. 3. Median values for free and sulfoconjugated catecholamines in the umbilical artery of 20 human newborns. The different ratios of free to sulfoconjugated catecholamines are probably the result of different affinities of free catecholamines to the enzyme phenolsulfotransferase

amines in human newborns and to investigate the role of sulfoconjugation (Schwab et al. 1985).

In the first assay we measured the free part of total catecholamines (Peuler and Johnson 1977). The sulfoconjugated catecholamines were then transformed to the free form by adding arylsulfatase (type VI, Biosigma) and we measured total catecholamines in a second assay. The amount of sulfoconjugated catecholamines was calculated by substracting the amount of free from the amount of total catecholamines (Johnson et al. 1980). We investigated 20 newborns (gestational age 39–41 weeks) after uncomplicated vaginal delivery. Apgar values at 1 min after birth ranged between 6 and 10. The mean fetal arterial pH was 7.22 ± 0.07. Blood samples were separately obtained from the umbilical artery and vein.

Figure 3 shows the median values of free and sulfoconjugated catecholamines in the umbilical artery. A wide range of free NE (950–74000 pg/ml), E (40–7500 pg/ml), and DA (30–650 pg/ml) occurs due to differences in individual delivery stress. Quantitatively, the predominant free catecholamine is NE (median 5170 pg/ml), which is in the same range as sulfoconjugated DA. Most of the total DA is sulfoconjugated and biologically inactive. The different ratios of free to sulfoconjugated catecholamines may be explained by varying affinities of free catecholamines to the enzyme phenolsulfotransferase. The affinity seems to be highest in free DA, lower in free E, and lowest in free NE.

The correlation of free to sulfoconjugated catecholamines shows highly significant results for all three catecholamines: increasing amounts of free catecholamines lead to increased levels of sulfoconjugated catecholamines. This means that even in severe hypoxic stress, free catecholamines are also rapidly inactivated by sulfoconjugation.

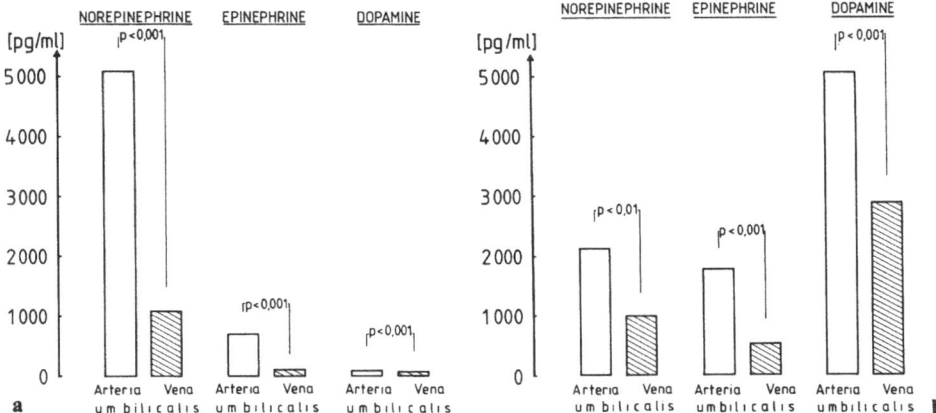

Fig. 4a, b. Median values of free and sulfoconjugated catecholamines in the umbilical artery in comparison to the umbilical vein. The arterial values of all three catecholamines *(open bars)* are significantly higher than the venous levels *(hatched bars)* due to placental extraction

Role of the Placenta in Catecholamine Inactivation

We measured plasma levels of free and sulfoconjugated catecholamines in the umbilical artery and vein. As shown in Fig. 4 the arterial values of all three catecholamines were significantly higher than the levels in the umbilical vein. The median value of free NE was five times and that in free E seven times as high in the arterial as in the venous vessels. Sulfoconjugated NE and DA were twice and E four times as high. Therefore, free and sulfoconjugated catecholamines in the umbilical cord blood originate from the fetus; the difference between arterial and venous plasma levels represents the amount which is due to placental extraction. The placenta has a huge capacity to inactivate free and sulfoconjugated catecholamines, probably by action of the enzymes catechol-*O*-methyltransferase and monoamine oxidase. Both enzymes are present in high concentrations in placental tissue (Castrén and Saarikoski 1974).

Placental extraction rates (percentages) of free sulfoconjugated catecholamines were calculated using the following formula:

$$\frac{C_{Art.\,umb.} - C_{vena\,umb.}}{C_{Art.\,umb.}} \times 100$$

The rates thus calculated were as follows:

Free catecholamines	Median	(Range)
Norepinephrine	78%	(51%–99%)
Epinephrine	84%	(9%–99%)
Dopamine	30%	(1%–83%)

Sulfoconjugated Catecholamines

Norepinephrine	60%	(0%–92%)
Epinephrine	65%	(34%–95%)
Dopamine	39%	(17%–94%)

No explanation could be found for the wide scattering of catecholamine extraction rates. In spite of this remarkable range, there was a highly significant correlation ($y = 0.7x + 21.6$, $r = 0.77$, $p < 0.001$) between the extraction rates of free NE and E. This suggests that there is no secretion of NE into the fetal blood during placental passage. This observation supports the view that the placenta has no sympathetic innervation, which agrees with morphological studies (Reilly and Russell 1977).

Conclusions

1. An oxygen saturation level of 15%–20% represents a critical threshold for severe acidosis and a marked increase of free catecholamines in the sheep fetus.
2. The elimination of catecholamines after brief hypoxia exhibits a biphasic course. There is fast elimination during the initial 10 min followed by a slower inactivation during the next 20 min.
3. In the healthy human fetus, delivery induces a marked increase of free and sulfoconjugated catecholamines. Quantitatively, the predominant catecholamines are free NE and sulfoconjugated DA. Plasma levels of free DA are very low, probably due to a high affinity of free DA to phenolsulfotransferase and rapid sulfoconjugation. The concomitant increase of free and sulfoconjugated catecholamines suggests that sulfoconjugation is an important inactivation pathway of free catecholamines.
4. The median placental extraction rates of catecholamines were 60%–80% for NE and E and 30%–40% for DA. Thus, human placenta has a huge capacity to inactivate free and sulfoconjugated catecholamines.

References

Alexander N, Yoneda S, Vlachakis ND, Maronde RF (1984) Role of conjugation and red blood cells for inactivation of circulating catecholamines. Am J Physiol 247: R203–R207

Castrén O, Saarikoski S (1974) The simultaneous function of catechol-*O*-methyltransferase and monoamine oxidase in human placenta. Acta Obstet Gynecol Scand 53: 41–47

Cohn HE, Sacks EJ, Heymann MA, Rudolph AM (1974) Cardiovascular responses to hypoxemia and acidaemia in fetal lambs. Am J Obstet Gynecol 120: 817–824

Claustre J, Serusclat P, Peyrin L (1983) Glucuronide and sulfate catecholamine conjugates in rat and human plasma. J Neural Transm 56: 265–278

Johnson GA, Baker CA, Smith RT (1980) Radioenzymatic assay of sulfate conjugates of catecholamines and dopa in plasma. Life Sci 26: 1591–1598

Jones CT, Robinson RO (1975) Plasma catecholamines in foetal and adult sheep. J Physiol 248: 15-33

Lawson EE, Brown ER, Torday JS, Mandansky DL, Taeusch HW (1978) The effect of epinephrine on tracheal fluid flow and surfactant efflux in fetal sheep. Am Rev Respir Dis 118: 1023-1026

Paulick R, Kastendieck E, Weth B, Wernze H (1987) Metabolische, kardiovaskuläre und sympathoadrenale Reaktionen des Feten auf eine progrediente Hypoxie - tierexperimentelle Untersuchungen. Z Geburtshilfe Perinatol 191: 130-139

Paulick R, Schwab O, Kastendieck E, Wernze H (1988) Plasma free and sulfoconjugated catecholamines during acute hypoxia in the sheep fetus - relation to cardiovascular parameters. J Perinat Med (in press)

Peuler J, Johnson GA (1977) Simultaneous single isotope radioenzymatic assay of plasma norepinephrine, epinephrine and dopamine. Life Sci 21: 625-636

Reilly FD, Russell PT (1977) Neurohistochemical evidence supporting an absence of adrenergic and cholinergic innervation in the human placenta and umbilical cord. Anat Rec 188: 277-285

Schwab O, Wernze H, Paulick R, Kastendieck E (1985) Circulating free and sulfate-conjugated catecholamines in the umbilical vessels as related to fetal stress. Acta Endocrinol 108 [Suppl 267]: 60-61

Sperling MA, Ganguli S, Leslie N, Landt K (1984) Fetal perinatal catecholamine secretion: role in perinatal glucose homeostasis. Am J Physiol 247: E69-E74

Tipton KF (1973) Biochemical aspects of monoamine oxidase. Br Med Bull 29: 116-119

Vandongen R (1984) The significance of sulfate-conjugated catecholamines in man. Neth J Med 27: 129-135

Walters DV, Olver RE (1978) The role of catecholamines in lung liquid absorption at birth. Pediatr Res 12: 239-242

Wang PC, Buu NT, Kuchel O, Genest J (1983) Conjugation pattern of endogenous plasma catecholamines in human and rat. J Lab Clin Med 101: 141-151

Placental Hormone Transfer

W. Moll[1]

Some maternal-fetal interrelationships are evident. The growing conceptus induces deep changes in maternal circulation; fetal growth is strongly correlated with maternal weight. Signals are exchanged between the fetoplacental and the maternal organisms. It is possible that special signals adjust the uteroplacental arteries, allowing for continuous adaptation of uteroplacental blood flow to fetal growth, even if we still have not found conclusive evidence that fetal hormones are involved in the regulation of the uterine arterial lumen.

Findings on the placental transfer of hormones appear confusing. There are reports that the placental membrane has different permeabilities for transfer from the mother to the fetus (maternal-fetal transfer) and transfer from the fetus to the mother (fetal-maternal transfer). It has been reported that the placental membrane secretes diffusible steroid exclusively to one side of the placental membrane, e.g. oestradiol to the maternal circulation (Walsh and McCarty 1981). There are reports that maternal-fetal transfer may be the same as maternal-fetal transfer even it quite different concentrations prevail (Kittinger 1974), or that even more hormone diffuse from a fetus with lower concentration than from the mother with a higher concentration (Mitchell et al. 1981). Thus, placental hormone transfer seems to be not yet understood. In view of these statements a theoretical treatment of placental transfer of hormones seems justified. I would like to present a theoretical account of the relationships between hormonal transfer across the placenta and its parameters.

Hypothesis 1: Fetal, Maternal and Placental Metabolic Clearance Rates and Placental Transfer Clearance Rates Define the Concentrations of Transferred Hormones

Most of the work done on placental transfer is theoretically based on the compartment analysis of Gurpide (1972). Gurpide's work has been stimulating, demonstrating that a number of fundamental key values of hormonal distribution, such as maternal and fetal hormone production and unidirectional flows, may be derived from measurements of specific activities during infusion of radioactive into the fetal and

---

[1] University of Regensburg, Universtätsstrasse, 8400 Regensburg, FRG

The Endocrine Control of the Fetus
Ed. by W. Künzel and A. Jensen

maternal organisms. I think that the consideration of a more substantial model which is more restricted in one sense but more explicative on the other hand may be useful, since it shows the importance of the metabolic clearance rate and the placental clearance rate for the hormone concentrations and may permit evaluation of these parameters.

The pregnant organism may be considered as a system of apposed circulatory units, the maternal unit and the fetal unit, brought in contact with each other by the placenta (see Fig. 1a). Each unit consists of parallel stream beds, one of them mediating maternal-fetal exchange, the placental vascular bed. The source of hormones is located on one of these parallel stream beds or, in the case of placental hormones, in the placental stream bed itself. Metabolic removal of hormones takes place predominantly in the liver. It is significant that the liver is situated at different strategic positions in the fetal and the maternal circulation: in the fetus, the liver is partially (in some species entirely) in series with the placental vascular bed, whilst in the maternal circulation it is parallel.

In the following section I shall derive the essential parameters determining the level of maternal hormones in the fetal circulation. Let us consider the fate of hormones produced in the maternal circulation. A non-placental hormone is produced in one of the vascular beds parallel to the placental vascular bed. It is admixed to the blood of other veins, cleared in other parallel vascular beds, predominantly in the hepatic circulation, and transferred by the placenta. The maternal metabolic removal rate R_m is proportional to the maternal concentration c_{ma}. The maternal metabolic removal rate divided by concentration gives the maternal metabolic clearance rate MCR_m:

$$R_m = c_{ma} \cdot MCR_m.$$

The rate of the diffusional transfer across the placenta (T_p) is proportional to the maternal arterial concentration minus the fetal arterial concentration c_{fa}. The rate of placental transfer divided by the fetal/maternal arterial concentration difference gives the placental transfer clearance rate PCR_t:

$$T_p = (c_{ma} - c_{fa}) \cdot PCR_t.$$

The hormone transferred is broken down by the metabolism on the fetal side at a rate which is proportional to the arterial concentration c_{fa}. The fetal metabolic removal rate (R_f) divided by the arterial concentration gives the fetal metabolic clearance rate MCR_f:

$$R_f = c_{fa} \cdot MCR_f.$$

The metabolic clearance rates and the placental transfer clearance rate represent a system of conductances which are partially in series and partially in parallel. For a given rate of hormone production the ratio and the arrangement of conductances define the concentrations of the hormone in the two compartments and the hormone flow rates. The hormone flow rates as given by concentrations and clearance rates are analogous to water as given by pressures and conductances. The flow of

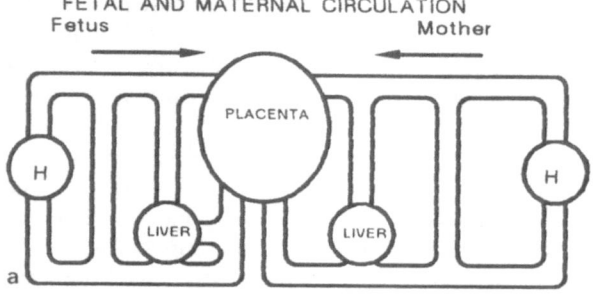

FETAL AND MATERNAL CIRCULATION

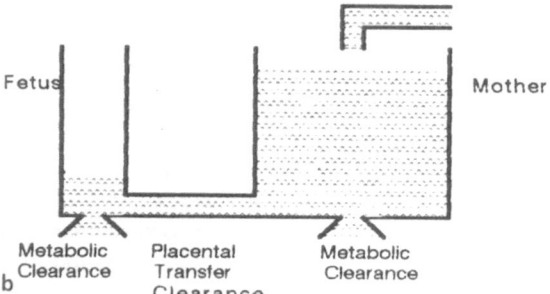

MODEL FOR MATERNAL – FETAL TRANSFER

Fig. 1. a Diagram of the maternal and fetal circulatory systems. **b** Model for the flow of maternal hormone to the fetus

maternal hormones can thus be compared schematically (see Fig. 1b) by a flow of water from a tap, representing the hormone source, into a big container (the maternal circulation) which it leaves through a sink, the luminal diameter of which represents the maternal metabolic clearance rate, and a tube, the luminal diameter of which represents the placental clearance rate. The flow reaches a second smaller container (the fetal circulation). This container has sink with a smaller lumen (the lower fetal metabolic clearance rate). The diameter of the lumen of the sink depends on the position of the liver and the distribution of umbilical blood flow (see below). The pressures are determined by the flow rate from the tap and the ratio and arrangement of the conductances just as, in a comparable way, the hormone concentrations are determined by the secretion rate and the respective conductances. A similar model can be constructed for the hormones produced in the fetus. The model demonstrates that the ratio of placental transfer clearance to metabolic clearance rates on the other side of the placenta determines the concentrations of the transferred hormones.

Using this model, the pertinent parameters of hormonal concentrations, metabolic clearance rates and placental clearance rate, can be calculated from the concentrations prevailing when labelled hormones are infused into the maternal and fetal organism at known rates. Maternal-fetal and fetal-maternal placental clearances can be separately determined from the concentrations measured when labelled hormones are infused into the maternal and fetal circulation. The maternal-fetal placental transfer clearance rate is given by

$$PCR_t = I_f \cdot MF/(FF \cdot MM - MF \cdot FM)$$

and the fetal-maternal placental transfer clearance rate by

$$PCR_t = I_m \cdot FM/(FF \cdot MM - MF \cdot FM),$$

where I_m and I_f are the rates of infusion of labelled hormone into the maternal and fetal circulation, FF and FM the fetal and maternal concentrations of the hormone infused into the fetal circulation and MM and MF the maternal and fetal concentrations of hormones infused into the maternal blood.

The strategic position of the fetal liver implies that the maternal hormone that is transferred is immediately presented to the hepatic metabolism. The metabolic clearance rate is speeded up in proportion to the ratio of umbilical venous and central arterial concentrations.

Hypothesis 2: Placental Transfer Clearance is Dependent upon Placental Permeability–area Product, Placental Blood Flow, and Binding by Plasma Proteins

Placental diffusional clearance is limited by placental blood flow, placental metabolism and placental permeability. Ignoring placental metabolism for the sake of simplicity, the limitation imposed by permeability is indicated by the ratio of the product of permeability times area to the product of blood flow times binding capacity. If this ratio is 1, transfer in a countercurrent system is equally limited by permeability and blood flow. Placental blood flow is around $1 \text{ ml} \times \text{min}^{-1} \text{ g}^{-1}$ placenta or higher (for references, see Moll 1981).

The *placental permeability-area product* for the various hormones is not known; however, on the basis of information on the permeability for other substances (see Moll 1978) we may reach some rough estimations. Steroid hormones are lipid-soluble molecules which are expected to permeate the cell membranes in the placental barrier rapidly. When red blood cells loaded with radioactive oestradiol are mixed with saline, equilibrium is reached with a half-life of less than 2 s at 273 K. At body temperature the half-life is presumably less than 1 s. That means that oestradiol rapidly crosses the red cell membrane (Fig. 2). This indicates that, indeed, leaving placental metabolism out of account, the transport of oestradiol, and possibly of other steroids too, is limited by the thickness and the area of the placental membrane in a similar way to the transport of oxygen which is considered to be predominantly flow-limited.

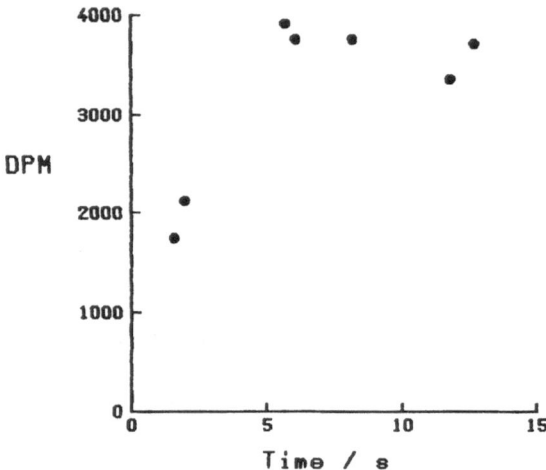

Fig. 2. Release of [³H]oestradiol by human red blood cells. The red cells were loaded with [³H]oestradiol and mixed with saline at 273 K. During constant rapid stirring cell-free saline was drawn from the stirred solution using the filtration technique of Dalmark and Wieth (1972). (Unpublished results obtained by W. Moll in collaboration with K. Schnell, Regensburg, 1987)

For water-soluble molecules like amino acids and hormones derived from amino acids that have a molecular weight of around 150 daltons, placental permeability-area product is around one-tenth of placental blood flow. For these hormones the concentrations transferred are determined by the ratio of permeability-area product to fetal metabolic clearance rate. The same holds true for peptide hormones.

Hormone binding in plasma affects placental clearance by reducing the concentration gradient of free (diffusible) hormone. The reduction is due to

(a) the reduction of free hormone in the chemical equilibrium
(b) finite velocity of dissociation during placental exchange.

Binding occurs for steroid hormones and peptides. The degree of binding of steroids is high. Even in the guinea pig, for which no specific binding globulins are known, only one out of 20 oestrogen molecules is diffusible. This is shown in Fig. 3, where on diffusible oestrogen molecule is shown among 20 oestrogen-albumin complexes. In other species, including humans, the proportion of diffusible steroid is even smaller (Dunn et al. 1981).

Whether the rate of dissociation is a limiting factor seems to depend upon the nature of the binding protein. In the guinea pig albumin seems to be the only binding protein. The fraction of free oestradiol is around 5% in maternal and fetal plasma (W. Moll, unpublished observations, 1986). The equilibrium constant for association is 2×10^4 M^{-1} in plasma containing 1 mM albumin. Rate constants of association of several tightly bound ligands were found to range from $1-40 \times 10^6$ M^{-1} s^{-1} (for references see Pardridge 1987), i.e. in the range of the rate of association of oxygen to haemoglobin. According to these values we obtain for the rate of dissociation a figure of 50 s^{-1} or higher. This is in the range of the dissociation con-

Fetus   Membrane   Mother

Fig. 3. Schematic diagram illustrating the diffusion of oestradiol in the guinea pig placenta. The filled circles indicate albumin-oestradiol complexes. Only one out of 20 oestradiol molecules is free and may cross the placental barrier as indicated by the diffusion pathway

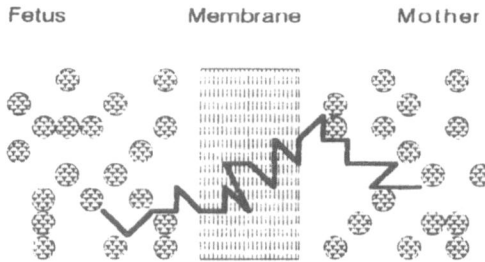

stant of oxygen from haemoglobin, which is not considered to limit oxygen release. Indeed, Pardridge and Mietus (1979) demonstrated that steroid hormones bound to albumin are transferred to tissues. So, apparently, the rate of dissociation of oestrogen and probably also that of other steroids from albumin is not a limiting factor of placental hormonal transfer. On the other hand, the maximum rate constant for dissociation of specifically bound hormone is approximated to be $0.1 \times s^{-1}$ or less; the rate of dissociation from the globulins may well limit placental hormone transfer.

Hypothesis 3: The Placenta is an Organ of Metabolic Breakdown As Well As of Diffusional Exchange

In the placenta, diffusion occurs simultaneously with metabolism. Placental metabolism deeply affects transfer. Fig. 4a shows schematically the placental membrane with cytoplasm between the cell membranes on the fetal and maternal side. We must distinguish two quantities of hormone. One crosses both membranes; this is the transfer determining the true placental diffusional clearance. Another quantity crosses only the first membrane and is metabolized in the cell. The cellular metabolism has two different effects: first, it offers additional conductance and increases the total flow across first membrane, i.e. the amount of hormone taken up by the placenta is larger than the transfer. Secondly, it lowers the intracellular concentration and the transfer across the second membrane. The situation is shown schematically in Fig. 4b: the placental membrane is shown as a container with a hole, the conductance of which represents the cell metabolic clearance rate, and two passages connected to greater container on either side, representing the permeability-area product of the two side membranes of the placental membrane. Fig. 4c indicates schematically uptake and transfer across this system when the permeability-area products of the two sides are equal. It can be seen that the uptake increases with the placental metabolic rate while the transfer falls with increasing metabolic clearance rate of the placental cells. Thus, placental transfer depends on the diffu-

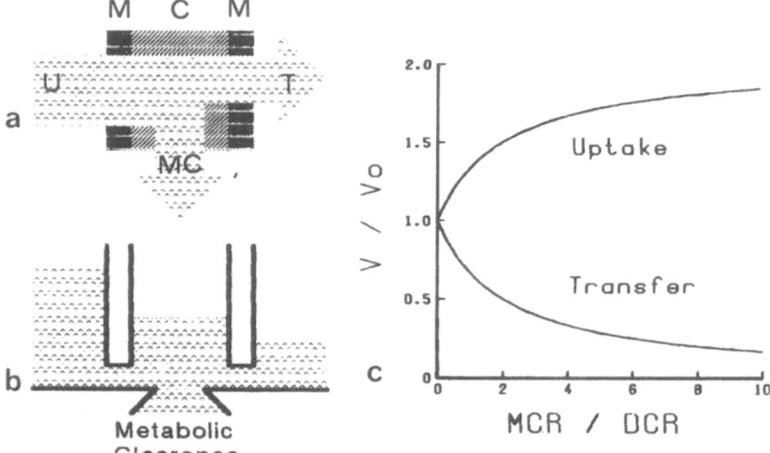

Fig. 4a–c. a Schematic representation of uptake *(U)*, transfer *(I)* and metabolic clearance *(MS)* in a placenta consisting of cell membranes *(M)* and cytoplasm *(C)*. **b** Model illustrating the intracellular concentration as a function of the permeability-area product of the two sides of the membrane and the metabolic clearance rate in the cells. **c** Uptake and transfer across the system related to the ratio of placental metabolic clearance rate and placental transfer clearance rate when the permeability-area products of the cell membranes are assumed to be equal. *M*, membrane; *C*, cytoplasm; *U*, uptake; *T*, transfer; *V/V_o*, relative transport; *MCR/DCR*, metabolic clearance rate/diffusional clearance rate

sional clearance, metabolic clearance and the ratio of diffusional resistances before and after the site of metabolism.

Placental metabolism seems to be especially pertinent for catecholamines. Saarikoski (1974) demonstrated radioactivity in fetal blood after injection of radioactive [³H]-noradrenaline into the mother. However, only 10% of the radioactivity in the umbilical vein was still in noradrenaline and no radioactive hormone could be demonstrated in the fetal arteries.

Hypothesis 4: The Strategic Position of the Fetal Liver Serves to Protect Fetal Autonomy

The fetal liver is positioned partly in series with the placenta (see Fig. 1a). It receives a high share of umbilical blood flow, i.e. a higher portion of cardiac output, than the postnatal liver. This results in high metabolic clearance rates and low hormone concentrations in the fetus before birth, after which metabolic clearance rate is decreased and plasma concentrations are increased (Mitchell et al. 1981). In fetal sheep, even at early gestation, metabolic clearance rate of aldosterone is around

25% of maternal metabolic clearance rate, while the ratio of body weights is lower than 1:20 (Wintour et al. 1980). The clearance rate is especially high in species where all of the umbilical blood crosses the liver, as in the guinea pigs (Girard et al. 1983). In guinea pigs the clearance rate of the cortisol is almost as high as in the mother when the tracer is infused into the jugular vein (Giry and Delost 1979).

The fetal liver, positioned in between the placenta and the fetus, forms a controlling barrier as it does between the gut and the body in the adult. Hepatic protection against maternal hormones is especially effective in species where the ductus venous is closed, as in the guinea pig. In the placental vessels the maternal hormones reach higher concentrations.

Hypothesis 5: Fetomaternal Placental Clearance Rate Appears to be Higher than Maternofetal Placental Clearance Rate

In Table 1 values for the placental diffusional clearance rate are presented, as determined from the published radioactive concentrations during infusion of radioactive hormones into the maternal and fetal circulation. The rate of placental clearance to the fetus equals the transfer to the fetus divided by the maternal/fetal arterial concentration difference. Because of the fetal liver function, maternal-fetal placental clearance is lower than fetal-maternal placental clearance. The liver removes part of the hormone transferred before it enters the fetal arteries.

In rhesus macaque monkeys, the placental clearance rate of progesterone is around 25% of the blood flow, according to data of Ducsay et al. (1985); the placenta is certainly not impermeable to progesterone. Placental clearance of aldosterone in sheep is around 80 ml in the maternal-fetal direction and about 110 ml $\times$ min^{-1} in the fetal-maternal direction. In the guinea pig, maternal-fetal placental clearance is around 0.5 ml $\times$ min^{-1}, while fetal-maternal placental clearance

Table 1. Placental clearance of steroids

Steroid	Species	Placental clearance (ml $\times$ min^{-1})		Reference
		Mat.-Fet.	Fet.-Mat.	
Aldosterone	Sheep	80	111	Wintour et al. 1980
Oestradiol	Rhesus monkey	0	7	Walsh and McCarty 1981
Cortisol	Rhesus monkey	5	17	Mitchell et al. 1981
Progesterone	Rhesus monkey	11	17	Ducsay et al. 1985
Cortisol	Guinea pig	0.2	2.5	Dalle and Delost 1979
Aldosterone	Guinea pig	2	6	Giry and Delost 1979

is 3-4 times higher. The table shows that the various placentae allow for steroid transfer. In part, the differences in maternal-fetal and fetal-maternal clearance rate are related to differing binding in maternal and fetal blood and by strategic position of the liver.

Hypothesis 6: Fetoplacental Hormones May Reach Effective Concentrations in the Uterine Vein

Maternal metabolic clearance is usually high when compared to the placental clearance. So maternal concentrations of fetal hormones are only a small percentage of fetal concentrations of fetal hormones. In the uterine vein, however, fetoplacental hormones may reach levels near to those in fetal blood, when placental clearance is high and placental metabolism is low. In the guinea pig, in whom placental oestradiol metabolism is minimal, the oestradiol concentration in the uterine vein rises steeply when oestradiol is injected into the fetal jugular vein. The concentration of oestradiol in the uterine vein was found to rise by 200 pg/ml when the fetal concentration was increased by 400 pg/ml (Scholl 1984). In the guinea pig, however, no net transfer occurs: the concentration of free oestrogen is the same in fetal and maternal plasma (Scholl 1984). The data show, however, that fetal steroid hormones may reach effective concentrations in placental vessels if produced at a high rate in the fetal organism and *may* play a role in the control of uteroplacental blood flow. The level of the hormone is a function of the permeability-area product and the placental metabolism.

References

Dalle M, Delost P (1979) Foetal-maternal production and transfer of cortisol during the last days of gestation in the guinea pig. J Endocrinol 82: 43-51

Dalmark M, Wieth JO (1972) Temperature dependence of chloride, bromide, iodide, thiocyanate and salicylate transport in human red cells. J Physiol 224: 583-610

Ducsay CA, Stanczyk FZ, Novy MJ (1985) Maternal and fetal production rates of progesterone in rhesus macaques: placental transfer and conversion to cortisol. Endocrinology 117: 1253-1258

Dunn JF, Nisula BC, Rodbard D (1981) Transport of steroid hormones: bindings of 21 endogenous steroids to both testosterone-binding globulin and corticoid-binding globulin in human plasma. J Clin Endocrinol Metab 53: 58-68

Girard H, Klappstein S, Bartag I, Moll W (1983) Blood circulation and oxygen transport in the fetal guinea pig. Dev Physiol 5: 181-193

Giry J, Delost P (1979) Placental transfer of aldosterone in the guinea-pig during late gestation. J Steroid Biochem 10: 541-547

Gurpide E (1972) Mathematical analysis for the interpretation of in vivo tracer infusion experiments. Acta Endocrinol 158: 26-43

Kittinger GW (1974) Feto-maternal production and transfer of cortisol in the rhesus *(Macaca mulatta)*. Steroids 23: 229-243

Mitchell BF, Seron-Ferre M, Hess DL, Jaffe RB (1981) Cortisol production and metabolism in the late gestation rhesus monkey fetus. Endocrinology 108: 916-924

Moll W (1978) Physiological and pathophysiological aspects of the placenta. In: Beller FK, MacGillivray I (eds) Hypertensive disorders in pregnancy. Springer, Berlin Heidelberg New York, pp 50-56

Moll W (1981) Physiologie der maternen placentaren Durchblutung. In: Becker V, Schiebler TH, Kubli F (eds) Die Placenta des Menschen. Thieme, Stuttgart, pp 172-194

Pardridge WM (1987) Plasma protein-mediated transport of steroid and thyroid hormones. Am J Physiol 252: E157-E164

Pardridge WM, Mietus LJ (1979) Transport of protein-bound steroid hormones into liver in vivo. Am J Physiol 237: E367-E372

Saarikoski S (1974) Fate of noradrenaline in the human foeto-placental unit. Acta Physiol Scand [Suppl] 421: 1-82

Scholl D (1984) Oestradiol-Konzentrationen und placentäre Oestradiol-Produktion beim Meerschweinchen. Thesis, University of Regensburg, Regensburg

Walsh SW, McCarty MS (1981) Selective placental secretion of estrogens into fetal and maternal circulations. Endocrinology 109: 2152-2159

Wintour EM, Coghlan JP, Hardy KJ, Lingwood BE, Rayner M, Scoggins BA (1980) Placental transfer of aldosterone in the sheep. J Endocrinol 86: 305-310

Endocrine Control of Circulation II – Neuronal Mechanisms

Hormonal Control of Fetal Breathing

G. S. DAWES[1]

After birth breathing movements are linked closely to metabolism, and there is small scope for hormonal control. Before birth, however, the situation is different. Fetal breathing movements are normally associated with episodes of low voltage electrocortical activity and rapid eye movements in sheep. In man they are also episodic in character, and are modulated by episodic sleep rhythms as identified by external ultrasound measurements. The natural history of fetal breathing suggests that there may be three circumstances in which hormones influence the control of breathing, directly or indirectly. These are during diurnal variation, shortly after food ingestion by the mother and immediately preceding parturition.

Diurnal Variation

Diurnal variation in fetal breathing movements was first observed in sheep (Boddy et al. 1973; Dawes 1973a; Dawes and Robinson 1976). In 1978 Patrick et al. (1978b) gave a thorough account of fetal breathing movements in man and drew attention to the diurnal and postprandial variations. We had speculated that the diurnal variation in sheep might be related to maternal changes (the most likely), or of exogenous origin (sound, light or temperature effects) or endogenous, due to inherent fetal rhythms. The last seemed unlikely since it was known that in some species diurnal variation was not established until weeks after birth. In 1979 Patrick et al. described a maternal endocrine diurnal rhythm, and we guessed that this might relate to the diurnal variation in human fetal heart rate (Visser et al. 1982) as well as breathing movements.

Recently Arduini, Rizzo and their colleagues (Arduini et al. 1986a, b, 1987) have made some interesting new observations on this phenomenon in man. In ten women near term, over a period of 24 h, they measured fetal activity, defined as fetal heart rate variation greater than 10 beats per minute associated with movements and cardiac accelerations. They found that it had a highly significant negative correlation ($r = -0.87$) with maternal plasma cortisol concentrations. There was a positive correlation with maternal plasma adrenocorticotrophic hormone (ACTH) ($r = 0.80$). The high values suggest a direct relationship between fetal activity and

---

[1] Charing Cross Sunley Research Centre, Lurgan Avenue, London W6 8LW, UK

The Endocrine Control of the Fetus
Ed. by W. Künzel and A. Jensen
© Springer-Verlag Berlin Heidelberg 1988

maternal hormonal values. These authors also demonstrated that the fetus of a woman who had been adrenalectomised and subjected to pituitary irradiation for Cushing's syndrome, and was receiving 50 mg cortisone twice daily, had no diurnal variation in activity. Finally, they showed that administration of 8 mg triamcinolone three times a day for 3 days at 35 weeks' gestation abolished the diurnal variation in fetal activity as compared with a control group. Diurnal variation returned by 38 weeks' gestation. Hence there is a prima facie case for believing that diurnal variation in fetal activity, a normal phenomenon in the evening towards the end of the pregnancy, is directly associated with changes in maternal hormonal concentrations of ACTH or cortisol. We are left with the problem of whether it is one or other of these two hormones, or another, which crosses the placenta and affects the fetal behavioural pattern. It is notable that in the early evening fetal activity is increased and may become continuous for periods up to some hours at a time. This could well be due to modulation of the hypothetical sleep cycle generator, located in sheep somewhere above the inferior colliculus. One of the candidate hormones is progesterone, and this possibility should be investigated, since Walsh et al. (1984) have reported diurnal variations in fetal plasma progesterone concentrations in rhesus monkeys near term. We should also note that Arduini and his colleagues' observations relate to fetal activity rather than to breathing movements. Patrick et al. (1981) found that fetal breathing was suppressed in six pregnant women receiving synthetic glucocorticoid treatment for various diseases. Both activity and breathing movements in human fetuses may therefore be modulated by changes in maternal hormones, possibly acting through the sleep cycle generator.

We should also consider whether the fetal hypothalamus and pituitary are involved, directly or indirectly, with the central control of fetal breathing movements. Bennett et al. (1986) have reported that thyrotropin-releasing hormone (TRH) injected into the lateral ventricles of fetal lambs causes an immediate switch to low voltage electrocortical activity and initiation of fetal breathing of increased rate and depth persisting through the next epoch of high voltage electrocortical activity. Bennett et al. (1987) have recently presented evidence that corticotrophin-releasing factor (CRF) also is a potent stimulus to fetal breathing. However, transection of the brain rostral to the superior colliculus did not alter the normal episodic character of fetal breathing movements, even when the hypothalamus was destroyed (Dawes et al. 1983). Mean tracheal pressure amplitude, inspiratory time and modal breath interval were unaltered. There was a small increase in sensitivity CO_2 and 'deep inspiratory efforts' were no longer detected. Subsequently Walker and Young (1985) reported that fetal hypophysectomy, by the trans-sphenoidal approach (to avoid damage elsewhere in the brain), also did not alter the normal episodic pattern of fetal breathing before term. However, it does not appear from the accounts so far available that a search was made to determine whether or not a diurnal rhythm was still present.

Postprandial Variation

There is good evidence in man that meals and oral or intravenous administration of glucose are all followed by an increase in fetal breathing movements (Boddy et al. 1975; Patrick et al. 1978 a; Natale et al. 1978). The maximal effect on breathing succeeds the peak maternal plasma glucose concentration by about an hour. The phenomenon has been described as early as 24 weeks' gestation (Nijhuis et al. 1985). We do not have a satisfactory animal analogue of this phenomenon. In sheep, hypoglycaemia (spontaneous or by maternal insulin administration) is associated with a decrease in the incidence of fetal breathing movements, but infusion of glucose to raise the fetal plasma glucose above normal limits is not associated with an increased prevalence of breathing movements. In man Patrick et al. (1981) observed in their six patients treated with synthetic glucocorticoid throughout pregnancy that the postprandial increase in fetal breathing movements was still present (the overnight increase being suppressed). This is good evidence that the postprandial variation is effected by a different mechanism. That it is already active by 24 weeks, before the earliest sign of fetal behavioural sleep states can be detected (which is at 28 weeks), is further evidence of another mechanism. It seems unlikely that hyperglycaemia could act directly, as this would imply that glucose supply is rate-limiting to the control of fetal breathing in the human brain.

Parturition

Fetal breathing movements were reported to cease before the onset of parturition in sheep (Dawes 1973 b). Similar observations were made in man within the next few years (Richardson et al. 1979; Carmichael et al. 1984). There was a decreased incidence of breathing in fetuses within 3 days of spontaneous parturition. After induction, no episodes of fetal breathing were observed once accelerated labour had begun. It has also been proposed that the continuation of normal episodic fetal breathing movements can be used to discriminate in suspected preterm labour between those women who will continue to term from those who will deliver within the next few days (Castle and Turnbull 1983). Prostaglandin synthetase inhibitors were found to induce continuous fetal breathing movements (Kitterman et al. 1979) by a central action (Koos 1985). It has been suggested that, since PGE_2 arrests fetal breathing movements (Kitterman et al. 1983), and since prostaglandin secretion in the fetal membranes, uterus and cervix is associated with the onset of normal labour, the effect might be due to prostaglandin release. It is a well-established fact that fetal behavioural patterns other than breathing continue episodically through labour, commonly until the second stage. It is therefore likely that the diminution or arrest of fetal breathing movements prior to the onset of labour is through a mechanism acting directly on the medulla rather than through the hypothetical sleep cycle generator. This would fit with the known actions of prostaglandins. However, though Wallen et al. (1985) found an inverse correlation between plasma PGE_2 concentrations and the incidence of breathing in fetal lambs, its low value ($r = -0.45$) suggests that other factors are involved.

Other Variations

There are some other natural phenomena for which we have as yet no adequate mechanistic explanation. For instance, abnormal prolonged episodes of regular fetal breathing, lasting many hours at a time, are usually seen before death in fetal lambs (Patrick et al. 1976). Similar prolonged episodes of vigorous fetal breathing have been observed by Walker and Davies (1986) during recovery from hyperthermia, when core temperature and arterial CO_2 tension had already returned to normal values; they extended through both low and high voltage electrocortical activity. In this respect they resemble the effects of prostaglandin synthetase inhibitors (described above) or of serotonin agonists such as 5-hydroxytryptophan (Quilligan et al. 1981) and 5-methoxy-N,N-dimethyltryptamine (5-MDMT; R. A. Ward, O. S. Bamford, and G. S. Dawes, unpublished observations. 5-MDMT is active in much smaller doses than 5-hydroxytryptophan: 0.1–1 mg i.a.). The effects of 5-MDMT persisted after transection of the brain stem or denervation of the systemic arterial chemoreceptors. Hence, the sites of action on fetal breathing of both the prostaglandin synthetase inhibitors and serotonin agonists are directly on the medulla. They are possible candidates for what we now recognise as likely to be complex phenomena, independent of and additional to the classic control of respiratory movements.

To complete the picture, there is evidence of dopaminergic, α_2-adrenergic, γ-aminobutyric acid and endorphin effects on fetal breathing movements (Table 1), which can also be affected by anaesthesia or ethyl alcohol acting either indirectly (e. g.

Table 1. Dopaminergic, α_2-adrenergic, γ-aminobutyric acid (GABA) and endorphin effects on fetal breathing movements

Class	Agonist	Antagonist	Site of action	Agonist action on breathing	Reference
Dopaminergic	Apomorphine	Haloperidol	Above inferior colliculus	Stimulation	Bamford et al. (1986a)
α_2-Adrenergic	Clonidine	Idazoxan	Medulla	Inhibition	Bamford et al. (1986b)
GABA	Muscimol	Picrotoxin	Not known	Inhibition	Johnston and Gluckman (1983)
Endorphin	Morphine	Naloxone	Not known	Inhibition	Olsen and Dawes (1985)

pentobarbitone, through the sleep cycle generator) or directly. While there are thus a number of potential mechanisms by which fetal breathing movements can be modulated by circulating hormones, the normal function of these putative neuro-transmitter systems is still obscure.

Infusions of catecholamines (e.g. adrenaline) can cause a small increase in the incidence of fetal breathing movements, but the effects on metabolism, the cardio-vascular and endocrine systems are relatively much greater (Jones and Ritchie 1978a, b), so that such effects as there are may be secondary.

In summary, in late fetal life breathing movements are episodic and not yet connected to metabolic requirements. There are several diurnal and other variations in fetal breathing, possibly hormonal in character, but whose mechanisms are still obscure. Since the lung is not yet an organ of gaseous exchange there can be no short-term value for survival, but in the long term there is good evidence that fetal breathing movements are necessary to pulmonary development. There also remains the possibility that prenatal practice is required to determine the development of the central control of respiration.

Acknowledgement. I am indebted to Prof. G.C. Liggins for drawing my attention to the paper by Walsh et al. (1984).

References

Arduini D, Rizzo G, Giorlandino G, Valensise H, Dell'Acqua S, Romanini C (1986a) Modifications of ultradian and circadian rhythms of fetal heart rate after fetal-maternal adrenal gland suppression: a double blind study. Prenat Diagn 6: 409–417

Arduini D, Rizzo G, Parlati E, Dell'Acqua S, Romanini R, Mancuso S (1986b) Loss of circadian rhythms of fetal behaviour in a totally adrenalectomized pregnant woman. Gynecol Obst Invest 23: 226–229

Arduini D, Rizzo G, Parlati E, Dell'Acqua S, Mancuso S, Romanini R (1987) Are the fetal heart rate patterns related to fetal maternal-endocrine rhythms at term of pregnancy? J Fetal Med 6: 53–57

Bamford OS, Dawes GS, Ward RA (1986a) Effects of apomorphine and haloperidol in fetal lambs. J Physiol 377: 37–47

Bamford OS, Dawes GS, Denny R, Ward RA (1986b) Effects of the α_2-adrenergic agonist clonidine and its antagonist idazoxan on the fetal lamb. J Physiol 381: 29–37

Bennett L, Gluckmann PD, Johnston BM (1986) The central effects of thyrotrophin-releasing hormone (TRH) on fetal breathing movements and electrocortical activity. International Union of Physiological Sciences Satellite Symposium: Fetal physiology – cellular and systems approaches, p 55

Bennett L, Johnston BM, Vale W, Gluckman PD (1987) Corticotrophin releasing factor (CRF) is a potent stimulus to fetal breathing. Society for the Study of Fetal Physiology, 14th annual meeting, poster 1

Boddy K, Dawes GS, Robinson J (1973) A 24-hour rhythm in the foetus. In: Cross KW, Dawes GS, Comline R, Nathanielsz P (eds) Foetal and neonatal physiology: Proceedings of the Sir Joseph Barcroft Centenary Symposium. Cambridge University Press, Cambridge, pp 63–66

Boddy K, Dawes GS, Robinson JS (1975) Intrauterine fetal breathing movements. In: Gluck L (ed) Modern perinatal medicine. Year Book Publishers, Chicago, p 381

Carmichael LK, Campbell K, Natale R, Patrick J, Richardson B (1984) Decrease in human fetal breathing movements prior to spontaneous labour at term. Am J Obstet Gynecol 148: 675–682

Castle BM, Turnbull AC (1983) The presence or absence of fetal breathing movements predicts the outcome of preterm labour. Lancet ii: 471–473

Dawes GS (1973a) Revolutions and cyclical rhythms in prenatal life: fetal respiratory movements rediscovered. Pediatrics 51: 965–971

Dawes GS (1973b) Foetal physiology and the onset of labour. Mem Soc Endocrinol 25–36

Dawes GS, Robinson JS (1976) Rhythmic phenomena in prenatal life. Prog Brain Res 45: 383–390

Dawes GS, Gardener WN, Johnston BM, Walker DW (1983) Breathing in fetal lambs: the effects of brain stem section. J Physiol 335: 535–553

Johnston BM, Gluckman PD (1983) GABA-mediated inhibition of breathing in the late gestation sheep fetus. J Dev Physiol 5: 353–360

Jones CT, Ritchie JWK (1978a) The cardiovascular effects of circulating catecholamines in fetal sheep. J Physiol 285: 381–393

Jones CT, Ritchie JWK (1978b) The metabolic and endocrine effects of circulating catecholamines in fetal sheep. J Physiol 285: 395–408

Kitterman JA, Liggins GC, Clements JA, Tooley WH (1979) Stimulation of breathing movements in fetal sheep by inhibitors of prostaglandin synthesis. J Dev Physiol 1: 453–466

Kitterman JA, Liggins GC, Fewell JE, Tooley WH (1983) Inhibition of breathing movements in fetal sheep by prostaglandins. J Appl Physiol 54: 687–692

Koos BJ (1985) Central stimulation of breathing movements by prostaglandin synthetase inhibitors. J Physiol 362: 455–466

Natale R, Patrick J, Richardson B (1978) Effects of human maternal venous plasma glucose concentrations on fetal breathing movements. Am J Obstet Gynecol 132: 36–41

Nijhuis JG, Jongsma HW, Crijns IJMJ, de Valk IMGM, van der Valden JWHJ (1985) Incidence of fetal breathing movements after glucose intake by the mother at 24 and 28 weeks gestation. Society for the Study of Fetal Physiology, 12th annual meeting, Haifa, July 1985, communication 34

Olsen GD, Dawes GS (1985) Morphine-induced depression and stimulation of breathing movements in the fetal lamb. Jones C, Nathanielsz P (ed) The physiological development of the fetus and newborn. Academic, London, pp 633–638

Patrick J, Dalton KJ, Dawes GS (1976) Breathing patterns before death in fetal lambs. Am J Obstet Gynecol 125: 73–78

Patrick J, Fetherston W, Vick H, Voegelin R (1978a) Human fetal breathing movements and gross fetal body movements at weeks 34–35 of gestation. Am J Obstet Gynecol 130: 693–699

Patrick J, Natale R, Richardson B (1978b) Patterns of human fetal breathing activity at 34–35 weeks gestational age. Am J Obstet Gynecol 132: 507–513

Patrick J, Challis J, Natale R, Richardson B (1979) Circadian rhythms in maternal plasma cortisol, estrone, estradiol and estriol at 34–35 weeks gestation. Am J Obstet Gynecol 135: 791–798

Patrick J, Challis J, Campbell K, Carmichael L, Richardson B, Tevaarwerk G (1981) Effects of synthetic glucocorticoid administration on human fetal breathing movements at 34–35 weeks gestational age. Am J Obstet Gynecol 139: 324–328

Quilligan EJ, Clewlow F, Johnston BM, Walker DW (1981) Effect of 5-hydroxytrytophane on electrocortical activity and breathing movements of fetal sheep. Am J Obstet Gynecol 141: 271–275

Richardson B, Natale R, Patrick J (1979) Human fetal breathing activity during induced labour at term. Am J Obstet Gynecol 133: 247–255

Visser GHA, Goodman JDS, Levine DH, Dawes GS (1982) Diurnal and other cyclic variations in human fetal heart rate near term. Am J Obstet Gynecol 142: 535–544

Walker DW, Davies AN (1986) Effects of hyperthermia on fetal breathing movements. J Dev Physiol 8: 485–497

Walker DW, Young IR (1985) Changes in breathing movements and electrocortical activity after trans-sphenoidal hypophysectomy of fetal lambs. Society for the Study of Fetal Physiology, 12th annual meeting, Haifa, July 1985, poster 6

Wallen LD, Clyman RI, Kitterman JA (1985) Breathing movements and prostaglandin E_2 concentrations in fetal sheep. Society for the Study of Fetal Physiology, 12th annual meeting, C7

Walsh SW, Stanczyk FZ, Novy MJ (1984) Daily hormonal changes in the maternal, fetal, and amniotic fluid compartments before parturition in a primate species. J Clin Endocrinol Metab 58: 629–639

Development of Reflex Control of the Fetal Circulation

A. M. WALKER[1]

Introduction

The cardiovascular system is the first organ system to reach a functional state in the embryo, and reflex control on the fetal circulation is evident from at least 0.6 of term. During fetal development, a remarkable functional feature of the cardiovascular system is the high level of pumping of the heart, despite its structural immaturity. Associated with high resting levels of cardiovascular performance, the fetus has no capacity to increase cardiac output in response to stress such as hypoxia. Consequently, autonomic nervous system control of the heart and circulation assumes an important role in the fetal defence against stress in utero, principally through maintenance and redistribution of the available cardiac output.

Receptor Mechanisms

Baroreceptors

During fetal life, functional baroreceptors have been demonstrated by measurements of phasic electrical activity in baroreceptor afferent nerves. This activity is synchronous with the arterial pulse (Ponte and Purves 1973). At 0.6 of term in the fetal lamb, elevations and small reductions of arterial pressure cause the heart rate to change in the opposite directions (Macdonald et al. 1980), signifying that a functional baroreflex control of the heart rate exists at this early state of development. However, baroreflex responses show substantial variability, which is not explained by changes in the sleep state of the fetus, the presence or absence of breathing movements or limb movements, or changes of blood gases (Dawes et al. 1980). With advancing gestation baroreflex control of the heart rate matures, as evidenced by an increasing proportion of positive responses to brief evaluations of arterial pressure, and by an increasing sensitivity of these responses quantified by relating the degree of heart rate slowing to the extent of blood pressure elevation (Shinebourne et al.

[1] Monash University Centre for Early Human Development, Monash Medical Centre, 246 Clayton Road, Clayton, Victoria 3168, Australia

The Endocrine Control of the Fetus
Ed. by W. Künzel and A. Jensen
© Springer-Verlag Berlin Heidelberg 1988

1972). Other animal studies have not been able to show gestational changes of sensitivity, and this question is unresolved (Dawes et al. 1980.

There is considerable variation among species in the age at which baroreceptor reflex responses are functional. At birth, species differences correspond roughly to the general maturity of the animal. For example, cardiac slowing in response to elevation of arterial pressure is less well developed in the maternally dependent newborn puppy than in the more active and independent newborn lamb (Vatner and Manders 1979). Nevertheless, the majority of species, including humans, exhibit low sensitivity of baroreceptor reflexes at birth and progressive postnatal maturation to adult levels (Gootman et al. 1979; Vatner and Manders 1979; Dawes et al. 1980). Differences in sensitivity of the baroreceptors per se do not appear to explain low responsiveness of the reflex arc at birth or the changes after birth. Thus, carotid baroreceptor responsiveness quantified by discharge in the afferent nerve (%/mm Hg) is greater in the fetal lamb than the 1-month-old lamb (Blanco et al. 1985) and equivalent in newborn and adult rabbits (Tomomatsu and Nishi 1982). Heart rate responses to postural tilting in term and preterm infants may signify well developed baroreflex control of the heart from 33 weeks' gestational age (Finley et al. 1984). Responses (tachycardia in response to head-up tilt and bradycardia with head-down tilt) were not different in term and preterm groups, nor in active and quiet sleep. Possibly, baroreflex control of heart rate is not well developed at earlier ages (Waldman et al. 1979), and may be transiently depressed in the first few days after birth (Young and Holland 1958). Clearly, the exact nature of baroreflex control in the newborn needs further study, taking account of central modification of afferent baroreceptor traffic, sleep state, and the maturity of efferent components of the reflex arc.

Mechanoreceptors in the atria, the ventricles, and the pulmonary artery are potential sites of cardiovascular reflexes which have not been systematically investigated in the fetus. Atrial or ventricular receptors causing reflex bradycardia may be implicated in the fetal response to lowered arterial pressure (Oberg 1976). When fetal blood pressure is reduced by hemorrhage or by venous occlusion, tachycardia only occurs when the pressure change is small (Macdonald et al. 1980; Walker et al. 1983). With larger pressure falls, the transient increase in heart rate is reversed and bradycardia occurs. Following administration of atropine the bradycardia disappears and the classical inverse relationship between heart rate and blood pressure of the adult circulation is unmasked. These studies show that in fetal hypotension, sympathetic acceleration of the heart is counteracted by increased vagal tone, and they emphasize that failure to demonstrate an adult form of response in the fetus cannot be interpreted to mean that receptor function is absent, nor that other components of the reflex arc are not effective.

Baroreceptor activation can be expressed as changes in venous and arterial tone and cardiac contractility, in addition to changes in heart rate. The sensitivity of the baroreceptor-heart rate reflex does not necessarily reflect the capacity of the arterial baroreceptors to control blood pressure. For example, in adult life, dissociation between the control of heart rate and the control of blood pressure occurs in exercise and hypertensive states (Ludbrook et al. 1980). Whether fetal baroreflexes play an important role in regulating fetal blood pressure can be assessed by examining blood pressure variability after arterial baroreceptor denervation (Yardley et al. 1983; Itskovitch et al. 1983). Surgical denervation of the carotid sinus, carotid artery, and aortic arch baroreceptor regions in fetal lambs results in effective denervation of the arterial baroreceptors and strikingly increases the natural variability of arterial pressure and heart rate. Coefficients of variation of mean arterial pressure measured over 24-h periods (standard deviation of the mean arterial pressure over the mean value of the mean arterial pressure) are twice as high in barodenervated fetuses as in intact fetuses. This response to denervation in the lamb fetus is surprisingly similar to the effect of barodenervation in active unanesthetized adult animals and supports the suggestion that arterial baroreceptors have a natural role in regulating fetal arterial pressure.

Chemoreceptors

Aortic chemoreceptors are active in the mature lamb fetus (A. M. Walker 1984). These receptors respond to asphyxia and to chemical stimulation in a way similar to adult receptors, producing bradycardia, increased arterial pressure, and sympathetic vasoconstriction. Because the aortic chemoreflex is activated by small falls of arterial oxygen tension, it has been proposed as the first line to defense against arterial hypoxemia in fetal lambs (Dawes et al. 1969). Fetal carotid body chemoreceptors by contrast, were only marginally responsive.

These studies in anesthetized lambs support the long-held view that aortic chemoreceptors are active and play an important role in fetal cardiovascular regulation, wheras carotid chemoreceptors are inactive. Historically, the absence of fetal breathing movements in the anesthetized fetus and the failure to stimulate fetal respiratory activity by hypoxemia and carotid body stimulation supported this view (Purves 1981). However, recent animal studies suggest that the carotid bodies are more active prenatally than previously believed; moreover, these become transiently inactive at birth when arterial oxygen tension increases, and only slowly reset into the adult range (Blanco et al. 1984). The unanesthetized fetus responds to moderate hypoxemia with bradycardia, not with tachycardia, as seen in studies of anesthetized animals (Walker et al. 1979). This response is due to vagal activation and should be distinguished from the bradycardia which results from profound hypoxemia (PO_2 less than 10 Torr), which is not prevented by parasympathetic blockade (Lewis et al. 1980). Because fetal arterial pressure usually increases, the bradycardia has regularly been explained as a baroreflex response. However, hypoxemic bradycardia is probably not due to baroreceptor stimulation, as arterial pressure does not increase in all hypoxemic fetuses, and the extent by which heart rate slows in different animals is the same whether or not arterial pressure rises (Walker et al. 1979). Secondly, brief episodes of fetal hypoxemia produced by occluding the uterine artery in sheep do not cause arterial pressure to increase, but the heart rate slows (Parer et al. 1980). Finally, pretreatment with α-adrenergic blocking agents prevents the hypoxemia-induced hypertension without affecting the fall in heart rate (Lewis et al. 1980). Carotid chemoreceptor stimulation is probably the primary cause of the increased vagal activity and bradycardia which accompany fetal hypoxemia, as the primary effect of carotid body stimulation is to slow the heart via activation of the medullary cardioinhibitory center and vagal outflow (de Burgh Daly 1972). Aortic chemoreceptor stimulation, by contrast, accelerates the heart (Sleight 1974).

Autonomic Nervous System

Natural development of autonomic control of the heart begins during fetal life in many animal species and in humans (Pappano 1977). During ontogenesis the postsynaptic components of the neuroeffector process (the transmitter receptor and the target organ effector mechanism) appear before the presynaptic component (the efferent autonomic nerve). Thus, responsiveness of the heart to neurotransmitters develops long before effective innervation is found. The sympathetic and parasympathetic cardiac innervation follow the same developmental sequence. Receptor responsivity is established simultaneously, but sympathetic nerves appear and effective neurotransmission starts later in the sympathetic system than in the parasympathetic system. Because cholinergic neurotransmission develops before adrenergic innervation, a potential for autonomic nervous imbalance exists during development (Pappano 1977; Vlk and Vincenzi 1977). However, as cholinergic and β-adrenergic receptors appear at about the same time in ontogenesis, other sources of catecholamines could protect the developing fetus from unopposed vagal activity.

The gestational development of basal autonomic tone is readily understood using the concept of the *intrinsic heart rate,* the natural frequency of discharge of the pacemaker cells in the sinoatrial node. The progressive natural reduction of fetal heart rate throughout gestation can be ascribed to two significantly different processes (Fig. 1). In early gestation the natural decline in heart rate reflects a reduction in the intrinsic rate (phase 1). Isolated fetal atria show an age-dependent decrease in their rate of beating prior to 15 weeks' gestation (D. Walker 1975), and as the isolated fetal heart does not respond to ganglion-stimulating drugs at this time, it is unlikely that the fetal heart is subject to tonic vagal inhibition (D. Walker 1975). A second developmental period is characterized by growth of nervous influence on the heart. In this period (Fig. 1, phase 2) the reduction of the normal heart rate can be ascribed to growth in vagal inhibition which predominates over a smaller tonic sympathetic stimulation (Walker et al. 1978). Intrinsic heart rate measured after administration of both atropine and propranolol is constant during developmental phase 2. In the human fetus this phase is from 15 weeks to term (Schifferli and Caldeyro-Barcia 1973); in the fetal lamb, it is from 0.7 of term to term (Walker et al. 1978; Bell et al. 1986). Nonneurogenic changes become important once more in determining the general pattern of heart rate development after birth. The intrinsic

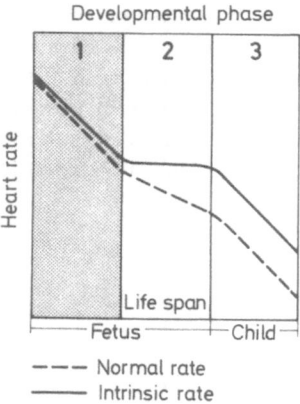

Fig. 1. Schematic representation of phases in the development of heart rate during fetal and postnatal life

heart rate falls from birth throughout early childhood in humans and in neonatal life in lambs (Walker 1984), with predominant vagal inhibition lowering the heart rate below the intrinsic rate. Detailed longitudinal studies have identified progressive slowing after a peak rate at approximately 1 month of age (Harper et al. 1976). Transient increases of newborn heart rate above fetal values which occur in the 1st h after birth in humans (Bustos et al. 1975) and during the first few days of life in lambs (Walker et al. 1978) are not shown in Fig. 1.

Concentrations of norepinephrine can be used as an index of the magnitude of sympathetic innervation in developing myocardium because the heart stores of norepinephrine are located predominantly within sympathetic nerves (Friedman 1972). Assessed in this way, myocardial innervation is incomplete at term and preterm birth because concentrations in the fetus are much lower than in the adult. Significant growth of cardiac sympathetic innervation continues after birth in many animal species and it would be unwise to assume that the human fetal heart is fully innervated at birth. Comparative studies show that adrenergic innervation is better developed in newborn animals that are relatively more independent at birth (Pappano 1977). The effectiveness of the immature sympathetic nervous system in the fetus at birth may be enhanced by supersensitivity of the target organ, due to suppressed uptake of the neurotransmitter and by high circulating catecholamine levels (Friedman 1972). Plasma catecholamine concentrations in various animal fetuses increase over the last quarter of gestation, with adrenaline concentration showing the sharpest rise, reflecting the increasing epinephrine : norepinephrine concentration ratio in the fetal adrenal gland (Comline and Silver 1966; Jones 1980). Anesthesia, gentle handling and restraint stress cause large increases of catecholamines, especially of epinephrine (Buhler et al. 1978). In the unanesthetized lamb fetus basal values are similar to those found in adult animals. From 0.7 of term there is a gradual rise which becomes rapid 2–3 days prior to delivery. Umbilical arterial catecholamine levels exceed umbilical venous levels in the basal state and during hypoxemia, reflecting the major role of the placenta in the clearance of the fetal plasma catecholamines (Jones 1980). At parturition, beginning with the onset of labor (Eliot et al. 1981), there is a marked increase of sympathetico-adrenal activity. In the human newborn, catecholamines are further increased when there is evidence of fetal distress (Bistolotti et al. 1983). The preterm sheep fetus shows a more substantial increase of plasma catecholamines at birth than the full-term fetus (Padbury et al. 1985). In contrast, the associated hemodynamic changes (increased heart rate and blood pressure) are less pronounced in the preterm fetus, and the metabolic responses are blunted, with lower blood glucose and free fatty acid concentration. These blunted physiological responses in the face of a greater catecholamine surge probably represent both reduced sympathetic innervation and immature target tissue receptor-effector mechanisms in the preterm lamb.

Responses to Stress

Cardiovascular responses of the fetal lamb to hypoxemia vary with the age of the lamb (Walker et al. 1979). Between 0.6 and 0.8 of term heart rate and blood pressure are not significantly changed during hypoxemia, but in older fetuses bradycardia and hypertension occur. The increasing pressor response in older fetuses parallels the increasing reactivity of the vascular neuroeffector mechanism (Assali et al. 1978). Plasma catecholamines are elevated to similar levels in younger and older fetuses (Jones and Robinson 1975) and these are sufficient to sustain cardiac output, myocardial flow and cerebral flow during hypoxia. However, efferent sympathetic activity is required to increase arterial pressure and to effect the normal redistribution of cardiac output (Iwamoto et al. 1983). Moreover, combined α- and β-adrenergic blockade is lethal for the fetus if given during hypoxia (Parer et al. 1978). Absence of fetal heart rate changes during hypoxemia does not indicate failure of autonomic nervous system activation. Selective pharmacological blockade of the opposing vagal and sympathetic divisions of the autonomic nervous system unmasks considerable activation during hypoxemia (Walker et al. 1979). Prior to 0.8 of term increased sympathetic stimulation during hypoxemia is matched by an increased vagal inhibition and heart rate is not changed, but later in gestation a larger increment in vagal inhibition causes bradycardia (Fig. 2).

Opposing, augumented activity of the sympathetic and parasympathetic influences on the heart is also found in acute fetal hypotension (Walker et al. 1983), fetal hemorrhage (Macdonald et al. 1980) and in the chronically growth-retarded and hypoxemic fetal lamb (Llanos et al. 1980), in which plasma catecholamines are elevated (Jones 1980). This pattern is not seen after birth, and the possible benefits for the fetus have not been fully elucidated. Clearly there is an important role for sympathetic activation during hypoxemia in supporting ventricular output in the fetus (Walker et al. 1985). After the application of β-adrenergic blockade in the fetus, unopposed parasympathetic activation during hypoxemia results in falls of heart rate and left and right ventricular output. In the newborn, significant increases of heart

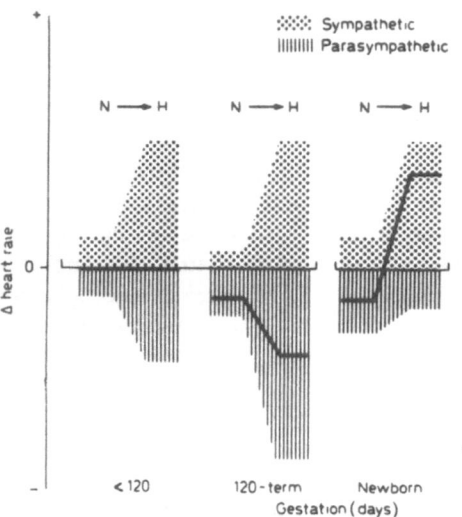

Fig. 2. Autonomic nervous influences on the heart rate of normoxemic *(N)* and hypoxemic *(H)* fetal and newborn lambs. The height and depth of shaded areas represent the opposing sympathetic and parasympathetic chronotropic influences. The zero line represents the intrinsic rate of the heart when isolated from autonomic nervous influence. The heavy line represents the actual heart rate in relation to the intrinsic rate. Up until 120 days (0.8) of gestation, increased sympathetic stimulation during hypoxemia is matched by an increasing parasympathetic inhibition – the heart rate is not changed. Later in gestation, sympathetic influence increases as before, but a larger increment in parasympathetic inhibition is reflected in a net bradycardia. In the newborn, augmentation of sympathetic influence plus a small contribution from parasympathetic withdrawal results in pronounced tachycardia. (From Walker et al. 1979)

rate and cardiac output occurring in hypoxemia are reduced by propranolol injection. Shortening of the pre-ejection period (PEP) consistent with inotropic stimulation during hypoxemia is abolished by propranolol injection in the fetus, signifying an important role of the β-adrenergic system in augmenting fetal inotropy. Parasympathetic activation opposing sympathetic activation appears to be important for the efficiency of the fetal cardiac response to hypoxemia (Fig. 3). Left ventricular output in fetal lambs does not increase in control experiments, when heart rate was unchanged, nor in the presence of atropine, when large increases of heart rate occurred under unopposed sympathetic drive (Fig. 3 a). This was in contrast to neonatal lambs, in which tachycardia was effective in augmenting left ventricular output in hypoxemia both in the unblocked state and in the presence of vagal blockade (Fig. 3 b). As spontaneous heart rate changes are positively correlated with ventricular output changes in the normoxemic fetus (Rudolph and Heymann 1974), the ineffectiveness of tachycardia in augmenting fetal cardiac output in hypoxemia conditions may reflect an imbalance between myocardial oxygen supply and de-

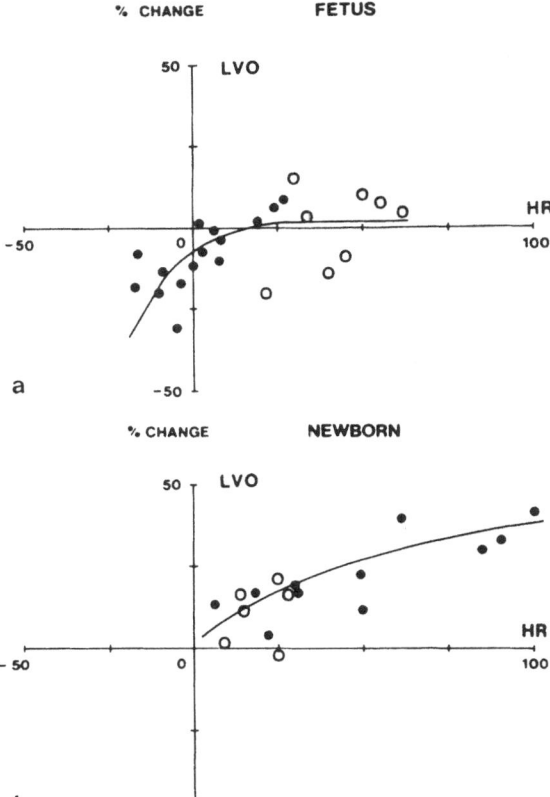

Fig. 3. Relationship between changes in aortic flow (left ventricular outputs *LVO*) and heart rate *(HR)* during hypoxemia in control experiments (●) and in the presence of parasympathetic blockage with atropine (○). **a** In fetal lambs, left ventricular output is sensitive to falls in heart rate in the control experiments (●) but is not increased during unopposed tachycardia in the presence of atropine (○). **b** In neonatal lambs left ventricular output is augmented by tachycardia in both conditions. (From Walker et al. 1985)

mand. Fetal myocardial metabolism remains aerobic during moderate hypoxemia, and myocardial oxygen consumption is maintained by increased myocardial blood flow (Fisher et al. 1982). If the fetus is unable to increase myocardial blood sufficiently to meet extra oxygen requirements associated with tachycardia (Cohn et al. 1980), oxygen availability may impose a limit on ventricular performance, even in mild hypoxemic or hypotensive stress. If so, parasympathetic activation would provide an important benefit to the fetus by minimizing myocardial oxygen consumption for a given level of cardiac output.

Sleep and Cardiovascular Control

Development of Sleep Activity Cycles

In humans and other animal species behavioral states develop during gestation. Distinct behavioral states are easily recognizable in neonates with a gestational age of 35 weeks or more. Several states may be recognized based upon observation of behavioral features, supplemented by physiological recording:

1. Active sleep (called rapid eye movement sleep, REM)
2. Quiet sleep (called non-REM, NREM, or slow-wave sleep)
3. Indeterminate sleep
4. Wakefulness.

In preterm infants of less than 35 weeks' gestation, cycles of rest and activity, regular and irregular breathing, and other behavioral features, as seen in older infants, are present but alternate independently. When these become more coincident after 35 weeks and alternate in synchrony, they are recognized as true behavioral states (Casaer and Devlieger 1984). In utero, stable behavioral states can also be recognized using real time ultrasonic imaging of the fetus from 36 weeks' gestational age onwards. At birth, total sleep time is at a lifetime maximum in most species and humans.

The brain of the fetal lamb matures early in utero and the cortex might influence the cardiovascular system after approximately 0.5 of term. After 120 days (0.8) of gestation the electrocorticogram (ECoG) becomes patterned into high-voltage slow activity (HVSA) and low-voltage fast activity (LVFA). Using measurements of fetal ECoG together with other measures of fetal activity, criteria have been established for recognizing fetal behavioral states corresponding to active wakefulness (AW), REM sleep, and quiet or NREM sleep. The ECoG pattern of HVSA is recognized as NREM sleep; the pattern of LVFA comprises REM sleep and AW, but conclusive evidence for the identity of these fetal ECoG states with sleep states after birth is lacking.

Cardiovascular Control in Sleep

Baroreceptor-heart rate reflex responses have been examined in relation to ECoG and fetal breathing and limb movements in mature fetal lambs (Dawes et al. 1980). It was anticipated that the previously observed variability of baroreflex responses in fetal lambs (Shinebourne et al. 1972; Maloney et al. 1977) might be explained by changes in fetal activity state. However, only small differences were found in heart rate slowing in response to arterial pressure elevations during LVFA and HVSA, and no association with breathing or limb movements was appearant. In agreement with Dawes et al. (1980), our studies in fetal and newborn lambs show no baroreflex sensitivity differences between fetal HVSA and LVFA, or between newborn REM and NREM sleep (Walker et al. 1986). On the other hand, increased gain (greater change in heart rate per unit change in arterial pressure) has been reported in response to hypertension and hemorrhagic hypotension during REM sleep in lambs (Fewell and Johnson 1984; Fewell et al. 1984). Conversely, smaller increases of systemic vascular resistance during hemorrhagic hypotension during REM suggest diminished responsiveness in the baroreflex pathway (Fewell et al. 1984). Further study is needed to resolve these questions relating to cardiovascular control in sleep before and after birth, as studies in the adult show a significant effect of sleep state cycling on the cardiovascular system.

Alternating patterns of basal heart rate and heart rate variations are closely related to cyclic changes in fetal behavior (Visser 1984). In the fetal sheep, basal heart rate is higher during HVSA and lower during LVFA, in a similar pattern to NREM and REM sleep states of the newborn (Ruckebusch et al. 1977). Variations in fetal heart rate and blood pressure are greater during fetal breathing movements occurring in LVFA, and these increase further following baroreceptor denervation (Itskovitz et al. 1983), supporting the concept that nervous control differs in low voltage and high voltage ECoG states in fetal life. Examination of autonomic tone using selective pharmacological blockade reveals that heart rate is slower during LVFA in the fetal lamb because sympathetic tone is less, and because parasympathetic tone is greater (Walker et al. 1985). Study of the cyclic variations in autonomic activity during hypoxemic stress presents difficulty, because acute hypoxemia modifies the ECoG cycle itself, increasing HVSA at the expense of LVFA (Harding et al. 1981). However, study of chronic hypoxemic lambs reveals accentuated depression of heart rate during LVFA, due to exaggerated parasympathetic tone (Walker et al. 1985).

Conclusion

Reflex control of the fetal circulation is evident from very early on in gestation and plays an important role in the fetal defense against hypoxia and other stresses (e.g., hypotension) which threaten oxygen delivery to developing fetal tissues. Low sensitivity of fetal baroreflexes and chemoreflexes is apparently not explained by diminished responsivity of receptors, which are active in fetal life. Rather, there may be modification of afferent activity by the central nervous system to effect a pattern of cardiovascular control which is uniquely appropriate for the fetus, combined with immaturity of the afferent pathways and/or target tissue receptor-effector mechanisms.

References

Assali NS, Brinkman CR, Wood R, Dandavino A, Nuwayid B (1978) Ontogenesis of the autonomic control of cardiovascular functions in the sheep. In: Longo LD, Reneau DD (eds) Fetal and newborn cardiovascular physiology. Garland, New York, pp 47-91

Bell AW, Kennaugh JM, Battaglia FC, Makowski EL, Meschia G (1986) Metabolic and circulatory studies of the fetal lamb at mid-gestation. Am J Physiol 250: E538-E544

Bistoletti P, Nylund L, Lagercrantz H, Hjendahl P, Strom H (1983) Fetal scalp catecholamines during labour. Am J Obstet Gynecol 147 (7): 785-788

Blanco CE, Dawes GS, Hanson MA, McCooke HB (1984) The response to hypoxia of arterial chemoreceptors in fetal sheep and new-born lambs. J Physiol (Lond) 351: 25-37

Blanco CE, Dawes GS, Hanson MA, McCooke HB (1985) Studies of carotid baroreceptor afferents in fetal and newborn lambs. In: Jones CT, Mott JC, Nathanielsz PW (eds) Physiological development of the fetus and newborn. Academic, London, pp 595-598

Buhler HU, Da Prada M, Haefely W, Picotti GB (1978) Plasma adrenaline, noradrenaline and dopamine in man and different animal species. J Physiol (Lond) 276: 311-320

Bustos R, Bejar R, Arrojave H, Jacomo AJD, Burghi M, Ramirez F, Cordano MC, Burbelo V, Guayasamin O, Minetti MA, Guemberena L, Caldeyro-Barcia R (1975) Heart rate in fetuses and neonates in normal conditions and with mild depression. J Perinat Med 3: 172-179

Casaer P, Devlieger H (1984) The behavioural state in human perinatal life. J Dev Physiol 6 (3): 187-194

Cohn HE, Piasecki GJ, Jackson BT (1980) The effect of fetal heart rate on cardiovascular function during hypoxemia. Am J Obstet Gynecol 138: 1190-1199

Comline RS, Silver M (1966) Development of activity in the adrenal medulla of the foetus and newborn animals. Br Med Bull 22: 16-20

Dawes GS, Duncan SL, Lewis BV, Merlet CL, Owen-Thomas JB, Reeves JT (1969) Cyanide stimulation of the systemic arterial chemoreceptors in foetal lambs. J Physiol (Lond) 201: 117-128

Dawes GS, Johnston BM, Walker DW (1980) Relationship of arterial pressure and heart rate in fetal, newborn and adult sheep. J Physiol (Lond) 309: 405-417

de Burgh Daly M (1972) Interaction of cardiovascular reflexes. In: Gilliland I, Francis J (eds) Scientific basis medicine annual review. Athlone, London, pp 307-332

Eliot RJ, Klein AH, Glatz TH, Nathanielsz PW, Fisher DA (1981) Plasma norepinephrine, epinephrine and dopamine concentrations in maternal and fetal sheep during spontaneous parturition and in premature sheep during cortisol-induced parturition. Endocrinology 108: 1678-1682

Fewell JE, Johnson P (1984) Acute increases in blood pressure cause arousal from sleep in lambs. Brain Res 311: 259-265

Fewell JE, Williams BJ, Hill DE (1984) Behavioural state influences the cardiovascular response to hemorrhage in lambs. J Dev Physiol 6: 339-48

Finley JP, Hamilton R, Mackenzie MG (1984) Heart rate response to tilting in newborns in quiet and active sleep. Biol Neonate 45: 1-10

Fisher DJ, Heymann MA, Rudolph AM (1982) Fetal myocardial oxygen and carbohydrate consumption during acutely induced hypoxemia. Am J Physiol 242: H657-H661

Friedman WF (1972) The intrinsic physiologic properties of the developing heart. In: Friedman WF, Lesch M, Sonnenblick EH (eds) Neonatal heart disease. Grune and Stratton, New York, pp 21-49

Gootman PM, Buckley NM, Gootman N (1979) Postnatal maturation of neural control of the circulation. Rev Perinat Med 3: 1-72

Harding R, Poore ER, Cohen GL (1981) The effect of brief episodes of diminished uterine blood flow on breathing movements, sleep states and heart rate in fetal sheep. J Dev Physiol 3: 231-243

Harper RM, Hoppenbrouwers T, Sterman MB, McGinty DJ, Hodgman J (1976) Polygraphic studies of normal infants during the first six months of life. I. Heart rate and variability as a function of state. Pediatr Res 10: 945-951

Itskovitz J, La Gamma EF, Rudolph AM (1983) Baroreflex control of the circulation in chronically instrumented fetal sheep. Circ Res 52: 589-596

Iwamoto HS, Rudolph AM, Mirkin BL, Keil LC (1983) Circulatory and humoral responses of sympathectomized fetal sheep in hypoxemia. Am J Physiol 245: H767-H772

Jones CT (1980) Circulating catecholamines in the fetus: their origin, actions and significance. In: Parvez H, Parvez S (eds) Biogenic amines in development. Elsevier/North Holland, Amsterdam, pp 63-86

Jones CT, Robinson RO (1975) Plasma catecholamines in foetal and adult sheep. J Physiol (Lond) 248: 15-33

Lewis AB, Donovan M, Platzker ACG (1980) Cardiovascular responses to autonomic blockade in hypoxemic fetal lambs. Biol Neonate 37: 233-242

Llanos AJ, Green JR, Creasy RK, Rudolph AM (1980) Increased heart rate response to parasympathetic and β-adrenergic blockade in growth retarded fetal lambs. Am J Obstet Gynecol 136: 808-813

Ludbrook J, Mancia G, Zanchetti A (1980) Does the baroreceptor-heart rate reflex indicate the capacity of the arterial baroreceptors to control blood pressure? Clin Exp Pharmacol Physiol 7: 499-503

Macdonald AA, Rose J, Heymann MA, Rudolph AM (1980) Heart rate response of fetal and adult sheep to hemorrhage stress. Am J Physiol 239: H789-H793

Maloney JE, Cannata J, Dowling MH, Else W, Ritchie BC (1977) Baroreflex activity in conscious fetal and newborn lambs. Biol Neonate 31: 340-350

Oberg B (1976) Overall cardiovascular regulation. Ann Rev Physiol 38: 537-570

Padbury JF, Diakomanolis ES, Hobel CJ, Perlman A, Fisher DA (1981) Neonatal adaptation: sympathoadrenal response to umbilical cord cutting. Pediatr Res 15: 1483-1487

Padbury JF, Polk DH, Newnham JP, Lam RW (1985) Neonatal adaptation: greater sympathoadrenal response in preterm than full-term fetal sheep at birth. Am J Physiol 248: E443-E449

Pappano AJ (1977) Ontogenic development of autonomic neuroeffector transmission and transmitter reactivity in embryonic and fetal hearts. Pharmacol Rev 29: 3-33

Parer JT, Krueger TR, Harris JL, Reuss ML (1978) Autonomic influences in umbilical circulation during hypoxia in fetal sheep. Abstracts of Scientific Papers, EJ Quilligan Symposium, San Diego

Parer JT, Krueger TR, Harris JL (1980) Fetal oxygen consumption and mechanisms of heart rate response during artificially produced late decelerations of fetal heart rate in sheep. Am J Obstet Gynecol 136: 478-482

Ponte J, Purves MJ (1973) Types of different nerve activity which may be measured in the vagus nerve of the sheep fetus. J Physiol (Lond) 229: 1490-1496

Purves MJ (1981) The neural control of respiration before and after birth. Rev Perinat Med 4: 299-336

Ruckebusch Y, Gaujoux M, Eghbali B (1977) Sleep cycles and kinesis in the foetal lamb. Electroencephalogr Clin Neurophysiol 42: 226-237

Rudolph AM, Heymann MA (1974) Cardiac output in the fetal lamb: the effects of spontaneous and induced changes of heart rate on right and left ventricular output. Am J Obstet Gynecol 124: 183-192

Schifferli PY, Caldeyro-Barcia R (1973) Effects of atropine and β-adrenergic drugs on the heart rate of the human fetus. In: Boreus LO (ed) Fetal pharmacology. Raven, New York, pp 259-279

Shinebourne EA, Vapaavouri EK, Williams RL, Heymann MA, Rudolph AM (1972) Development of baroreflex activity in unanaesthetised foetal and newborn lambs. Circ Res 31: 710-718

Sleight P (1974) Neural control of the cardiovascular system. In: Oliver MF (ed) Modern trends in cardiology, vol 3. Butterworth, London, pp 1-43

Tomomatsu E, Nishi K (1982) Comparison of carotid sinus baroreceptor sensitivity in newborn and adult rabbits. Am J Physiol 243: H546-H550

Vatner SF, Manders WT (1979) Depressed responsiveness of the carotid sinus reflex in conscious newborn animals. Am J Physiol 237: H40-H43

Visser GH (1984) Fetal behaviour and the cardiovascular system. J Dev Physiol 6: 215-224

Vlk J, Vincenzi FF (1977) Functional autonomic innervation of mammalian cardiac pacemaker during the perinatal period. Biol Neonate 31: 19-26

Waldman S, Krauss AN, Auld PAM (1979) Baroreceptors in preterm infants. Their relationship to maturity and disease. Dev Med Child Neurol 21: 714-722

Walker AM (1984) Physiological control of the fetal cardiovascular system. In: Beard RW, Nathanielsz PW (eds) Fetal physiology and medicine. The basis of perinatology, 2nd Edition, Marcel Dekker, Inc. Publishers, New York

Walker AM, Cannata J, Dowling MH, Ritchie B, Maloney JE (1978) Sympathetic and parasympathetic control of heart rate in unanesthetised fetal and newborn lambs. Biol Neonate 33: 135-143

Walker AM, Cannata JP, Dowling MH, Ritchie BC, Maloney JE (1979) Age-dependent pattern of autonomic heart rate control during hypoxia in fetal and newborn lambs. Biol Neonate 35: 198-208

Walker AM, Cannata JP, Ritchie BC, Maloney JE (1983) Hypotension in fetal and newborn lambs; different patterns of reflex heart rate control revealed by autonomic blockade. Biol Neonate 44: 358-365

Walker AM, Berger PJ, Cannata J, Horne R, Maloney JE (1984) Autonomic control of heart state during electrocortical activity cycles in chronically hypoxaemic fetal lambs. Proc Society for the Study of Fetal Physiology, C6

Walker AM, Cannata JP, Ritchie BC, Maloney JE (1985) Different patterns of reflex heart rate control during stress in fetal and neonatal lambs: implications for ventricular function. In: Physiological development of the fetus and newborn, [Jones CT, Mott JC, Nathanielsz PW (eds)] Academic, London, pp 395-399

Walker AM, Horne RSC, de Preu N, Berger P (1986) The baroreflex in sleep in newborn lambs. Proc Austral Perinat Soc, Brisbane

Walker D (1975) Functional development of the autonomic innervation of the human fetal heart. Biol Neonate 25: 31-43

Walker DW (1984) Peripheral and central chemoreceptors in the fetus and newborn. Annu Rev Physiol 46: 687-703

Yardley RW, Bowes G, Wilkinson M, Cannata JP, Maloney JE, Ritchie BC, Walker AM (1983) Increased arterial pressure variability after arterial baroreceptor denervation in fetal lambs. Circ Res 52: 580-588

Young IM, Holland WW (1958) Some physiological responses of neonatal arterial blood pressure and pulse rate. Br Med J ii: 276-278

Comparison of the Effects of Anti-NGF and Thyroidectomy on Sympatho-Adrenal Function in Fetal Lambs

D. W. WALKER[1]

It is now well known that neuronal development is dependent on a number of survival factors, and on certain growth factors that are specific for particular neuron types. Of these, nerve growth factor (NGF) is the best known, and the requirement of sensory, sympathetic and central cholinergic neurons for this substance for both development and maintenance has been described.

There is also evidence of an important interaction between thyroid hormones and NGF in neural tissue. In neonatal and adult mice administration of thyroxine (T_4) increases the concentration of NGF in the brain (Walker et al. 1979, 1981). Also, both NGF and T_4 promote the assembly of microtubes within neurons and increase the activity of Na^+/K^+-ATPase (Levi-Montalcini et al. 1974; Levi et al. 1975; Fellous et al. 1979; Varon and Skaper 1983).

Although there is a great deal of morphological and biochemical evidence for the roles of NGF and thyroid hormones in sympatho-adrenal development, the functional consequences of removing the actions of these growth factors before birth is not well understood. In this presentation, the effects of treating fetal lambs with antibodies to NGF or of removing the thyroid gland will be described. Some similarities in the effects of the two procedures on sympatho-adrenal function will be discussed.

Anti-NGF

Prenatal exposure of rats and rabbits to anti-NGF of maternal origin (auto-immune immuno-sympathectomy) resulted in reduction of tissue noradrenaline concentrations and impaired development of sympathetic and sensory neurons in the superior cervical and dorsal root ganglia respectively (Gorin and Johnson 1979, 1980; Johnson and Gorin 1980; Padbury et al. 1986). In these experiments the antibody was present in the mother before and during the whole of the gestational period, and was presumably present at all stages of neurogenesis and differentiation in the conceptus because of the ready transmission of maternal antibodies through the placenta in these species (Brambell 1970). In view of this it is perhaps surprising to find that there was only partial inhibition of catecholamine synthesis in the sympa-

---

[1] Department of Physiology, Monash University, Clayton, Victoria 3168, Australia

The Endocrine Control of the Fetus
Ed. by W. Künzel and A. Jensen
© Springer-Verlag Berlin Heidelberg 1988

thetically innervated tissues. At 31 days' gestation in the rabbits the noradrenaline content of the lung and heart was reduced by 32% and 46% respectively (Padbury et al. 1986). There was no effect on the adrenal content of catecholamines. If it can be assumed that the catecholamines present in the tissues of immuno-sympathecto-mized rabbit fetuses are of neural origin, these results imply that NGF permits full and appropriate development of sympathetic neurons but is not an obligatory growth factor. However, some of the catecholamines may be localized in dispersed chromaffin tissues which, like the adrenal medulla, is thought to be relatively insen-itive to the actions of anti-NGF. Indeed, para-aortic chromaffin tissue in the auto-immunosympathectomized rabbits contained increased concentrations of catechol-amines (Padbury et al. 1986). This indicates the importance of determining the origin of catecholamines within an organ in normal and treated animals, since there may be an important interrelationship between neural, adrenal and extra-adrenal sites of production.

Aloe et al. (1981) noted that newborn rates given anti-NGF during fetal life were sluggish and apathetic, and showed poor regulation of body temperature. Apart from this, little is known of the functional effects of fetal immunosympathectomy. Therefore, the effects of treating fetal sheep by direct intramuscular injection of an-ti-NGF was determined (Schuijers et al. 1986, 1987). Because of limitations on the amount of material available, it was not possible to treat fetal lambs with anti-NGF throughout gestation, and the effects of a single treatment at 80 days' gestation have been examined. At this age most of the sympathetically innervated organs contain only 14%–34% of the noradrenaline concentrations present at the end of gestation (Schuijers et al. 1987), indicating the incompleteness of sympathetic innervation at 0.55 of gestation in the sheep. However, it should be remembered that, at any age, sensory and sympathetic neurons have different and changing requirements for NGF, so that treatment at earlier or later stages of gestation might yield quite differ-ent results.

When studied at 120–135 days' gestation, fetal lambs treated with anti-NGF showed no significant change of arterial pressure, heart rate or plasma catechol-amines when made hypoxic by giving the ewe 9% O_2 in N_2 to breathe (Schuijers et al. 1986). This contrasts with the effect of hypoxia in control fetuses, where there was an increase of arterial pressure and bradycardia and a six- to ten-fold increase of plasma noradrenaline concentrations. Since most of the increase of plasma nor-adrenaline which occurs during hypoxia in untreated fetuses is of adrenal origin (Jones et al. 1984), the results with the anti-NGF-treated fetuses suggest that no adrenal release of catecholamine occurred in response to hypoxia. Anti-NGF treat-ment did not alter the concentrations of catecholamines in the adrenal gland. It is possible that the lack of response to hypoxia in the treated fetuses was due to (a) impairment of splanchnic nerve function, (b) an effect of anti-NGF on central catecholamine pathways, which prevented relay of the hypoxic signal to the pre-ganglionic sympathetic outflow, or (c), loss of chemosensitivity or afferent nerve function from the aortic chemoreceptors. Thus, it is possible that the anti-NGF-treated fetuses did not 'sense' the hypoxic conditions. The absence of bradycardia at the onset of hypoxia in the treated fetuses is further evidence that the aortic chemoreflex had been impaired.

The anti-NGF treatment was associated with a reduction in the catecholamine concentrations in some but not all of the organs of the thorax and abdomen (Schuijers et al. 1987). Concentrations of noradrenaline (ng/g wet weight) were reduced by more than 60% in the thymus, thyroid, atrium, lung, liver and kidney, but, curiously, there was no change in the spleen, an organ which has a rich sympathetic innervation. While there was 75% and 95% reduction in the concentration of noradrenaline and adrenaline respectively in the jejunum, there was no changes in the ileum, colon or bladder. The rostrocaudal gradient of effect may indicate greater sensitivity of more anterior sympathetic ganglia to the actions of anti-NGF. However, as mentioned above, it will be important to establish whether the remaining catecholamine was localized in neural or non-neural stores.

Anti-NGF treatment also had effects on noradrenaline concentrations in the brain, with reductions of more than 50% occurring in the thalamus, hypothalamus, hippocampus, medulla and cerebellum (Schuijers et al. 1987). The adrenaline concentrations in the hypothalamus and medulla were also reduced by over 75%. It might be expected that hypothalamic and brainstem functions mediated by monoaminergic cell groups may have been altered. It should also be noted that there is substantial evidence that NGF has actions on cholinergic neurons within the brain (Korsching et al. 1985), so that the anti-NGF treatment may have had more widespread actions not detected by measurement of catecholamines alone.

Thyroidectomy

The probable interaction between NGF and thyroid hormones in maturation of the nervous system has been reviewed by Fisher et al. (1982). Effects of thyroid deficiency on sympathetic neuron development include delayed development of tyrosine hydroxylase and choline acetyltransferase activities in the superior cervical ganglion (Black 1974), whereas the development of sympatho-adrenal function can be accelerated when neonatal rats are given excess thyroid hormones (Lau and Slotkin 1979, 1980).

To determine the importance of thyroid hormones on sympatho-adrenal development in sheep, the thyroid glands were removed from 12 fetal lambs at 80 days' gestation, and they were then studied at 125–135 days' gestation. Every thyroidectomized (TX) fetus was growth-retarded and had an immature appearance, as described previously (Hopkins and Thorburn 1972). The TX fetuses had a significantly lower arterial pressure than age-matched controls (34 ± 0.11 mm Hg vs. 45 ± 0.2 mm Hg, $p < 0.05$), which may be caused by a decrease of cardiac contractility due to a decrease in activity of the myocardial ATPase (Hoh and Egerton 1979). Changes in β-receptor activity following thyroidectomy in lambs have also been implicated (Breall et al. 1984).

In these TX fetuses hypoxia did not result in a significant change of arterial pressure, heart rate or plasma catecholamines. Thyroidectomy did not alter the concentrations of catecholamines in the adrenal gland or of noradrenaline in any of the thoracic or abdominal organs. Thus, the absence of a hypoxic response might be due to impairment of either the chemoreceptors and their afferent fibres, or of the central relay and transmission of the hypoxic stimulus to the sympathetic efferent outflow. Intravenous infusion of tyramine revealed that noradrenaline could be displaced from the postganglionic neuronal sites, although the increase of plasma noradrenaline was significantly less in the TX fetuses (1.9-fold) compared to the control fetuses (9.2-fold). Arterial pressure increased from 33 ± 0.2 mm Hg to 56 ± 0.4 mm Hg ($p < 0.05$) when tyramine was infused, and heart rate *increased* from 156 ± 7 beats/min to 215 ± 10 beats/min ($p < 0.05$), whereas in the control fetuses there was bradycardia during the tyramine-induced hypertension (see also Schuijers et al. 1986).

Thyroidectomy and treatment with anti-NGF appear to have altered the sympatho-adrenal responses of the fetal lamb in similar ways. In both groups of fetuses there was a decrease in the amount of noradrenaline available for release by tyramine from postganglionic neuronal stores. In both groups the bradycardia response to hypertension and hypoxia was absent. Arterial hypoxia did not increase plasma catecholamine concentration in the anti-NGF-treated or TX fetuses. These effects may be due to an impairment of sensory nerve fibres which relay the effects of hypoxia and hypertension to the vagal and sympathetic motor fibres, and/or to impairment of transmission in autonomic ganglia. The similarity of the functional deficits produced by the two procedures suggest that the synthesis and actions of NGF may depend upon the availability of thyroid hormones. If it is assumed that the catecholamine content of an organ is related to the density of sympathetic innervation, then thyroidectomy apparently does not restrict the growth of sympathetic nerves, although less transmitter is released from them by tyramine.

References

Aloe L, Cozzari C, Calissano P, Levi-Montalcini R (1981) Somatic and behavioral postnatal effects of fetal injections of nerve growth factor antibodies in the rat. Nature 291: 413–415

Black IR (1974) The role of the thyroid in the growth and development of adrenergic neurons in vivo. Neurology 24: 377 (abstract)

Brambell FWR (1970) The transmission of passive immunity from the mother to young. In: Neuberger A, Tatun EL (eds) Frontiers of biology, vol 18. Elsevier, New York

Breall JA, Rudolph AM, Heyman MA (1984) Role of thyroid hormone in postnatal circulatory and metabolic adjustments. J Clin Invest 73: 1418–1424

Fellous A, Lennon A, Francon J, Nunez J (1979) Thyroid hormones and microtubule assembly in vitro during brain development. Eur J Biochem 101: 365–376

Fisher DA, Hoath S, Lakshmanan J (1982) The thyroid hormone effects on growth and development may be mediated by growth factors. Endocrinol Exp 16: 259–271

Gorin PD, Johnson EM (1979) Experimental autoimmune model of nerve growth factor deprivation: effects on developing peripheral sympathetic and sensory neurons. Proc Natl Acad Sci USA 76: 5382-5386

Gorin PD, Johnson EM (1980) Effects of exposure to nerve growth factor antibodies on the developing nervous system of the rat: an experimental autoimmune approach. Dev Biol 80: 313-323

Hoh JFY, Egerton LJ (1979) Action of tri-iodothyronine on the synthesis of rat ventricular myosin isoenzymes. FEBS Lett 101: 143-148

Hopkins PS, Thorburn GD (1972) The effects of foetal thyroidectomy on the development of the ovine foetus. J Endocrinol 54: 55-60

Johnson EM, Gorin PD (1980) Dorsal root ganglia neurones are destroyed by exposure in utero to maternal antibody to nerve growth factor. Science 210: 916-918

Jones CT, Roebuck MM, Walker DW (1984) The effects of adrenal medullary activity on the sensitivity to catecholamine stimulation of the sympathetic system of fetal sheep. In: Usdin E, Carlsson A, Dahlström A, Engel J (eds) Catecholamines: basic and peripheral mechanisms. Liss, New York, pp 121-128

Korsching S, Auburger G, Heumann R, Scott J, Thoenen H (1985) Levels of nerve growth factor and its mRNA in the central nervous system of the rat correlate with cholinergic innervation. EMBO J 4: 1389-1393

Lau C, Slotkin TA (1979) Accelerated development of rat sympathetic neurotransmission caused by neonatal triiodothyronine administration. J Pharm Exp 208: 485-490

Lau C, Slotkin TA (1980) Maturation of sympathetic neurotransmission in rat heart. II. Enhanced development of presynaptic and postsynaptic components of noradrenergic synapses as a result of neonatal hyperthyroidism. J Pharm Exp 212: 126-130

Levi A, Cimino M, Mercanti D, Chen JS, Calissano P (1975) Interaction of nerve growth factor with tubulin, studies in binding and induced polymerization. Biochim Biophys Acta 339: 50-60

Levi-Montalcini R, Revoltella R, Calissano P (1974) Microtubule proteins in the nerve growth factor mediated response. Recent Prog Horm Res 30: 635-669

Padbury JF, Lam RW, Polk DH, Newnham JP, Lakshmanan J, Fisher DA (1986) Autoimmune sympathectomy in fetal rabbits. J Dev Physiol 8: 369-376

Schuijers JA, Walker DW, Browne CA, Thorburn GD (1986) Effect of hypoxemia on plasma catecholamine concentrations in intact and immunosympathectomized fetal lambs. Am J Physiol 251: R893-R900

Schuijers JA, Walker DW, Browne CA, Thorburn GD (1987) Peripheral and brain tissue catecholamine content in intact and anti-NGF treated fetal sheep. Am J Physiol 252: R7-R12

Varon S, Skaper SD (1983) The Na^+,K^+ pump may mediate the control of nerve cells by nerve growth factor. Trends Biochem Sci 8: 22-25

Walker P, Weichsel ME, Guo SM, Fisher DA (1979) Thyroxine increases nerve growth factor concentration in adult mouse brain. Science 204: 427-429

Walker P, Weil ML, Weichsel ME, Fisher DA (1981) Effect of thyroxine on nerve growth factor concentration in neonatal brain. Life Sci 28: 1777-1787

Fetal Chemical Sympathectomy:
Circulatory and Hormonal Responses to Hypoxemia

H. S. Iwamoto[1, 2], A. M. Rudolph[1-4], B. L. Mirkin[5], and L. C. Keil[6]

Introduction

Delivery of an adequate supply of oxygen to the fetus is essential for normal growth and development. When oxygen delivery to the fetus is reduced moderately, adjustments in the circulation and metabolism, mediated by various neural, hormonal, and reflex mechanisms, occur to ensure survival. The purpose of this presentation is to consider the role of the sympathetic nervous system in the fetal response to hypoxemia. Because most of the currently available information relating to this topic has been obtained from sheep fetuses, this discussion will be largely confined to these studies.

Fetal Responses to Acute Hypoxemia

The typical fetal response to an acute hypoxemic episode, induced by decreasing the fraction of oxygen in the inspired air delivered to its mother, is hypertension, bradycardia, and a redistribution of blood flow to peripheral organs (Cohn et al. 1974; Peeters et al. 1979). The hypertension and bradycardia are in part mediated by chemoreceptor-reflex stimulation of α-adrenergic and vagal mechanisms (Dawes et al. 1969; Itskovitz and Rudolph 1982). Total output of blood from the heart and blood flow to the umbilical-placental circulation do not change significantly with moderate hypoxemia. However, blood flow to the brain, heart, and adrenal increase, and oxygen delivery to these organs is maintained, while blood flow and oxygen delivery to the lungs decrease (Cohn et al. 1974; Peeters et al. 1979; Sheldon

[1] Cardiovascular Research Institute, University of California San Francisco, San Francisco CA 94143, USA
[2] Department of Physiology, University of California San Francisco, San Francisco CA 94143, USA
[3] Department of Pediatrics, University of California San Francisco, San Francisco CA 94143, USA
[4] Department of Obstetrics, Gynecology, and Reproductive Sciences, University of California San Francisco, San Francisco CA 94143, USA
[5] Department of Pharmacology, University of Minnesota, Minneapolis MN 55455, USA
[6] NASA-Ames Research Center, Moffett Field CA, USA

The Endocrine Control of the Fetus
Ed. by W. Künzel and A. Jensen
© Springer-Verlag Berlin Heidelberg 1988

et al. 1979). In response to severe hypoxemia or to hypoxemia combined with acidemia, blood flow to the kidneys, gastrointestinal tract, and musculoskeletal and cutaneous (peripheral) vascular beds also decreases (Cohn et al. 1974; Peeters et al. 1979). In addition to these regional changes in blood flow, the pattern of blood flow through shunt pathways in the fetal circulation is altered so that a greater proportion of oxygenated blood returning to the fetal body in the umbilical vein bypasses the hepatic microvasculature through the ductus venosus and preferentially perfuses the heart and brain (Reuss and Rudolph 1980).

A number of hormonal systems in the fetus respond to acute hypoxemia (Rurak 1978; Boddy et al. 1974; Wardlaw et al. 1981; Robillard et al. 1981; Stark et al. 1982). Vasopressin has been shown by several investigators to increase during acute hypoxemia (Robillard et al. 1981; Rurak 1978; Stark et al. 1982). Administration of vasopressin to fetuses in amounts that mimic the response to hypoxemia produces bradycardia, hypertension, and a redistribution of blood flow to organs, responses that are in many ways similar to those of fetuses made hypoxemic acutely (Iwamoto et al. 1979). Although this suggests that vasopressin participates in the fetal response to hypoxemia, the exact roles of vasopressin and other hormones have not yet been determined.

Possible Role of the Sympathetic Nervous System in the Fetal Response to Acute Hypoxemia

The sympathoadrenal system is clearly involved in the fetal response to acute hypoxemia or asphyxia. Plasma norepinephrine and epinephrine concentrations increase in response to hypoxemia. The magnitude of the increase is proportional to the decrease in fetal arterial oxygen tension (Comline et al. 1965; Cohen et al. 1982), is reproducible (Lewis et al. 1984b), and increases with gestational age (Robillard et al. 1981). Studies of adrenal medullary secretion of norepinephrine and epinephrine in acute and chronic preparations of fetal sheep have shown the adrenal response to be dynamic, proportional to stimulus intensity, and increasingly sensitive to repeated stimulation (Comline and Silver 1961; Comline et al. 1965; Cohen et al. 1984). The sympathetic nervous system also contributes to the plasma norepinephrine response, presumably as a result of increased spillover from nerve terminals with the increase in activity, because chemically and immuno-sympathectomized fetal sheep have smaller plasma norepinephrine responses to hypoxemia than do intact (Iwamoto et al. 1983; Lewis et al. 1984a; Schuijers et al. 1986). In addition, the extraadrenal chromaffin tissue in the fetus can in certain instances contribute to the plasma norepinephrine response to hypoxemia (Jones et al. 1984).

To determine the extent to which the sympathoadrenal system participates in the response to hypoxemia, several investigators have administered inhibitors of ganglionic transmission or adrenergic receptor blocking agents to hypoxemic fetuses (Jones et al. 1984; Reuss et al. 1982; Cohn et al. 1982; Court et al. 1984). The results from these studies suggest that α-adrenergic mechanisms are important for maintaining arterial pressure and blood flow to the placenta, heart, brain, and adrenal by constricting the splanchnic and pulmonary vascular beds (Reuss et al. 1982), and β-adrenergic mechanisms are important primarily for maintaining heart rate and output of blood flow from the heart (Cohn et al. 1982; Court et al. 1984). One of the disadvantages of these studies is that it is not possible to determine the relative importance of the sympathetic nervous system, as distinct from the adrenal medulla, in the response to hypoxemia.

To determine the role of the sympathetic nervous system, 6-hyroxydopamine (6-OHDA), a synthetic catecholamine that selectively destroys sympathetic nerve terminals, was administered to fetal sheep to produce chemical sympathectomy. The responses of intact and chemically sympathectomized fetal sheep to hypoxemia were compared.

Methods

Experiments were carried out in fetal sheep at 121–145 days' gestation. Polyvinyl catheters were placed in the ascending and descending aorta, and the inferior and superior venae cavae of the fetus, as well as in the amniotic fluid cavity, descending aorta, and inferior vena cava of the ewe. In one group of 11 fetuses on the 2nd and 4th days after surgery, 6-OHDA HCl (50 mg/kg body wt., Sigma) was infused intravenously. Experiments were carried out at least 2 days after the second dose of 6-OHDA. Fetal blood pressures and heart rate were monitored continuously. Combined ventricular output and blood flow to fetal organs were measured using the radionuclide-labeled microsphere technique as described in detail previously (Iwamoto et al. 1983). Arterial blood samples were obtained for determination of pH, PCO_2, PO_2, and plasma catecholamine and vasopressin concentrations. Acute hypoxemia in the fetus and ewe was induced by giving the ewes a gas mixture of 3% CO_2, 9% O_2 in N_2 to breathe. Measurement were repeated 15 min and 40 min after the induction of hypoxemia. Plasma concentrations of vasopressin were determined by radioimmunoassay, and catecholamines were determined by electrochemical detection following separation by HPLC. Additional details of methods have been published (Iwamoto et al. 1983). All values are expressed as mean ± SD.

Results and Discussion

Effects of 6-Hydroxydopamine on the Fetal Circulation

6-OHDA was administered to 11 fetal sheep at 117–142 days gestational age. Intravenous administration of 6-OHDA to fetal sheep acutely displaces norepinephrine from sympathetic nerve terminals and causes marked hypertension and tachycardia

Table 1. Effect of chemical sympathectomy and of hypoxemia on arterial blood

	Before 6-OHDA	After 6-OHDA	Hypoxia
pH	7.39 ± 0.03	7.37 ± 0.05	7.31 ± 0.06[a]
PO_2 (Torr)	21 ± 3	22 ± 3	13 ± 2[a]
PCO_2 (Torr)	45 ± 5	45 ± 4	43 ± 3
Hemoglobin (g/dl)	9.1 ± 2.1	8.9 ± 1.9	9.5 ± 1.8[a]
Blood O_2 saturation (%)	57 ± 10	52 ± 9	20 ± 7[a]

'Before 6-OHDA' values were obtained on the 2nd postoperative day prior to exposure to 6-OHDA; 'After 6-OHDA' values were obtained following recovery from the acute effects of 2 doses of 6-OHDA described. 'Hypoxia' values were obtained from the 11 chemically sympathectomized fetuses made hypoxemic for 40 min. $n = 11$. Values are expressed as mean ± SD.

[a] Significant change from previous values, $p < 0.05$.

(H. S. Iwamoto and A. M. Rudolph, unpublished observations; Lewis et al. 1984a; Tabsh et al. 1982). For this reason, it was necessary to administer 6-OHDA slowly while monitoring arterial blood pressure and heart rate. The degree of sympathectomy achieved by 6-OHDA was assessed by measuring the arterial blood pressure response to an injection of tyramine, an indirectly acting adrenergic agonist. Prior to exposure to 6-OHDA, tyramine, injected in amounts of 200–300 µg/kg body wt., increased arterial pressure by 9 ± 5 Torr and decreased heart rate by 20 ± 14 beats/min. After 1–2 injections of 6-OHDA, tyramine elicited no response and the norepinephrine content in portions of myocardium decreased by 67%–80%. Sympathectomy in these fetuses was judged to be complete.

Following recovery from the acute effects of 6-OHDA, arterial pH, PO_2, PCO_2, hemoglobin concentration, and blood O_2 saturation were similar to those values prior to exposure to 6-OHDA (Table 1). Arterial blood pressure was 48 ± 6 Torr, and heart rate was 178 ± 28 beats/min, values that are not significantly different from those in intact fetuses. These results are similar to those of Tabsh et al. (1982) but differ slightly from those of Lewis et al. (1984a), who reported a significant increase in mean arterial blood pressure. This discrepancy many be related to the observation that the animals in the studies by Lewis et al. (1984a) had slightly, but not significantly, greater plasma catecholamine concentrations following 6-OHDA treatment. Chemical sypathectomy renders fetuses more sensitive to α- and β-adrenergic stimulation (Tabsh et al. 1982), which suggests the development of end-organ supersensitivity following removal of functional innervation. The supersensitivity to α-adrenergic stimulation with phenylephrine or norepinephrine has been reported to persist for up to 5 months after birth in animals sympathectomized in utero (Lumbers et al. 1980). The small increase in circulating catecholamines in the studies of Lewis et al. (1984a) may be responsible for the increased mean arterial pressure.

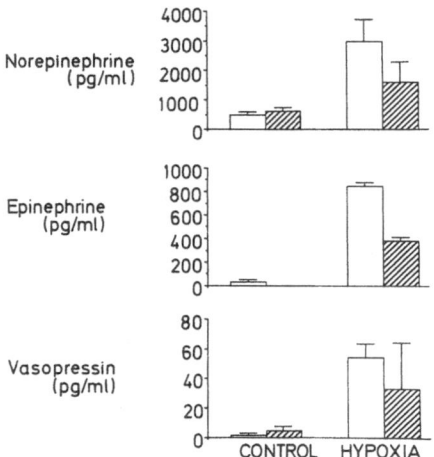

Fig. 1. Response of intact (□) and chemically sympathectomized (▨) fetal sheep to hypoxia. Chemically sympathectomized fetuses had undetectable plasma epinephrine concentrations but normal norepinephrine and vasopressin concentrations. Acute hypoxemia increases concentrations of these hormones, but the responses in intact fetuses were greater

Effect of 6-Hydroxydopamine on the Fetal Response to Hypoxemia

Plasma Catecholamine and Vasopressin Concentrations

In intact fetal sheep, plasma norepinephrine concentrations range from 120–500 pg/ml, plasma epinephrine concentrations from 7–50 pg/ml, and plasma vasopressin concentrations from 3–5 pg/ml during the last 3 weeks of gestation (Palmer et al. 1984; Iwamoto et al. 1979). In some chemically sympathectomized fetal sheep, the plasma concentrations of norepinephrine, epinephrine, and vasopressin were 630 ± 10, <100, and 4.5 ± 3.5 pg/ml, respectively, during the control period, but norepinephrine and epinephrine were not detectable in all fetuses (Fig. 1). Acute hypoxemia increased these values significantly in both intact and sympathectomized fetuses. However, the responses of the sympathectomized fetuses were smaller than those of intact fetuses. In a similar series of studies by Lewis et al. (1984a), the increase in plasma norepinephrine and epinephrine concentration produced by umbilical cord compression was smaller in chemically sympathectomized than in intact fetuses. However, the plasma catecholamine response to hypoxemia was abolished in immunosympathectomized fetuses, even though basal norepinephrine concentrations were significantly increased (Schuijers et al. 1986).

Arterial Blood Pressure and Heart Rate

When the chemically sympathectomized fetuses were made hypoxemic acutely, mean arterial pressure did not change significantly (51 ± 8 Torr), in contrast to the increase (48 ± 4 to 57 ± 6 Torr) that occurred in intact fetuses. Heart rate decreased transiently in both groups of animals, but returned to control values by 40 min. Similar results were obtained by Schuijers et al. (1986), who studied immunosympa-

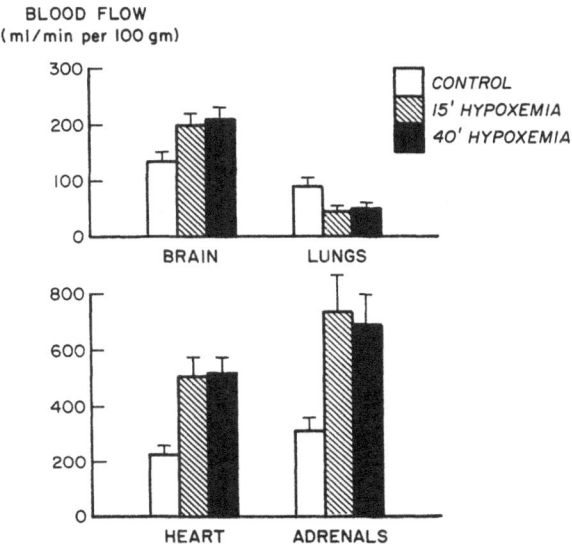

Fig. 2. Acute hypoxemia significantly increased blood flow to the heart, brain, and adrenals ($p<0.05$) and decreased blood flow to the lungs ($p<0.05$) in chemically sympathectomized fetal sheep. $n=11$, values are mean $\pm$ SE

thectomized fetuses. In contrast, Lewis et al. (1984a) found that hypoxemia resulting from umbilical cord compression was associated with rapid development of hypertension and bradycardia in chemically sympathectomized fetuses. In the studies of Lewis et al., the hypertension may have resulted directly from the mechanical effects of cord compression and may be unrelated to arterial hypoxemia per se.

Combined Ventricular Output and Organ Blood Flow

During the control period in chemically sympathectomized fetuses, combined ventricular output was 486 ± 76 ml$\cdot$min$^{-1}\cdot$kg^{-1} umbilical-placental blood flow was 213 ± 51 ml$\cdot$min$^{-1}\cdot$kg^{-1}, and blood flow to the fetal body (total output minus umbilical-placental blood flow) was 273 ± 54 ml$\cdot$min$^{-1}\cdot$kg^{-1}. These values are comparable to previously reported values obtained from intact fetuses at a similar stage of development (Cohn et al. 1974). Acute hypoxemia had no significant effect on any these values.

Despite the absence of changes in total output, acute hypoxemia was associated with changes in blood flow to certain organs. Blood flow to the cerebral, myocardial, and adrenal vascular beds increased significantly from 133 ± 14 to 208 ± 20, 215 ± 27 to 519 ± 55, and 291 ± 49 to 690 ± 110 ml$\cdot$min$^{-1}\cdot$100 g^{-1}, respectively, while that to the pulmonary circulation decreased from 94 ± 16 to 50 ± 7 ml$\cdot$

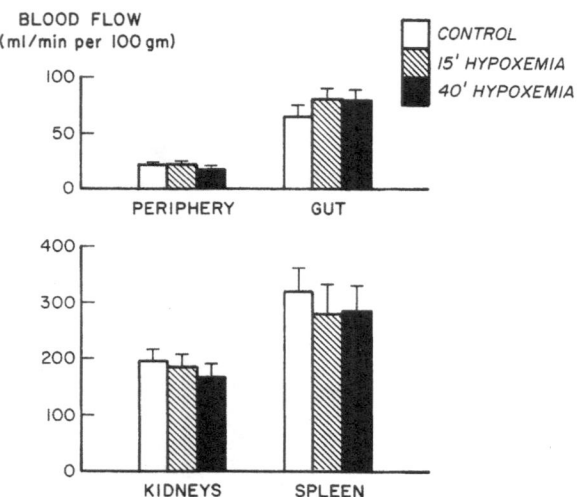

Fig. 3. Acute hypoxemia significantly increased blood flow to the gastrointestinal tract ($p <$ 0.05), decreased renal blood flow ($p < 0.05$), and did not alter peripheral or splanchnic blood flows in chemically sympathectomized fetal sheep. $n = 11$, values are mean + SE

$\min^{-1} \cdot 100 \, g^{-1}$ (Fig. 2). These responses also occur in intact fetuses and are largely intact in those pretreated with propranolol and phenoxybenzamine (Cohn et al. 1982; Cohn et al. 1974; Peeters et al. 1979; Reuss et al. 1982). The changes in blood flow to these organs with hypoxia are probably a response to local changes in oxygen tension or content.

By contrast, blood flow to the peripheral, gastrointestinal and splenic vascular beds did not decrease as it does in intact fetuses (Cohn et al. 1974); in fact, gastrointestinal blood flow increased significantly from 62 ± 8 to $80 \pm 9 \, ml \cdot min^{-1} \cdot 100 \, g^{-1}$ (Fig. 3). These results suggest that the sympathetic nervous system normally mediates the vasoconstrictor response to hypoxemia in these vascular beds. However, α-adrenergic inhibition with phenoxybenzamine abolishes only the vasoconstrictor response in the gastrointestinal and splenic vascular beds (Reuss et al. 1982). Thus, the absence of vasoconstriction in the peripheral circulation may also be related to the reduced vasopressin response to hypoxemia in the sympathectomized fetal sheep. In some intact (Reuss et al. 1982; Robillard et al. 1981; Iwamoto et al. 1985; Cohn et al. 1974; Cohn et al. 1974; Peeters et al. 1979) and in the chemically sympathectomized fetuses, acute hypoxemia decreases blood flow to the kidneys significantly. Renal nerve stimulation in fetal sheep stimulates an α-adrenergically mediated vasoconstriction, a β_2-adrenergically mediated vasodilatation, and an overall decrease in renal blood flow velocity (Robillard et al. 1987). In the chemically sympathectomized fetuses either 6-OHDA had no effect on renal sympathetic nerve function and renal vascular response to hypoxemia or it increased the sensitivity of the renal vasculature, predominantly to α-adrengeric stimulation.

Conclusions

These studies show, by comparing the responses of intact and chemically sympathectomized fetuses to acute hypoxemia, that the sympathetic nervous system plays a role in the fetal response to hypoxemia, specifically by constricting peripheral vascular beds and increasing arterial blood pressure. These effects may be mediated directly or indirectly through an influence on vasopressin secretion. The results from these studies should be interpreted as indicating the *minimum* extent to which the sympathetic nervous system participates in the response because other regulatory mechanisms compensate for the absence of normal sympathetic nerve function. In addition, available evidence demonstrates that the chemically sympathectomized fetus becomes supersensitive to α- and β-adrenergic stimulation. Thus the increase in plasma catecholamine concentrations, although smaller than that which occurs in intact fetuses, may have profound circulatory and metabolic effects.

When comparing the results from similar studies of chemically or immuno sympathectomized fetal sheep, (Lewis et al. 1984a; Schuijers et al. 1986), several differences become evident that indicate that administration of 6-OHDA or anti-nerve growth factor antiserum produces many changes in addition to destruction of the peripheral nervous system. Among these are the development of increased responsiveness to exogenous catecholamines; increases, decreases or no changes in basal and stimulated catecholamine concentrations; maintenance or increased in arterial blood pressure and heart rate despite apparent destruction of innervation believed to be important for tonic regulation. Future studies in this area should involve investigation of several aspects of neural function and response.

References

Boddy K, Jones CT, Mantell C, Ratcliffe JG, Robinson JS (1974) Changes in plasma ACTH and corticosteroids of the maternal and fetal sheep during hypoxia. Endocrinology 94: 588–591

Cohen WR, Piasecki GJ, Jackson BT (1982) Plasma catecholamines during hypoxemia in fetal lambs. Am J Physiol 243: R520–R525

Cohen WR, Piasecki GK, Cohn HE, Young JB, Jackson BT (1984) Adrenal secretion of catecholamines during hypoxemia in fetal lambs. Endocrinology 114: 383–390

Cohn HE, Sacks EJ, Heymann MA, Rudolph AM (1974) Cardiovascular responses to hypoxemia and acidemia in fetal lambs. Am J Obstet Gynecol 120: 817–824

Cohn HE, Piasecki GJ, Jackson BT (1982) The effect of β-adrenergic stimulation on fetal cardiovascular function during hypoxemia. Am J Obstet Gynecol 114: 810–816

Comline RS, Silver M (1961) The release of adrenaline and noradrenaline from the adrenal glands of the foetal sheep. J Physiol (Lond) 156: 424–444

Comline RS, Silver IA, Silver M (1965) Factors responsible for the stimulation of the adrenal medulla during asphyxia in the foetal lamb. J Physiol (Lond) 178: 211–238

Court DJ, Parer JT, Block BSB, Llanos AJ (1984) Effects of beta-adrenergic blockade on blood flow distribution during hypoxemia in fetal sheep. J Dev Physiol 6: 349–358

Dawes GS, Duncan SL, Lewis BV, Merlet CL, Owen-Thomas JB, Reeves JT (1969) Hypoxaemia and aortic chemoreceptor function in foetal lambs. J Physiol (Lond) 201: 105-116

Itskovitz J, Rudolph AM (1982) Denervation of arterial chemoreceptors and baroreceptors in fetal lambs in utero. Am J Physiol 242: H916-H920

Iwamoto HS, Rudolph AM (1985) Metabolic responses of the kidney in fetal sheep: effect of acute and spontaneous hypoxemia. Am J Physiol 249: F836-F841

Iwamoto HS, Rudolph AM, Keil LC, Heymann MA (1979) Hemodynamic responses of the sheep fetus to vasopressin infusion. Circ Res 44: 430-436

Iwamoto HS, Rudolph AM, Mirkin BL, Keil LC (1983) Circulatory and humoral responses of sympathectomized fetal sheep to hypoxemia. Am J Physiol 254: H767-H772

Jones CT, Ritchie JWK (1983) The effects of adrenergic blockade on fetal response to hypoxia. J Dev Physiol 5: 211-222

Jones CT, Roebuck MM, Walker DW (1984) The effects of adrenal medullary activity on the sensitivity of catecholamine stimulation of the sympathetic system in fetal sheep. In: Catecholamines: basic and peripheral mechanisms. Liss, New York, pp 121-128

Lewis AM, Wolf WJ, Sischo W (1984a) Fetal cardiovascular and catecholamine responses to hypoxemia after chemical sympathectomy. Pediatr Res 18: 318-322

Lewis AB, Wolf WJ, Sischo W (1984b) Cardiovascular and catecholamine responses to successive episodes of hypoxemia in the fetus. Biol Neonate 45: 105-111

Lumbers ER, Stevens AD, Alexander G, Stevens D (1980) The cardiovascular responses of conscious newborn lambs treated in utero with 6-hydroxydopamine. J Dev Physiol 2: 139-149

Palmer SM, Oakes GK, Lam RW, Oddie TH, Hobel CJ, Fisher DA (1984) Catecholamine physiology in the ovine fetus. 1. Gestational age variation in basal plasma concentrations. Am J Obstet Gynecol 149: 420-425

Peeters LLH, Sheldon RE, Jones MD Jr, Makowski EL, Meschia G (1979) Blood flow to fetal organs as a function of arterial oxygen content. Am J Obstet Gynecol 135: 637-646

Reuss ML, Rudolph AM (1980) Distribution and recirculation of umbilical and systemic venous blood flow in fetal lambs during hypoxia. J Dev Physiol 2: 71-84

Reuss ML, Parer JT, Harris JL, Krueger TR (982) Hemodynamic effects of alpha-adrenergic blockade during hypoxia in fetal sheep. Am J Obstet Gynecol 142: 410-415

Robillard JE, Weitzman RE, Burmeister L, Smith FG Jr (1981) Developmental aspects of the renal response to hypoxemia in the lamb fetus. Circ Res 48: 128-138

Robillard JE, Nakamura KT, Wilkin MK, McWeeny OJ, DiBona GF (1987) Ontogeny of renal hemodynamic response to renal nerve stimulation in sheep. Am J Physiol 252: F605-F612

Rurak DW (1978) Plasma vasopressin levels during hypoxemia and the cardiovascular effects of exogenous vasopressin in fetal and adult sheep. J Physiol (Lond) 277: 341-357

Schuijers JA, Walker DA, Browne CA, Thorburn GD (1986) Effect of hypoxemia on plasma catecholamines in intact and immunosympathectomized fetal lambs. Am J Physiol 251: R893-R900

Sheldon RE, Peeters LLH, Jones MD Jr, Makowski EL, Meschia G (1979) Redistribution of cardiac output and oxygen delivery in the hypoxemic fetal lamb. Am J Obstet Gynecol 135: 1071-1078

Stark RI, Wardlaw SL, Daniel SS, Husain MK, Sanocka UM, James LS, Vande Wiele RL (1982) Vasopressin secretion induced by hypoxia in sheep: developmental changes and relationship to β-endorphin release. Am J Obstet Gynecol 143: 204-215

Tabsh K, Nuwayhid B, Murad S, Ushioda E, Erkkola R, Brinkman CR III, Assali NS (1982) Circulatory effects of chemical sympathectomy in fetal, neonatal, and adult sheep. Am J Physiol 243: H113-H122

Wardlaw SL, Stack RI, Daniel S, Frantz AG (1981) Effects of hypoxia on β-endorphin and β-lipotropin release in fetal, newborn and maternal sheep. Endocrinology 108: 1710-1715

Dynamics of Circulatory Centralization and Release of Vasoactive Hormones During Acute Asphyxia in Intact and Chemically Sympathectomized Fetal Sheep*

A. Jensen[1] and U. Lang

Introduction

Acute fetal asphyxia caused by reduction in uterine blood flow both increases sympathetic activity and decreases blood flow to peripheral organs, including that to the skin (Jensen and Künzel 1980; Paulick et al. 1985; Jensen et al. 1985b, 1987b). These changes, which result in a circulatory centralization, characteristic of fetal and neonatal shock (Jensen et al. 1987a), are associated with increased mortality and morbidity in the perinatal period (Jensen and Schumacher 1987). Therefore, in an effort to reduce perinatal morbidity and mortality, it has been suggested to supplement monitoring of the heart rate by that of skin blood flow-dependent variables to detect fetal circulatory centralization early during labour (Jensen and Künzel 1980; Paulick et al. 1985).

Experimental studies in sheep have confirmed that skin blood flow is an index of fetal shock, because during asphyxia decreased blood flow to the fetal skin correlates closely with both decreased blood flow to other peripheral organs and increased sympathetic activity (Jensen et al. 1985b, 1987a, 1987b). However, neither the time course of changes in blood flow to the skin and to other organs nor the mechanisms that govern this time course are clear.

Only recently has it become possible to study the dynamics, i.e. the time course of circulatory centralization during acute asphyxia using a modified isotope-labelled microspheres technique (Rudolph and Heymann 1967; Jensen et al. 1985a, 1987c). First results on blood flow changes during prolonged acute asphyxia at 1-min intervals show that peripheral blood flow, including that to the skin and scalp, falls rapidly, because the vessels in peripheral organs constrict. The rapidity and uniformity of this peripheral vasoconstriction, which is accompanied by a decrease in fetal oxygen consumption (Jensen 1986; Jensen et al. 1987c; A. Jensen and A. M. Rudolph, unpublished observations), suggest that the sympathoneuronal system may reflexly cause these initial blood flow changes during acute asphyxia and hence may be important in fetal short-term adaptation to asphyxia.

The present investigation was designed to test this hypothesis and to further validate skin blood flow as an index of fetal circulatory centralization. We first studied

* This investigation was supported by the Deutsche Forschungsgemeinschaft (Je 108/4-1).
[1] Department of Obstetrics and Gynaecology, Justus-Liebig-Universität Giessen, Klinikstr. 32, D-6300 Giessen, Federal Republic of Germany

The Endocrine Control of the Fetus
Ed. by W. Künzel and A. Jensen
© Springer-Verlag Berlin Heidelberg 1988

the time course of changes in both organ blood flow and vasoactive hormone concentrations during and after arrest of uterine blood flow for 2 min. Secondly, to assess the influence of the sympathoneuronal system on the time course of circulatory centralization separately from the influence of the adrenal medulla upon it, we administered 6-hydroxydopamine to a second group of fetuses to destroy selectively their sympathetic nerve terminals (Iwamoto et al. 1983) and examined the circulatory and hormonal responses of these sympathectomized fetuses using the same protocol.

Material and Methods

Animal Preparation

Eleven fetal sheep near term were chronically prepared (term is at 147 days). All ewes were anaesthetized by subarachnoid or epidural injection of 2-4 ml of 1% (w/v) tetracaine HCl at the lower spine and were operated on under sterile conditions. Polyvinyl catheters were placed in a maternal iliac artery and vein through tibial vessels. The ewe's abdominal wall was opened in the midline and a snare was placed around the descending aorta below the renal artery, which was required in the experiment to arrest uterine and ovarian blood flow. Care was taken not to include any nerves. Through a small uterine incision fetal catheters were inserted under local anaesthesia (xylocaine 1%, w/v) via a tibial vein and artery of each hindleg and advanced to the caudal vena cava and abdominal aorta, respectively. Through a second incision additional catheters were placed in both a carotid and a brachial artery. The uterus was closed around another catheter in the amniotic cavity after the fluid lost had been replaced by warm saline (0.9%, w/v). All catheters were filled with heparin (1000 IU/ml) and plugged. Catheters and snare were passed subcutaneously to the ewe's flank where they were exteriorized and protected by a pouch sewn to the skin. The fetal sheep received antibiotics daily, injected intravenously into the ewe and into the amniotic cavity (4 g cefotaxime and 80 mg gentamycin). These were allowed to recover for 2 days before being studied.

Experimental Protocol

Five intact and six sympathectomized normoxic unanaesthetized chronically catheterized fetal sheep (weight 3.2 ± 0.5 kg) were studied in utero at 125-135 days' gestation. The experiments, one on each fetus, were conducted on the 3rd day after the operation. Sympathectomy was performed on the 2nd day after operation by infusing 6-hydroxydopamine (46.1 ± 6 mg/kg fetal weight) over 1-2 h (Iwamoto et al. 1983). This was accompanied by an increase in mean arterial blood pressure from 42 ± 4 mmHg to a maximum of 83 ± 5 mmHg. To determine whether sympathectomy was complete the pressure response to tyramine (200 µg/kg) was measured before and after 6-hydroxydopamine infusion.

On the day of the experiment fetal and maternal catheters were flushed with warm saline (0.9%, w/v, 39 °C), and continuous recording of fetal and maternal descending aortic pressure and heart rate was begun, while the ewe was standing quietly in a metabolic cage. To prevent the ewe's sitting down during constriction of the snare, each thigh was put into an adjustable sling. Thus the ewe's pelvis was suspended without obstruction of uterine blood flow. After approximately 2 h control recording, blood samples were obtained from the descending aorta to measure blood gases, oxygen saturation of haemoglobin, acid-base balance and plasma concentrations of norepinephrine, epinephrine, vasopressin ($n=9$) and angiotensin II ($n=9$).

To determine the time course of circulatory centralization during and after acute asphyxia and the effects of sympathectomy on it, blood flow to fetal organs and the distribution of combined ventricular output was measured by injecting six batches of differently isotope-labelled microspheres (^{141}Ce, ^{114}Ind, ^{113}Sn, ^{103}Ru, ^{95}Nb and ^{46}Sc, 16 µm diameter, New England Nuclear) into the inferior vena cava (Rudolph and Heymann 1967; Heymann et al. 1977). Asphyxia was caused by arrest of uterine blood flow with the snare. The first five microsphere injections were made at short intervals (Jensen et al. 1985a, 1987c), i.e. 75 s before (control measurement " −1"), at 1 and 2 min during arrest of uterine blood flow, and at 1, 2 and 28 min after release of the snare (recovery period; Fig.1a,b). In one intact fetus blood flows were measured at −1, 1, 2, 3 and 4 min only. Reference blood samples were withdrawn from both a carotid and a femoral artery at a rate of 2.5 ml/min during control and asphyxia over 270 and 75 s respectively. This volume of blood was simultaneously replaced by maternal blood maintained at 39 °C in a waterbath.

Measurements

Fetal heart rate and arterial blood pressure, amniotic fluid pressure, and maternal arterial blood pressure were recorded on a polygraph (Hellige, FRG). Complete occlusion of the ewe's descending aorta by the snare throughout the 2-min study period was checked by the fall in pressure distal to the obstruction (Fig.1a,b). Before each blood flow measurement a blood sample was withdrawn from the descending aorta and analysed for blood gases, acid-base balance (Technicon BG IA, FRG; OSM 2, Radiometer, Denmark), lactate (Roche), vasopressin, angiotensin II (RIA), and catecholamine concentrations. Plasma concentrations of catecholamines were determined by reversed phase ion-pair high-performance liquid chromatography with electrochemical detection (HPLC-ECD), with a detection limit of 5 pg, an intra-assay variance below 5% and an inter-assay variance below 10%. A detailed description of the HPLC-ECD catecholamine assay is given by Jelinek and Jensen elsewhere in this book.

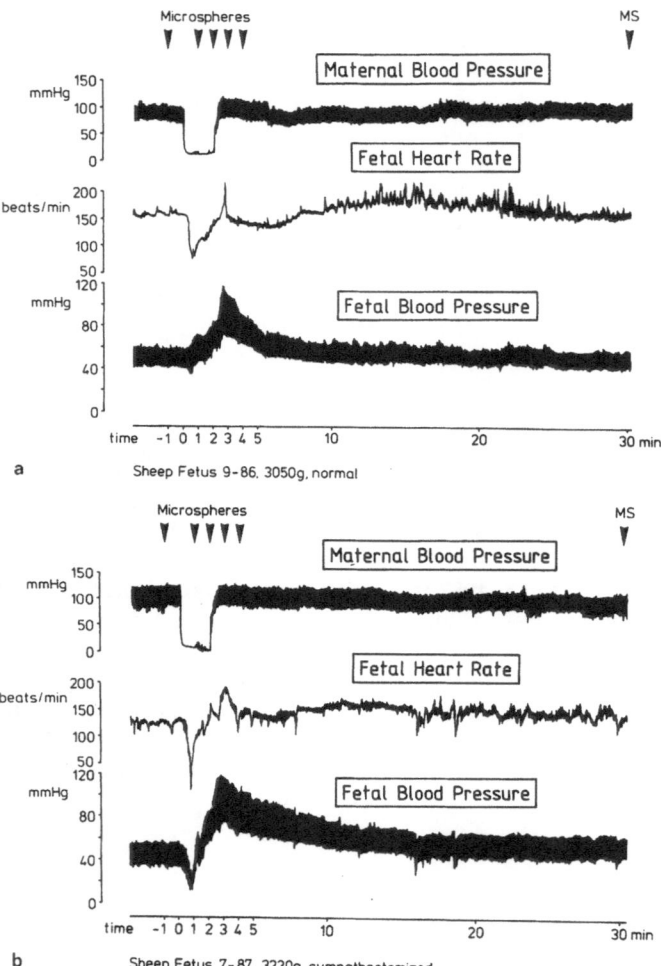

Fig. 1 a, b. Experimental protocol for studying the dynamics of circulatory centralization in
a intact and **b** chemically sympathectomized fetal sheep near term during and after acute asphyxia caused by arrest of uterine blood flow for 2 min. The distribution of combined ventricular output and blood flow to fetal organs was measured by injecting isotope-labelled microspheres (16 μm diam.) at −1, 1, 2, 3, 4 and 30 min. Uterine blood flow was arrested by
constricting a snare around the maternal aorta below the renal artery, as indicated by the fall
in maternal arterial blood pressure distal to the obstruction (*top* tracing). Note that in intact fetuses (**a**) arterial blood pressure increases progressively during asphyxia (*bottom* tracing),
whereas in sympathectomized fetuses (**b**) arterial blood pressure falls transiently, reaching a
nadir at 1 min of asphyxia

After the experiment a lethal dose of sodium pentobarbitone was given to the ewe and the fetuses were perfused with 300 ml of formalin (15%, w/v, saline). Fetal organs and cotyledons were weighed and placed in vials, which were filled to the same height to reduce variations in geometry. The intestines were separated from the mesentery, opened and cleared of contents. Paired organs (lungs, kidneys and adrenals) were counted separately, as were the right and left sides of the cerebrum and of the brain stem parts. No significant preferential streaming of microspheres was found. Specimens of the skin and muscle were taken from the hips and shoulders of each side; there were no differences in blood flow between the intact side and the side on which the femoral or brachial arteries were catheterized. Upper and lower carcass were carbonized and aliquots placed in vials.

The solid-state semiconductor gamma counter [Ge (Li)] used had a high energy resolution of about 3 keV and was connected to a multichannel (2048) pulse height analyser (Nuclear Data Inc.). The results were normalized with respect to time and sample weight, punched on paper tape, transferred to a disc file and processed with a programmable computer (PDP 11/45) (Winkler et al. 1982).

Calculations

Fetal combined ventricular output and blood flow to the various organs were calculated from both counts of the injected nuclide recovered in fetal organs and placenta and counts in the appropriate reference samples, and from the withdrawal rate of the reference sample (Rudolph and Heymann 1967; Heymann et al. 1977). Portal venous blood flow was calculated by summing the actual blood flow to all gastrointestinal organs, including stomach, small and large gut, mesentery, pancreas and spleen. The percentage of combined ventricular output distributed to a given organ (%) was calculated by dividing absolute blood flow to that organ (ml/min) times 100 by combined ventricular output (ml/min). The vascular resistance was calculated by dividing arterial blood pressure (corrected for amniotic fluid pressure) by blood flow and was expressed in $mmHg/ml \times min$ per 100 g of tissue; blood flows lower than $1 \, ml/min \times 100$ g were treated as $1 \, ml/min \times 100$ g.

Blood flow changes associated with changes in the electrocortical state of activity, e.g. those observed in both brain stem and gastrointestinal organs on transition from high to low voltage (Jensen et al. 1986), were not taken into account, because they are small as compared to those observed during arrest of uterine blood flow (Jensen et al. 1985a, 1987c).

Results are given as means $\pm$ SE. The data were analysed statistically by two-way multivariate analysis of variance for repeated measures and by Student's t tests. To obtain appropriate p values in a repeated measures design a Bonferroni correction was applied (Miller 1966).

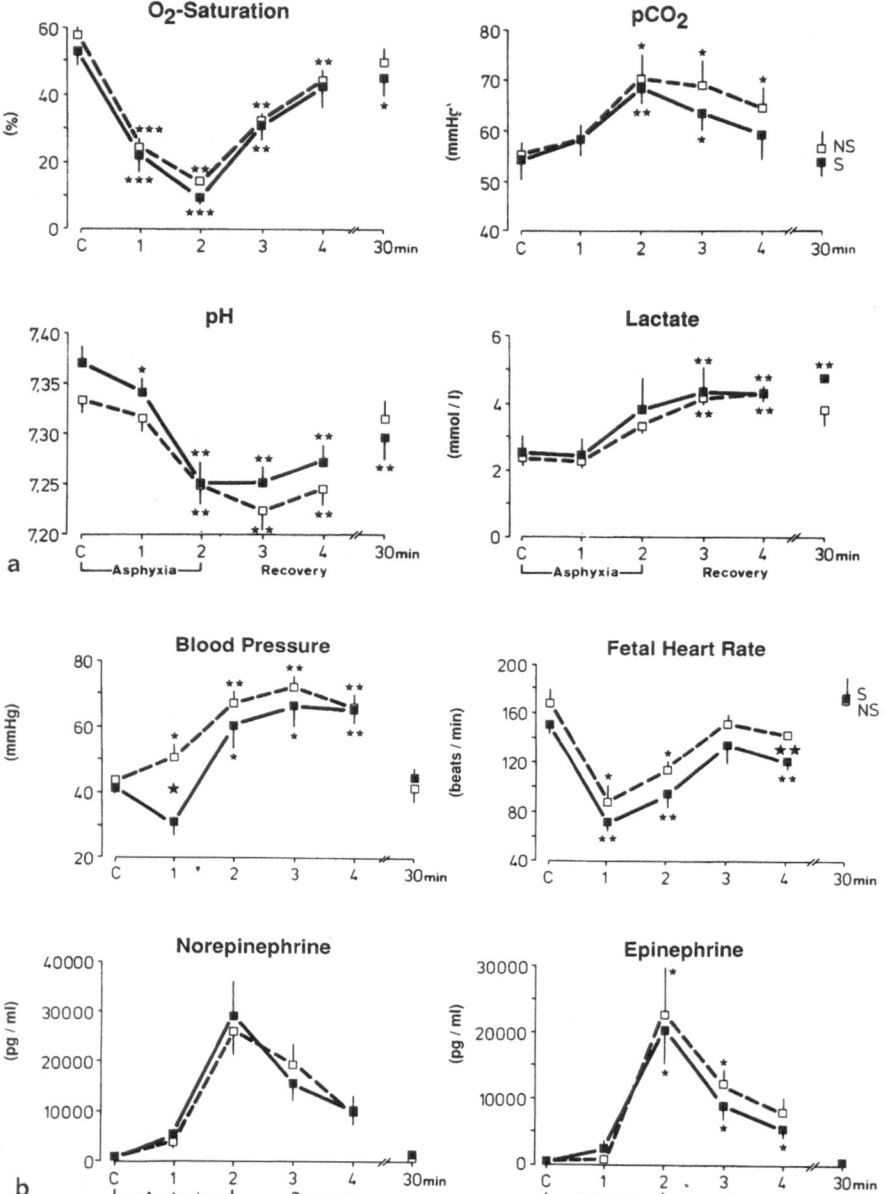

Fig. 2 a, b. Changes in **a** arterial O_2 saturation of haemoglobin, pCO_2, pH, and lactate concentrations and **b** in arterial blood pressure, heart rate and norepinephrine and epinephrine concentrations before, during and after acute asphyxia caused by arrest of uterine blood flow for 2 min in intact (--- □ ---) and sympathectomized (— ■ —) fetal sheep near term. Note that, unlike in intact fetuses, in sympathectomized fetuses the arterial blood pressure decreases after 1 min of asphyxia (see Fig. 1 b). Note also that there were no significant differences in plasma catecholamine concentrations between groups, suggesting that plasma norepinephrine and epinephrine derive largely from the adrenal medulla, which is not affected by sympathectomy

Results

Dynamics of Circulatory Centralization in Intact Fetuses

In the control period heart rate, arterial blood pressure, blood gases, pH and lactate, norepinephrine and epinephrine concentrations were in the normal range for chronically catheterized fetal sheep near term. Arrest of uterine blood flow for 2 min decreased fetal heart rate, arterial O_2 saturation of haemoglobin and pH and increased fetal arterial blood pressure, PCO_2 and lactate, norepinephrine and epinephrine concentrations (Fig. 2a, b).

After 1 min of asphyxia combined ventricular output fell by 35% ($p \leq 0.05$); however, the portion of it distributed to placenta and heart increased by 49% and 336%, respectively, whereas that to both adrenals and total brain did not change (Fig. 3a). The distribution of the combined ventricular output to various parts of the brain and its change during asphyxia varied (Fig. 3b). The portion distributed to both spinal medulla and midbrain increased progressively after 1 and 2 min asphyxia, that to the cerebrum and cerebellum did not change, and that to the choroid plexus decreased (Fig. 3b).

After 2 min asphyxia combined ventricular output was reduced by 42% ($P \leq 0.01$; Fig. 3a), but its redistribution towards central organs resulted in significant increases in blood flow to the heart (300%, $p \leq 0.001$), midbrain (81%, $p \leq 0.05$), medulla (148%, $p \leq 0.05$), no change in blood flow to the adrenals, cerebrum, and cerebellum, and a decrease in that to the choroid plexus (-86%, $p \leq 0.001$).

In the immediate recovery period, i.e. 1 and 2 min after arrest of uterine blood flow had been released, arterial blood pressure was still high (Fig. 2b), blood flows to the heart and brain stem gradually returned towards normal, those to the cerebrum, cerebellum and adrenals increased, and blood flow to the choroid plexus remained poor (-66%, $p \leq 0.01$). At 30 min, i.e. after 28 min recovery, these values were no different from control values.

In all peripheral organs of the upper and lower body segment, including carcass (-72%, $p \leq 0.001$), skeletal muscle, total gastrointestinal tract and portal vein (-95%, $p \leq 0.001$), kidneys, spleen, scalp, and body skin, the proportion of combined ventricular output and actual blood flow decreased to very low values during asphyxia, and recovered gradually thereafter (Figs. 4, 5).

The time course and the direction of changes in both distribution of combined ventricular output and blood flow to peripheral organs during acute asphyxia were very similar to those observed in the scalp and body skin, indicating that skin blood flow is an index of circulatory redistribution (Figs. 4, 5), as suggested previously (Jensen and Künzel 1980; Jensen et al. 1985a, 1987a, 1987b). However, there were some differences between blood flows to the two cutaneous areas. First, blood flow to the scalp did not decrease as rapidly as that to the body skin (Fig. 4), resulting in a significant blood flow difference ($p \leq 0.01$) between the two cutaneous areas after 1 min of asphyxia. Furthermore, after 2 min of asphyxia, blood flow to the scalp

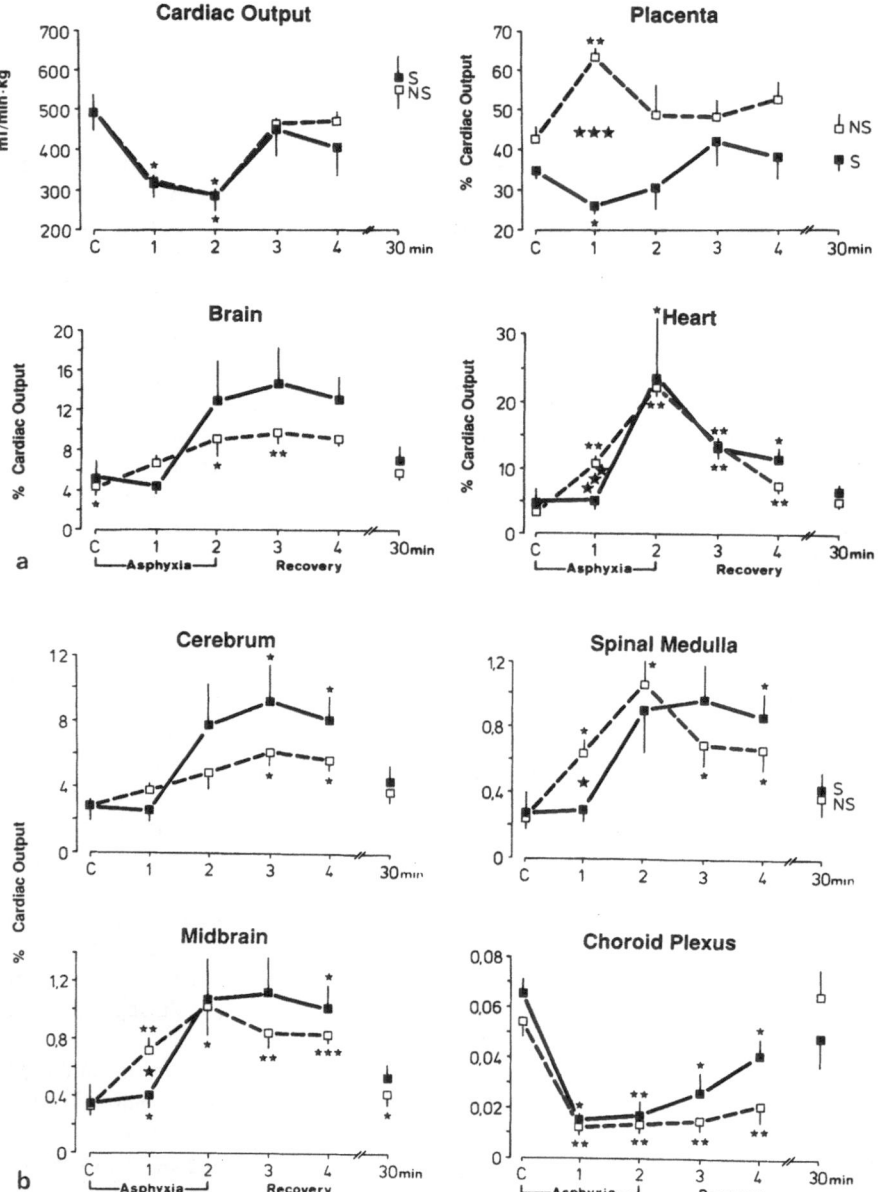

Fig. 3a, b. Changes in combined ventricular output (ml/min per kg) and in its proportional distribution (%) to **a** the placenta, total brain and heart and **b** the cerebrum, mid brain, spinal medulla and choroid plexus before, during and after acute asphyxia caused by arrest of uterine blood flow for 2 min in intact (--- □ ---) and sympathectomized (— ■ —) fetal sheep near term. ★ $p \leq 0.05$; ★★ $p \leq 0.01$; ★★★ $p \leq 0.001$; *small asterisks,* difference vs. control; *large asterisks,* difference between groups

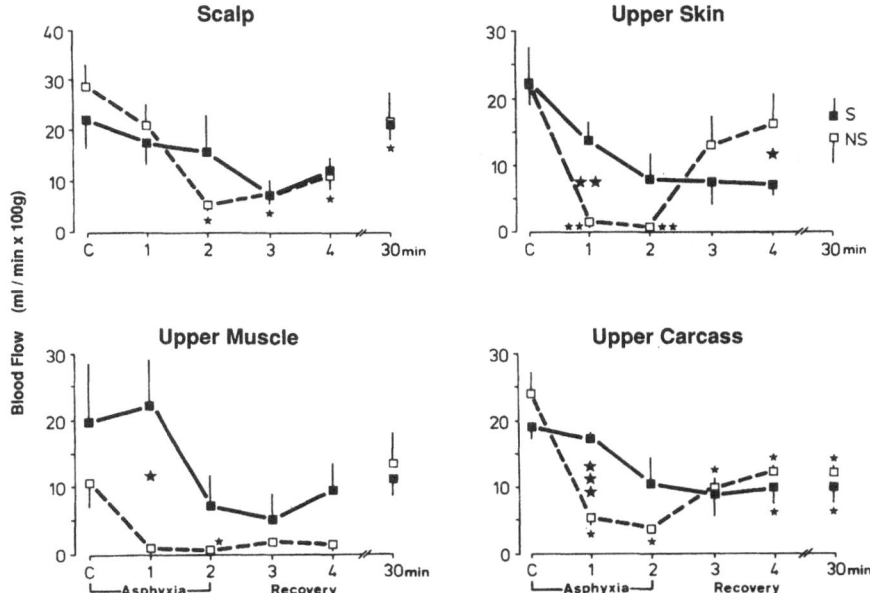

Fig. 4. Changes in blood flow (ml/min × 100 g tissue) to the scalp, skin, skeletal muscle and carcass of the upper body segment before, during and after acute asphyxia caused by arrest of uterine blood flow for 2 min in intact (--- □ ---) and sympathectomized (− ■ −) fetal sheep near term. Note that in intact fetuses the blood flow changes in the scalp and skin during asphyxia reflect those in the carcass and gastrointestinal tract (see Fig. 5). ★ $p \leq 0.05$; ★ ★ $p \leq 0.01$; ★ ★ ★ $p \leq 0.001$; *small asterisks,* difference between groups

did not fall as much as that to the body skin ($p \leq 0.001$). Finally, scalp blood flow did not recover after asphyxia as rapidly as the blood flow to the body skin: it was reduced significantly not only after 1 and 2 min, but also after 28 min of recovery (Fig. 4). These differences in both time course and magnitude of blood flow changes during and after acute asphyxia, which have been observed previously (Jensen et al. 1985 a, 1987 c), suggest that vascular control in the scalp may be different from that in the body skin.

Dynamics of Circulatory Centralization in Sympathectomized Fetuses

There were no significant differences between intact and sympathectomized fetuses with regard to blood gases, acid-base balance, and plasma concentrations of nor-epinephrine and epinephrine (Fig. 2 a, b). However, unlike in intact fetuses, in which mean arterial blood pressure increased progressively during asphyxia (Fig. 1 a), in sympathectomized fetuses blood pressure decreased transiently (−25 %) with de-

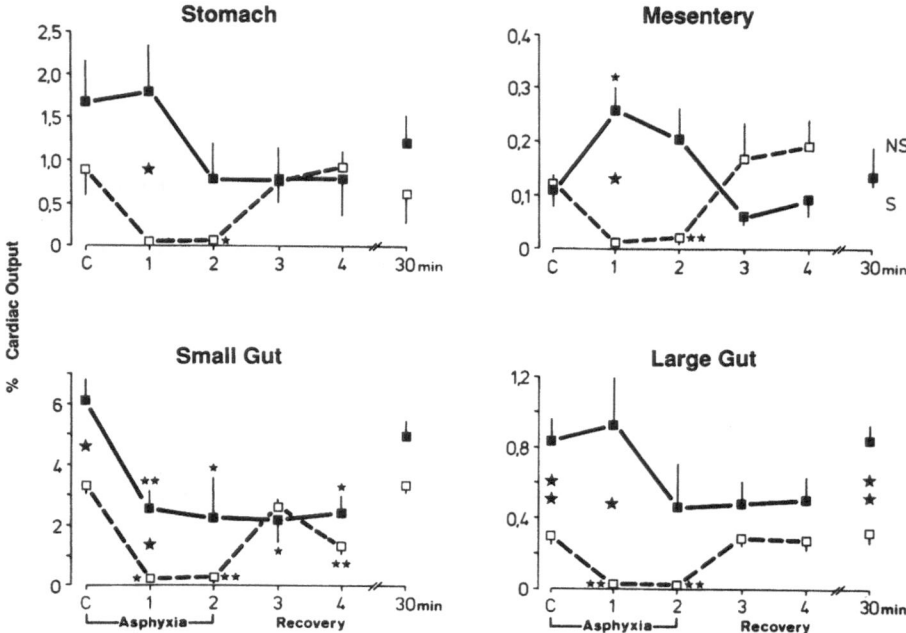

Fig. 5. Changes in the percentage combined ventricular output distributed to the stomach, mesentery and small and large intestines before, during and after acute asphyxia caused by arrest of uterine blood flow for 2 min in intact (--- □ ---) and sympathectomized (— ■ —) fetal sheep near term. ★ $p \le 0.05$; ★★ $p \le 0.01$; ★★★ $p \le 0.001$; *small asterisks*, difference vs. control; *large asterisks*, difference between groups

creasing heart rate (Fig. 1 b), resulting in a significant difference between the groups at 1 min of asphyxia (Fig. 2 b). Furthermore, vasopressin concentrations (-55%, $p \le 0.05$) and heart rate (-15%, $p \le 0.05$) were lower in sympathectomized than in intact fetuses after 2 min of asphyxia and of recovery.

In the control period adrenal blood flow (97%, $p \le 0.05$; not illustrated) and the portion of combined ventricular output distributed to the total gastrointestinal tract, including stomach, small and large intestines, mesentery, spleen and pancreas, and hence that to the portal vein, were significantly higher (78%, $p \le 0.05$) in sympathectomized than in intact fetuses (Fig. 5).

After 1 and 2 min of asphyxia combined ventricular output decreased progressively by 36% and 43% respectively ($p \le 0.001$), and recovered thereafter, and there were no differences between groups (Fig. 3 a). However, there were a number of differences between intact and sympathectomized fetuses in both distribution of cardiac output and actual blood flow after 1 min of asphyxia (Fig. 3 a).

First, unlike in intact fetuses, the portion of cardiac output distributed to the total carcass increased from 31% at control to 48% at 1 min ($p \le 0.01$; Fig. 6) and that

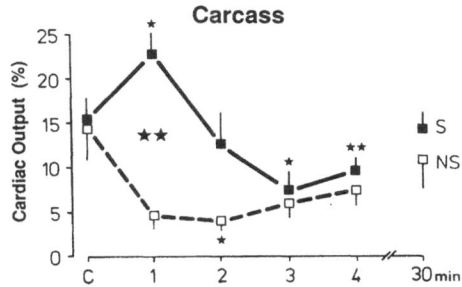

Fig. 6. Changes in the percentage of combined ventricular output distributed to the carcass of the upper body segment before, during and after acute asphyxia caused by arrest of uterine blood flow for 2 min in intact (--- □ ---) and sympathectomized (— ■ —) fetal sheep near term. Note that in intact fetuses the percentage of combined ventricular output distributed to the upper and lower carcass decreases at 1 min of asphyxia, whereas it increases in sympathectomized fetuses, in which the percentage distributed to the placenta is decreased (see Fig. 3 a). ★ $p \le 0.05$; ★★ $p \le 0.01$ ★★★ $p \le 0.001$; *small asterisks*, difference vs. control; *large asterisks*, difference between groups

to the placenta decreased from 35% to 26% ($p \le 0.01$; Fig. 3 a). Secondly, the portion of combined ventricular output directed to the adrenals and to peripheral organs, including those of the gastrointestinal tract (Fig. 5) and portal vein, kidneys and spleen, was significantly higher, and that to both heart and brain stem was lower in sympathectomized fetuses (Fig. 3).

A detailed analysis of the distribution of the combined ventricular output to various parts of the fetal brain and spinal cord after 1 min of asphyxia is given in Fig. 7. It shows that both brain stem and spinal cord are predominantly affected by the redistribution of the combined ventricular output towards carcass and gastrointestinal organs caused by ablation of the peripheral sympathetic nervous system. Sympathectomy reduced specifically the portion of combined ventricular output distributed to parts of the brain stem from the mid medulla rostrally to the hypothalamus and that to the upper spinal enlargement (Fig. 7).

As a result of the difference in the distribution of combined ventricular output between intact and sympathectomized fetuses at 1 min of asphyxia, blood flow to the placenta was maintained in the previous and decreased in the latter group. Furthermore, blood flow to the adrenals and to peripheral organs, including those of the gastrointestinal tract, carcass and body skin, but not scalp, were higher and those to the heart and brain stem parts were lower after sympathectomy.

The effects of chemical sympathectomy on the time course and magnitude of fetal circulatory responses to acute asphyxia were even more conspicuous when the calculated vascular resistance was evaluated. In intact fetuses, vascular resistance of most of the peripheral organs increased to very high values after 1 min of asphyxia and decreased thereafter (Fig. 8). This rapid peripheral vasoconstriction was blunted by sympathectomy, resulting in significant differences between the two groups in

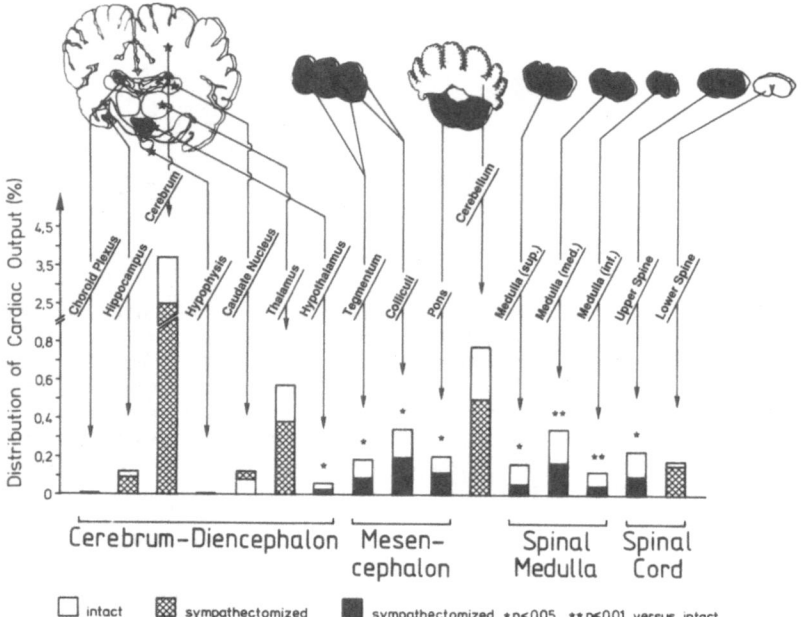

Fig. 7. Changes in the percentage of combined ventricular output distributed to various parts of the brain and spinal cord after 1 min of asphyxia, caused by arrest of uterine blood flow in intact *(open bars)* and sympathectomized (*black bars* if significantly different, *cross-hatched bars* if not) fetal sheep near term. Note that in sympathectomized fetuses the percentage of combined ventricular output distributed to a number of brain stem parts and to the upper spinal enlargement is significantly reduced at 1 min of asphyxia (*black bars* correspond to *black areas* shown above the graph)

the gastrointestinal organs, carcass, skeletal muscle, and body skin at both 1 and 2 min of asphyxia (Fig. 8). Only in the scalp, in which vascular resistance did not change after 1 min, but increased after 2 min of asphyxia, were there no differences between intact and sympathectomized fetuses (Fig. 8a).

Unlike the adrenals, in which vascular resistance was lower at control, throughout asphyxia and after 1 min of recovery, there were no significant group differences in the heart, brain and lungs (not illustrated).

Summary and Conclusions

We conclude that in intact fetal sheep near term, arrest of uterine blood flow for 2 min causes a rapid circulatory centralization at the expense of blood flow to peripheral organs, including skin and scalp. The decreases in blood flow to the fetal

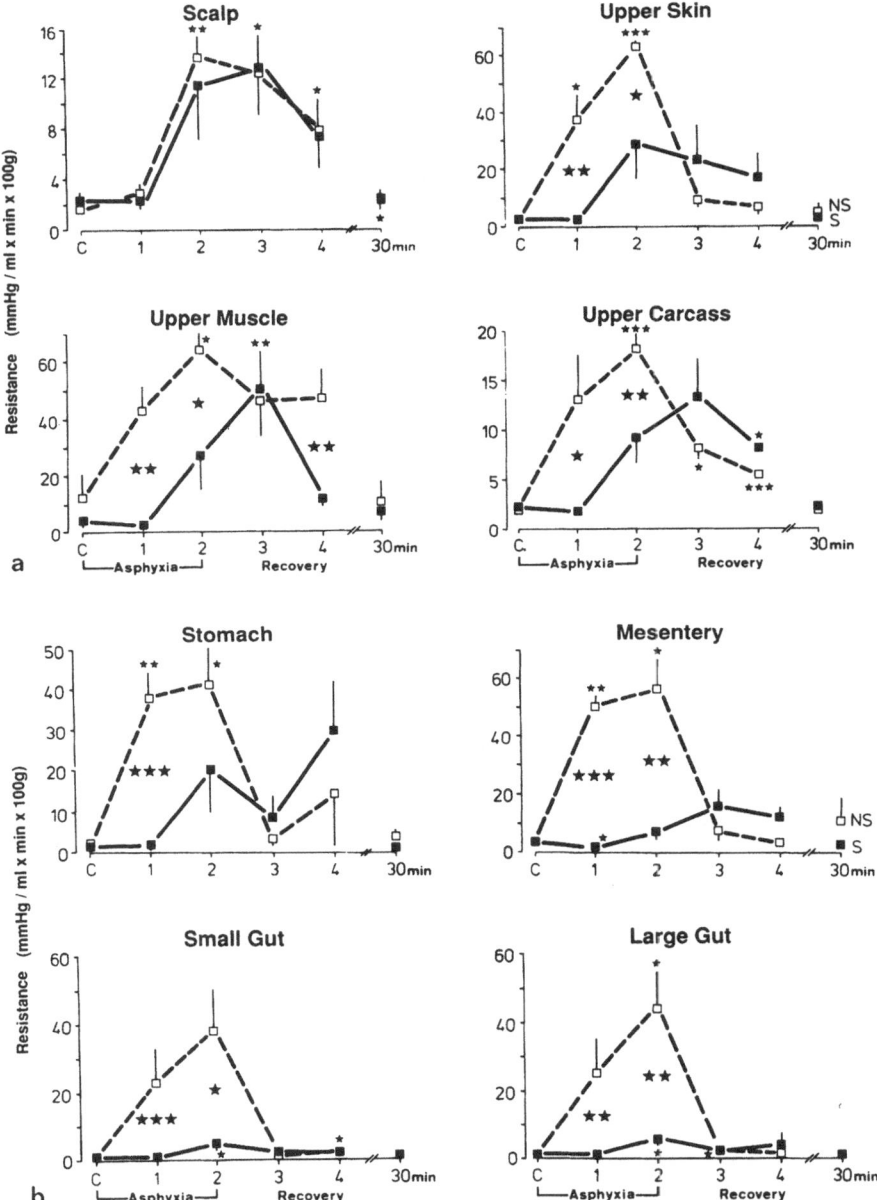

Fig. 8 a, b. Changes in vascular resistance in **a** the scalp, skin, skeletal muscle and carcass of the upper body segment and **b** the stomach, mesentery and small and large intestines before, during and after acute asphyxia caused by arrest of uterine blood flow for 2 min in intact (--- □ ---) and sympathectomized (— ■ —) fetal sheep near term. Note that in intact fetuses the changes in vascular resistance in the scalp and skin reflect those in the carcass and gastrointestinal tract. Note also that sympathectomy blunts the increase in vascular resistance during asphyxia in the carcass and, even more, in the gastrointestinal tract, but not in the scalp. ★ $p \leq 0.05$; ★★ $p \leq 0.01$; ★★★ $p \leq 0.001$; *small asterisks*, difference vs. control; *large asterisks*, difference between groups

skin and scalp during asphyxia and their increases after asphyxia reflect the blood flow changes in the principal peripheral organs and hence indicate both the development of and the recovery from fetal circulatory centralization caused by acute asphyxia.

From our studies on chemical sympathectomy we further conclude that in fetuses near term the initial phase of circulatory centralization, which increases blood flow to the heart and brain stem, is highly dependent on the sympathoneuronal system, because activation of peripheral sympathetic neurons increases arterial blood pressure by increasing vascular resistance in the carcass and gastrointestinal tract. This rapid peripheral vasoconstriction is necessary to counteract the fall in combined ventricular output during acute asphyxia and the pressure – related adverse effects it has on both flow blood and delivery of oxygen and substrates to the central fetal organs. Thus, the sympathoneuronal system is important for fetal short-term adaptation to and intact survival of asphyxia.

Acknowledgements. We thank S. and J. Jelinek, B. Maid, R. Achenbach, O. Adam, and H.-J. Klönne for their technical assistance. We are also indebted to Dr. B. Winkler and Prof. Dr. W. Schaper, Max Planck Institute for Experimental Cardiology, Bad Nauheim, FRG, for the blood flow analyses.

References

Heymann MA, Payne BD, Hoffman JIE, Rudolph AM (1977) Blood flow measurements with radionuclide - labeled particles. Prog Cardiovasc Dis 20: 55-79

Iwamoto HS, Rudolph AM, Mirkin BL, Keil LC (1983) Circulatory and humoral responses of sympathectomized fetal sheep to hypoxemia. Am J Physiol 245: H767-H772

Jensen A (1986) Die Zentralisation des fetalen Kreislaufs. Thesis, University of Giessen

Jensen A, Künzel W (1980) The difference between fetal transcutaneous PO_2 and arterial PO_2 during labour. Gynecol Obstet Invest 11: 249-264

Jensen A, Schumacher R (1987) Intrakranielle Blutungen: Ursache, Diagnostik und Bedeutung: Gynäkologe 20: 52-64

Jensen A, Künzel W, Hohmann M (1985a), Dynamics of fetal organ blood flow redistribution and catecholamine release during acute asphyxia. In: Jones CT, Nathanielsz PW (eds) The physiological development of the fetus and newborn. Academic, London, pp 405-410

Jensen A, Künzel W, Kastendieck E (1985b) Repetitive reduction of uterine blood flow and its influence on fetal transcutaneous PO_2 and cardiovascular variables. J Dev Physiol 7: 75-87

Jensen A, Bamford OS, Dawes GS, Hofmeyr GJ, Parkes MJ (1986) Changes in organ blood flow between high and low voltage activity in fetal sheep. J Dev Physiol 8: 187-194

Jensen A, Hohmann M, Künzel W (1987a) Redistribution of fetal circulation during repeated asphyxia in sheep: Effects on skin blood flow, transcutaneous PO_2, and plasma catecholamines. J Dev Physiol 9: 41-55

Jensen A, Künzel W, Kastendieck E (1987b) Fetal sympathetic activity, transcutaneous PO_2, and skin blood flow during repeated asphyxia in sheep. J Dev Physiol 9: 337-346

Jensen A, Hohmann M, Künzel W (1987c) Dynamic changes in organ blood flow and oxygen consumption during acute asphyxia in fetal sheep. J Dev Physiol 9: 543-559

Miller RG Jr (1966) Simultaneous statistical interference. McGraw-Hill, New York

Paulick R, Kastendieck E, Wernze H (1985) Catecholamines in arterial and venous umbilical blood placental extraction, correlation with fetal hypoxia, and transcutaneous partial oxygen pressure. J Perinat Med 13: 31–42

Rudolph AM, Heymnn MA (1967) The circulation of the fetus in utero: methods for studying distribution of blood flow, cardiac output and organ blood flow. Circ Res 21: 163–184

Winkler B, Stämmler G, Schaper W (1982) Measurement of radioactive tracer microsphere blood flow with NaJ (T1)- and Ge-well type detectors. Basic Res Cardiol 77: 292–300

Pancreatic Blood Flow, Insulin Concentrations, and Glucose Metabolism in Intact and Chemically Sympathectomized Fetal Sheep During Normoxaemia and Asphyxia*

U. LANG, W. KÜNZEL, and A. JENSEN[1]

Introduction

The sympathetics play an important part in the regulation of fetal and neonatal insulin secretion and glucose metabolism during normoxaemia and asphyxia (Dawes 1969). Their humoral effects are well documented: exogenous or endogenous catecholamines, e.g. those released during hypoxaemia and asphyxia, changed both insulin and glucose concentrations as well as glucose consumption, in part through α-adrenoceptor mediated mechanisms (Feige et al. 1976; Bassett 1977; Jones and Ritchie 1978; Jones et al. 1983; Jensen et al. 1987b). However, even though there is a rich sympathetic innervation of both pancreatic blood vessels and pancreatic islets (Van Campenhout 1927; Simard 1935; Hickson 1970), little is known about indirect or direct neuronal effects of the sympathetics on fetal insulin secretion and glucose metabolism.

To determine whether pancreatic sympathetic neurones are involved in the regulation of insulin secretion and glucose metabolism we studied insulin and glucose concentrations and pancreatic blood flow in intact and chemically sympathectomized normoxaemic fetal sheep near term. Since sympathectomy was performed by infusing 6-hydroxydopamine (Iwamoto et al. 1983), which selectively destroys sympathetic nerve terminals by depletion of norepinephrine stores without affecting the adrenal medulla, we also addressed the question of whether a potential change in insulin secretion caused by ablation of sympathetic neurones can be reversed by increased plasma concentrations of endogenous catecholamines. Therefore, the fetuses were also studied during and after acute asphyxia caused by arrest of uterine blood flow for 2 min.

---

* This investigation was supported by the Deutsche Forschungsgemeinschaft (Je 108/4–1).
[1] Department of Obstetrics and Gynaecology, Justus-Liebig-Universität Giessen, Klinikstr. 32, D-6300 Giessen, Federal Republic of Germany

---

The Endocrine Control of the Fetus
Ed. by W. Künzel and A. Jensen
© Springer-Verlag Berlin Heidelberg 1988

Material and Methods

Experimental Protocol

The preparation of the animals, the experimental protocol, including sympathecto-my (Iwamoto et al. 1983), and the modified microsphere method used for blood flow measurements (Jensen et al. 1987a) are reported in detail by Jensen and Lang elsewhere in this book.

Briefly, six intact (weight 3050 ± 390 g) and six sympathectomized (weight 3295 ± 575 g) normoxaemic unanaesthetized chronically catheterized fetal sheep were studied in utero at 125–135 days gestation (term is 147 days). After approximately 2 h of control recordings of physiologic variables, in the absence of uterine contractions both blood samples were drawn from the descending aorta and pancreatic blood flow was measured by injecting a batch of isotope-labelled microspheres into the inferior vena cava. Then uterine blood flow was arrested for 2 min by constricting a snare and the blood flow measurements were repeated at 1, 2, 3, 4 and 30 min. Blood samples for insulin determinations were drawn at 0, 4, 10 and 30 min, those for measurements of glucose and lactate concentrations, blood gases, oxygen saturation of haemoglobin, acid-base balance and plasma concentrations of norepinephrine and epinephrine at 0, 1, 2, 3, 4, 5, 10 and 30 min.

Measurements

Fetal heart rate and arterial blood pressure, amniotic fluid pressure, and maternal arterial blood pressure were recorded on a polygraph (Hellige, FRG). Blood samples were analysed for insulin, glucose concentration (Beckmann glucose analyser, FRG), lactate concentration (Roche lactate analyser), blood gases, acid-base balance (Technicon BG IA, FRG; OSM 2, Radiometer, Denmark) and norepinephrine and epinephrine concentrations. A detailed description of the HPLC-ECD catecholamine assay is given by Jelinek and Jensen elsewhere in this book. Fetal insulin concentrations were measured by radioimmunoassay (Pharmacia Insulin RIA 100, Pharmacia Diagnostics, Uppsala, Sweden) with a detection limit of 2 µU/ml and an intra-assay variance of 6% (Prof. Dr. W. Bretzel, Department of Internal Medicine, University of Giessen, FRG). The assay standard was calibrated against research standard A for insulin, human, for immuno-assay 66/304 from the WHO International Laboratory for Biological Standards.

To determine pancreatic blood flow before, during and after acute asphyxia six batches of differently isotope-labelled microspheres ([141]Ce, [114]Ind, [113]Sn, [103]Ru, [95]Nb and [46]Sc, 16 µm diameter, New England Nuclear) were injected into the inferior vena cava (Rudolph and Heyman 1967; Heymann et al. 1977). The first five microsphere injections were made at short intervals (Jensen et al. 1985, 1987a), i.e. 75 s before arrest of uterine blood flow (control period), at 1 and 2 min during arrest

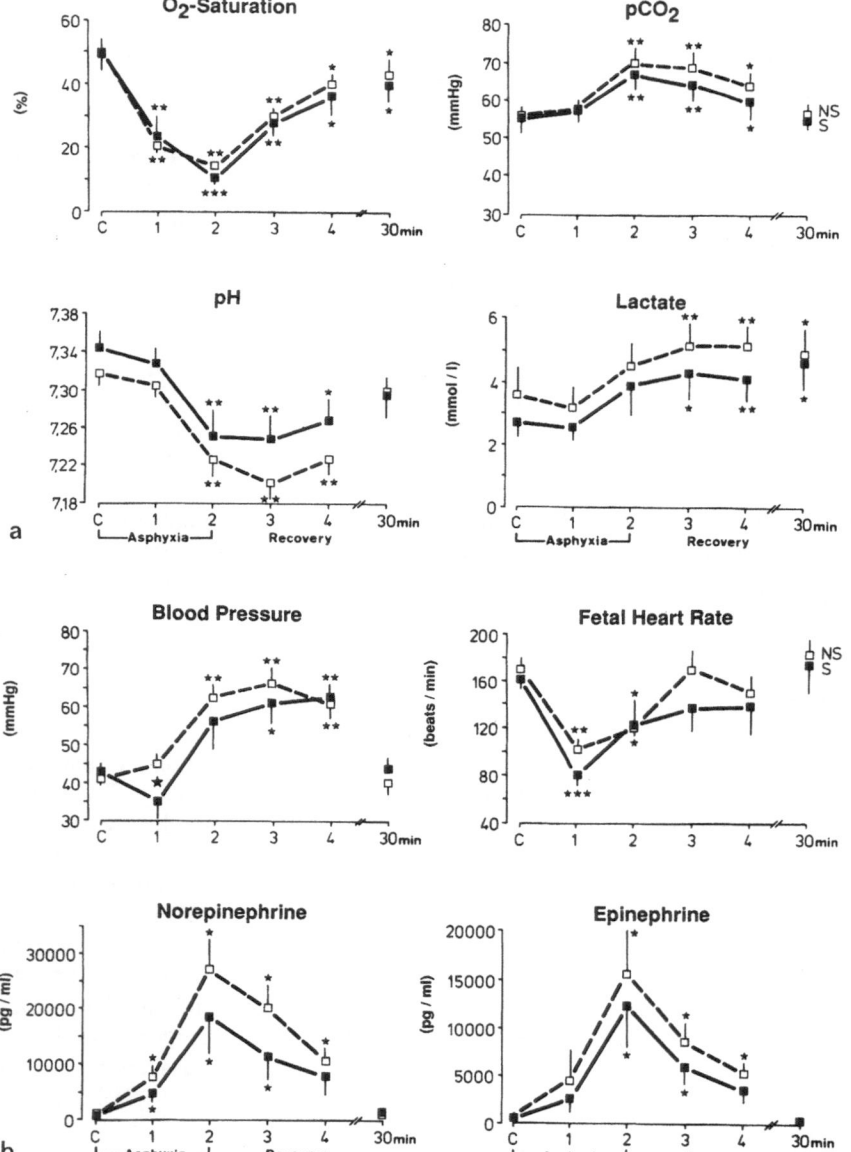

Fig. 1a, b. Changes in arterial O_2 saturation of haemoglobin and concentrations of pCO_2, pH and lactate (**a**), and in arterial blood pressure, heart rate and concentrations of norepinephrine and epinephrine (**b**) before, during and after acute asphyxia caused by arrest of uterine blood flow for 2 min in intact (- - - □ - - -) and sympathectomized (- ■ -) fetal sheep near term. Note that there were no significant differences in these variables between groups, except for blood pressure, which increased in intact, but decreased in sympathectomized fetuses after 1 min of asphyxia. ★ $p < 0.05$, ★★ $p < 0.01$, ★★★ $p < 0.001$; *small asterisks* different vs control, *large asterisk* difference between groups and at 1, 2 and 28 min after release of the snare (recovery period). Reference blood samples were withdrawn from both a carotid and a femoral artery at a rate of 2.5 ml/min during control and asphyxia over 270 and 75 s respectively. This volume of blood was simultaneously replaced by maternal blood maintained at 39 °C in a waterbath.

Calculations

The pancreatic vascular resistance was calculated by dividing arterial blood pressure (corrected for amniotic fluid pressure) by blood flow and was expressed in mmHg/ml × min per 100 g of tissue; blood flows below 1 ml/min × 100 g were treated as 1 ml/min × 100 g.

Changes in pancreatic blood flow associated with changes in the electrocortical activity on transition from high to low voltage (Jensen et al. 1986) were not taken in-

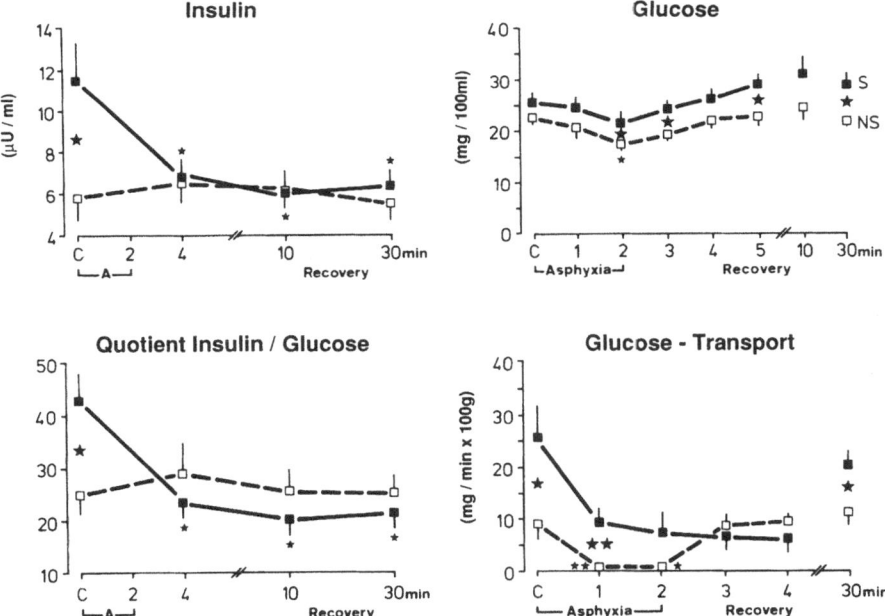

Fig. 2. Changes in insulin and glucose concentrations, insulin/glucose quotient, and glucose transport to the pancreas before, during and after acute asphyxia caused by arrest of uterine blood flow for 2 min in intact (- - - □ - - -) and sympathectomized (— ■ —) fetal sheep near term. Note that under resting conditions insulin concentrations and the insulin/glucose quotient are significantly higher in sympathectomized than in intact fetuses, indicating that sympathetic neurones are involved in the regulation of insulin concentrations by changing the sensitivity of the pancreatic islets to glucose. Note also that these changes are reversed by asphyxia when catecholamine concentrations are high (see Fig. 1b). ★ $p < 0.05$, ★ ★ $p < 0.01$; *small asterisks* different vs control, *large asterisks* differences between groups

to account, because they are small as compared to those observed during arrest of
uterine blood flow (Jensen et al. 1985, 1987a).

Results are given as means ± SE. The data were analysed statistically using two-
way multivariate analysis of variance for repeated measures and Student's *t* tests.
To obtain appropriate *p* values in a repeated measures design, a Bonferroni correction was applied (Miller 1966).

Results

In the control period, glucose and lactate concentrations, blood gases, pH, norepinephrine and epinephrine concentrations, heart rate and arterial blood pressure
were in the normal range for resting chronically catheterized fetal sheep near term,
and there were no significant differences between intact and sympathectomized fetuses (Figs. 1 a b, 2). However, insulin concentrations, the insulin/glucose quotient,

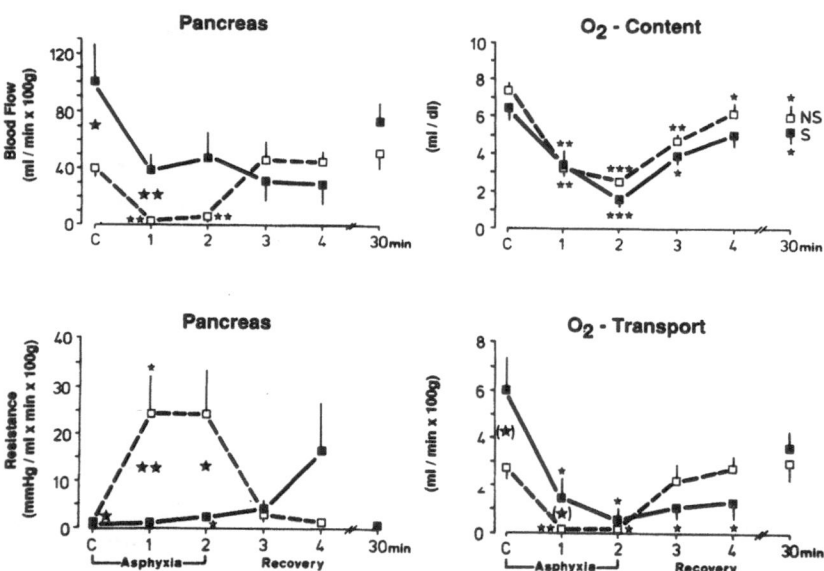

Fig. 3. Changes in pancreatic blood flow and vascular resistance, arterial O_2 content and pancreatic oxygen transport before, during and after acute asphyxia caused by arrest of uterine
blood flow for 2 min in intact (- - - □ - - -) and sympathectomized (- ■ -) fetal sheep near
term. Note that under resting conditions blood flow and oxygen transport to the pancreas are
significantly higher in sympathectomized than in intact fetuses, suggesting that sympathetic
neurones in part may act indirectly on the regulation of insulin concentrations and glucose
metabolism by changing pancreatic blood flow (see Fig. 2). ★ $p < 0.005$, ★ ★ $p < 0.01$, ★ ★ ★
$p < 0.001$; *small asterisks* different vs control, *large asterisks* differences between groups

pancreatic blood flow and the delivery of oxygen and glucose to the pancreas were higher and pancreatic vascular resistance was lower in sympathectomized than in intact fetuses (Figs. 2, 3). Interestingly, there was a close positive correlation between glucose and insulin concentrations in sympathectomized but not in intact fetuses (Fig. 4).

Arrest of uterine blood flow for 2 min caused asphyxia. Glucose concentrations, arterial O_2 saturation of haemoglobin and pH decreased, and fetal blood pCO_2, lactate, norepinephrine and epinephrine concentrations increased (Figs. 1 a, b, 2). There were no significant differences between groups. But, unlike in intact fetuses, in which mean arterial blood pressure increased progressively during asphyxia, in sympathectomized fetuses blood pressure decreased transiently (-25%) with decreasing heart rate, resulting in a significant difference between the groups at 1 min of asphyxia (Fig. 1 b). Furthermore, glucose concentrations at 2, 3, 5 and 30 min, but not at 4 and 10 min, were significantly higher in sympathectomized than in intact fetuses. (Fig. 2).

During asphyxia in sympathectomized fetuses, insulin concentrations, the insulin/glucose quotient, pancreatic blood flow and the delivery of oxygen and glucose to the pancreas decreased from high contol values, so that there were no significant differences between these and intact fetuses at 4 min (Figs. 2, 3). The decrease in pancreatic blood flow and the increase in vascular resistance observed in intact fetuses after 1 and 2 min asphyxia was blunted by sympathectomy, resulting in significant differences between the groups in these two variables (Fig. 3).

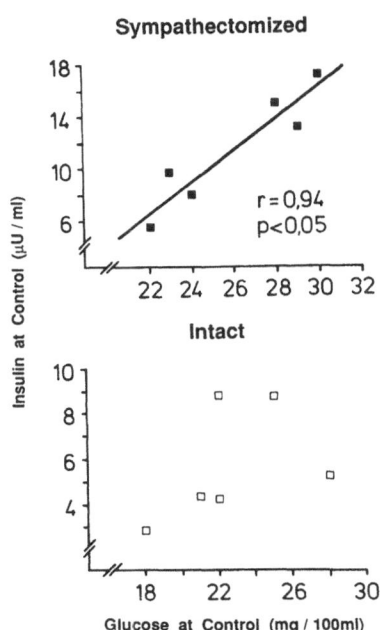

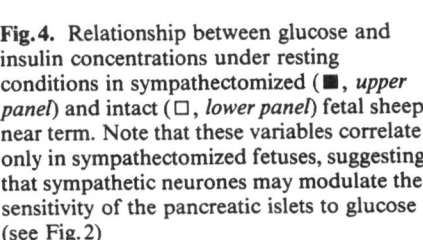

Fig. 4. Relationship between glucose and insulin concentrations under resting conditions in sympathectomized (■, *upper panel*) and intact (□, *lower panel*) fetal sheep near term. Note that these variables correlate only in sympathectomized fetuses, suggesting that sympathetic neurones may modulate the sensitivity of the pancreatic islets to glucose (see Fig. 2)

Summary and Conclusions

In unanaesthetized fetal sheep near term under resting conditions, chemical sympathectomy increases insulin concentrations, the insulin/glucose quotient, pancreatic blood flow, and the delivery of both oxygen and glucose to the pancreas. These changes are reversed in part during asphyxia when plasma catecholamine concentrations are high. Thus, we conclude that pancreatic sympathetic neurones *and* circulating catecholamines are involved in the regulation of fetal insulin concentrations and glucose metabolism through direct and/or indirect mechanisms.

Acknowledgements. We thank S. and J.Jelinek, B.Maid, R.Achenbach, O.Adam, and H.-J.Klönne for their technical assistance. We are also indepted to Prof. Dr. W.Bretzel, Department of Internal Medicine, University of Giessen, FRG, for the insulin analyses.

References

Bassett JM (1977) Glucagon, insulin and glucose homeostasis in the fetal lamb. Ann Rech Vet 8: 362–373
Dawes GS (1969) Foetal and neonatal physiology. Year Book Medical, Chicago, pp 141–149
Feige A, Künzel W, Cornely M, Mitzkat HJ (1976) Die Beziehung zwischen Glukosekonzentration und Säure-Basenstatus im maternen und fetalen Blut während der Geburt. Zeitschrift für Geburtshilfe und Perinatologie 180: 106–111
Heymann MA, Payne BD, Hoffman JIE, Rudolph AM (1977) Blood flow measurements with radionuclide - labeled particles. Prog Cardiovasc Dis 20: 5579
Hickson JDC (1970) The secretory and vascular response to nervous and hormonal stimulation in the pancreas of the pig. J Physiol 269: 299–322
Iwamoto HS, Rudolph AM, Mirkin BL, Keil LC (1983) Circulatory and humoral responses of sympathectomized fetal sheep to hypoxemia. Am J Physiol 245: H767–H772
Jensen A, Künzel W, Hohmann M (1985) Dynamics of fetal organ blood flow redistribution and catecholamine release during acute asphyxia. In: Jones CT, Nathanielsz PW (eds) The physiological development of the fetus and newborn. Academic, London, pp 405–410
Jensen A, Bamford OS, Dawes GS, Hofmeyer GJ, Parkes MJ (1986) Changes in organ blood flow between high and low voltage activity in fetal sheep. J Dev Physiol 8: 187–194
Jensen A, Hohmann M, Künzel W (1987a) Dynamic changes in organ blood flow and oxygen consumption during acute asphyxia in fetal sheep. J Dev Physiol 9: 543–559
Jensen A, Künzel W, Kastendieck E (1987b) Fetal sympathetic activity, transcutaneous PO_2, and skin blood flow during repeated asphyxia in sheep. J Dev Physiol 9: 337:346
Jones CT, Ritchie JWK (1978) The metabolic and endocrine effects of circulation catecholamines in fetal sheep. J Physiol 285:395–408
Jones CT, Ritchie JWK, Walker D (1983) The effects of hypoxia on glucose turnover in the fetal sheep. J Dev Physiol 5: 223–235
Miller RG Jr (1966) Simultaneous statistical interference. McGraw-Hill, New York
Rudolph AM, Heyman MA (1967) The circulation of the fetus in utero: methods for studying distribution of blood flow, cardiac output and organ blood flow. Circ Res 21: 163–184
Simard LC (1935) Les complexes sympathetico - insulaires du pancreas de l'homme adulte. CR Soc Biol (Paris) 119: 27–28
Van Campenhout E (1927) Contribution a l'étude de l'histogenese du pancreas chez quelques mammiferes. Les complexes sympathico-insulaires. Arch Biol (Liège) 37: 121
Winkler B, Stämmler G, Schaper W (1982) Measurement of radioactive tracer microsphere blood flow with NaJ (T1)- and Ge-well type detectors. Basic Res Cardiol 77: 292–300

Distribution of Catecholamines in Central and Peripheral Organs of the Fetal Guinea Pig During Normoxaemia, Hypoxaemia, and Asphyxia*

J. Jelinek and A. Jensen[1]

Introduction

During intra-uterine hypoxia and asphyxia the centralization of the fetal circulation improves fetal survival. Together with simultaneous metabolic changes, supplies of both oxygen and substrates to the essential organs are maintained (Elsner et al. 1969; Cohn et al. 1974; Jones 1977; Peeters et al. 1979; Sheldon et al. 1979; Reuss et al. 1982; Jensen et al. 1985, 1987a; Jensen and Lang, this volume).

Catecholamines, through their ability to stimulate specific adrenergic receptors, are important for many cardiovascular and metabolic responses to oxygen deficiency. Their release is caused by stimulation of the sympathoneuronal and sympathomedullary system. Therefore, plasma concentrations of catecholamines provide a valid index of short-term changes in sympathetic activity (Kopin 1978), which is increased during both hypoxia and asphyxia (Comline and Silver 1961; Comline et al. 1965; Jones and Robinson 1975; Cohen et al. 1982, 1984; Lewis et al. 1982, 1984; Gu et al. 1985; Paulick et al. 1985; Jensen et al. 1987b).

There are also close interrelations between tissue concentrations of catecholamines and the innervation of organs by the sympathetics. Norepinephrine concentrations in organs correspond to the density of noradrenergic neurons (Euler 1954; Greenberg and Lind 1961; Iversen 1967; Mandel et a. 1975). Norepinephrine concentrations therefore serve as an index of maturity of the sympathetic innervation during ontogenesis (Phillippe 1983).

Epinephrine is mainly localized in the adrenal medulla. Comparatively little is found in extra-adrenal chromaffin cells, e.g. in para-aortic ganglia and in the heart, spleen and kidneys. The fetal tissue concentrations of epinephrine depend on the activity of phenyl-ethanolamine-N-methyltransferase (PNMT), an enzyme which develops slowly during ontogenesis (Iversen 1967; Mandel et al. 1975).

The sympathetic nervous system of guinea pig fetuses near term, which are frequently used for perinatal investigations, is not well understood (Karki et al. 1962). Due to problems in studying fetal guinea pigs at rest, little is known about the function of the sympathetic nervous system during hypoxaemia and asphyxia in this

* This investigation was supported by the Deutsche Forschungsgemeinschaft (Je 108/1; Je 108/4-1).
[1] Department of Obstetrics and Gynaecology, Justus-Liebig-Universität Giessen, Klinikstr. 32, D-6300 Giessen, Federal Republic of Germany

animal model. Attempting to gain a better understanding of the responses of the sympathetic nervous system to oxygen deficiency in the fetal guinea pig, we studied plasma and tissue concentrations of catecholamines during normoxaemia, hypoxaemia and asphyxia. Specifically, we were interested in the following questions:

1. What are the concentrations of norepinephrine and epinephrine in central and peripheral organs of the fetal guinea pig during normoxaemia?
2. How do catecholamines change during hypoxaemia and asphyxia in both plasma and tissue?
3. How do plasma catecholamine concentrations correlate with tissue catecholamines?

Material and Methods

Experimental Protocol

Twenty-two guinea pigs (Purbright white) between days 59 and 61 of pregnancy (term is at 68 days) were anaesthetized with 0.15 ml of a 2% solution of xylazin (Rompun, Bayer AG, Leverkusen, FRG) and 10 mg/100 g ketamine HCl (Ketanest, Parke-Davis, Munich, FRG). After uterotomy fetal blood was obtained by cardiopuncture (Graham and Scothorne 1970) through the intact membranes for measurements of plasma catecholamines, blood gases and acid-base balance. After the blood sample was drawn the fetus was sacrificed by decapitation. In about 6 min 14 brain fractions and 15 organs were removed, frozen in liquid nitrogen and stored at −70 °C for subsequent analysis of tissue catecholamines by high-performance liquid chromatography. At the end of the experiment a lethal dose of barbiturate [Thiopental (Lentia), Hormon-Chemie, Munich, FRG] was given to the mother.

Measurements of Calculations

Plasma and tissue catecholamines were measured by reversed-phase ion-pair (RP-IP) high-performance liquid chromatography with electrochemical detection [HPLC-ECD (HPLC system by Waters Associates, Milford, Mass., USA; electrochemical detector LC 4A by Bioanalytic Systems, West Lafayette, Ind., USA)]. The potential of the glassy carbon working electrode was set at 0.7 V versus the Ag/AgCl reference electrode. The stationary phase was a microparticle (5 μm) C_{18} column RCM 100, I.D. 8 mm (Waters Associates). The mobile phase was composed of 10% (v/v) methanol, 0.1 M citric acid, 0.1 M sodium acetate (Merck, Darmstadt, FRG), 1 mM dibutylamine (Merck-Schuchard, Hohenbrunn, FRG) and 5 mM sodium octyl sulphonate (PIC B8, Waters Associates). The pH of the mobile phase was 4.65.

The frozen tissue specimens were homogenized (Mikrodismembrator II, Braun, Melsungen, FRG), and the tissue catecholamines were extracted with 0.4 M perchloric acid (Merck) containing antioxidants (Wagner et al. 1982).

Plasma and dissolved tissue catecholamines were isolated and preconcentrated by aluminum adsorption under alkaline conditions (pH 8.6) according to the method of Anton and Sayre (1962). The recovery of norepinephrine, epinephrine, dopamine and the internal standard 3,4-dihydroxybenzylamine (Sigma Chemie, Deisenhofen, FRG) was 70%-75%.

The intra-assay variance of the extraction and measurement of norepinephrine, epinephrine and dopamine was below 5%, the inter-assay variance was below 12%. The sensitivity of the HPLC-ECD system was 5 pg.

In the blood pH, O_2 saturation of haemoglobin, PO_2, PCO_2 and base excess were determined (blood gas analyser BGA II A, Technicon, Bad Vilbel, FRG; haemoxymeter OSM 2, Radiometer, Copenhagen, Denmark).

All values were divided into three groups, normoxaemic, hypoxaemic and asphyxic, using O_2 saturation of haemoglobin and blood pH as criteria. For statistical evaluation the variables were tested by analysis of variance and linear regression analysis. Comparisons between groups were made by t test, considering the whole variance (computer programme based upon Tatsuoka 1971). Logarithmic transformation was performed when necessary.

Results and Discussion

Blood Gases, Acid-Base Balance, and Catecholamines

The normoxaemic fetuses showed values for O_2 saturation, blood gases, pH (Fig. 1) and base excess (not shown) which were in the range of those known for example from chronically prepared fetal lambs near term (Cohn et al. 1974, 1982; Jones 1977; Lorijn and Longo 1980; Fisher et al. 1982; Lewis et al. 1982; Adamson et al. 1984; Court et all. 1984).

In hypoxaemic fetuses O_2 saturation of haemoglobin and PO_2 were reduced to about 50%. There were no significant changes in PCO_2, base excess or pH.

In addition to hypoxaemia in asphyxic fetuses there was an increase in PCO_2 and a decrease in base excess and pH, consistent with a combined respiratory and metabolic acidosis.

In accordance with the fairly physiological blood gas and acid-base values in the normoxaemic group, the plasma catecholamine concentrations of these fetuses were low (Fig. 2).

As a result of hypoxaemia, plasma norepinephrine and epinephrine concentrations increased threefold and 80-fold, respectively. The quotient of norepinephrine and epinephrine fell from about 13 to 4.

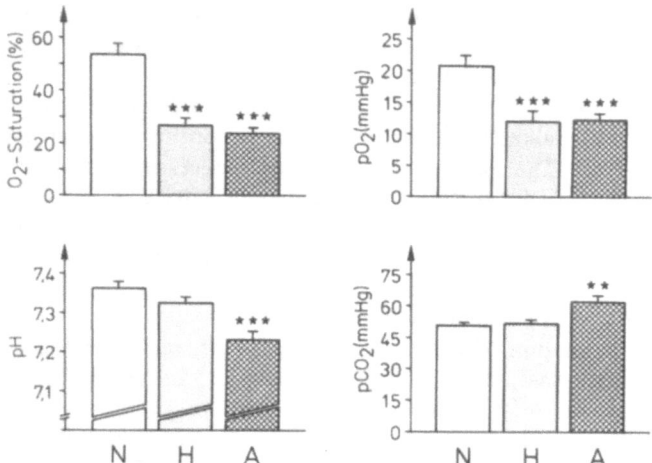

Fig. 1. O_2 saturation of haemoglobin, blood gases and pH of fetal guinea pigs near term during normoxaemia (N, $n = 5$), hypoxaemia (H, $n = 6$) and asphyxia (A, $n = 11$). Means $\pm$ SE ($\star\star$ $p \leq 0.01$; $\star\star\star$ $p \leq 0.001$)

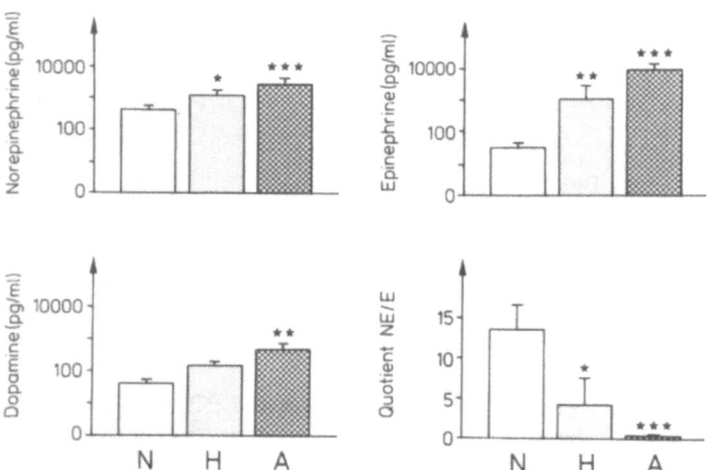

Fig. 2. Plasma catecholamine concentrations of fetal guinea pigs near term during normoxaemia ($n = 5$), hypoxaemia ($n = 6$) and asphyxia ($n = 11$). NE/E, norepinephrine/epinephrine. Means $\pm$ SE ($\star$ $p \leq 0.05$; $\star\star$ $p \leq 0.01$; $\star\star\star$ $p \leq 0.001$)

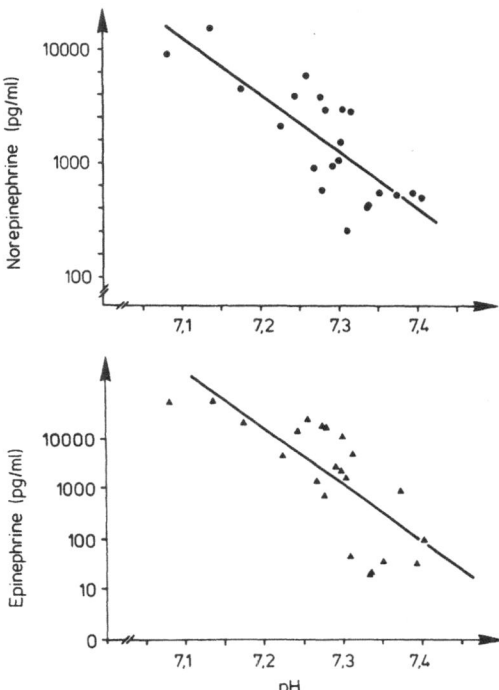

Fig. 3. Linear correlations between the blood pH and the log-norepinephrine and log-epinephrine concentrations in the plasma of fetal guinea pigs near term ($p < 0.001$)

In asphyxic fetuses the plasma catecholamine concentrations increased further. The norepinephrine/epinephrine quotient decreased to below 1.

The tremendous increase in plasma catecholamine concentrations was closely correlated to changes in blood gases and acid-base balance, e.g. the pH (Fig. 3).

Tissue Concentrations of Norepinephrine

In Fig. 4 the norepinephrine concentrations in various organs of the fetal guinea pig during normoxaemia are shown. As expected, the adrenal concentration of norepinephrine was much higher than that in any other organ. In fact, it was 33 times higher than, for example, in the heart. High concentrations - between 800 and 400 ng/g tissue - were also detected in the spleen, large and small gut, and pancreas. The norepinephrine concentrations in the kidney, brown adipose tissue, liver, skin and muscle were comparatively low. In the lung and, especially, in the placenta norepinephrine was hardly detectable. The minimal amount of norepinephrine in the placenta can be explained by both absence of sympathetic innervation (Iversen 1967) and high content of catecholamine metabolizing enzymes [monoamine oxi-

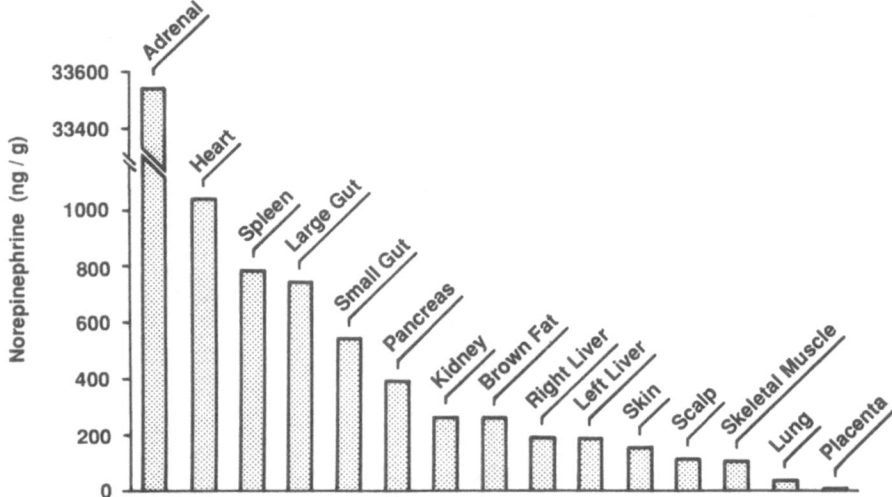

Fig. 4. Distribution of norepinephrine (ng/g fresh tissue weight) in organs of fetal guinea pigs near term during normoxaemia ($n = 5$)

dase, catecholamine-O-methyltransferase (Iisalo and Castrén 1967; Castrén and Saarikoski 1974; Chen et al. 1974; Saarikoski 1974)]. Indeed, the placenta can account for 50% and more of the plasma catecholamine clearance in the fetus (Jones and Rurak 1976).

During normoxaemia the regional distribution and the concentrations of norepinephrine in organs of the fetal guinea pig are in the range of those described for other animal species (Euler 1954; Anton and Sayre 1962; Iversen 1967; Mandel et al. 1975).

In some fetal organs norepinephrine concentrations depended on the degree of oxygen deficiency. The changes during hypoxaemia and asphyxia were variable. We observed an increase of norepinephrine concentrations in lung and skin and a decrease in concentrations in the left liver lobe, pancreas and scalp. Both increase and decrease of norepinephrine concentrations correlated with changes in blood gases and acid-base balance (Fig. 5).

During hypoxaemia and asphyxia there were also correlations between norepinephrine concentrations in the tissue and those in the plasma, particularly in those organs in which norepinephrine decreased, e. g. in the left liver lobe (Fig. 5). This might be related to sympathetic overflow (Folkow et al. 1967).

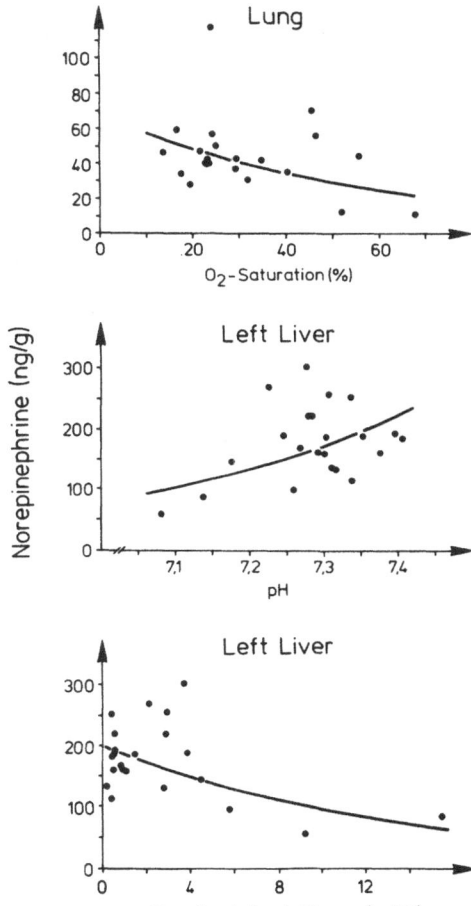

Fig. 5. Correlations between O_2 saturation, blood pH and plasma norepinephrine concentrations and norepinephrine concentrations in the lung and left liver lobe of fetal guinea pigs near term ($p < 0.05$)

Tissue Concentrations of Epinephrine

The distribution of epinephrine concentrations in various organs of the fetal guinea pig during normoxaemia is shown in Fig. 6. Considering the absolute values, which amount to nearly 460 000 ng/g tissue, epinephrine is almost exclusively localized in the adrenal. The concentration found in the fetal adrenal is in good agreement with that described for the adult guinea pig (Chatterjee and Gosh 1982). The distribution of epinephrine in the other organs is different to the pattern of the distribution observed for norepinephrine. The highest concentrations were found in the pancreas, followed by those found in the kidney, heart and brown adipose tissue. With levels below 10 ng/g tissue, the concentrations in the skeletal muscle, spleen, skin, liver,

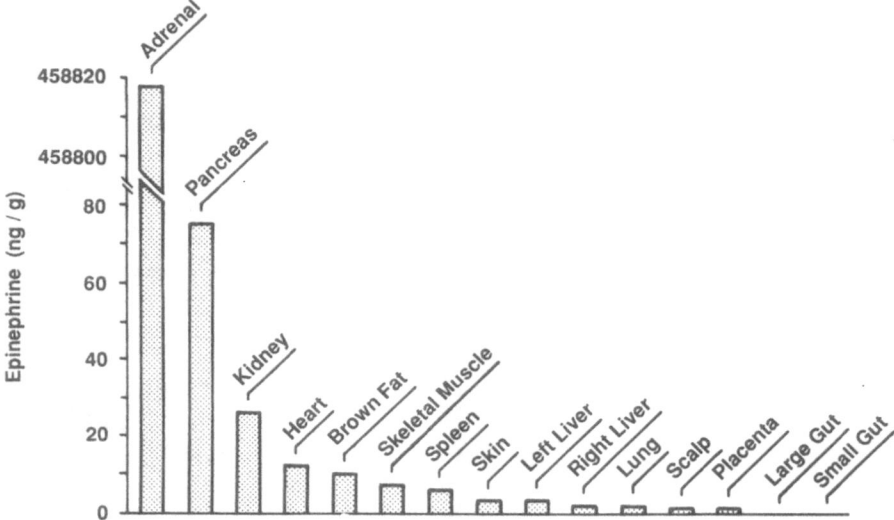

Fig. 6. Distribution of epinephrine (ng/g fresh tissue weight) in organs of fetal guinea pigs near term during normoxaemia ($n = 5$)

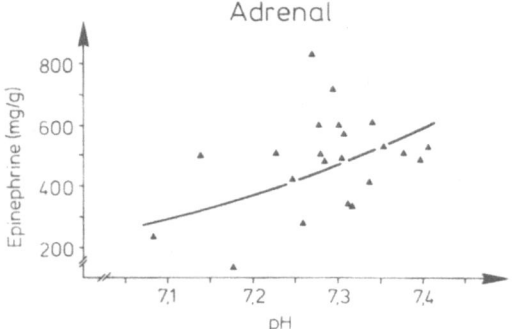

Fig. 7. Correlation between blood pH and epinephrine concentrations in the adrenal of fetal guinea pigs near term ($p < 0.05$)

lung, scalp and placenta were very low. In both large and small gut epinephrine was undetectable.

During hypoxaemia and asphyxia epinephrine concentrations in the tissue changed, for example, in the adrenal, heart, lung, right liver lobe and scalp. In the adrenal, epinephrine concentrations decreased with decreasing blood pH (Fig. 7), and this decrease correlate with increasing plasma epinephrine concentrations (not illustrated), indicating a depletion of adrenal epinephrine stores during asphyxia. In the heart, lung, right liver lobe and scalp tissue epinephrine concentrations increased during hypoxaemia and asphyxia (Fig. 8).

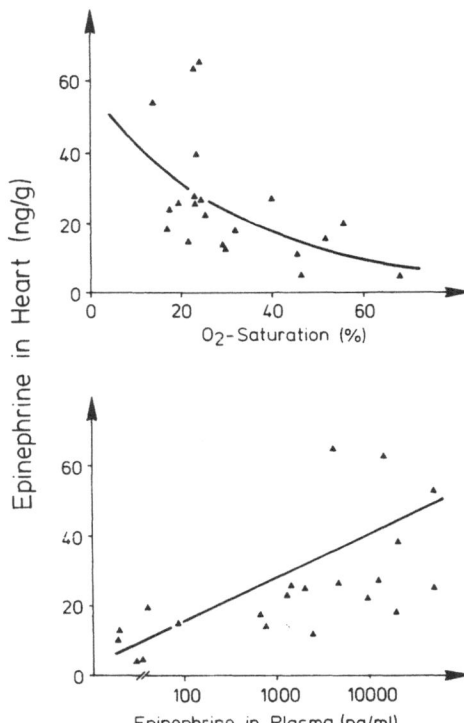

Fig. 8. Correlations between O_2 saturation and plasma epinephrine concentrations and epinephrine concentrations in the heart of fetal guinea pigs near term ($p < 0.01$)

Correlations between epinephrine plasma and tissue concentrations were observed for the heart and scalp. In the heart, for instance, there was a linear increase of tissue epinephrine with increasing concentrations of plasma epinephrine, suggesting epinephrine uptake from the blood (Fig. 8).

Relationship Between Tissue Concentrations of Norepinephrine and Epinephrine

The changes in norepinephrine and epinephrine concentrations in fetal organs during hypoxaemia and asphyxia discussed so far result in changing relationships between the two catecholamines, as indicated by the norepinephrine/epinephrine ratio. There were significant drops in the norepinephrine/epinephrine ratio in the scalp, heart, liver and kidney during hypoxaemia and asphyxia. Only in the adrenal did this ratio increase, supporting the view that the adrenal is depleted of epinephrine (Fig. 9). The changes in norepinephrine/epinephrine ratio in the adrenal, heart and left liver lobe were closely correlated with the decrease in O_2 saturation and blood pH (Fig. 10).

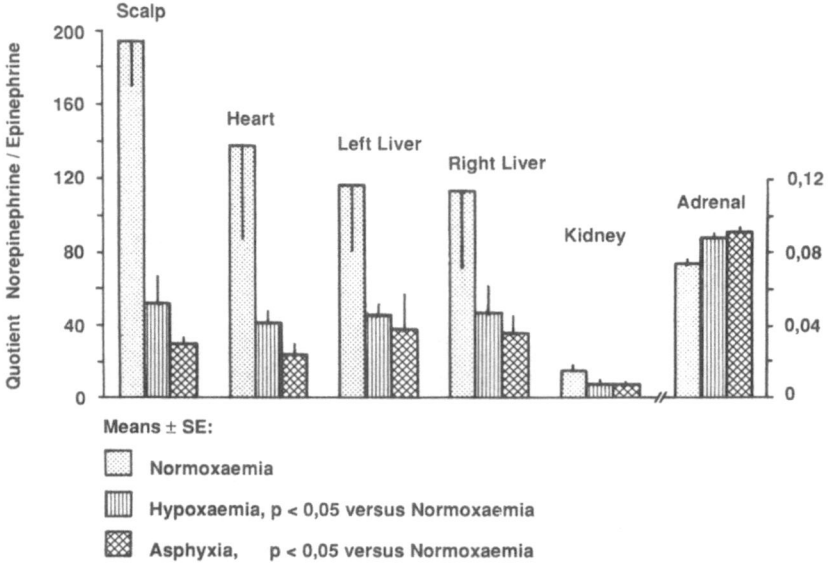

Fig. 9. Norepinephrine/epinephrine quotients in organs of fetal guinea pigs near term during normoxaemia ($n = 5$), hypoxaemia ($n = 6$) and asphyxia ($n = 11$). Means ± SE

Catecholamine Concentrations in the Brain

The catecholamine concentrations in the brain have to be regarded separately because of the blood-brain barrier, which is impermeable for catecholamines.

The norepinephrine distribution in various brain parts of the fetal guinea pig during normoxaemia is shown in Fig. 11. The highest concentrations were found in the hypothalamus, medulla, pons, thalamus, lamina tecti and tegmentum. Lower concentrations of norepinephrine were detected in cerebellum, hippocampus and caudate nucleus. The lowest levels were those in the cortex and bulbus olfactorius. This pattern of distribution agrees well with that from other mammals.

During acute hypoxaemia and asphyxia only small changes in norepinephrine concentrations in the brain could be observed, supporting earlier studies of brain catecholamines in rats subjected to acute hypoxia (Davis and Carlsson 1973; Brown et al. 1975).

Epinephrine was undetectable in any of the brain parts examined. This can be explained by the absence of PNMT, the catalysing enzyme of epinephrine formation, in the brain of the guinea pig (Fuller and Hemrick-Luecke 1983). Similarly, epinephrine could not be detected in brains of hypoxaemic or asphyxic fetuses, suggesting that the blood-brain barrier is impermeable for epinephrine during moderate hypoxaemia and asphyxia.

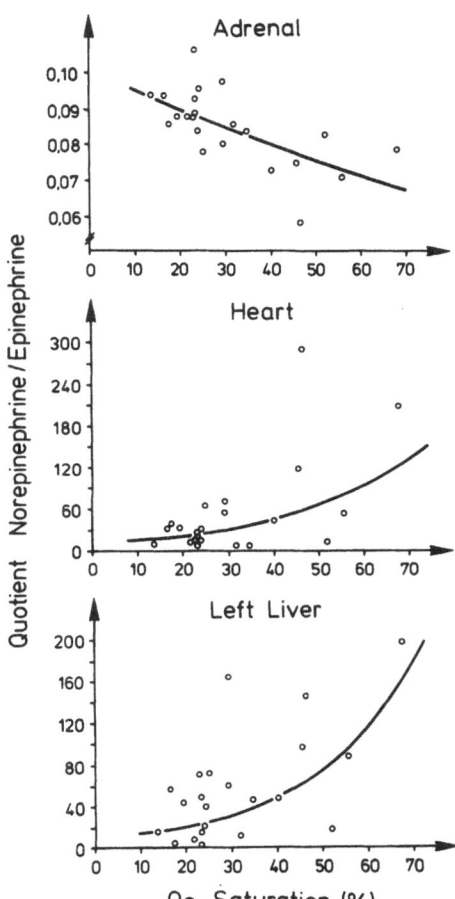

Fig. 10. Correlations between O_2 saturation and norepinephrine/epinephrine quotients in the adrenal, heart and left liver lobe of fetal guinea pigs near term ($p < 0.05$)

Conclusions

1. The concentrations of norepinephrine and epinephrine in central and peripheral organs of normoxaemic fetal guinea pigs near term agree well with those of adult guinea pigs, suggesting a relative maturity of the fetal sympathetic nervous system.
2. During hypoxaemia and asphyxia, norepinephrine and epinephrine concentrations change in various organs in different directions.
3. These changes relate to plasma catecholamine concentrations in different ways in different organs. We therefore conclude that in the fetal guinea pig near term, the

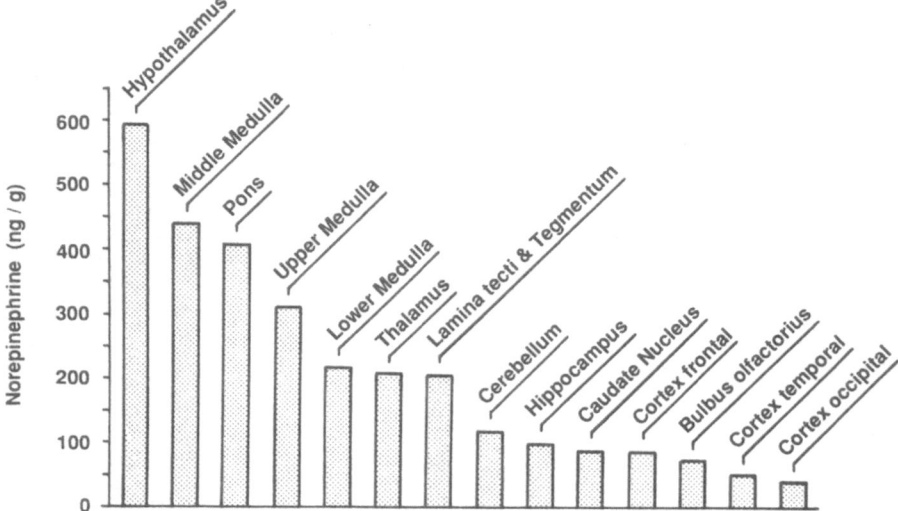

Fig. 11. Distribution of norepinephrine (ng/g fresh tissue weight) in 14 brain parts of fetal guinea pigs near term during normoxaemia ($n = 5$)

changes of organ-bound catecholamines during hypoxaemia and asphyxia and the interrelations with changes in plasma catecholamine concentrations may reflect circulatory and metabolic responses governed by the sympathetics during fetal adaptation to oxygen deprivation.

Acknowledgements. We thank S. Jelinek, R. Achenbach, O. Adam and K. Valentin for their technical assistance.

References

Adamson SL, Patrick JE, Challis JRG (1984) Effects of naloxone on the breathing, heart rate, glucose and cortisol responses to hypoxia in the sheep fetus. J Dev Physiol 6: 495–507

Anton AH, Sayre DF (1962) A study of the factors affecting the aluminum oxide-trihydroxyindole procedure for the analysis of catecholamines. J Pharm Exp Ther 138: 360–375

Brown RM, Kehr W, Carlsson A (1975) Functional and biochemical aspects of catecholamine metabolism in brain under hypoxia. Brain Res 85: 491–509

Castrén O, Saarikoski S (1974) The simultaneous function of catechol-O-methyltransferase and monoamine oxidase in human placenta. Acta Obstet Gynecol Scand 53: 41–47

Chatterjee S, Ghosh A (1982) Influence of experimental scurvy on adrenomedullary catecholamine levels in birds and mammals. Cell Mol Biol 28 (4): 401–403

Chen CH, Klein DC, Robinson JC (1974) Catechol-O-methyltransferase in rat placenta, human placenta and chorion carcinoma grown in culture. J Reprod Fertil 39: 407–410

Cohen WR, Piasecki GJ, Jackson BT (1982) Plasma catecholamines during hypoxemia in fetal lamb. Am J Physiol 243: R520-R525

Cohen WR, Piasecki GJ, Cohn HE, Young JB, Jackson BT (1984) Adrenal secretion of catecholamines during hypoxemia in fetal lamb. Endocrinology 114 (2): 383-390

Cohn HE, Sacks EJ, Heymann MA, Rudolph AM (1974) Cardiovascular responses to hypoxemia and acidemia in fetal lambs. Am J Obstet Gynecol 120 (6): 817-824

Cohn HE, Piasecki GJ, Jackson BT (1982) The effect of β-adrenergic stimulation on fetal cardiovascular function during hypoxemia. Am J Obstet Gynecol 144: 810-816

Comline RS, Silver M (1961) The release of adrenaline and noradrenaline from the adrenal glands of the foetal sheep. J Physiol 156: 424-444

Comline RS, Silver IA, Silver M (1965) Factors responsible for the stimulation of the adrenal medulla during asphyxia in the foetal lamb. J Physiol 178: 211-238

Court DJ, Parer JT, Block BSB, Llanos AJ (1984) Effects of beta-adrenergic blockade on blood flow distribution during hypoxemia in fetal sheep. J Dev Physiol 6: 349-358

Davis JN, Carlsson A (1973) Effect of hypoxia on tyrosine and tryptophan hydroxylation in unanaesthetized rat brain. J Neurochem 20: 913-915

Elsner R, Hammond DD, Parker HR (1969) Circulatory responses to asphyxia in pregnant and fetal animals: a comparative study of Weddell seals and sheep. Yale J Biol Med 42: 202-217

Euler US von (1954) Adrenaline and noradrenaline. Distribution and action. Pharmacol Rev 6: 15-22

Fisher DJ, Heymann MA, Rudolph AM (1982) Fetal myocardial oxygen and carbohydrate metabolism in sustained hypoxemia in utero. Am J Physiol 243: H959-H963

Folkow B, Häggendal J, Lisander B (1967) Extent of release and elimination of noradrenaline at peripheral adrenergic nerve terminals. Acta Physiol Scand [Suppl] 307: 5-38

Fuller RW, Hemrick-Luecke SK (1983) Species differences in epinephrine concentration and norepinephrine N-methyl-transferase activity in hypothalamus and brain stem. Comp Biochem Physiol 74C (1): 47-49

Graham RW, Scothorne RJ (1970) Calcium homeostasis in the foetal guinea pig. Q J Exp Physiol 55: 44-53

Greenberg RE, Lind J (1961) Catecholamines in tissues of the human fetus. Pediatrics 27: 904-911

Gu W, Jones CT, Parer JT (1985) Metabolic and cardiovascular effects on fetal sheep of sustained reduction of uterine blood flow. J Physiol 368: 109-129

Iisalo E, Castrén O (1967) The enzymatic inactivation of noradrenaline in human placental tissue. Ann Med Exp Biol Fenn 45: 253-257

Iversen LL (1967) The uptake and storage of noradrenaline in sympathetic nerves. Cambridge University Press, Cambridge

Jensen A, Künzel W, Hohmann M (1985) Dynamics of fetal organ blood flow redistribution and catecholamine release during acute asphyxia. In: Jones CT, Nathanielsz PW (eds) The physiological development of fetus and newborn. Academic, London, pp 405-410

Jensen A, Hohmann M, Künzel W (1987a) Redistribution of fetal circulation during repeated asphyxia in sheep: effects on skin blood flow, transcutaneous PO2, and plasma catecholamines. J Dev Physiol 9: 41-55

Jensen A, Künzel W, Kastendieck E (1987b) Fetal sympathetic activity, transcutaneous PO2, and skin blood flow during repeated asphyxia in sheep. J Dev Physiol 9: 337-346

Jones CT (1977) The development of some metabolic responses to hypoxia in the foetal sheep. J Physiol 266: 743-762

Jones CT, Ritchie JWK (1983) The effect of adrenergic blockade on the fetal response to hypoxia. J Dev Physiol 5: 211-222

Jones CT, Robinson RO (1975) Plasma catecholamines in foetal and adult sheep. J Physiol 248: 15-33

Jones CT, Rurak D (1976) The distribution and clearance of hormones and metabolites in the circulation of the foetal sheep. Q J Exp Physiol 61: 287-295

Karki N, Kuntzman R, Brodie BB (1962) Storage, synthesis, and metabolism of monoamines in the developing brain. J Neurochem 9: 53-58

Kopin I (1978) Plasma catecholamines: a brief overview. In: Usdin E, Kopin I, Barchas J (eds) Catecholamines: basic and clinical frontiers, vol 1. Pergamon, pp 897-902

Lewis AB, Evans WN, Sischo W (1982) Plasma catecholamine responses to hypoxemia in fetal lambs. Biol Neonate 41: 115-122

Lewis AB, Wolf JW, Sischo W (1984) Cardiovascular and catecholamine responses to successive episodes of hypoxemia in the fetus. Biol Neonate 45: 105-111

Lorijn RHW, Longo LD (1980) Norepinephrine elevation in the fetal lamb: oxygen consumption and cardiac output. Am J Physiol 239: R115-R122

Mandel P, Mack G, Goridis C (1975) Function of the central catecholaminergic neuron: synthesis, release and inactivation of transmitter. In: Friedhoff AF (ed) Catecholamines and behavior. Plenum, New York, pp 1-40 (Basic neurobiology, vol 1)

Paulick R, Kastendieck E, Wernze H (1985) Catecholamines in arterial and venous umbilical blood: placental extraction, correlation with fetal hypoxia, and transcutaneous partial oxygen tension. J Perinat Med 13: 31-42

Peeters LLH, Sheldon RR, Jones MD Jr, Makowski EL, Meschia G (1979) Blood flow to fetal organs as a function of arterial oxygen content. Am J Obstet Gynecol 135 (5): 637-646

Phillippe M (1983) Fetal catecholamines. Am J Obstet Gynecol 146 (7): 840-855

Reuss ML, Parer JT, Harris JL, Krueger TR (1982) Hemodynamic effects of alpha-adrenergic blockade during hypoxia in fetal sheep. Am J Obstet Gynecol 142 (4): 410-415

Saarikoski S (1974) Fate of norepinephrine in the human foetoplacental unit. Acta Physiol Scand [Suppl] 421: 1-82

Sheldon RE, Peeters LLH, Jones MD Jr, Makowski EL, Meschia G (1979) Redistribution of cardiac output and oxygen delivery in the hypoxemic fetal lamb. Am J Obstet Gynecol 135 (8): 1071-1078

Tatsuoka MM (1971) Multivariate analysis. Wiley, New York

Wagner J, Vitali P, Palfreyman MG, Zraika M, Huot S (1982) Simultaneous determination of 3,4-dihydroxyphenylalanine, 5-hydroxytryptophan, dopamine, 4-hydroxy-3-methoxyphenylalanine, norepinephrine, 3,4-dihydroxyphenylacetic acid, homovanillic acid, serotonin, and 5-hydroxyindoleacetic acid in rat cerebrospinal fluid and brain by high performance liquid chromatography with electrochemical detection. J Neurochem 38 (5): 1241-1254

Endocrine Control of
the Fetus – Basic Mechanisms

Biological Timekeeping During Pregnancy and the Role of Circadian Rhythms in Parturition*

L. D. Longo[1] and S. M. Yellon

Abstract

A considerable body of evidence suggests that both the hour of onset of labor and the hour of birth in humans, as well as many animal species, is related to time of day. In humans, over 30 studies have demonstrated such a relationship, with the peak hours of onset of labor and of birth between 2300 and 0400 hours. In day-active animals a similar pattern exists, while in night-active species the reverse holds. Underlying the 24-h periodicity in onset of labor and birth time may be a rhythm in uterine myometrial activity. In rhesus macaques, uterine contractions show a 24-h rhythmicity with peak activity between 2300 and 0300 hours. The mechanism responsible for daily rhythms in uterine activity is proposed to reflect 24-h patterns of endocrine secretion. Several hormones in maternal and fetal circulation, including melatonin, cortisol (mother only), dehydroepiandrosterone (fetus only), and progesterone, have also been shown to demonstrate 24-h rhythms, although not all these hormones have been studied in any one species. Evidence in sheep demonstrates that the circadian melatonin pattern mediates the effect of photoperiod and controls the neuroendocrine system regulating reproduction. Extending this concept to the mechanism timing parturition, we raise the possibility that the 24-h melatonin pattern conveys information about photoperiod and synchronizes various maternal and fetal endocrine rhythms to the light-dark cycle. We hypothesize that the temporal coordination of these endocrine and uterine rhythms with environmental photoperiod plays a key role in the initiation of parturition.

* This study was supported in part by US Public Health Service grant HD 03807 and a grant from the Department of Pediatrics, Loma Linda University School of Medicine.
[1] Division of Perinatal Biology, Departments of Physiology, Obstetrics and Gynecology, and Pediatrics, School of Medicine, Loma Linda University, Loma Linda, CA 92350, USA

The Endocrine Control of the Fetus
Ed. by W. Künzel and A. Jensen
© Springer-Verlag Berlin Heidelberg 1988

Introduction

> "Most men are begotten in the night, most animals in the day;
> but whether more persons have been born in the night or the day,
> were a curiosity undecidable ..."
>
> Sir Thomas Browne (Browne circa 1672)

The question of whether there exists a periodicity in the time of human birth, and if so at what hour the peak number of deliveries occur, has been of interest since at least the 1600s, when the physician-philosopher Sir Thomas Browne speculated on the matter. The popular idea that the great proportion of births occurs at night has been discussed in numerous works, including several nineteenth and early twentieth century obstetrical texts (for instance, see Williams 1922). Considerable evidence supports the concept of a daily rhythm in the onset of labor and the timing of delivery in humans and animals, but mention of the subject is absent in contemporary obstetrical literature (Aladjem 1980; Pritchard et al. 1985; Reid et al. 1972). In an effort to examine this topic in some depth we will consider the following questions:

(a) What is the relationship between time of day and periodicity in labor and delivery in humans?
(b) What is the evidence for such timing in subhuman primates and other mammals?
(c) What are the associations between daily hormone rhythms and the onset of labor?
(d) What are possible causal mechanisms of the rhythmicity in time of birth?
(e) What is the biological significance of this phenomenon?

Relationship Between Birth and Time of Day in Humans

During the past century more than three dozen authors have examined the question of whether there exists a diurnal rhythm in the time of the onset of labor and/or delivery. Table 1 lists these reports by date, locality, sample size, the peak and nadir for labor onset and/or delivery time, and author. Danz and Fuchs in 1848 were the first to report that the most frequent time for delivery was between 0400 and 0500 hours, while the least frequent period was between 1100 and 1200 hours. As can be seen from Table 1, with few exceptions subsequent studies have confirmed that a disproportionate number of births occur between 0100 and 0500 hours compared to other clock times. These reports also indicate that the least frequent hours for delivery are from 1400 to 1800 hours.

The upper panel of Fig. 1 is a composite of hourly onset of labor as reported in six studies with a total number of 114,910 cases (Charles 1953; Gauquelin 1967; Henry 1932; Málek et al. 1962; Schlegel et al. 1966; Shettles 1960). The peak hours

Table 1. Timing of parturition in humans

Year	City/Country	n	Onset of labor	Births (hour) Peak	Nadir	Reference
1848	Schmalkalden	1 000		0400–0500	1100–1200	Danz & Fuchs
		809		0400–0500	0700–0800	Berlinski[a]
				1300–1400	3000–2100	
1855	Berlin	14 036		0100–1200	1700–1800	Viet[a]
1868	Breslau		2100–2400	0000–0300		Spiegelberg
1905	Washington, DC	4 000		2200–0600	1400–2200	White
1907	Baltimore	1 508		0800–0900	1300–1400	Lynch[a]
1909	New York	39 000		2300–2400	1800–1900	Knapp[a]
1910	Norway					Horn
1932	New York, ex-	93 230		0300–0600	1200–1500	DePorte[a]
	cluding NYC	(3 398)		(1500–1800)	(0600–0900)[b]	
	Marseille	1 000	2100–0300	0200–0300	1300–1400	Henry[a]
1933	Switzerland			0200–0500	1400–1900	Jenny[a]
	Leipzig			0200–0500	1400–1900	Richter
1935	Kiel			0200–0500	1400–1900	Kirchhoff[a]
1936	Frankfurt	26 707		0200–0400		Guthmann & Bienhüls[a]
1937	England	32 224		0300–0600	1200–1500	Hill
		(1 272)		(0900–1200)	(0000–0300)[b]	
1938	New York,	82 077		0000–1200	1200–2400	Yerushalmy
	excluding NYC	(2 288)		(1500–1800)	(0300–0600)[b]	
1938	Czechoslovakia					Boháč
1940	London	2 225		1600–1800	0800–1000	Spiller
1946						Hosemann
1949	Freiburg					Kaiser & Maurath
1949	Stuttgart	5 308				Wurster
1952	Prague	69 875	0100–0200			Málek
	England	4 031		0300–0400	1900–2000	Simpson
				1000–1100		
1953	Birmingham	16 000[c]	0100–0300	0300	2000	Charles[a]
1954						Málek
1956	5 Hospitals in Maine, New York, Pennsylvania	33 215		0500–0600	1900–2000	King[a]
	Oklahoma City			0300–0700	1500–1600	Points[a]
1959	Indiana	115 950		0800–0900	1500–1600	Calhoun et al.[a]
1960	Richmond VA	10 469[c]		1400–1500	0800–0900	King[a]
		13 266[d]		0200–0300	1900	
	New York City	4 154	0200–0500			Shettles
1961						Kaiser
1962	(Review of 10 previous reports)	(601 222)		0100–0400	1600–2000	Kaiser[a] & Halberg
	Czechoslovakia	92 590	0100–0200			Málek et al.
1964	Zurich	10 000				Rippmann
			Labor intensity			Schlegel et al.
1966		1 800	0200			
1967	New York City	4 023		09000–1200	2100–2400	Erhardt et al.
	Strasbourg		0100–1200			Gauquelin
1967	Budapest	43 638		0400–0700	2000–2200	Orban & Czeizel
1972			2100–1300	0100–0700		Jolly

[a] Studies used in computing data for Fig. 1.
[b] Data for stillbirths in parentheses
[c] White
[d] Black

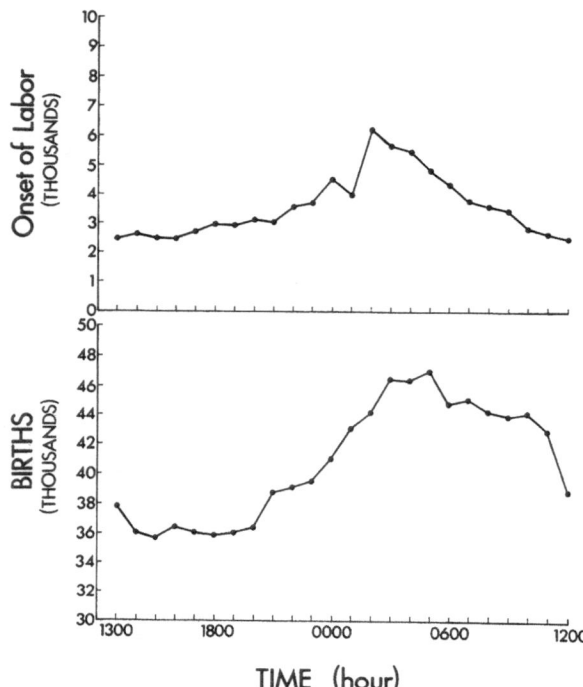

Fig. 1. *Upper panel* The number of patients with on-set of labor at a given hour; composite of six studies with total number of 114,910. *Lower panel* The number of births by hour of day summarized for 23 studies with total number of 997,158 (see text and Table 1 for details)

for the onset of labor are 2100–0300 hours. The lower panel of Fig. 1 shows the hourly frequency of natural births for a total of 997,168 cases (see Table 1). A similar 24-h periodicity is strikingly apparent in each of the larger studies for time of birth (Hill 1937; Calhoun et al. 1959; De Porte 1932; Guthmann and Biehüls 1936; King 1956; Knapp 1909; Orban and Czeizel 1967; Viet 1855; Yerushalmy 1938). Statistically, both the 0100–0500 hour peak and 1400–1800 hour nadir are significantly different from the mean values during the 24-h cycle ($p < 0.001$ by both the χ^2 and Kolmogorov-Smirov one-sample test).

Because the duration of labor is 4–6 h in multiparous patients and 8–10 h in those that are nulliparous (Pritchard et al. 1985, p.315), one would expect that the onset of labor would be phase-shifted 5–10 h prior to the peak delivery time. As seen in Fig. 1, however, such is not the case. The discrepancy between actual and expected initiation of labor may actually provide evidence to support a true nocturnal rhythm in human births. Several large studies have reported that the duration of labor is significantly shorter during the period of midnight to 0600 hours than at other times of day (Charles 1953; Schlegel et al. 1966). Although some subjective bias exists in the detection of the exact time of onset of uterine contractions, this cannot account for the definite nighttime peak in onset of this event (Jolly 1972).

In a large obstetrical service a variety of factors could influence the hour of delivery other than natural causes, including: artificial induction, chemical augmentation, operative interference, and social considerations that might alter the normal pattern of delivery. In fact, this may have been a factor in the lack of a night-day difference in birth frequency in a few reports (King 1960; Rippmann 1964; White 1905), but not for the vast remainder of studies which exclusively focused on natural onset of labor and normal delivery.

The question of whether parity influences time of delivery has also been examined in several studies, with conflicting results. For instance, Charles (1953) and Schlegel and colleagues (1966) noted a stronger correlation with delivery between midnight and 0400 hours in nulliparous than in multiparous patients. Other workers report the reverse (Erhardt et al. 1967), or no such parity difference (Gauquelin 1967; Málek et al. 1962). Neither infant sex (Gauquelin 1967; Points 1956a) nor prematurity (Kaiser and Halberg 1962) are correlated with a significant difference from other deliveries as to time of day. For stillbirths, a number of studies indicate a peak delivery incidence during the early to mid-afternoon, almost 12 h out of phase with the early-hours normal spontaneous liveborn eminence (Charles 1953; Erhardt et al. 1967; DePorte 1932; Yerushalmy 1938).

Additional considerations include possible weekly or monthly cyclicity in the hour of onset of labor and birth. The day of the week on which births most commonly occur has been variously reported to be Sunday (Boháč 1938), Tuesday (Calhoun et al. 1959), and Wednesday (Málek et al. 1962). Sunday was reported to be a nadir in the latter two studies (Calhoun et al. 1959; Málek et al. 1962). In one report season of the year was noted to show distinct variations in numbers of deliveries, with the highest rates in August and a nadir in April (Points 1956a). However, other authors (Málek et al. 1962; White 1905) report no such relationship. These conflicting reports of weekly or seasonal rhythms are in contrast to the repeatedly consistent finding that human births are correlated with time of day.

Evidence for Circadian Timing of Birth in Subhuman Primates and Other Mammals

Table 2 presents data on the timing of parturition in a number of mammals including subhuman primates. Almost without exception, the peak hour for birth is at night or during the early morning hours in animals that are day-active; a time that is during the rest phase of the 24-h day. By contrast, in animals that are active at night (i. e., nocturnal), births occur during the day (Jolly 1972; Plaut et al. 1970). The validity of these observations might be questioned since in some studies observations were made at 3-h intervals (Lincoln and Porter 1976; Lindahl 1964; Murakami et al. 1987) while others used various apparatuses to record birth time automatically (Merton 1937; Mitchell and Yochim 1970). Despite methodological differences, however, there is striking consistency in the reports.

Table 2. Timing of parturition in animals

Species	Gest. length (days)	n	Births (hour)		Comments	References
			Peak	Nadir		
Rodents						
Mouse (albino)	19	164	0000–0400 1600–2000	0400–0800		Merton 1937
	~ 20	1,202	0100–0300 1400–1600	0800–2300	When light reversed	Svorad & Šáchová 1959
		21	1700–0900			Porter 1972
		167			Phase-shifted light with no effect on duration of pregnancy	
Rat (nocturnal)						
Sprague-Dawley	23	82	0900–1500			Murakami et al. 1987
		50	Reversal		Light reversed	
		40	Random Reversal		Constant darkness Reversed photoperiod	
		17	Reversal		Reversed photoperiod, SCN lesions	
Wistar	22–23	66	0900–1500		Biphasic with some delivered at night	
		77	1000, 1300			Lincoln & Porter 1976
					Phase-shifted	Bosc & Nicolle 1980
Holtzman		34	During light period		2:22 Light:dark 14:10 Light:dark 22:2 Light:dark Pinealectomized – no effect	Mitchell & Yochim 1970
		11	0900–01300			Reppert 1983
		13	2100–0100		Phase reversed	
Hamster (nocturnal)						
Chinese	20.5	2,500+	Late afternoon			Yerganian 1958
Horse		501	2200–2300	0900–1600		Rossdale & Short 1967
			2200–0600			Bane 1961
					With shorter nights in summer, the number of foals during darkness increased	Zwoliński & Siudiński 1965
Pigs (Yorkshire)		410	0200–0400			Deakin & Fraser 1935
Sheep		1270	1500–1800 0900–1200	2100–0300		Lindahl 1964
		230	0000–0400			Wallace 1949 (quoted by Lindahl 1964)

Table 2. (continued)

Species	Gest. length (days)	n	Births (hour)		Comments	References
			Peak	Nadir		
Primates (diurnal)						
Saimiri sciureus (squirrel monkey)		1	0130			Takeshita 1961-62
		2	"Early a.m." 2200–0415			Bowden et al. 1967
		10	"All at night"			Hopf 1967
Macaca mulatta (rhesus monkey)						Hartman & Tinklepaugh 1932
Macaca nemestrina (pig-tailed monkey)		12	Phase-shifted, 11 of 12 delivered during artificial night			Jensen & Bobbitt 1967
Presbytis entrellus (Hanuman langurs)		3	Between midnight and dawn			Sugiyama 1965

Jolly (1972) has extensively reviewed the literature on birth hour in primates. Several dozen studies strongly demonstrate peak birth hour during the sleeping period (Jolly 1972). Among the day-active primates, uterine contractile activity and frequency of birth is much greater at night than during the day (Bowden et al. 1967; Christen 1968; Hartman and Tinklepaugh 1932; Hopf 1967; Jensen and Bobbitt 1967; Kirschsofer 1960; Kuehn et al. 1965; Lorenz and Heinemann 1967; Lucas et al. 1937; Sugiyama 1965). For night-active prosimians, this pattern is reversed (Doyle et al. 1967, 1971; Petter-Rousseaux 1964). In contrast, among the great apes no time of day birth rhythmicity has been reported; however, the number of observations was very small (Jolly 1972).

Control of the time of day of birth by photoperiod (light–dark cycles) has been examined by several groups. In mice subjected to a 12-h phase shift, peak birth times changed from 0100–0300 hours to 1400–1600 hours, coincident with the new light period (Svorad and Šáchová 1959). In rats (a nocturnal species), numerous studies have demonstrated that changes in the light–dark cycle midway through pregnancy correspondingly shift the time of birth cycle (Bosc and Nicolle 1980; Mitchell and Yochim 1970; Plaut et al. 1970; Reppert 1983; Sherwood et al. 1983). However, light changes made 1–2 days before parturition had no effect on birth time (Lincoln and Porter 1976). The failure to change birth time after an acute photoperiod shift suggests the mechanism linking birth to time of day requires an adaptation greater than 2 days. Adaptation periods of more than a few days are indicative of circadian biological timekeeping (Reppert 1983). As an aside, in rats the duration of pregnancy is not affected by photoperiod shifts (Porter 1972; Reppert 1983) and factors such as maternal parity or litter size do not seem to influence the hour of birth (Lincoln and Porter 1976).

Additional support for the role of light in the birth process is provided by a study in which rats placed on a 2-h light-22-h dark regimen delivered during the relatively brief light period (Mitchell and Yochim 1970). Finally, in the only known photoperiod study of its kind in the pig-tailed macaque, time of parturition was shifted after only about 10 days of a 180° reversal of lighting (Jensen and Bobbitt 1967).

The circadian system is further implicated as the mechanism initiating birth by two recent studies in the rat. Experimental lesions of the maternal suprachiasmatic nuclei (the SCN is believed to be a "master" oscillator generating many circadian rhythms) caused the hour of birth to become random (Reppert et al. 1987; Murakami et al. 1987). Elimination of circadian melatonin rhythms is among the many effects of SCN lesions (Moore 1978). However, pinealectomy in the Holtzman rat did not seem to affect birth time (Mitchell and Yochim 1970). The absence of a daily pattern of melatonin in circulation does not necessarily preclude a role for the pineal gland of its melatonin rhythm in the parturition process. In the absence of a primary zeitgeber, other entraining signals may influence the timing of birth. Nonetheless, these initial findings need to be confirmed.

Association Between Daily Hormone Rhythms and the Mechanism of Birth

The question arises, what, if any, relation has the circadian periodicity of births to do with hormonal changes associated with labor and delivery? Parturition results from a combination of factors and mechanisms finely tuned to result in a healthy offspring (Liggins 1979; Thorburn and Challis 1979). Although the factors which determine the onset of labor are poorly understood, a number of hormonal systems are involved including: fetal pituitary (Gordon and Sherwood 1982), fetal adrenal (Bosc and Nicolle 1979, 1980; Liggins 1979), prostaglandins (Dukes et al. 1974; Strauss et al. 1975), catecholamines, relaxin (Downing and Sherwood 1985), ovarian steroids (Catalá and Deis 1973; Thorburn and Challis 1979), and other hormonal interrelations between the mother and fetus (Sherwood et al. 1985). Through maturation of its hypothalamic-pituitary-adrenal-placental and extraembryotic membrane axis the fetus is hypothesized to initiate the onset of labor and birth (Liggins 1979; Thorburn and Challis 1979). In turn, the mother may, through hormonal interrelations, provide information to define the exact hour for parturition, barring unforeseen complications.

Uterine Activity Rhythms.

Studies in the pregnant rhesus macaque have demonstrated daily rhythms in uterine activity during late gestation (Ducsay et al. 1983; Taylor et al. 1983). Peak uterine myometrial activity occurred at approximately midnight and gradually evolved into labor and delivery.

Rhythms in uterine activity were also reported by Harbert and Spisso (1980, 1981), but their data are difficult to interpret for several reasons. Average number of contractions per minute had no relationship to hourly contraction frequency or time of day. Moreover, no mention was made of photoperiod treatment prior to the actual days of experimentation. Finally, and perhaps most importantly, only 9 of 15 monkeys gave birth at night and 4 of those were premature (146–157 day's gestation). Therefore, this report is difficult to compare with recent studies which clearly define nocturnal rhythms in uterine activity (Ducsay et al. 1983; Taylor et al. 1983).

For women, similar daily rhythms in myometrial activity are also presumed to occur (Málek 1952; Tamby Raja and Hobel 1983). However, in the sheep, a day-night difference in myometrial contractility has yet to be defined (see Fig. 2e). Ultimately, the relation between daily uterine activity rhythms and the uterine contractions of labor remains to be established. The significance of the rhesus data is that it raises the hypothesis that parturition is an extension or amplification of these normally occurring rhythms in uterine activity.

Hormonal Rhythms

In an effort to elucidate the factors involved in the generation of the nocturnal activity rhythm, profiles of several hormones in the mother and fetus have been examined. In the pregnant rhesus monkey (blood collected during the day, 0700–1000 hours, or evening, 1700–2100 hours), daytime elevations of cortisol, estradiol, and estrone were detected in maternal plasma, while progesterone was higher in the evening (Challis et al. 1980). By comparison, in fetal circulation no significant day-evening cortisol difference was found (Challis et al. 1980). Higher concentrations were found for dehydroepiandrosterone sulfate (DHEAS) and progesterone in the fetus in the evening; estrogens were not measured. This evening peak in DHEAS has since been confirmed (Serón-Ferré et al. 1983; Walsh et al. 1984).

More frequent blood samples (every 3 h) taken from fetal and maternal circulations have been obtained in the rhesus monkey by Walsh and colleagues (1984). In the fetus, a nocturnal rhythm of DHEAS in circulation paralleled increases in uterine activity (peak at 2300–0300 hours; Fig. 2c, f). Changes in DHEAS concentration were accompanied by significant parallel changes in the circulating concentration of progesterone in both mother and fetus (Fig. 2d). However, cortisol concentrations showed no significant diurnal changes in the fetus. In maternal blood, cortisol concentrations peaked between 0600 and 0900 hours (Fig. 2b), while cortisol in amniotic fluid peaked later, between 0900 and 1200 hours. The changes in estrogens,

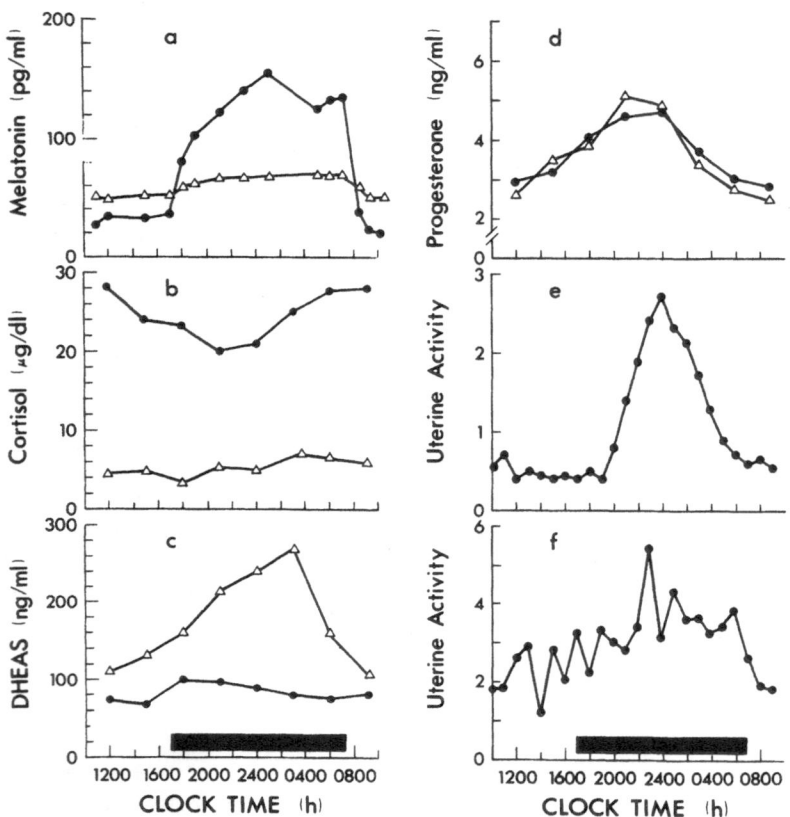

Fig. 2 a–f. Fluctuations in several hormones in maternal and fetal blood by time of day for 24 h. ●————● maternal; △————△ fetal. **a** Melatonin concentrations in sheep. **b** Cortisol concentrations in rhesus monkeys. **c** Dehydroepiandrosterone sulfate concentrations in monkeys. **d** Progesterone concentrations in monkeys. **e** Uterine activity pattern in monkeys. **f** Uterine activity pattern in sheep. See text for details

progestins, and corticosteroids are hypothesized to play a key role in the onset of parturition due to their effect on myometrial contractility and stimulation of prostaglandin production (Thorburn and Challis 1979; Ducsay et al. 1983).

Another possibly relevant hormone to consider regarding circadian activity is the pineal indoleamine, melatonin. In many seasonally breeding species, the nocturnal melatonin rhythm is part of an endogenous biological clock mediating information about day length to time the onset of puberty and the annual adult reproductive cycle. To determine whether timekeeping persists during pregnancy, Yellon and Longo (1987) studied the pattern of circulating melatonin in sheep during the last trimester of gestation. In the chronically catheterized ewe and fetus ($n = 6$) circulating melatonin concentrations were determined over a 48-h period (every 1–4 h) at

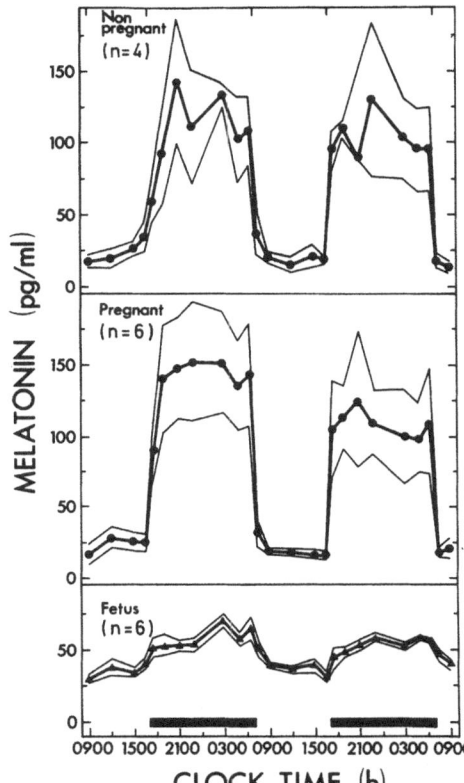

Fig. 3. Pattern of melatonin in circulation over 48 h in nonpregnant ewes *(top)* and in pregnant ewes *(middle)* and their fetuses *(bottom)* at 120 days' gestation. Mean (± SE) plasma melatonin concentrations are plotted with respect to clock time. The *Dark bar* indicates night (10:14 light:dark period; lights off at 1700 hours)

about 120, 127, and 135 days' gestation. As shown in Fig. 3, a typical rhythm was present in the pregnant ewes; plasma melatonin was low during the day and increased throughout the night.

For the fetus, a 24-h pattern of melatonin was also present in the circulation (Fig. 3), but it differed from the adult melatonin rhythm in two important respects. First, the amplitude of the fetal melatonin pattern was remarkably low, due, in part, to a high daytime baseline and a modest 50% rise at night. Second, the rise in plasma melatonin was phase-delayed 0.5–1.5 h, extending into the light period unlike the adult rhythm. Despite these concerns, information about photoperiod in seasonal breeders is currently hypothesized to be conveyed by the duration of increased circulating melatonin rather than its amplitude or phase (Goldman and Darrow 1983; Yellon et al. 1986).

Evidence already mentioned in the sheep indicates that the duration of the plasma melatonin rise is similarly maintained in fetal and maternal circulations. In pregnant sheep, without exception, a greater than fivefold increase in circulating

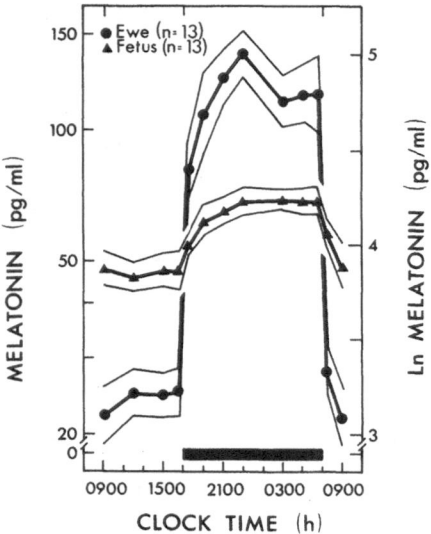

Fig. 4. Natural log of plasma melatonin concentrations (mean ± SE) in the pregnant ewe and fetus during the last trimester of gestation. Blood samples were collected over a 48-h period on 13 separate occasions (120 days, $n = 6$; 126 days, $n = 4$; 135 days, $n = 3$). After log transformation to normalize variance, a two-way ANOVA with replication was performed; the second 24-h sampling period was treated as a replicate (see text for further details). The *dark bar* indicates night (10:14 light:dark period; lights off at 1700 hours)

melatonin concentrations occurred at night. In fact, even the actual birth process does not affect the 24-h pattern of melatonin in circulation (Fig. 4). This observation does not suggest that pregnancy impairs the mechanism generating a circadian rhythm in pineal melatonin secretion. Whether the processing of this photoperiodic message is influenced by the marked endocrinologic changes of pregnancy is as yet unresolved.

Preliminary findings in this study (Fig. 5) suggest that photoperiod and, presumably, melatonin rhythms are important for timing parturition because in sheep that went to term, births occurred at night at 0130 hours ± 2.6 h (mean ± SE; $n = 3$). The nocturnal occurrence of birth strongly indicates that the mechanism of photoperiodic time measurement is sustained in the pregnant sheep.

The presence of a 24-h melatonin pattern in fetal circulation does not necessarily indicate that it is endogenously generated by the fetus. Melatonin rapidly crosses the placenta in a variety of species (Reppert 1985) including the sheep (S. M. Yellon and L. D. Longo, 1987, unpublished observations). Perhaps placental transfer of melatonin from maternal circulation may account for the 24-h pattern and its presence *in toto* in the fetus.

In addition, the neural regulation of melatonin secretion may not yet be established in the fetus. Even though the enzymes synthesizing indoles in the fetal pineal

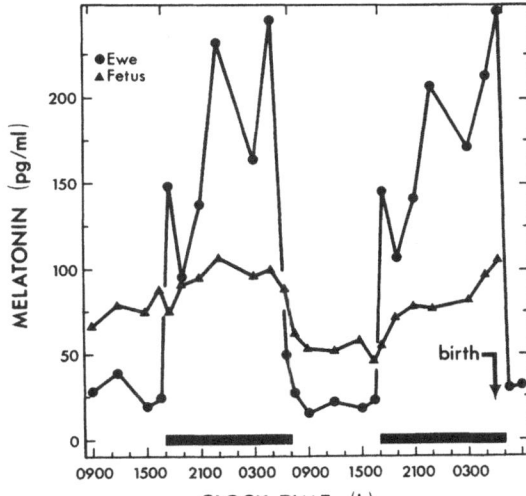

Fig. 5. Melatonin pattern in maternal and fetal circulations during the time period immediately preceding birth at 139–140 days' gestation in a pregnant sheep. (Details as in Fig. 3 legend)

gland are active prior to birth (Kennaway et al. 1977), the lack of appropriate innervation may account for elevated daytime melatonin concentrations in fetal versus maternal circulation. A direct central nervous system connection between the pineal gland and the subcommissural organ has been reported for fetuses of several species (Moller et al. 1975). Also, the absence of normal melatonin rhythms in neonatal lambs much before 10 weeks of age (Yellon and Foster 1986) raises the possibility that the sympathetic innervation of the pineal gland has not developed prior to birth. Thus, it remains to be determined whether an endogenous circadian melatonin rhythm is generated by the fetus.

Mechanisms Underlying the Relation of Birth to Time of Day

Anecdotal accounts have for several centuries suggested that human labor and delivery are more frequent at night. The data presented in Table 1 and Fig. 1 indicate that such beliefs are not unfounded. Additional evidence strongly supports the idea that animals give birth during their resting phase: at night or during early morning hours for those that are day-active and during the day for those that are night-active. Nonetheless, several potential problems regarding the data need to be considered.

For both humans and the several animal species the problem of avoiding bias in data selection is vital (Charles 1953; Erhardt et al. 1967; Rippmann 1964). In recording the onset of labor, one must rely on the testimony of the mother and be certain that one's criteria for onset is consistent. For delivery *per se* the criterion is well

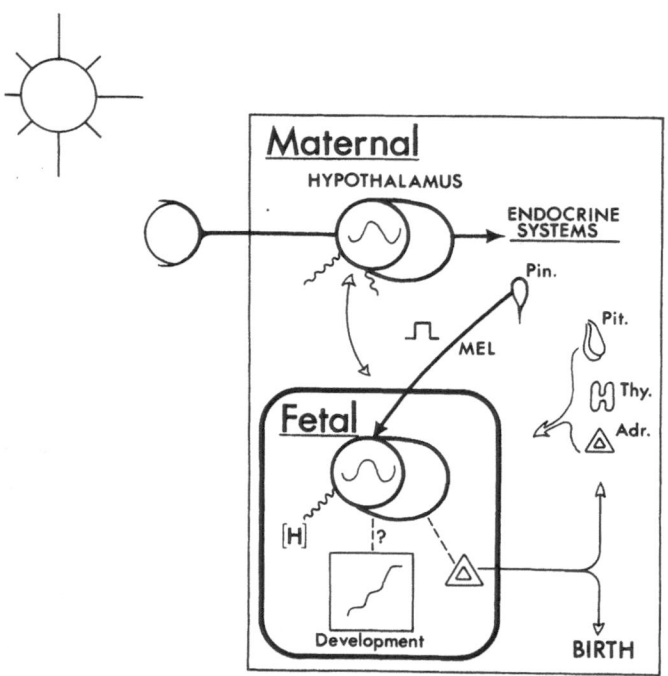

Fig. 6. Possible maternal-fetal melatonin relationship, and the relations to other hormone systems and the onset of parturition. *Adr.*, adrenal *[H]*, hormonal concentration; *MEL*, melatonin; *Pin.*, pineal; *Pit.*, pituitary; *Thy.*, thyroid

defined. In determining the times of normal, spontaneous, vaginal delivery, cases complicated by artificial induction, chemical augmentation, and operative interference must be excluded. These factors, in fact, undoubtedly skewed the results of some studies (Rippmann 1964).

Other environmental and biological considerations may also modulate the process of parturition. Variability in geographic latitude, climate, temperature, prior photoperiod experience, light intensity, and work schedules could be of potential significance for comparing data among studies. Furthermore, diversity in the gene pool could affect the time of day of birth in select populations; more inbred groups may concentrate peak time of labor or delivery. Finally, appropriate data analysis is needed to discriminate between a true rhythm and random periodicity. Kaiser and Halberg (1962) explored various methods of data evaluation of the frequency-time relationship for human birth. These included: curve fitting, Fourier analysis, variance spectrum analysis, periodic regression, the so-called periodogram, and a so-called circadian quotient. These techniques disclosed a prominent peak at the 24-h periodicity.

In mammals, photoperiodic regulation of the time of birth is strongly suggested for several species (see Reppert et al. 1987). Photic information passes from the retina to the suprachiasmatic nucleus (SCN) of the anterior hypothalamus (Fig. 6). Studies in rodents suggest that the SCN (which serves as a circadian pacemaker or biological clock) of the fetus and newborn is entrained to the light-dark cycle. Reppert and Schwartz (1983), using 2-deoxyglucose autoradiography, demonstrated that, prior to innervation of the SCN by the retinohypothalamic pathway, the fetal clock is entrained to environmental lighting conditions. Entrainment of the neural oscillator in the fetal SCN probably occurs via the maternal circadian system because the developing biological clock is synchronized by the maternal SCN which, in turn, is entrained by ambient lighting (for review, see Reppert 1985). Moreover, it is clear that biological timekeeping is critical for timing the initiation of parturition, because maternal SCN lesions abolish normal circadian timing of birth (Reppert et al. 1987).

Consideration of these issues highlights our lack of understanding of the complex mechanisms which account for the precise time of birth. Many authors, enumerated in Table 1, have considered the phenomenological relationship of birth frequency to time of day. Knapp (1909) postulated that maternal activity during the daylight hours stimulated the onset of labor, with nocturnal delivery. Others have proposed a "nervous" mechanism (Málek 1952) or that in the early morning hours the attending obstetrician might stimulate labor and delivery so that he could prepare for a new day in the office (DePorte 1932; King 1960; Málek 1952).

During recent years several authors have stressed the role of hormones in the initiation of parturition (Jolly 1972; Murakami et al. 1987; Thorburn and Challis 1979). The presence of a 24-h melatonin pattern in fetal circulation as noted by Yellon and Longo (1987) leads to consideration of a role of melatonin in gestation and parturition. The results raise the possibility that the 24-h melatonin pattern conveys information about the photoperiod and synchronizes various maternal and fetal endocrine rhythms to the light-dark cycle. Coordination of maternal hormone secretion would ensure that the fetal circadian system is provided with information about the prevailing environmental photoperiod.

Biological Significance

The observation that birth frequency in both humans and many animal species is related to time of day probably has a number of implications. First, one must recognize that the maternal-fetal organism is structured in time as well as space. Ostensibly, the temporal relation of parturition to the resting period has survival value for the several animal species if not humans (Jolly 1972; Kaiser and Halberg 1962). Also, it is of interest that so many endocrine systems display circadian rhythms, of which some may be unique for pregnancy, i.e., progesterone, DHEAS, (Walsh et al. 1984). Such phasic changes also may have diagnostic and therapeutic value in health and disease.

Certainly photoperiodic timekeeping in the fetus and its importance for parturition cannot be ignored because, at least in some species, it is the fetus that appears to initiate the process of birth (Liggins 1979; Lincoln and Porter 1976; Thorburn and Challis 1979). These findings raise questions about the mechanism through which the fetus, either directly or indirectly through the mother, recognizes photoperiod cues in the environment to influence the onset of parturition.

Another question regards the role of light in entraining these endocrine rhythms. Because the hormonal, uterine activity, and birth frequency rhythms are so tightly linked to day-night cycles, it appears clear that light is a key factor. A large body of evidence has demonstrated the dependence of pineal melatonin activity on an intact retinal-suprachiasmatic nucleus-pineal pathway, providing the interface for light information in the environmental physiology. However, the exact role of melatonin, if any, in influencing the other hormonal rhythms and the timing of parturition is unclear. At present, the specific target site for melatonin action and mechanism through which memory of time is conveyed is unknown. Although it is probably not the factor that initiates the onset of labor, the current working hypothesis is that melatonin plays a key role in entraining the endocrine-uterine system.

References

Aladjem S (ed) (1980) Obstetrical practice. Mosby, St. Louis

Bane A (1961) Förlossningen hos sto och des förebud. Medlemsbl Sveriges Vet Forb 10: 253–261

Berlinski (1848) Dissertatio de nascentium morientiumque numero. Rostock

Boháč V (1938) Týdenní rythmus porodů. Stat Obzor 17–18

Bosc MJ, Nicolle A (1979) Effect of stress on the course of labor and parturition time in normal or adrenalectomized rats. Ann Biol Anim Bioch Biophys 19: 31–44

Bosc MJ, Nicolle A (1980) Influence of photoperiod on the time of parturition in the rat. I. Effect of the length of daily illumination on normal or adrenalectomized animals. Reprod Nutr Develop 20: 735–745

Bowden D, Winter P, Ploog D (1967) Pregnancy and delivery behavior in the squirrel monkey *(Saimiri sciureus)* and other primates. Folia Primatol (Basel) 5: 1–42

Browne T (circa 1672) A letter to a friend upon occasion of the death of his intimate friend. In: Keynes G (ed) The Works of Sir Thomas Browne, vol 1. Faber and Gwyer, London, p 169

Calhoun RA, Harvey VK, Jr, Holwager KE (1959) An analysis of 1957 Indiana births by time of day and day of week. J Indiana State Med Assoc 52: 1992–1996

Catalá S, Deis RP (1973) Effect of oestrogen upon parturition, maternal behaviour and lactation in ovariectomized pregnant rats. J Endocrinol 56: 219–225

Challis JRG, Socol M, Murata Y, Manning FA, Martin CB Jr (1980) Diurnal variations in maternal and fetal steroids in pregnant rhesus monkeys. Endocrinology 106: 1283–1288

Charles E (1953) The hour of birth. A study of the distribution of times of onset of labour and of delivery throughout the 24-hour period. Br J Prev Soc Med 7: 43–59

Christen A (1968) Haltung und Brutbiologie von *Cebuella.* Folia Primatol (Basel) 8: 41–49

Danz CF, Fuchs CF (1848) Physisch-medicinische Topographie des Kreises Schmalkalden. Gesellschaft zur Beförderung der gesamten Naturwissenschaften zu Marburg 6: 205–206

Deakin A, Fraser EB (1935) Fecundity and nursing capacity of large Yorkshire sows. Sci Agriculture 15: 458–462

DePorte JV (1932) The prevalent hour of stillbirth. Am J Obstet Gynecol 23: 31–37

Downing SJ, Sherwood OD (1985) The physiological role of relaxin in the pregnant rat. I. The influence of relaxin on parturition. Endocrinology 116: 1200–1205

Doyle GA, Pelletied A, Bekker T (1967) Courtship, mating and parturition in the lesser bush-baby *(Galago senegalensis moholi)* under semi-natural conditions. Folia Primatol (Basel) 7: 169–197

Ducsay CA, Cook MJ, Walsh SW, Novy MJ (1983) Circadian patterns and dexamethasone-induced changes in uterine activity in pregnant rhesus monkeys. Am J Obstet Gynecol 45: 389–396

Dukes M, Chester R, Atkinson P (1974) Effect of oestradiol and prostaglandin F_{2a} on the timing of parturition in the rat. J Reprod Fertil 38: 325–334

Erhardt CL, Nelson FG, McMahon M (1967) Hour of birth. NY State J Med 67: 421–423

Gauquelin M (1967) Note sur le rythme journalier du début des douleurs de l'accouchement. Gynecol Obstet (Paris) 66: 229–236

Goldman BD, Darrow JM (1983) The pineal gland and mammalian photo-periodism. Neuro-endocrinology 37: 389–396

Gordon WL, Sherwood OD (1982) Evidence that luteinizing hormone from the maternal pituitary gland may promote antepartum release of relaxin, luteolysis, and birth in rats. Endocrinology 111: 1299–1310

Guthmann H, Bienhüls M (1936) Wehenbeginn, Geburtsstunde und Tageszeit. Monatsschr Geburts Gynäkol 103: 337–348

Harbert GM, JR, Spisso KR (1980) Biorhythms of the primate uterus *(Macaca mulatta)* during labor and delivery. Am J Obstet Gynecol 138: 686–696

Harbert GM, Jr, Spisso KR (1981) Effect of adrenergic blockade on dynamics of the pregnant primate uterus *(Macaca mulatta)*. Am J Obstet Gynecol 139: 767–780

Hartman CG, Tinklepaugh OL (1932) Weitere Beobachtungen über die Geburt beim Affen *Macacus rhesus*. Arch Gynaekol 149: 21–37

Henry J-R (1932) Sur les heures de début du travail et de l'accouchement. Marseille Med 1: 203–205

Hill AB (1937) Principles of medical statistics. Lancet, London, pp 90–91

Hopf S (1967) Notes on pregnancy, delivery, and infant survival in captive squirrel monkeys. Primates 8: 323–332

Horn J (1910) Naar i døgnet begynder og naar afsluttes den spontane fødsel? Norsk Mag Laegevidenskab 8: 630–631

Hosemann H (1946) Über die Dauer der Geburt. Klin Wochenschr 71: 181–184

Jenny E (1933) Tagesperiodische Einflüsse auf Geburt und Tod. Schweiz Med Wochenschr 63: 15–17

Jensen GD, Bobbitt RA (1967) Changing parturition time in monkeys *(Macaca nemestrina)* from night to day. Lab Anim Care 17: 379–381

Jolly A (1972) Hour of birth in primates and man. Folia Primatol (Basel) 18: 108–121

Kaiser F, Maurath J (1949) Kreislaufdynamische 24-Stunden-Rhythmik beim Menschen. Klin Wochenschr 27: 659–662

Kaiser IH (1961) Circadian aspects of birth and development. In: Circadian systems. Report of the 39th Ross Conference on Pediatric Research. Ross Laboratories. Columbus, Ohio, pp 33–34

Kaiser IH, Halberg F (1962) Circadian periodic aspects of birth. Ann NY Acad Sci 98: 1056–1068

Kennaway DJ, Matthews CD, Seamark RF, Phillipou G, Schilthuis M (1977) On the presence of melatonin in the plasma of foetal sheep. J Steroid Biochem 8: 559–563

King PD (1956) Increased frequency of births in the morning hours. Science 123: 985–986

King PD (1960) Distortion of the birth frequency curve. Am J Obstet Gynecol 79: 399–400

Kirchhoff H (1935) Unterliegt der Wehenbeginn kosmischen Einflüssen? Zentralbl Gynaekol 59: 134–144

Kirschsofer R (1960) Einige Verhaltensbeobachtungen an einem Guereza-Jungen *Colobus polykomos kikuyensis* unter besonderer Berücksichtigung des Spiels. Z Tierpsychol 17: 506–514

Knapp CB (1909) The hour of birth. Bull Lying - In Hosp City New York 6: 69–74

Kuehn RE, Jensen GD, Morrill RK (1965) Breeding *Macaca nemestrina:* a program of birth engineering. Folia Primatol (Basel) 3: 251–262

Liggins GC (1979) Initiation of parturition. Br Med Bull 35: 145–150

Lincoln DW, Porter DG (1976) Timing of the photoperiod and the hour of birth in rats. Nature 260: 780–781

Lindahl IL (1964) Time of parturition in ewes. Anim Behav 12: 231–234

Lorenz R, Heinemann H (1967) Beiträge zur Morphologie und körperlichen Jugendentwicklung des Spring-tamarin *Callimico geldii.* Folia Primatol (Basel) 6: 1–27

Lucas NS, Hume EM, Smith HH (1937) The breeding of the common marmoset in captivity. Proc Zool Soc Lond 107: 205–211

Lynch FW (1907) Hour of birth. A discussion as to the hour at which birth most often occurs. Surg Gynecol Obstet 5: 677–680

Málek J (1952) The manifestation of biological rhythms in delivery. Gynaecologia (Basel) 133: 365–372

Málek J (1954) Der Einfluß des Lichtes und der Dunkelheit auf den klinischen Geburtsbeginn. Gynaecologia (Basel) 138:401–405

Málek J, Šebasta J (1949) Vliv některých faktorů na denní rytmus klinického začátku porodu. Stat Obzor 29: 124–129

Málek J, Budinsky J, Budinska M (1950) Analyse du rhythme journalier du début clinique de l'accouchement. Rev Fr Gynecol Obstet 45: 222–226

Málek J, Malý V, Masák J (1958) Vztah denního rytmu klinického začátku porodu k délce porodu. Sb Lek Fak Karlovy Univ 60: 13–23

Málek J, Gleich J, Malý V (1962) Characteristics of the daily rhythm of menstruation and labor. NY Acad Sci 98: 1042–1055

Merton H (1937) Studies on reproduction in the albino mouse. I. The period of gestation and the time of parturition. R Soc Edinburgh Proc 58: 80–97

Mitchell JA, Yochim JM (1970) Influence of environmental lighting on duration of pregnancy in the rat. Endocrinology 87: 472–480

Moller M, Mollgard K, Kimble JE (1975) Presence of pineal nerve in sheep and rabbit fetuses. Cell Tissue Res 158: 451–459

Moore RY (1978) The innervation of the mammalian pineal gland. Prog Brain Res 4: 1–29

Murakami N, Abe T, Yokoyama M, Katsume A, Kuroda H, Etoh T (1987) Effect of photoperiod, injection of pentobarbitone sodium or lesion of the suprachiasmatic nucleus on prepartum decrease of blood progesterone concentrations of time of birth in the rat. J Reprod Fertil 79: 325–333

Orban VG, Czeizel E (1967) Der Tagesrhythmus der Geburten. Gynaecologia (Basel) 163: 173–178

Petter-Rousseaux A (1964) Reproductive physiology and behavior of the Lemuroidea. In: Buettner-Janusch (ed) Evolutionary and genetic biology of the primates, vol 2. Academic, New York, pp 91–132

Plaut SM, Grota LJ, Ader R, Graham CW III (1970) Effects of handling and the light-dark cycle on time of parturition in the rat. Lab Anim Care 20: 447–453

Points TC (1956a) Twenty-four hours in a day. Obstet Gynecol 8: 245–248

Points TC (1956b) Day after day. Obstet Gynecol 8: 748–752

Porter DG (1972) The light regimen and gestation length in the mouse. J Reprod Fertil 28: 9–14

Pritchard JA, MacDonald PC, Gant NF (1985) Williams obstetrics. Appleton-Century-Crofts, Norwalk, Conn

Reid DE, Ryan KJ, Benirschke K (1972) Principles and management of human reproduction. Saunders, Philadelphia

Reppert SM (1983) Time of birth in the rat is gated to the daily light-dark cycle by a circadian mechanism. Pediatr Res 17: 154A

Reppert SM (1985) Maternal entrainment of the developing circadian system. Ann NY Acad Sci 453: 162-169

Reppert SM, Henshaw D, Schwartz WJ, Weaver DR (1987) The circadian-gated timing of birth in rats: disruption by maternal SCN lesions or by removal of the fetal brain. Brain Res 403: 398-402

Reppert SM, Schwartz WJ (1983) Maternal coordination of the fetal biological clock in utero. Science 220: 969-971

Richter J (1933) Die geburtshilflich-gynäkologische Tierklinik der Universität Leipzig in den Jahren 1927-1931. Berl Tierarztl Wochenschr 33: 517-521

Rippmann ET (1964) Die zeitliche Verteilung von 10000 Geburten auf die 24-Stunden des Tages. Gynaecologia (Basel) 158: 31-34

Rossdale PD, Short RV (1967) The time of foaling of thoroughbred mares. J Reprod Fertil 13: 341-343

Schlegel VL, Stembera ZK, Porkorný J (1966) Tagesrhythmus der Wehentätigkeit unter der Geburt. Gynaecologia (Basel) 162: 185-196

Serón-Ferré M, Taylor NF, Rotten D, Koritnik DR, Jaffe RB (1983) Changes in fetal rhesus monkey plasma dehydroepiandrosterone sulfate: relationship to gestational age, adrenal weight and preterm delivery. J Clin Endocrinol Metab 57: 1173-1178

Sherwood OD, Downing SJ, Golos TG, Gordon WL, Tarbell MK (1983) Influence of light-dark cycle on antepartum serum relaxin and progesterone immunoreactivity levels and on birth in the rat. Endocrinology 113: 997-1003

Sherwood OD, Downing SJ, Rieber AJ, Fraley SW, Bohrer RE, Richardson BC, Shanks RD (1985) Influence of litter size on antepartum serum relaxin and progesterone immunoreactivity levels and on birth in the rat. Endocrinology 116: 2554-2562

Shettles LB (1960) Hourly variation in onset of labor and rupture of membranes. Am J Obstet Gynecol 79: 177-179

Simpson AS (1952) Are more babies born at night? Br Med J 2: 831

Spiegelberg O (1868) Bericht über die Leistungen der gynäkologischen Klinik und Poliklinik and der Universität zu Breslau in den Studienjahren vom October 1865 bis ebendahin 1967. Monatsschr Geburtskd Frauenkr 32: 267-307

Spiller V (1940) An inquiry into the hour of birth. Br Med J 1: 435

Strauss JF, III, Sokoloski J, Caploe P, Duffy P, Mintz G, Stambaugh RL (1975) On the role of prostaglandins in parturition in the rat. Endocrinology 96: 1040-1045

Sugiyama Y (1965) Behavioral development and social structure in two troops of Hanuman langurs *(Presbytis entellus)*. Primates 6: 213-247

Svorad D, Šáchová V (1959) Periodicity of the commencement of birth in mice and the influence of light. Physiol Bohemoslov 8: 439-442

Takeshita H (1961-1962) On the delivery behavior of squirrel monkeys *(Saimiri sciurea)* and a mona monkey *(Cercopithecus mona)*. Primates 3: 59-72

Tamby Raja RL, Hobel CJ (1983) Characterization of 24-h uterine activity in the second half of the human pregnancy. Proceedings of the 30th Annual Meeting of the Society for Gynecologic Investigation 17-20 March 1983, Washington, DC, p 280

Taylor NF, Martin MC, Nathanielsz PW, Serón-Ferré M (1983) The fetus determines circadian oscillation of myometrial electromyographic activity in the pregnant rhesus monkey. Am J Obstet Gynecol 146: 557-567

Thorburn GD, Challis JRG (1979) Endocrine control of parturition. Physiol Rev 59: 863-918

Viet G (1855) Beiträge zur geburtshülflichen Statistik. Monatsschr Geburtskd S: 344-381

Wallace LR (1949) Parturition in ewes and lamb mortality. Sheepfarming Annual. Massey Agricultural College, Palmerston North, New Zealand, pp 5-24

Walsh SW, Ducsay CA, Novy MJ (1984) Circadian hormonal interactions among the mother, fetus and amniotic fluid. Am J Obstet Gynecol 150: 745-753

White CS (1905) The relation of conception and birth to season and hour. Am J Obstet 52: 527-532
Williams JW (1922) Obstetrics. A textbook for the use of students and practitioners, 4th edn. Appleton, New York, p 251
Wurster K (1949) Untersuchung zur Frage tagesperiodischer, lunarer und meteorobiologischer Einflüsse auf den Wehenbeginn. Zentralbl Gynäkol 2: 159-168
Yellon SM, Foster DL (1986) Development of the pineal melatonin rhythm and its role in normal and delayed puberty in the female lamb. In: Gupta D, Reiter RJ (eds) The pineal gland during development: from fetus to adult. Croom Helm, London, pp 153-165
Yellon SM, Longo LD (1987) Melatonin rhythms in fetal and maternal circulation during pregnancy in the sheep. Am J Physiol 252: E799-E802
Yerganian G (1958) The striped-back or Chinese hamster, Cricetulus griseus. J Natl Cancer Instit 20: 705-721
Yerushalmy J (1938) Hour of birth and stillbirth and neonatal mortality rates. Child Dev 9: 373-378
Zwoliński J, Siudiński S (1965) Dobowy rozklad wyźrebień u klaczy. Med Wet 21: 614-616

Ontogenetic Changes in Olfactory Bulb Neural Activities of Pouched Young Tammar Wallabies

F. Ellendorff[1, 3], C. H. Tyndale-Biscoe[2], and R. Mark[1]

Introduction

Within the wide field of fetal physiology, fetal neuroendocrinology has sprouted only relatively recently (Ellendorff et al. 1984a, b). Direct neurophysiological approaches to brain control systems are at the seeding stage at best, and the difficulties encountered in approaching the fetal brain in placental animals directly in situ are numerous (Konda et al. 1979). Major problems are due to maternal influences, and these can be avoided by using non-placental animals, such as the chick (Grossmann et al. 1986a, b). Another alternative non-placental model is provided by marsupials. This presentation will use the pouched young tammar wallaby as a model to study early neonatal maturation. The olfactory system was chosen because it is well exposed anatomically in the marsupial, because of some experience with the olfactory system by one of the authors (Reinhardt et al. 1981; MacLeod 1979, 1983) and because of the close functional relationship which exists between the olfactory bulb and the pituitary-gonadal system (MacLeod et al. 1979; Merriam et al. 1977).

Materials and Methods

Thirty-three tammar wallabies *(Macropus eugenii)* originated from the department breeding colony of the CSIRO-Wildlife Research at Canberra.

Twenty-five pouched young (PY) of various ages [1–70 days in pouch (PY 1–70), $n=15$; PY 71–140, $n=10$] were taken from the maternal pouch either on the day of experiment or after incubation at 37 °C for no more than 24 h after removal from the mother. The youngest pouched animal used was PY 1, the oldest was PY 135. In addition, eight male adolescent wallabies (1.7–3.0 kg) were used. The

---

[1] Department of Behavioural Biology, The Australian National University, Research School of Biological Sciences, PO Box 475, Canberra City, ACT 2601, Australia
[2] Division of Wildlife Research, Commonwealth Scientific and Industrial Organization, Lynetham, ACT, Australia
[3] Institut für Tierzucht und Tierverhalten (FAL), Mariensee, 3057 Neustadt 1, FRG (Address of correspondence)

The Endocrine Control of the Fetus
Ed. by W. Künzel and A. Jensen
© Springer-Verlag Berlin Heidelberg 1988

age of the pouched young was estimated from their weight and naso-occipital measurements. All pouched young were anaesthetized with urethane (400 mg-1.4 g per kilogram body weight) intraperitoneally. Adults received 60 mg/kg pentobarbital in addition as pretreatment. Wallabies up to PY 70 were then fixed into cubical Silastic moulds made in several different sizes to fit the various age groups. The mould was subsequently placed into a rat stereotaxic frame. An advanced type moulding device has since been described for the chicken embryo by Grossmann and Ellendorff (1986a) and their report can be consulted for details. The silastic mould allowed fixation of the soft skull without exerting undue pressure on the brain. Older pouched young and adults were fixed directly into a rat or a cat stereotaxic frame, respectively. All pouched young were surrounded by cotton wool and placed into a glass cylinder. Body temperature was maintained at 37 °C by a heating lamp. This is of particular importance, since at lower temperatures younger pouched young strongly reduce functional activities including respiration.

The skin above the skull was retracted over the area of the olfactory bulb and the skull was carefully removed either by forceps or by rongeurs, or, in adults, a hole was drilled, for electrode insertion. The dura mater was pierced or cut over the medial quadrant of the olfactory bulb, which protruded considerably. In very young pouched young care must be taken to keep the hole small.

A micropipette containing a glass-fibre filament (1-2 μm tip diameter) and filled with 4 M NaCl was introduced at an anterior-to-posterior angle of 45° and advanced under microscopic vision to near the surface of the olfactory bulb. A hydraulic microdrive was used to advance further until contact was made with the surface and then until neurons were encountered. Extracellular unit activity was recorded, using conventional amplifying, processing and storage equipment. Data were recorded on tape for later analysis.

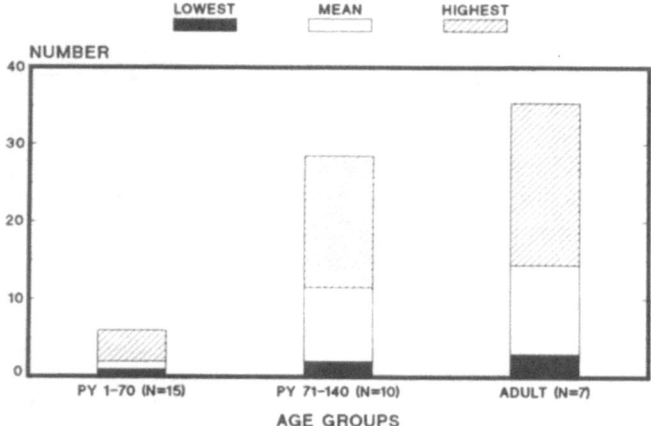

Fig. 1. Distribution of number of neurons encountered in three age groups of tammar wallabies

In five animals, three aged PY 44, PY 103, and PY 135 and two adolescents, field potentials were recorded. To this end a second, fork-shaped stainless steel electrode (tip separation 1.5 mm) was introduced posterior to the olfactory bulb into the lateral olfactory tract and adjusted until maximum field amplitude occurred with the recording electrode on the surface of the bulb. As the electrode was advanced, the field was recorded to obtain an estimate of the depth and function of the mitral cell layer.

Upon appearance of a neuron, firing rates were recorded and the shape of the action potential was assessed. In several cases where a neuron was stable enough, an olfactory test was made. Either a small cotton swab dipped in eucalyptus oil was placed near the nostrils or air was passed by. Changes in firing rates were observed.

Results

General Neuronal Activities

A total of 189 neurons were obtained from the 32 tammars used. In 15 animals aged PY 1-PY 70 only 14 cells were recorded (Fig. 1). From six animals (PY 1-PY 38) not a single neuron could be maintained. The first neuron was obtained in an animal aged PY 39; it fired at a rate of 5.7 Hz. Of the 13 neurons recorded in animals between PY 41 and PY 70, seven were stable, six were transient. Characteristic of neurons at this age is the low amplitude, instability and, usually, axo-somatic spike (A/B wave in Fig. 2) separation, indicative of slow initial segment conduction veloc-

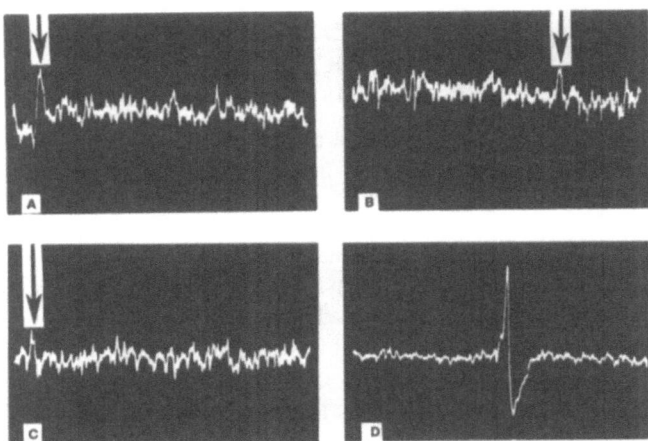

Fig. 2 A-D. Axo-somatic spike separation. Typical units in young pouched young: Here, PY 44. **A-C** Hard to recognize and to process; *arrows* indicate the spikes. **D** "Normal" unit

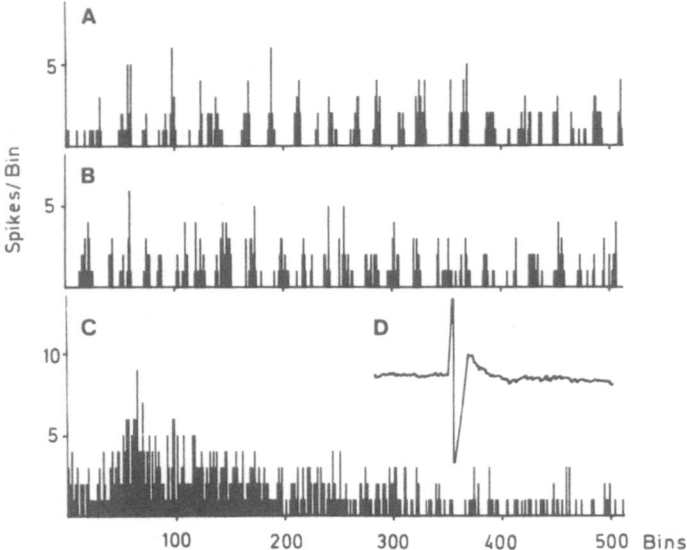

Fig. 3. Continuous record (**A, B**) and interval histogram (**C**) of a neuron from an animal aged PY 81. Redrawn action potential in (**D**). Bin width in **C**: 1 ms

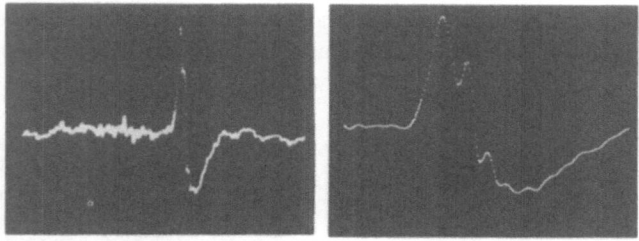

Fig. 4. Neuron from an animal aged PY 81: Axo-somatic spike separation at two different speeds

ity. The absence of recordable neurons at early developmental ages does not imply that no neurons are present: typically, higher frequency discharges, acoustically similar to those of neurons approaching the electrode and then dying, were recognized in the background. But low-amplitude neurons were often lost as soon as they were encountered.

In the second half of the pouched young age range, PY 71–PY 140, 95 neurons were recorded from ten animals (Figs. 1, 3). Interestingly, every single animal yielded between two and 21 neurons that fired at rates between less than 0.1 Hz and 4.0 Hz. Over 90% of the neurons in which firing rates could be established fired at

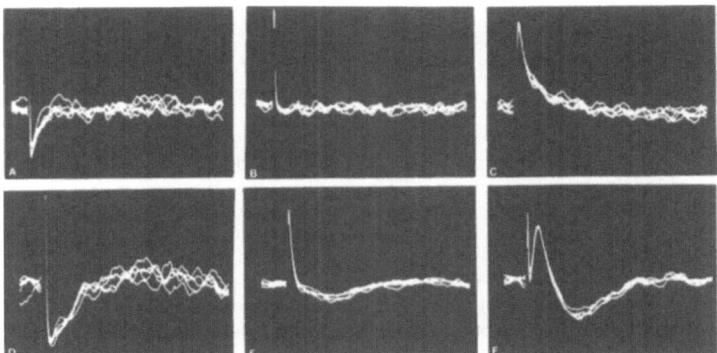

Fig. 5. Olfactory bulb field in an animal aged PY 103 (A–C) and in an adult (D–F). Field potentials recorded in the lower mitral layers are shown. A 4320 μm; B 4410 μm; D 5650 μm; E 7610 μm; F 8000 μm. 15 V, 5 sweeps. **B,E:** zero isopotential line

below 2.0 Hz. Axo-somatic spike separation was still observed (Fig. 4), but did not seem to be as frequent as at the earlier age levels.

In adolescent tammars 80 neurons were recorded from eight animals. Between two and 17 neurons were counted per animal, which is similar to the numbers obtained from animals aged PY 70–PY 140. Firing rates at this age ranged between 0.2 Hz and 20.2 Hz; 66% of the neurons fired at rates above 2.0 Hz. A/B wave separations were present occasionally.

Field Potentials

The mitral cell layer depth was estimated in some animals from the depth at which the zero isopotential line was observed (Fig. 5). In the earliest usable field potential of an animal aged PY 44 the upper mitral layer occurred at 1920 μm down from the surface of the bulb; in animals at PY 103 and PY 135 the lower mitral layer was encountered at 4358 μm and 8300 μm, respectively. In two adults, the upper mitral layer was estimated at 2800 μm, the lower at 7400 μm and 9450 μm, respectively. Thus, field potential recordings in the PY 44 animal indicate that by then the mitral cell layer must be differentiated functionally and its upper layer is at a depth of nearly 2 mm.

Olfactory Stimulus

Fourteen neurons in six pouched young (PY 63, PY 81, PY 81, PY 100, PY 100, PY 135) were tested with a strong smell that they would normally encounter in the wild, eucalyptus oil. Five of the neurons responded by increased activity, the others

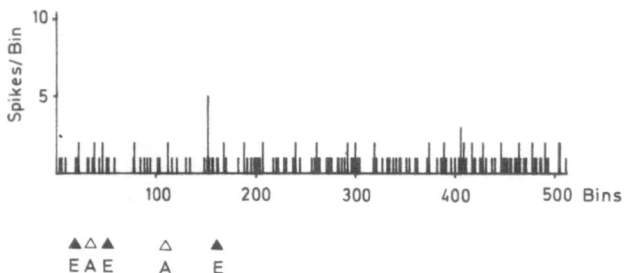

Fig. 6. Post-stimulus histogram of a neuron from an animal aged PY 81: no response to eucalyptus odour. *E*, eucalyptus; *A*, Air. Bin width 500 ms. Firing rate 1.4 Hz

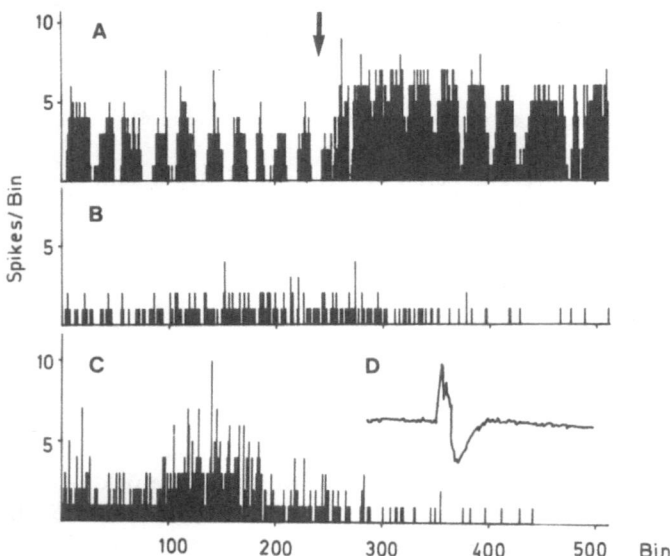

Fig. 7 A–D. Olfactory bulb neuron in PY 81 animal. **A** Spontaneous firing rate before and after *(arrow)* eucalyptus oil. B, C Interval histograms before (**B**) and after (**C**) eucalyptus oil. Bin
width in A/C 1 ms. Redrawing of the original action potential. Firing rate 4.05 Hz

did not (Fig. 6). The earliest change in neuronal firing was evoked in four neurons
of two animals aged PY 81, the next recording was possible in a PY 100 animal. In
the former, both relatively fast-firing (4 Hz), and slow (0.76 Hz), irregularly firing
neurons were affected. After stimulation, both firing rates were elevated and firing
interval distributions were markedly altered (Fig. 7), increasingly so upon repeated
exposure (Fig. 8). Thus, the PY tammar wallaby's olfactory system is sensitive to
strong odours at least from PY 81 on.

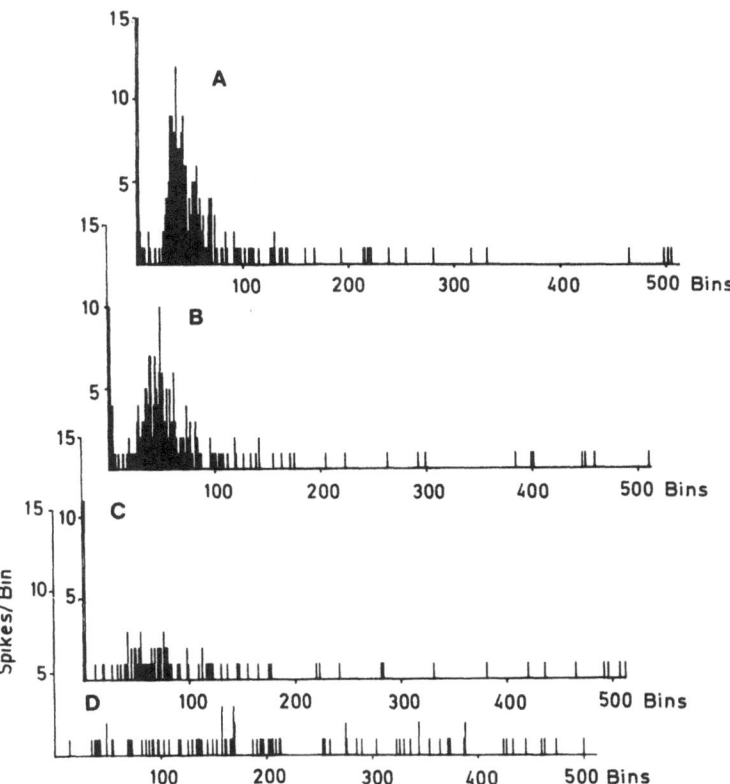

Fig. 8 A–D. Change in spike interval histograms of one neuron with successive exposure to eucalyptus smell. **A** Control; **B** first treatment; **C** second treatment; **D** third treatment. Bin width 1 ms

Conclusions

In a first attempt, early functional developmental changes in neuronal characteristics were established in a marsupial ontogenetic model, that has so far attracted relatively little attention, despite obvious advantages for invasive studies of brain development. In the pouched young tammar wallaby, olfactory bulb neuronal activity can be recorded from PY 39 on. Between PY 1–PY 70 low stability, low amplitude and axo-somatic spike separation are characteristic. Olfactory neurons from animals aged PY 71–PY 140 fire markedly slower than neurons from adolescent tammars. Evoked potential changes, encountering a zero isopotential zone at the presumptive mitral cell layer as early as PY 44, indicate that by then a functional differentiation of mitral cells has occurred. Responses of olfactory bulb neurons to a strong-smelling stimulus (eucalyptus oil) were evoked at PY 81. Attempts in a younger animal (PY 63) failed to alter firing rates.

References

Ellendorff F, Gluckman PD, Parvizi N (eds) (1984a) Fetal neuroendocrinology. Perinatology Press, Ithaca (Research in perinatal medicine, vol 2)
Ellendorff F, Parvizi N, Elsaesser F, Bruhn Th, Ponzilius KH (1984b) Ontogeny of neuroendocrine control of gonadotrophin secretion in the fetal pig. In: Ellendorff F, Gluckman PD, Parvizi N (eds) Fetal neuroendocrinology. Perinatology Press, Ithaca, pp 165-174 (Research in perinatal medicine, vol 2)
Grossmann R, Ellendorff F (1986a) Functional development of the prenatal brain. I. Recording of extracellular action potentials from the magnocellular system of the 18-day-old chicken embryo. Exp Brain Res 62: 635-641
Grossmann R, Ellendorff F (1986b) Functional development of the prenatal brain. II. Ontogeny of the hypothalamo-neurohypophyseal axis in the pre- and perinatal chicken brain. Exp Brain Res 62: 642-647
Konda N, Dyer RG, Bruhn T, Macdonald AA, Ellendorff F (1979) A method for recording single unit activity from the brains of foetal pigs in utero. J Neurosci Methods 1: 289-300
MacLeod N, Reinhardt W, Ellendorff F (1979) Olfactory bulb neurons of the pig respond to an identified steroidal pheromone and testosterone. Brain Res 164: 323-327
Merriam GR, Beitins IZ, Bode HH (1977) Father-to-son transmission of hypogonadism with anosmia. Kallmann's syndrome. Am J Dis Child 131: 1216-1219
Reinhardt W, Konda N, MacLeod N, Ellendorff F (1981) Electrophysiology of olfacto-limbic-hypothalamic connections in the pig. Exp Brain Res 43: 1-10
Reinhardt W, MacLeod NK, Ladewig J, Ellendorff F (1983) An electrophysiological study of the accessory olfactory bulb in the rabbit. II. Input-output relations as assessed from analysis of intra- and extracellular unit recordings. Neuroscience 10: 131-139

The Functional Maturation of the Somatotropic Axis in the Perinatal Period

P. D. Gluckman[1], N. Bassett, and K. T. Ball

Introduction

The regulation of growth differs between fetal and postnatal life. The growth of the fetus in late gestation is primarily determined by the availability of substrate and by the degree of maternal constraint on growth. The fetus does not grow to its maximal potential and the role of hormones in the regulation of fetal growth must be seen as that of mediating the interaction between limited substrate availability and the genetic drive to grow (see Gluckman and Liggins 1984; Gluckman 1986 for review). By contrast, after weaning the postnatal organism is able to grow to its genetic potential, because the availability of substrate is not normally limiting. Endocrine factors are thus acting to co-ordinate the growth of the postnatal organism to its maximum potential. It is therefore to be expected that the status of the somatotropic axis, the major endocrine axis involved in the regulation of growth, will differ between fetal and postnatal life.

In the mammalian fetus growth hormone concentrations are high, are not greatly influenced by nutritional factors, and appear to have only a very limited role in influencing fetal growth and development. Receptor immaturity appears to be the basis of this limitation. By contrast, postnatally, following differentiation of the growth hormone receptor, growth hormone has a major role in the regulation of postnatal growth; its concentrations are much lower and are strongly influenced by nutrition (see Gluckman et al. 1987 for review). The insulin-like growth factors which are believed to mediate the effects of growth hormone on somatogenic growth postnatally appear also to have this role in utero. However, as would be expected from the above discussion, they are regulated in utero not by growth hormone but by substrate availability and perhaps by placental factors. This review will focus on these issues with particular reference to the developing sheep, the species for which most data are available.

---

[1] Developmental Physiology Laboratory, Department of Paediatrics, University of Auckland, Private Bag, Auckland, New Zealand

The Endocrine Control of the Fetus
Ed. by W. Künzel and A. Jensen
© Springer-Verlag Berlin Heidelberg 1988

Circulating Growth Hormone

In all species studied growth hormone concentrations in the fetal circulation are considerably higher than in the adult (see Gluckman et al. 1981 for review). In the fetal sheep growth hormone is secreted in a markedly pulsatile manner, the amplitude of the pulses being 10- to 20-fold higher than in the mature sheep (Bassett and Gluckman 1986). The pulses are frequent, and consequently interpulse values are also much higher than postnatally. The ontogeny of the pulsatile pattern of release has been studied and shows that growth hormone concentrations are already high from at least 65 days of gestation (term is 147 days). Between 120 and 140 days of gestation growth hormone secretion rises; it then decreases during the last week before delivery. Fetal growth hormone secretion is sexually dimorphic, with higher concentrations in the male than female (Bassett and Gluckman 1986); analogous to the pattern described postnatally in some other species (Jansson et al. 1985). This sexual dimorphism presumably reflects sexual differentiation of the hypothalamus at the critical period, which in the sheep is between 60 and 80 days of gestation (Short 1974). The decrease in growth hormone secretion over the week prior to birth can be mimicked by glucocorticoid infusion to the fetus, suggesting that it is related to the prenatal glucocorticoid surge (Lowe et al. 1985). However, at birth growth hormone concentrations are still very elevated; the mean concentration in umbilical cord blood is 119 ± 35 ng/ml. The most dramatic feature is the rapid fall in growth hormone concentrations immediately after birth, to reach 35 ± 14 ng/ml within 1 h (Bassett and Gluckman 1986). In premature delivery the fall is also linked to birth, but it is not to such low levels, and there appears to be a more gradual decrease over some days to reach typical postnatal values. This is best documented in the human neonate (Cornblath et al. 1965).

Earlier it had been suggested that the decrease immediately linked to parturition might reflect kinetic factors (Bassett et al. 1970). We have studied the disappearance rate of ovine growth hormone immunoreactivity when the growth hormone was intravenously administered in late-gestation fetuses, neonatal lambs and an experimental preparation in which the events of birth were simulated in utero (Power et al. 1987a). The disappearance rate was studied using noncompartmental analysis and multiple exponential techniques. Simulated birth in utero was achieved in fetuses treated at least 24 h previously by sequentially cooling the fetus by means of a coil of tubing around it, oxygenating it to a partial oxygen pressure of approximately 100 Torr, then separating it from any placental influence by use of a snare around the umbilical cord. No change in the plasma clearance rate was observed (fetus 3.4 ± 0.2 ml/min·kg body weight; simulated birth 3.2 ± 0.2 ml/min·kg body weight; lamb 3.9 ± 0.1 ml/min·kg body weight). The secretory rate fell from 2.4 ± 0.2 µg/min in the fetus to 1.2 ± 0.1 µg/min following simulated birth and to 0.27 ± 0.02 µg/min in the neonatal lamb. Thus placental separation in some way leads to an immediate decrease in the secretory rate of growth hormone, and the fall in growth hormone levels at birth reflects a change in secretion, not in clearance.

There are several major questions to be addressed here: Why are fetal growth hormone concentrations so markedly elevated? Is growth hormone secretion in the fetus regulated in a manner analogous to that after birth? What factors determine the changing concentrations of growth hormone in the fetal circulation prior to birth? And what is the basis of the very rapid decrease in growth hormone secretion at birth?

Postnatally, secretion of growth hormone is primarily regulated by two hypothalamic peptides, growth hormone releasing factor (GRF) and somatostatin (SRIF). They interact on the somatotrope to create the typical pattern of low amplitude pulsatile growth hormone release seen postnatally. In turn, the release of GRF and SRIF appears dependent on a variety of neurotransmitters and extrahypothalamic influences. Growth hormone exerts negative feedback on its own secretion by both ultrashort and short feedback loops. There is increasing evidence that insulin-like growth factor-1 (IGF-1) can inhibit growth hormone secretion both at pituitary and hypothalamic level.

Fetal growth hormone secretion is clearly dependent on the fetal hypothalamus. Stereotactic lesions of the fetal mediobasal hypothalamus in the region of the cell bodies for GRF cause a rapid fall in growth hormone secretion to very nonpulsatile levels (Gluckman and Parsons 1985). Thus, the fall in growth hormone at birth cannot be simply due to the withdrawal of a direct placental stimulus, although GRF has recently been identified in the rat placenta (Baird et al. 1985). Synthetic GRF stimulates fetal release of growth hormone in the sheep from at least 70 days of gestation. The response is much greater than postnatally but shows some decrease in magnitude as gestation progresses (Gluckman 1984a). A number of other neurotransmitter agonists also stimulate fetal release of growth hormone; their actions are generally considered to be mediated by GRF. These include β-endorphin (Gluckman et al. 1980), cholinergic agonists (Gluckman et al. 1985) and serotinergic agonists (Marti-Henneberg et al. 1980). All induce considerably greater release of growth hormone prenatally than postnatally. This hyper-responsiveness is also seen postnatally in very altricial species such as rat (Szabo and Cuttler 1986) and suggests that one feature of the developing pituitary is hyper-responsiveness to GRF. This appears to change with maturation and in some species to continue to reduce after birth. The fetal pituitary is also responsive to SRIF. However, high-dose intravenous infusion of somatostatin to the fetal lamb only partially inhibits growth hormone release (Gluckman et al. 1979). Similarly, other inhibitory factors whose action may either be mediated by SRIF or be direct on the somatotrope fail to reduce fetal growth hormone release by more than 50%. These include dopaminergic, β-adrenergic and GABA-ergic agonists (see Gluckman 1984b for review). The only experiment in which fetal growth hormone can be reduced to the low levels typical of neonate is by the infusion of a fatty acid emulsion (Bassett and Gluckman 1987a). This has been shown to act at the level of the somatotrope by blocking responsivity to GRF. Thus we hypothesised that the fetal somatotrope, while excessively responsive to stimulatory secretagogues, was relatively resistant to inhibitory control. This gating would be an explanation for high fetal growth hormone concentrations, and a gradual change in the relative secretion of and response to GRF and SRIF might explain the changing pattern of growth hormone secretion in the fetus.

While these changes might be at the level of the GRF and/or SRIF receptor, recent developments in our understanding of the postreceptor signal transduction systems offer further possibilities. GRF is now known to affect growth hormone release via the Gs protein system and SRIF via the Gi system. An immaturity in the Gi system would offer one model by which the pituitary appears gated to stimulatory secretagogues and only partially responsive to inhibitory agents. Recent experimental evidence supports such a gating model. In vitro both GRF and SRIF modulate growth hormone secretion by the ovine (Silverman et al. 1987) and human fetal somatotrope (Goodyer et al. 1985). However, while SRIF can block the stimulatory effect of GRF in the postnatal pituitary it does not do so in the fetal pituitary (Goodyer et al. 1985). Similar observations are reported in the neonatal rat, where SRIF was less effective in suppressing the effect of GRF in vitro than in the mature rat (Cuttler et al. 1986). Thus the current hypothesis is that the high fetal growth hormone concentrations reflect primarily an immaturity at the level of the somatotrope rather than incomplete differentiation of hypothalamic neural mechanisms. Because the pituitary responds poorly to inhibitory factors, growth hormone itself and perhaps also placental lactogen do not provide an efficient feedback loop. Furthermore, the low levels of IGF-1 in the fetus, a consequence of an immaturity of growth hormone receptors, also reduce the possibility of negative feedback.

The above discussion suggests that there is a progressive change in the ability of the somatotrope to respond to its normal regulators. This may be accelerated by the prenatal glucocorticoid surge. However, the very rapid decrease in the secretion of growth hormone, at least in the sheep, must be explained in any model of the maturation of growth hormone secretion. This very rapid fall soon after birth remains to be studied in other species.

A series of experiments in our laboratory has addressed this question. Our initial hypothesis was that it is the rapid increase in free fatty acid levels at birth (see Gluckman et al., this volume, pp 300–305) that led to the fall, in growth hormone secretion at birth. Growth hormone is known to be markedly influenced by free fatty acids after birth and we had shown that an emulsion rich in free fatty acids markedly inhibited growth hormone secretion in the fetus (Bassett and Gluckman 1987a). In our simulated birth in utero model the fetus was first cooled and then ventilated with oxygen, and then the umbilical cord was snared. The level of free fatty acids rose somewhat with oxygenation and markedly following cord snare (Power et al. 1987b) and the fall in growth hormone levels paralleled these changes in free fatty acids. Furthermore, when the model was extended by ventilating with nitrogen before the oxygen and preceding the cord snare with a triiodothyronine infusion, growth hormone levels again fell in direct parallel to the rise in free fatty acids. These studies suggested that placental separation in some way allowed the onset of non-shivering thermogenesis and this rise in free fatty acids acted to inhibit growth hormone release. Alternatively, the placenta may secrete a factor which directly inhibits the ability of SRIF to act effectively on the somatotrope or enhances the action of GRF. Any such placental factor would of necessity have a very short half-life. Such a model has also been used to explain how placental separation induces the onset of breathing (Adamson et al. 1987), and, in the presence of cooling, the onset of non-shivering thermogenesis (Power et al. 1987b).

However, a further factor must be considered. Once growth hormone secretion has reduced following delivery, growth hormone levels remain low relative to fetal levels. In the simulated birth in utero models in which the fetus was cooled before cord snare, growth hormone levels again rose following the cessation of cooling even in the absence of any placental influence. One possibility is that catecholamines have a priming effect on the somatotrope, to irreversible reset the gating of the pituitary. During normal labour the fetus undergoes prolonged exposure to elevated catecholamine levels. This is a complex model, and clearly more experimental data are required both in vivo and in vitro.

Growth Hormone Action

Despite its very high levels, the role of growth hormone in the fetus appears relatively minor, at least in late gestation (Gluckman 1986). The human infant with congenital growth hormone deficiency is of normal size, and reduced hepatic glycogen appears to be the major metabolic consequence (see Gluckman et al. 1981 for review). Growth hormone appears to play some role in the proliferation of the pancreatic islet β-cell. This has been demonstrated in the hypertrophy seen in infants of diabetic mothers and recently in in vitro experiments (Swenne et al. 1987). Similarly in the sheep growth hormone appears to have some influence on insulin secretion and action (Parkes and Bassett 1985). Growth hormone has been suggested to have a key role in the differentiation of the pre-adipocyte (Morikawa et al. 1984), and it may be that it has a number of actions in early embryogenesis. The various observations of natural gestation or experiments in the laboratory which have led to the conclusion that growth hormone does not have a major role in fetal development are complicated by the possible role of placental lactogen and the increasing number of other placental peptides, including (in humans) a placental growth hormone (Frankenne et al. 1987), that all may act at the same receptor as pituitary growth hormone.

Nevertheless, the weight of evidence is that growth hormone does not play a major role in fetal somatic growth. Somatogenic receptors cannot be demonstrated in sheep liver until some days after birth (Gluckman et al. 1983). Their appearance coincides with the rise in plasma IGF-1 concentrations (Gluckman and Butler 1983) and in the appearance of the growth hormone-dependent, 150000-dalton, IGF-binding protein (Butler and Gluckman 1986). Thus there appears to be reasonable evidence that the receptor measurements correlate with functional observations and this explains the lack of major effect of growth hormone on fetal growth in this species. The ontogeny of the somatogenic receptors in most other species can be inferred from a variety of data. Generally they appear postnatally at about the time of weaning. This would correlate with the time when the organism's growth can from a teleological perspective be considered to be no longer substrate-limited. The recent observation that in fetal liver there are specific receptors for placental lactogen which mediate metabolic events, including glycogen deposition, separates somatogenic from other potential actions of the growth hormone family of hormones in the fetus (Freemark and Handwerger 1986). In humans the picture is not as clear as in sheep. Growth hormone-dependent growth is not seen until after birth, but the 150000-dalton, IGF-binding protein does appear in the fetal circulation and appears to be under pituitary control (D'Ercole et al. 1980). Receptors with high affinity for human growth hormone are reported in the fetal liver (Hill et al. 1987) and it may be that there is a heterogeneity of growth hormone receptors subserving different functions, as has been suggested from postnatal studies (Breier et al. 1988), and that only some subtypes are seen in utero. Some data in mice suggest that various growth hormone-dependent proteins show quite different ontogenic patterns (Berry et al. 1986). There may well be a different timetable of appearance in different tissues. For example, while growth hormone does not have receptors in liver in fetal lamb, it appears to play a dominant role in the mobilisation of fat in lamb in late gestation, implying the presence of receptors in this tissue (Stevens and Alexander 1986).

The Insulin-like Growth Factors

Neither IGF-1 nor IGF-2 appears to be under growth hormone regulation in fetal lamb (Gluckman and Butler 1985). IGF-1 levels are relatively low in the fetus. Generally the plasma levels have correlated with birth size, suggesting a role in fetal growth. This is reviewed at considerable length elsewhere (see Gluckman 1986). However, as the 150000-dalton binding protein is low in the fetus and IGF-1 circulates primarily in low molecular weight form (45000 daltons; Butler and Gluckman 1986), the secretory rate of IGF-1 in the fetus is likely to be very high (the half-life of the 45000-dalton form being much less than that of the 150000-dalton form in postnatal lamb; Hodgkinson et al. 1987).

In view of the generally held hypothesis that IGF-1 has autocrine, paracrine and endocrine actions, tissue IGF concentrations as well as plasma concentrations need to be studied. The placenta as well as other fetal tissues has been demonstrated to be a source of IGF-1 (Fant et al. 1985). IGF-2 levels are high in the fetus of the rat, guinea pig (Daughaday et al. 1986) and sheep (see Gluckman 1986). In humans altough mRNA levels of IGF-2 are high, expression in the fetus appears to be at a low level (Ashton et al. 1985). Thus, the role of IGF-2 is very unclear. In both intrauterine growth retardation in the sheep secondary to either carunclectomy (Bassett and Gluckman 1987b) or pancreatectomy (Gluckman et al. 1987b) and in the growth-retarded guinea pig (Jones et al. 1987), IGF-2 levels are elevated, suggesting that IGF-2 cannot be considered as a major fetal growth regulator. As IGF-2 is affected by metabolic factors in the fetus it may be that IGF-2 has a role in fetal metabolic processes. The high concentration of IGF-2 type receptors in the placenta of all species suggests that IGF-2 may have a specific role in either placental growth or function.

It is clear that substrate availability, either directly or, possibly, mediated by insulin, plays a major role in the regulation of IGF-1 and IGF-2 in the fetus (see Gluckman 1986). It has also been suggested that placental lactogen is a significant determinant of fetal IGF secretion. While there is some limited in vitro data to support this contention (Adams et al. 1983), direct in vivo data are lacking. Similarly, direct evidence that the IGFs play a role in either fetal growth or fetal metabolism remains to be obtained.

Acknowledgement. This research was funded by grants and fellowships from the Medical Research Council of New Zealand.

References

Adams SO, Nissley SP, Handwerger S, Rechler MM (1983) Developmental patterns of insulin-like growth factor-I and -II synthesis and regulation in rat fibroblasts. Nature 302: 150–152

Adamson SL, Richardson BG, Homan J (1987) Initiation of pulmonary gas exchange by fetal sheep in utero. J Appl Physiol 62: 989–998

Ashton IK, Zapf J, Einschenk I, Mackenzie IZ (1985) Insulin-like growth factors (IGF) 1 and 2 in human foetal plasma and relationship to gestational age and foetal size during mid-pregnancy. Acta Endocrinol 110: 558–563

Baird A, Wehrenberg WB, Bohlen P, Ling N (1985) Immunoreactive and biologically active growth hormone-releasing factor in the rat placenta. Endocrinol 117: 1598–1601

Bassett N, Gluckman PD (1986) The ontogenesis of pulsatile growth hormone release in the fetal and neonatal lamb. J Endocrinol 109: 307–312

Bassett NS, Gluckman PD (1987a) The effect of fatty acid infusion on growth hormone secretion in the ovine fetus: evidence for immaturity of pituitary responsiveness in utero. J Dev Physiol (in press)

Bassett NS, Gluckman PD (1987b) Insulin-like growth factors in experimental fetal growth retardation in the sheep. Endocrinology 120: 234A

Bassett JM, Thorburn GD, Wallace ALC (1970) Plasma growth hormone concentrations of foetal lamb. J Endocrinol 48: 251–263

Berry SA, Manthei RD, Seelig S (1986) Ontogenesis of growth hormone (GH) responsive hepatic gene products: lack of correlation with hepatic GH receptor content. Endocrinology 119: 2290–2296

Breier BH, Gluckman PD, Bass JJ (1988) The somatotrophic axis in young steers: influence of nutritional status and oestradiol 17-B on hepatic high and low affinity somatotrophic binding sites. J Endocrinol 6: 169–177

Butler JH, Gluckman PD (1986) Circulating insulin-like growth factor binding proteins in fetal neonatal and adult sheep. J Endocrinol 109: 333–338

Cornblath M, Parker MK, Reisner SH, Forbes AE, Daughaday WH (1965) Secretion and metabolism of growth hormone in premature and full term infants. J Clin Endocrinol 25: 209–218

Cuttler L, Welsh JB, Szabo M (1986) The effect of age on somatostatin suppression of basal growth hormone releasing factor stimulated and bibutyryl adenosine 3,5-monophosphate stimulated GH release from rat pituitary cells in monolayer culture. Endocrinoloy 119: 152–158

Daughaday WH, Yanow CE, Kapadia M (1986) Insulin-like growth factors I and II in maternal and fetal guinea pig serum. Endocrinology 119: 490–494

D'Ercole AJ, Willson DF, Underwood CE (1980) Changes in circulating form of serum somatomedin C during fetal life. J Clin Endocrinol Metab 51: 674–676

Fant ME, Munro HN, Moses AC (1985) An autocrine/paracrine role for insulin-like growth factor I in the regulation of human placental growth. Endocrinology 116: 17A

Frankenne F, Rentier-Delrue F, Scippo M-L, Martial J, Hennen G (1987) Expressions of the growth hormone variant gene in human placenta. J Clin Endocrinol Metab 64: 635–637

Freemark M, Handwerger S (1986) The glycogenic effects of placental lactogen and growth hormone in ovine fetal liver are mediated through binding to specific fetal ovine placental lactogen receptors. Endocrinology 118: 613–618

Gluckman PD (1984a) Changing responsiveness to growth hormone releasing factor in the perinatal lamb. J Dev Physiol 6: 509–515

Gluckman PD (1984b) Functional maturation of the neuroendocrine axis in the perinatal period: Studies of the somatotropic axis in the ovine fetus. J Dev Physiol 6: 301–312

Gluckman PD (1986) The role of pituitary hormones, growth factors and insulin in the regulation of fetal growth. In: Clarke JR (ed) Oxford reviews of reproductive biology, vol 8. Clarendon, Oxford, pp 1–60

Gluckman PD, Butler JH (1983) Parturition related changes in insulin-like growth factors-I and -II in the perinatal lamb. J Endocrinol 99: 223–227

Gluckman PD, Butler JH (1985) Circulating insulin-like growth factor-I and -II concentrations are not dependent on pituitary influences in the midgestation fetal lamb. J Dev Physiol 7: 411–420

Gluckman PD, Liggins GC (1984) The regulation of fetal growth. In: Beard R, Nathanielsz P (eds) Fetal physiology and medicine. Dekker, New York, pp 511–558

Gluckman PD, Parsons Y (1985) Growth hormone secretion in the fetal sheep following stereotaxic electrolytic lesioning of the fetal hypothalamus. J Dev Physiol 7: 25–36

Gluckman PD, Mueller PL, Kaplan SL, Rudolph AM, Grumbach MM (1979) Hormone ontogeny in the ovine fetus. III. The effect of exogenous somatostatin. Endocrinology 104: 974

Gluckman PD, Marti-Henneberg C, Kaplan SL, Li CH, Grumbach MM (1980) Hormone ontogeny in the ovine fetus. X. The effects of beta endorphin and naloxone on circulating growth hormone, prolactin and chorionic somatomammotropin. Endocrinology 107: 76–80

Gluckman PD, Grumbach MM, Kaplan SL (1981) The neuroendocrine regulation and function of growth hormone and prolactin in the mammalian fetus. Endocr Rev 2: 363–395

Gluckman PD, Butler JH, Elliott TB (1983) The ontogeny of somatotropic binding sites in ovine hepatic membranes. Endocrinology 112: 1607–1612

Gluckman PD, Villiger JW, Taylor KM (1985) Muscarinic influences on growth hormone secretion in the fetal and neonatal sheep: pharmacological studies and in vitro binding studies. J Dev Physiol 7: 411–419

Gluckman PD, Breier BH, Davis SR (1987a) The physiology of the somatotropic axis with particular reference to the ruminant. J Dairy Sci 70: 442–464

Gluckman PD, Butler JH, Comline R, Fowden A (1987b) The effects of pancreatectomy on the plasma concentrations of insulin-like growth factors 1 and 2 in the sheep fetus. J Dev Physiol 9: 79–88

Goodyer CG, Marcovitz S, Berezuik M, De Stephano C, Lefebvre Y (1985) In vitro modulation of GH secretion from early gestation human fetal pituitaries. In: Ellendorff F, Gluckman PD, Parvizi N (eds) Fetal neuroendocrinology. Perinatology Press, Ithaca, pp 209–211

Han VKM, D'Ercole AJ, Lund PK (1987) Cellular localization of somatomedin (insulin-like growth factor) messenger RNA in the human fetus. Science 236: 193–197

Hill DJ, Strain AJ, Freemark M (1987) Presence of specific binding sites for human placental lactogen and growth hormone on human fetal tissue membranes during second trimester Endocrinology 120: 236A

Hodgkinson SC, Davis SR, Burleigh D, Henderson HV, Gluckman PD (1987) Metabolic clearance of protein bound and free insulin-like growth factor-I in fed and starved sheep. J Endocrinol 115: 223–240

Jansson JO, Eden S, Isaksson O (1985) Sexual dimorphism in the control of growth hormone secretion. Endocr Rev 6: 128–150

Jones CT, Lafeber HN, Price DA, Parer JT (1987) Studies on the growth of the fetal guinea pig. Effect of reduction in uterine blood flow on the plasma sulphation-promoting activity and on the concentrations of insulin-like growth factors I and II. J Dev Physiol 9: 181–201

Lowe KC, Gluckman PD, Jansen CAM, Nathanielsz PW (1985) Plasma growth hormone concentrations in the chronically catheterized ovine fetus during spontaneous term delivery and premature delivery induced by continuous intravascular infusion of low doses of adrenocorticotropin (Synacthen ACTH1-24) or cortisol to the fetus. Am J Obstet Gynecol 154: 420–423

Marti-Henneberg C, Gluckman PD, Kaplan SL, Grumbach MM (1980) Hormone ontogeny in the ovine fetus. XI. The serotoninergic regulation of growth hormone (GH) and prolactin (PRL) secretion. Endocrinology 107: 1273–1277

Morikawa M, Green H, Lewis UJ (1984) Activity of human growth hormone and relaxed polypeptides on the adipose conversion of 3T3 cells. Mol Cell Biol 4: 228–231

Parkes MJ, Bassett JM (1985) Antagonism by growth hormone of insulin action in fetal sheep. J Endocrinol 105: 379–382

Power GC, Ball KT, Gluckman PD (1987a) Disappearance of growth hormone from plasma of fetal and newborn sheep. Am J Physiol 254: E318–E322

Power GC, Gunn TR, Johnston BM, Gluckman PD (1987b) Oxygen supply and the placenta limit thermogenic responses in fetal sheep. J Appl Physiol 63: 1896–1901

Short RV (1974) Sexual differentiation of the brain of the sheep. International Symposium on Sexual Endocrinology of the Perinatal Period. INSERM, vol 32, pp 121–142

Silverman BL, Bettendorf M, Kaplan SL, Grumbach MM, Miller WL (1987) Modulation of basal and growth hormone releasing factor induced secretion of growth hormone in cultured fetal and neonatal ovine pituitary cells. Endocrinology 120: 234A

Stevens D, Alexander G (1986) Lipid deposition after hypophysectomy and growth hormone treatment in the sheep fetus. J Dev Physiol 8: 139–145

Swenne I, Hill DJ, Strain AJ, Milner RDG (1987) Growth hormone regulation of somatomedin C/insulin-like growth factor I produciton and DNA replication in fetal rat islets in tissue culture. Diabetes 36: 288–294

Szabo M, Cuttler L (1986) Differential responsiveness of the somatotroph to growth hormone-releasing factor during early neonatal development in the rat. Endocrinology 118: 69–73

Endocrine Control of
the Fetal Growth

Placental Metabolism and Endocrine Effects in Relation to the Control of Fetal and Placental Growth

C. T. Jones[1], J. E. Harding, W. Gu, and H. N. Lafeber

Introduction

In the search for factors that will explain the close interrelationship between placental and fetal growth there have been relatively few candidates providing clear pathways of regulation (Gluckman 1986; Jones 1989). The strongest candidate has been placental lactogen, but convincing evidence that it is a clear placental signal regulating fetal growth is awaited. The other strong candidates, the insulin-like growth factors (IGFs), also await clear evaluation as far as a role in coordinating fetal-placental growth is concerned (Gluckman 1986; Jones 1989). They are produced by both fetal and placental tissue (D'Ercole et al. 1980; Fant et al. 1986; Gluckman 1986; Jones 1988) and their concentration changes in reponse to manipulations of prenatal growth rate (Gluckman 1986; Jones 1989; Jones et al. 1987; Vileisis et al. 1986), but more detailed information on their biochemical actions on placental and fetal tissues is required before their precise significance can be ascertained.

In the absence of much direct information on the endocrine regulation of fetal-placental interaction in the control of growth, are there are situations in which there are any useful correlates with growth control? Perhaps the most significant and consistent are the marked increases or reductions in prenatal growth rate that are usually associated with substantial chagnes in fetal plasma glucose and insulin concentration (Gluckman 1986; Jones 1987; Robinson et al. 1980; Susa et al. 1979). Hence, intra-uterine growth retardation is normally associated with hypoglycaemia and hypoinsulinaemia (Jones et al. 1984; Susa et al. 1979). This is a surprising fact, as the fetus is normally reduced in proportion to placental size under such circumstances and a normal blood glucose concentration might be expected. The starting point for the present studies was therefore to provide an explanation for the close relationship between fetal glycaemia and placental and fetal growth retardation and to evaluate the potential actions of some of the hormones involved.

---

[1] Nuffield Institute for Medical Research, University of Oxford, Oxford OX3 9DS, UK

The Endocrine Control of the Fetus
Ed. by W. Künzel and A. Jensen
© Springer-Verlag Berlin Heidelberg 1988

Materials and Methods

The study employed pregnant guinea pigs using techniques essentially as described elsewhere (Jones et al. 1987; Lafeber et al. 1984). In those experiments involving the use of doubly perfused placentas the methods used were essentially as described by Yudilevich and coworkers (Yudilevich et al. 1979).

Results

If fetal or placental growth is restricted experimentally by limiting either uterine blood flow or placental size, the close relationship between fetal and placental size is maintained (Jones et al. 1983; Robinson et al. 1979). This relationship may even by improved as placental transport reserve is lost through restriction of placental size (Harding et al. 1985; Fig. 1). Such a consistent interdependent relationship has implied that either substrate supply to the fetus controls fetal growth or fetal supply of substrate controls placental growth, or both. Indeed, the close interrelationship between placental and fetal handling of the substrate glucose would support such conclusions (Gu et al. 1987; Sparks et al. 1983).

The importance of concentrating on handling of glucose is demonstrated in the very close relationship between fetal plasma glucose concentration and fetal and placental growth under a range of experimental conditions (Fig. 2; Jones et al. 1984; Robinson et al. 1980). Hence, administration of glucose, as opposed to glucose reduction, increases both fetal growth rate and body size in very close proportion to the rise in plasma glucose concentration (Fig. 2).

Although an increase in fetal growth rate might be expected from a rise in fetal glucose concentration through maternal hyperglycaemia, it is surprising that growth retardation caused by reduction of uterine blood flow should be associated with fetal hypoglycaemia (Jones et al. 1983, 1984; Lafeber et al. 1984; Robinson et al. 1980). Hence, fetal plasma glucose level should be determined by the balance between plasma transport and fetal consumption, which in the simplest case might be considered to be determined by size. However, in growth-retarded insulin concentration (Jones et al. 1984; Robinson et al. 1980), indicative of reduced fetal plasma glucose turnover. Although there is evidence ot suggest that in growth-retarded fetuses glucose uptake may be normal or even increased (Owens et al. 1987), experiments in which glucose turnover has been measured with [^{14}C]glucose suggest the opposite (Jones et al. 1985; Table 1). By contrast, placental consumption of glucose from the fetus is increased in such conditions.

If, therefore, fetal glucose consumption is reduced in growth retardation, why are such fetuses hypoglycaemic? In view of the close relationship between fetal growth rate and fetal plasma glucose, the answer to this question may tell us something about an important mechanism underlying the control of prenatal growth rate.

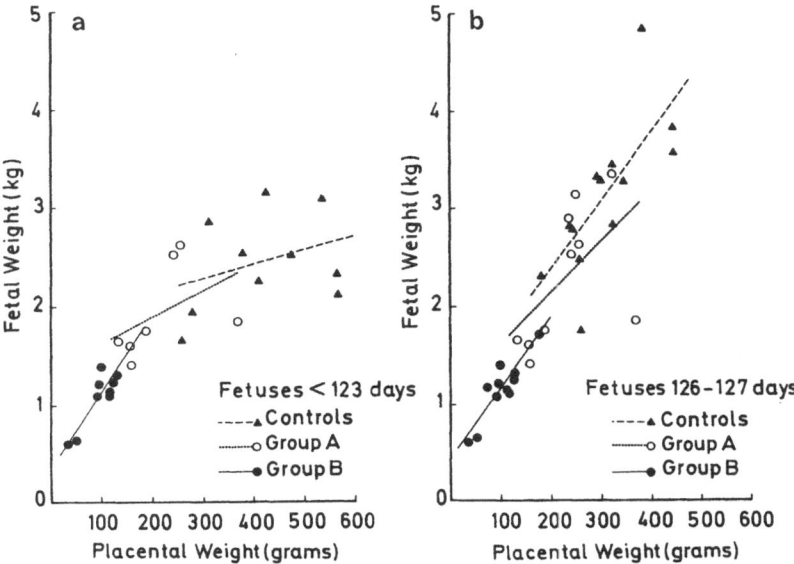

Fig. 1a, b. The relationship between fetal weight and placental weight in sheep at **a** less than 123 days' gestation and **b** 126–127 days' gestation. The fetuses were either of normal size (▲) or those (○, ●) for which the size of the placenta had been reduced surgically

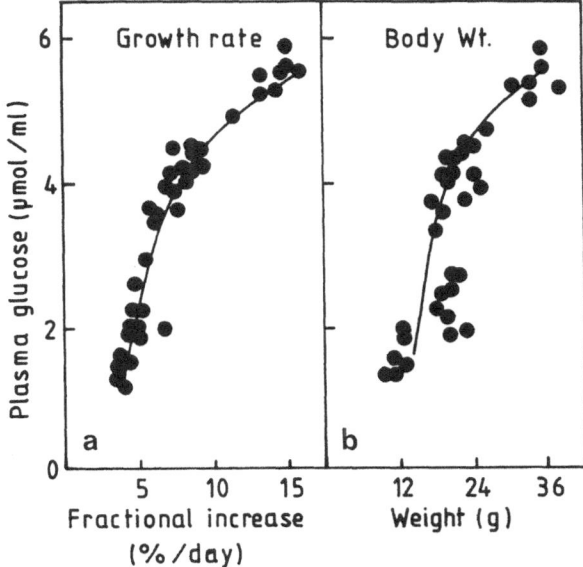

Fig. 2a, b. Relationship between fetal plasma glucose concentration and **a** growth rate and **b** fetal weight between 43 and 45 days' gestation in guinea pigs in which maternal plasma glucose was elevated by regular injection

Table 1. Fetal turnover and placental utilisation rates of glucose in normal and growth-retarded fetal sheep

	Normal	Retarded
Fetal weight (kg)	3.72 ± 0.31	2.28 ± 0.33
Fetal glucose uptake (μmol min^{-1} kg^{-1} fetus)	25.8 ± 4.3	18.7 ± 4.0
Utero-placental uptake from fetus (μmol min^{-1} kg^{-1} fetus)	41.7 ± 5.8	$68.3 \pm 4.7*$

The results are given as means $\pm$ SEM for 10–12 experiments.
* $p < 0.05$.
Growth retardation was produced in fetal sheep by carunclectomy as described previously (Jones et al. 1984). Glucose turnover and utero-placental consumption was measured using [^{14}C]glucose as outlined elsewhere (Gu et al. 1987)

The most likely explanation is either that placental capacity per unit mass to transport glucose is impaired or that glucose is being extracted at a uterine site other than fetal, which limits its availability to the fetus. In support of these two possibilities is the fact that, in fetal sheep made small by surgical reduction of the placenta, maternal glucose administration has an abnormally small effect on fetal plasma concentration (Harding et al. 1985). In addition, in growth-retarded fetal guinea pigs the relationship between placental blood flow and fetal plasma glucose concentration implies, at low rates of placental perfusion, that glucose transfer is relatively inefficient (Jones et al. 1983). Both of these observations tend to favour impairment of transport but neither proves it directly.

The transport capacity of placentas from growth-retarded fetuses was next evaluated, using the doubly-perfused isolated guinea pig placenta to measure unidirectional fluxes of glucose (Yudilevich et al. 1979). Such experiments gave no evidence for a reduction in the transport capacity of the small placenta from growth-retarded fetuses (Table 1), neither was there any clear evidence of a change in the rate of consumption of glucose or of its principle metabolites, CO_2 and lactate (Table 2).

This would imply that in growth retardation caused by restriction of maternal placental blood flow, placental transport and metabolism of glucose are undisturbed and cannot account for fetal hypoglycaemia and prenatal growth restriction. However, this interpretation is erroneous: because of the restriction of uterine blood flow, glucose supply to the uterus and placenta is substantially reduced (Table 3). In consequence, the fraction of this supply that must be extracted to maintain placental uptake increases by about 70% for the placenta of the growth-retarded fetus (Table 3). However, a major proportion of the glucose that is taken up by the placenta is used to meet its own metabolic demands (approximately two-thirds for the normal guinea pig placenta). Consequently, the maintained uptake of glucose by the placenta of the growth-retarded guinea pig constitutes such a substantial metabolic burden that there appears to be little glucose left over for the fetus (Table 3). It is thus the glucose requirement of the placenta that makes the fetus hypoglycaemia

Table 2. Unidirectional flux and metabolism of D-[ul^{14}C]glucose infused into the uterine circulation of doubly perfused placentas of normal and growth-retarded fetal guinea pigs of 60–62 days' gestation

	Normal	Retarded
Placental weight (g)	5.41 ± 0.83	2.26 ± 0.33
Fetal weight (g)	84.9 ± 9.2	34.6 ± 3.1
Unidirectional flux M to F (μmol min^{-1} g^{-1})	3.96 ± 0.71	3.60 ± 0.58
Rate of placental uptake of glucose (μmol min^{-1} g^{-1})	1.21 ± 0.53	1.36 ± 0.47
Placental conversion of glucose to lactate (μmol glucose min^{-1} g^{-1})	0.36 ± 0.14	0.40 ± 0.16
Placental conversion of glucose to CO_2 (μmol glucose min^{-1} g^{-1})	0.68 ± 0.20	0.70 ± 0.29

Results are given as means $\pm$ SD of 10–12 experiments. Placentas were perfused through umbilical and uterine circulations at 20 cm H_2O with Krebs' bicarbonate buffer containing 5 mM glucose on the uterine side and none on the umbilical. Flows were 5–10 ml min^{-1}. Leakage measured with L-[^{3}H]glucose was less than 6%

Table 3. Glucose supply to and consumption by normal and growth-retarded fetal guinea pigs of 60–63 days' gestation

	Normal	Retarded
Uterine blood flow (ml min^{-1})	17.3 ± 4.8	5.20 ± 2.14
Uterine blood glucose (μmol ml^{-1})	5.23 ± 1.25	5.05 ± 1.49
Glucose supply (μmol min^{-1})	90	26
Placental glucose uptake as fraction of supply (%)	7.1	12
Difference between M-to-F and F-to-M unidirectional flux of glucose (μmol min^{-1} g^{-1}	1.78 ± 0.59	1.39 ± 0.52
Net transfer of glucose as fraction of supply (%)	10.6	12.2
Fraction of supply remaining for fetus (%)	3.5	0.2
Umbilical plasma glucose (μmol ml^{-1})	2.6 ± 0.39	1.9 ± 0.33

Results are means $\pm$ SD of 10–13 experiments. Blood flow and glucose concentrations were determined in vivo as described previously (Harding et al. 1985). Net flux of glucose across the perfused placenta was calculated from the difference of maternal-to-fetal-transfer as outlined in Table 2 and fetal-to-maternal transfer as determined in separate experiments at the mean umbilical glucose concentrations in indicated above

and places severe limitation of fetal growth rate. This may explain the altered relationship between maternal placental blood flow and fetal plasma glucose concentration in conditions of severe growth restriction (Jones et al. 1983). It is consistent with the data in Table 1 for the growth-retarded fetal sheep and concurs with the view expressed elsewhere that the supply of glucose to the placenta is of paramount

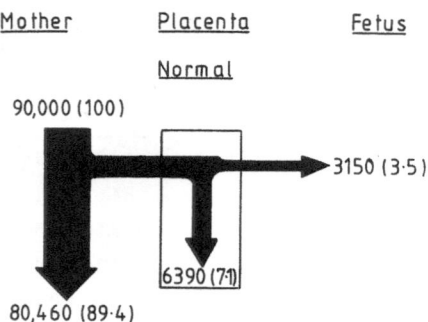

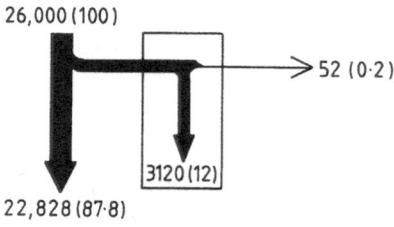

Flux across one placenta and to
one fetus = μmol glucose /min
(% of flux)

Fig. 3. Flux rates of glucose supply and utilization across one placenta and to one fetus in normal and growth-retarded fetal guinea pigs of 60–62 days' gestation. Figures are of μmol glucose min^{-1} (percentages in parentheses). Other details are as for Tables 2 and 3

importance to the extent that fetal glucose is used to a substantial degree to sustain it (Gu et al. 1987; Hay et al. 1984; Sparks et al. 1983). Hence, in growth retardation, the greater fractional extraction of glucose by the placenta will limit fetal growth, but also it is likely that limited availability of fetal glucose for the placenta will restrict placental growth, and this may be an important mechanism for matching the growth of the two.

The estimated pattern of glucose flux across the placenta to the normal and severely growth-retarded fetal guinea pig is summarized in Fig. 3. This illustrates the problem the placenta faces when uterine blood flow is reduced in needing to meet its requirements before those of the fetus.

This pattern of responses of the fetus and placenta to reduced substrate supply and hence depressed growth rate has been presented largely in terms of glucose availability without reference to adaptive endocrine mechanisms. Clearly, in intrauterine growth retardation or enhanced growth through maternal hypoglycaemia there are substantial changes in the fetal endocrine state (Jones et al. 1984; Robinson et al. 1980; Vileisis et al. 1986). This is particularly true for plasma insulin, glucagon, thyroxine [T$_4$], IGFs I and II, and prolactin (Jones et al. 1984, 1987; Vileisis

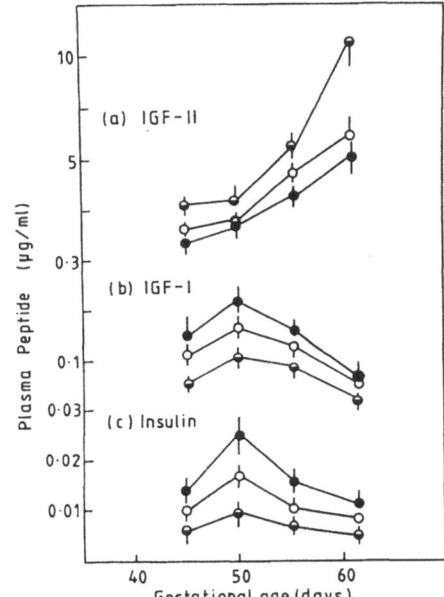

Fig. 4. Developmental changes in normal and growth-retarded fetuses of *(a)* plasma total IGF-II, *(b)* plasma total IGF-I and *(c)* plasma insulin. ●, Normal fetuses; ○, growth-retarded fetuses 40%–50% of normal weight; ◑ growth-retarded fetuses less than 40% of normal weight. Results are means ± SD of 5–14 experiments. (From Jones et al. 1987)

et al. 1986; Fig. 4). Do any of these changes have a role to play in modifying substrate supply to and handling by the placenta? Although both insulin and glucagon have marked effects upon glucose handling by the fetus, and in particular its liver, they were ineffective at causing a significant change in glucose metabolism by or transport across the perfused guinea pig placenta. This concurs with observations in the pregnant sheep, where insulin appears to have little effect upon placental glucose transfer, although no data on metabolism is available. Similarly, short-term exposure to T_4 also had little effect on the guinea pig placenta. IGF-I and IGF-II show dramatic changes during growth retardation, with the former rising and the latter falling. Moreover, IGF-II concentrations are particularly high in the fetus (Jones et al. 1987). Although the broad biological effects of the IGFs have been described – and in particular their ability to act as mitogenic agents, stimulate RNA and glycogen synthesis and enhance glucose oxidation (Zapf et al. 1984) – their importance under physiological conditions has yet to be evaluated. There has been a report that IGF-I incresaes glycogen synthesis, but technically that study was flawed (Freemark et al. 1985). Unfortunately IGF-II is not available in the quantities that would allow many in vitro studies and its effects on placental metabolism will not be reported here. It does in the perfused fetal liver increase glycogen deposition (Michael and Jones 1988). As IGF-II acts on the placental IGF-I receptor with an affinity (K_d of about 10^{-10}–10^{-9} M) comparable to that of IGF-I and probably at the same site on the receptor (Carsella et al. 1986), and as, moreover, at the IGF-II concentrations in the growth-retarded fetus guinea pig this receptor is likely

Table 4. The effect of IGF-I on glucose handling by the perfused placenta from guinea pig 60–62 days pregnant

	Control	IGF-I (1 μg/ml)
Unidirectional flux M to F (μmol min^{-1} g^{-1})	3.85 ± 0.82	3.50 ± 0.74
Rate of placental uptake of glucose (μmol min^{-1} g^{-1})	1.15 ± 0.36	1.81 ± 0.33*
Placental conversion of glucose to lactate (μmol glucose min^{-1} g^{-1})	0.31 ± 0.09	0.67 ± 0.15*
Placental conversion of glucose to CO_2 (μmol glucose min^{-1} g^{-1})	0.69 ± 0.11	0.89 ± 0.14*

Results are means $\pm$ SD of 10 experiments; details as for Table 2.
* $p < 0.05$

to be saturated (Jones et al. 1987), the effects of IGF-I at concentrations comparable with those of IGF-II have been evaluated.

In the perfused guinea pig placenta the addition of human$_r$ IGF-I to the fetal side of the placenta enhanced the fractional conversion of glucose to lactate and CO_2 and increased the fractional extraction of glucose by the placenta (Table 4).

Discussion

Although there is a wide family of hormones that have potential growth-regulating effects upon selected fetal tissues (Gluckman 1986; Jones 1987), a coordinating role in the control of prenatal growth has not been apparent. This is well exemplified by the relatively minor effect of substantial changes in fetal plasma growth hormone concentrations (Parkes et al. 1985). Of the hormones that show the closest relationship to growth, insulin has probably the most convincing case, particularly when prenatal growth rate is manipulated experimentally (Jones et al. 1984; Robinson et al. 1980), Hence it is not surprising that fetal plasma glucose concentration exhibits the best relationship to placental and fetal growth rate under such conditions. This implies that, to a large extent, growth rate is manipulated by substrate supply, which certainly fits with the relationship between blood flow and fetal weight distribution along the uterine arcade. However, the precise mechanisms of how substrate supply is related to fetal growth, except in so far as insulin concentration reflects that supply, are unclear.

The present experiments show that fetal and placental growth are intimately coupled through management of glucose balance between the two. Moreover, when glucose supply is limited through depression of uterine blood flow, fetal growth is restricted substantially because of the incresaed fractional extraction by the placenta. However, as the placenta obtains much of its glucose from the fetus, this is probably the cause of the fall in placental growth. There is little evidence that these processes are coordinated by changes in insulin concentration. However, IGF-II is also very sensitive to changes in the nutritional state of the fetus, and the elevation of its plasma concentration in growth retardation, caused by restricted uterine blood flow, could serve to increase placental fractional extraction from mother and fetus at a time of depressed supply. In that sense it could be considered as an important example of a hormone regulating fetal-placental interactions in substrate utilization in growth control. We await other examples.

References

Carsella SJ, Hay VK, D'Ercole AJ, Svoboda ME, van Wyk JJ (1986) Insulin-like growth factor II binding to the type I somatomedin receptor. Evidence for two high affinity binding sites. J Biol Chem 261: 9268–9273

D'Ercole AJ, Applewhite GT, Underwood LE (1980) Evidence that somatomedin is synthesized by multiple tissues in the fetus. Dev Biol 75: 315–328

Fant M, Munro H, Moses AC (1986) An autocrine/paracrine role for insulin-like growth factors in the regulation of human placental growth. J Clin Endocrinol Metab 63: 499–505

Freemark M, D'Ercole AJ, Handwerger S (1985) Somatomedin-C stimulates glycogen synthesis in fetal rat hepatocytes. Endocrinology 116: 2578–2582

Gluckman PD (1986) Hormones and fetal growth. Oxford Rev Reprod Biol 6: 1–60

Gu W, Jones CT, Harding JE (1987) Metabolism of glucose by fetus and placenta of sheep. The effects of normal fluctuations in uterine blood flow. J Dev Physiol 9: 369–389

Harding JE, Jones CT, Robinson JS (1985) Studies on experimental growth retardation in sheep. The effects of a small placenta in restricting transport to and growth of the fetus. J Dev Physiol 7: 427–442

Hay WW, Sparks JW, Battaglia FC, Meschia G (1984) Maternal-fetal glucose exchange: necessity of a 3 pool model. Am J Physiol 246: E528–E534

Jones CT (1989) The control of fetal growth. Rev Perinatal Med (in press)

Jones CT, Parer JT (1983) The effect of alterations in placental blood flow on the growth of and nutrient supply to the fetal guinea pig. J Physiol 343: 525–537

Jones CT, Lafeber HN, Roebuck MM (1984) Studies on the growth of the fetal guinea pig. Changes in plasma hormone concentration during normal and abnormal growth. J Dev Physiol 6: 461–472

Jones CT, Rolph TP, Lafeber HN, Gu W, Harding JE, Parer JT (1985) Experimental studies on the control of fetal growth. In: Jones CT, Nathanielsz PW (eds) Physiological development of the fetus and newborn. Academic, New York, pp 11–20

Jones CT, Lafeber HN, Price DA, Parer JT (1987) Studies on the growth of the fetal guinea pig. Effects of reduction in uterine blood flow on plasma sulphation promoting activity and on concentrations of insulin-like growth factors-I and -II. Dev Physiol 9: 181–201

Lafeber HN, Jones CT, Rolph TP (1984) Studies on the growth of the fetal guinea pig. The effects of ligation of the uterine artery on organ growth and development. J Dev Physiol 6: 441–459

Michael E, Jones CT (1988) Insulin, glucagon and insulin-like growth factor effects upon glycogen turnover in fetal rabbit liver. In: Jones CT (ed) Fetal and neonatal development. Perinatology Press, Ithaca (in press)

Owens JA, Falconer J, Robinson JS (1987) Effect of restriction of placental growth on fetal and utero-placental metabolism. J Dev Physiol 9: 225–238

Parkes MJ, Hill DJ (1985) The lack of growth-hormone dependant somatomedins or growth retardation in hypophysectomized fetal lambs. J Endocrinol 104: 193–199

Robinson JS, Kingston EJ, Jones CT, Thorburn GD (1979) Studies on experimental growth retardation in sheep. The effect of removal of endometrial caruncles on fetal size and metabolism. J Dev Physiol 1: 379–398

Robinson JS, Hart IC, Kingston EJ, Jones CT, Thorburn GD (1980) Studies on experimental growth retardation in sheep. The effects of reduction of placental size on hormone concentration in fetal plasma. J Dev Physiol 2: 239–248

Sparks JW, Hay WW, Meschia G, Battaglia FC (1983) Partition of nutrients to the placenta and fetus in sheep. Eur J Obstet Gynecol Reprod Biol 14: 331–340

Susa JB, McCormick KL, Widness JA, Singer DB, Oh W, Adamsons K, Schwartz R (1979) Chronic hyperinsulinaemia in the rhesus monkey: effects of physiological hyperinsulinaemia on fetal growth and composition. Diabetes 28: 1058–1063

Vileisis RA, D'Ercole AJ (1986) Tissue and serum concentrations of somatomedin C/insulin-like growth factor in fetal rats made growth retarded by uterine artery ligation. Pediatr Res 13: 126–130

Yudilevich DL, Eaton BM, Short AH, Leichtweiss HP (1979) Glucose carriers at maternal and fetal sides of the trophoblast in guinea pig placenta. Am J Physiol 237: C205–C212

Zapf J, Schmidt C, Froesch ER (1984) Biological and immunological properties of insulin-like growth factors (IGF) I and II. Clin Endocrinol Metab 13: 3–30

The Biosynthesis and Regulation of Fetal Insulin-like Growth Factors

V. R. Sara[1] and C. Carlsson-Skwirut

Introduction

The somatomedins or insulin-like growth factors (IGFs) are a family of growth-promoting peptide hormones. They consist of insulin-like growth factors 1 (IGF-1) and 2 (IGF-2) as well as their variant forms. As has been said in earlier reviews (Sara and Hall 1984; Sara 1987), the IGFs are believed to be major regulators of fetal growth. This hypothesis was initially based upon their growth-promoting action on fetal cells in vitro as well as the presence of IGF receptors in a wide variety of fetal tissues. As demonstrated by the studies summarized in Table 1, the IGFs have a potent action on fetal cell proliferation, protein and glycogen synthesis, and differentiation. These anabolic actions have been demonstrated in a wide variety of cell types from several species. Since the IGFs appear to be ubiquitous anabolic peptides, it is unlikely that they specifically influence cell differentiation. Rather this action may be due to an influence on the timing of cellular growth. Stimulation of proliferation may 'speed up' the developmental clock, pushing cells into their differentiating phase. With the recent availability of recombinant IGF-1 it has now been possible to confirm the growth-promoting actions of IGF-1 in vivo. For example, Philipps et al. (1988) have shown that the administration of IGF-1 to neonatal rats, but not of growth hormone (GH), stimulates their growth.

The biological action of the IGFs is mediated by their interaction with specific receptors on the plasma membranes of fetal cells. Two IGF receptors have been identified (Rechler and Nissley 1985). The type 1 or IGF-1 receptor is a glycoprotein (approximately 350000 daltons) consisting of 2 extracellular α-subunits (approximately 130000 daltons) which contain the hormone binding site and 2 transmembranal β-subunits (approximately 95000 daltons) with intrinsic tyrosine kinase activity. The type 2 or IGF-2 receptor is an approximately 250000-dalton monomeric transmembral protein whose mechanism of intracellular signal transduction is unclear. Both IGF-1 and IGF-2 receptors have been identified in a wide variety of fetal tissues, where their characteristics as well as predominance appear to be both development- and tissue-specific (Sara et al. 1983). Receptor responsiveness provides a cell-specific mechanism to regulate IGF bioactivity during development.

---

[1] Department of Psychiatry, Karolinska Institute, St. Göran's Hospital, Box 12500, 11281 Stockholm, Sweden

The Endocrine Control of the Fetus
Ed. by W. Künzel and A. Jensen
© Springer-Verlag Berlin Heidelberg 1988

Table 1. Biological action of IGF on fetal cells

Action	Species	Target	Reference
Cell proliferation	Human	Fibroblasts	Weidman and Bala 1980
			Conover et al. 1986
			Hill et al. 1986
		Cartilage	Ashton and Spencer 1983
		Brain	Sara et al. 1980
	Rat	Fibroblasts	Adams et al. 1983
		Calvaria	Canalis 1980
		Osteoblasts	Schmid et al. 1983a
		Brain	Sara et al, 1979b
			Lenoir and Honegger 1983
		Myoblasts	Ewton and Florini 1980
	Chick	Fibroblasts	Rechler et al. 1978
		Cartilage	Froesch et al. 1976
Protein synthesis	Rat	Calvaria	Canalis 1980
	Chick	Cartilage	Hall 1972
Glycogen synthesis	Rat	Osteoblasts	Schmid et al. 1983a
		Myoblasts	Ewton and Florini 1980
		Hepatocytes	Freemark 1986
Differentiation	Rat	Osteoblasts	Schmid et al. 1984
		Myoblasts	Ewton and Florini 1981
			Florini et al. 1986
		Oligodendrocytes	McMorris et al. 1986
	Chick	Myoblasts	Schmid et al. 1983b

The IGF-1 receptor mediates the growth-promoting action of the IGFs. The function of the IGF-2 receptor is as yet unclear, as the growth-promoting action of IGF-2 can be accounted for by interaction in the IGF-1 receptor. Only recently has any biological activity, namely glycogen synthesis in human hepatoma cells, been shown to be mediated by the IGF-2 receptor (Hari et al. 1987).

In more recent years it has become clear that the IGFs are synthesized in all fetal tissues, where they act as paracrine hormones on neighbouring cells to stimulate their growth (D'Ercole et al. 1980, 1986). In this review, we will concentrate on the identity of the fetal IGFs, their biosynthesis pathway in fetal cells, and the mechanisms which regulate their production. Our understanding of fetal IGF biosynthesis and its regulation is at present limited. With the application of techniques of molecular biology, however, this has become a rapidly expanding field, contributing to our knowledge of the basic principles of developmental regulation.

Structure

The first two IGFs to be characterized were IGF-1 and IGF-2, which were isolated from adult human plasma (Rinderknecht and Humbel 1978 a, b). Both IGF-1 and IGF-2 are single chain molecules with intrachain disulphide bridges consisting of 70 and 67 amino acids respectively. Their amino acid sequences have identical residues in 45 positions. As with proinsulin, an amino-terminal B region, a connecting C region and an A region can be defined in the sequences of the IGFs. In addition, the IGFs also contain a carboxyl-terminal D region not present in proinsulin. The similarities in primary structure between IGF-1 and IGF-2 as well as between the IGFs and proinsulin are restricted to the B and A regions. Subsequent characterization of somatomedins A and C has revealed these peptides to be identical to IGF-1 if possible deamidization differences are disregarded (Svoboda et al. 1980; Enberg et al. 1984). Variant forms of both IGF-1 and IGF-2 have been identified: for example, a larger form of IGF-2, with a substitution of three amino acids for one in posi-

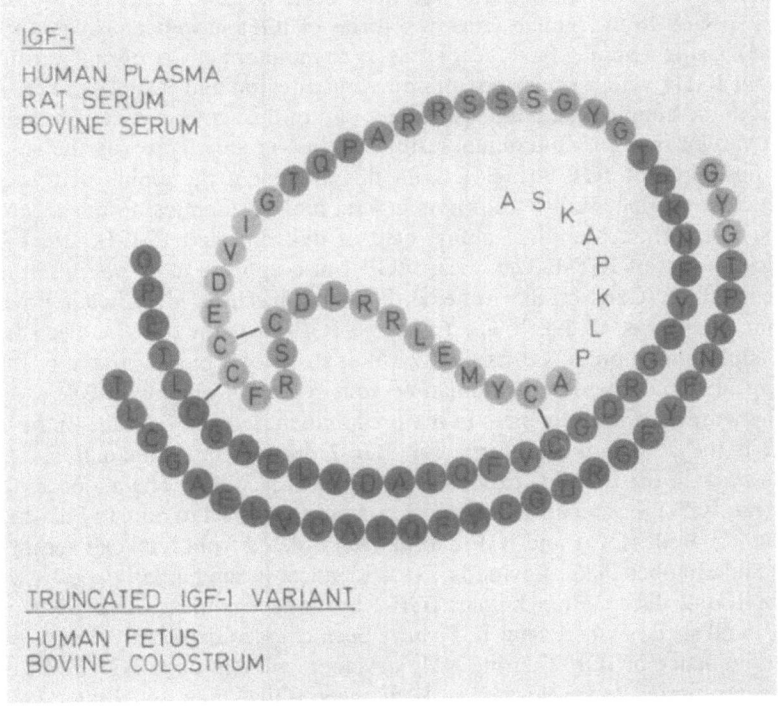

Fig. 1. The proven parts of the primary structure of the truncated IGF-1 variant compared to IGF-1. The source for purification of both peptides is given

Table 2. Fetal forms of IGF in various species

Species	Fetal IGF	Reference
Human	Variant IGF-1 IGF-2	Sara et al. 1986 Carlsson-Skwirut et al. 1987
Bovine	IGF-1 IGF-2	Honegger and Humbel 1986
Rat	IGF-2	Marquardt et al. 1981
Chicken	IGF-1-like peptide	Haselbacher et al. 1980
Sheep	IGF-2 acitivity	Gluckman and Butler 1985
Guinea pig	IGF-2 activity	Daughaday et al. 1986

tion 33 and a carboxyl-terminal extension of 21 amino acids, has also been purified from adult human serum (Zumstein et al. 1985). The IGFs, especially IGF-1, appear to be well conserved between different species.

The IGFs present during fetal life have been identified and characterized in several species. In the human fetus, two forms of IGFs have been isolated (Sara et al. 1986). Using a human fetal receptor assay to monitor activity during purification, a variant IGF-1 with a truncated amino-terminal region and IGF-2 were characterized from the human fetal brain. The larger part of the activity was attributed to the IGF-1 variant. The amino-terminal residue of this variant aligns with the amino acid in position 4 of IGF-1 (Fig. 1). With this alignment the amino acid sequence (1–29) of the variant IGF-1 from human fetal brain is identical to the sequence of IGF-1. The carboxyl-terminal amino acid of the truncated IGF-1 variant is also identical to that of IGF-1. The variant IGF-1 also appears to be present in human fetal circulation (Carlsson-Skwirut et al. 1987) and has been identified in bovine colostrum (Francis et al. 1986). The truncated IGF-1 variant has also been isolated from adult human brain (Carlsson-Skwirut et al. 1986). Thus, synthesis of the variant appears to continue in the human nervous system throughout life.

There are species differences in the predominant fetal IGF (Table 2). In the rat, IGF-2 is the dominant fetal form. Rat IGF-2 differs from human IGF-2 by five amino acids in the B and C regions (Marquardt et al. 1981). The major adult form in the rat, IGF-1, has an amino-terminal sequence identical to human IGF-1 (Rubin et al. 1982). Both IGF-1 and IGF-2 have been isolated from fetal calf serum (Honegger and Humbel 1986). Bovine IGF-1 is identical to the human peptide, whereas bovine IGF-2 differs from human IGF-2 by three amino acids in the C region. Thus, in all species where fetal IGFs have been characterized, only the rat displays a predominance of IGF-2 during early development. Although IGF-2 like activity has been reported to be elevated in both sheep (Gluckman and Butler 1985) and guinea pig (Daughaday et al. 1986), this activity remains to be identified. An IGF-1-like peptide has been partially purified from chick embryo liver (Haselbacher et al. 1980).

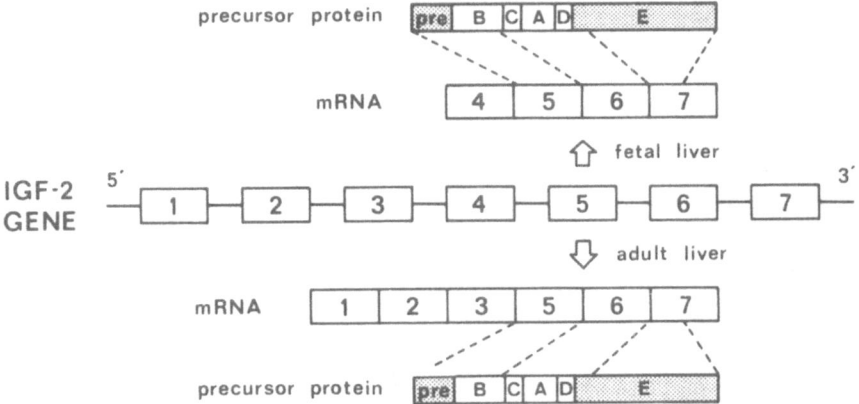

Fig. 2. Schematic representation of the IGF-2 gene, mRNA and protein product. Different transcription and splicing of the seven gene exons results in alternative 5'-untranslated regions in fetal and adult liver mRNA

Gene

Single gene loci for IGF-1 and IGF-2 have been mapped to the long arm of human chromosome 12 and the short arm of human chromosome 11, respectively (Bell et al. 1985). The IGF-2 gene is located in close proximity to the insulin gene. The IGF-1 gene consists of at least 5 exons (Rotwein 1986; Rotwein et al. 1986). The primary transcript of the IGF-1 gene can be alternatively spliced to result in two mRNAs encoding precursor proteins differing in their carboxyl-terminal amino acid extensions. Exon 1 contains 5'-nontranslated sequences and exons 2 and 3 encode the signal peptide, intact IGF-1 B, C, A and D regions, and the first 16 amino acids of the E region. Exons 4 and 5 contain the sequences encoding the rest of the E region together with 3'-nontranslated sequences. Differential splicing results in either IGF-1A (exons 1, 2, 3, 5) or IGF-1B (exons 1, 2, 3, 4) mRNA. The IGF-2 gene is composed of seven exons of which the first four are 5'-nontranslated; exons 5–7 contain the sequences encoding the IGF-2 precursor protein as well as 3'-nontranslated sequences (de Pagter-Holthuizen et al. 1986). Two different promoters have been identified in the IGF-2 gene, generating alternative transcripts containing either 5'-nontranslated exons 1–3 or 5'-nontranslated exon 4 (Fig. 2).

Biosynthesis

The genes for both IGF-1 and IGF-2 are expressed during fetal life. In humans, Scott et al. (1985) first demonstrated the presence of IGF-2 mRNA in a wide variety of fetal tissues where IGF-2 gene expression was markedly increased compared to the adult. Studies in our laboratory have demonstrated IGF-1 gene expression in human fetal brain where a 1.1 kb transcript, similar to that in the adult liver, was identified (Sandberg et al. 1987, unpublished material). Recently, Han et al. (1987) reported widespread IGF-1 and IGF-2 gene expression in the human fetus. Using in situ hybridization histochemistry, these authors localized IGF mRNA to connective tissue or to cells of mesenchymal origin. All fetal tissues examined, except cortex, showed hybridization to IGF-1A, IGF-1B and IGF-2 oligomers. There was no difference in the localization of the various IGF mRNAs, although IGF-2 appeared to be most abundant. These findings suggest mesenchymal cells as the primary site of IGF biosynthesis.

The majority of studies of IGF gene expression during development have been performed in the rat, where several mRNA species for both IGF-1 and IGF-2 have been identified. The relative abundance of these mRNA species varies during development and in different tissues (Lund et al. 1986). Soares et al. (1985) were the first to report tissue-specific IGF gene expression. In their study the size and abundance of IGF-2 transcripts in rat organs varied during development; however, tissue specificity remains to be confirmed. Similarly, the functional activity of transcripts whose size differences could represent incomplete RNA processing or polyadenylation remains to be demonstrated. It is clear, however, that in the rat, IGF-2 gene expression is developmentally regulated, abundant IGF-2 mRNA being present in the fetus, and, with the exception of the brain, very low amounts being present in the adult (Brown et al. 1986; Graham et al. 1986). Thus, IGF-2 gene expression becomes repressed during development in the rat. This appears to be regulated by distinct promoter regions of the gene, whose appearance is development specific (Soares et al. 1986).

IGF-2 gene expression appears to be similarly regulated in man. As illustrated in Fig. 2, two different promoters give rise to different IGF-2 transcripts during development. The transcript in the fetal liver contains exons 4–7 whereas that in the adult liver contains exons 1–3 together with exons 5–7 (de Pagter-Holthuizen 1987). Different promoter regions of the gene, whose expression is development-specific, generate these alternative 5'-nontranslated exons. Regulation of IGF-2 expression during development may thus result from interaction of regulatory proteins with these promoter regions. Presumably the 5'-untranslated regions are involved in the efficiency of translation of the mRNAs, so that their developmentally regulated appearance influences final precursor protein production.

It remains unclear whether IGF-1 gene expression is developmentally regulated. A tendency to greater expression in fetal than in adult tissues has been suggested in the rat (Lund et al. 1986). Identification of variant IGF-1 as the final protein product in the human fetus indicates development-specific processing from the IGF-1 gene. This may occur at several levels during biosynthesis, especially during RNA or precursor protein processing. Although two distinct IGF-1 mRNAs encoding different precursor proteins are generated from the gene, it remains to be determined whether the expression of these mRNAs is developmentally regulated. However, variant IGF-1 is unlikely to arise from alternative RNA processing, since the amino-terminal of IGF-1 is not an intron/exon hinge region (Jansen et al. 1983); it is more likely that it arises from specific post-translational modification of the precursor protein. Since variant IGF-1 has enhanced biological activity compared to IGF-1 (Sara et al. 1986), post-translational modification provides a mechanism to regulate potency during development. Since a single gene can express alternative protein products with differing biological action by alternative processing at the RNA or precursor protein level, regulation can occur at multiple sites in the biosynthetic pathway. Two distinct major pathways are believed to be used in the regulation of IGF expression during development. Whereas the major pathway for IGF-1 processing is proposed to occur at the precursor protein level, that for IGF-2 is proposed to occur at the RNA level. This implies separate mechanisms for the regulation of IGF-1 and IGF-2 production.

After synthesis of the preproIGF, studies in metabolically labelled cells suggest that the signal peptide is cleaved from the precursor by a microsomal peptidase and that processing of the prohormone occurs at the time of its secretion from the cell (Yang et al. 1985). Following their release from the cell, the IGFs are bound by carrier proteins, whose origin is as yet undetermined. Sara and Hall (1984) have proposed that the binding protein may be a transmembranal peptidase which proteolytically processes the prohormone and remains bound to the IGF. This would imply a distinct IGF-1 peptidase in the human fetus. The binding protein regulates the transport as well as binding of the IGFs to their target cells (Clemmons et al. 1986; de Vroede et al. 1986).

In fetal serum or in the extracellular fluid surrounding fetal cells, IGFs appear as a higher molecular weight complex of approximately 40000 daltons consisting of IGF associated with a binding protein of approximately 35000 daltons. Póvoa et al. (1984) have characterized a 35000-dalton binding protein from amniotic fluid where peak concentrations occur at midterm. This low molecular weight binding protein is not GH-regulated and predominates in the fetal circulation until the GH-dependent higher molecular weight binding complex of approximately 150000 daltons appears. In humans the switch over from the low to the high molecular weight binding protein occurs towards the end of gestation (D'Ercole et al. 1980). Similarly, in the rat, the lower molecular weight binding protein predominates until weaning, when a switch occurs over to the higher molecular weight GH-regulated binding complex (White et al. 1982). The timing of this switch over may depend upon maturation of GH-receptors and consequent prompting by the GH regulation system.

Apart from the influence of the binding proteins, there is a change in the circulating levels of IGFs during development. In the human, levels of both IGF-1 and IGF-2 immunoreactivity are low during fetal life, rising gradually with maturation (Sara and Hall 1984). Evidence suggests, however, a predominance of variant IGF-1 in the fetal circulation (Carlsson-Skwirut et al. 1986). The developmental pattern of circulating IGFs is best established in the rat, where IGF-2 appears as the predominant fetal form (Moses et al. 1980). Unlike in humans, IGF-2 levels are elevated in the fetal rat and fall with maturation until around weaning, when there is a switch over to predominance of IGF-1 (Sara et al. 1980a). This switch over occurs at the time that GH begins to regulate growth and IGF-1 production.

Regulation

The developmental pattern of IGFs and their carrier proteins in both human and rat is summarized in Fig. 3. The pattern of serum levels differs in rat and human. Gene expression is only reflected in rat serum IGF-2 levels; otherwise there seems to be no association between gene expression and serum levels. Instead, the latter more closely relates to changes in the binding proteins. This raises the possibility that serum IGF level reflects an inhibition of IGF degradation which occurs with the switch over to the higher molecular weight binding protein. The latter may be especially effective as a transport protein preventing IGF degradation in serum. This may imply a switch over from paracrine to endocrine function during development. In both species there seems to be a repression of IGF-2 expression with maturation. This occurs in parallel with the increase in GH-regulated IGF-1 and high molecular weight binding protein, which raises the possibility that they have an interactive feedback effect on IGF-2 gene expression.

The mechanisms which regulate fetal cell biosynthesis have yet to be elucidated. It is evident that regulation during fetal life differs from that at maturity (Sara and Hall 1984). For example, unlike in the adult; fetal IGF-1 production appears independent of pituitary growth hormone regulation, since serum levels are unaffected by anencephaly or hypophysectomy. GH begins to regulate IGF-1 production by

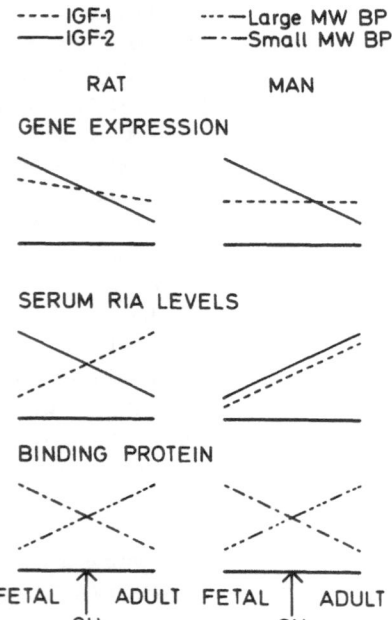

Fig. 3. Developmental pattern of IGF gene expression, serum immunoreactivity and binding proteins in rat and human. The onset of growth hormone regulation of production of IGF-1 and the high molecular weight binding protein is indicated

6 months of postnatal life in humans and 20 postnatal days in the rat. GH receptors are absent from human fetal liver and presumably develop at a larger maturational stage. The GH homologue, placental lactogen, stimulates IGF production and has been suggested to substitute for GH during fetal development (Adams et al. 1983). There is a positive relationship between insulin and IGF. Fetal hyperinsulinaemia is associated with elevated serum IGF levels (Hilla nd Milner 1980), whereas fetal pancreatectomy is associated with reduced IGF levels (Gluckman et al. 1987). It is unclear, whether insulin action is direct or mediated by enhanced nutrients to the cells. Other hormones such as thyroid and steroid hormones, which have been reported to influence postnatal IGF production, have not been investigated for an influence on fetal synthesis. Interestingly, in vitro studies suggest an interaction between IGFs and other growth factors, such as platelet-derived growth factor (Clemmons and Shaw 1983; Clemmons 1984) and epidermal growth factor (Richman et al. 1985). These findings raise the possibility that during development there is regulation between neighbouring cells. Such local cellular communication may provide means for the co-ordinated growth essential to early development.

Nutrition may prove to be the primary regulator of IGF production. The influence of nutrition is well established postnatally and overrides GH control (Soliman et al. 1986). Several studies have demonstrated nutritional influences on IGF production during early development. Undernutrition in the pre-weanling rat, for example, reduces IGF levels and is accompanied by growth retardation (Sara et al. 1979a). IGF-1 synthesis in the developing rat has been suggested to be primarily influenced by dietary protein intake (Prewitt et al. 1982), whereas IGF-2 synthesis relates to caloric intake (Philipps et al. 1987, unpublished material). An elevation in binding protein has also been observed during short-term starvation in neonatal rats, suggesting a protective mechanism to reduce IGF bioavailability (Philipps et al. 1987, unpublished material). Nutrition may prove to be the major regulator of IGF biosynthesis throughout life. This would provide for a local regulation of paracrine or autocrine IGF biosynthesis according to the availability of substrates to the cell. Higher neural control involving the hypothalamic-pituitary axis may only develop with maturation as a secondary regulator of IGF biosynthesis.

Conclusion

The IGFs have been characterized in fetal tissues where they are produced as paracrine hormones to regulate cellular growth. The expression of the IGF genes appears to be developmentally regulated at various levels in the biosynthetic pathway. The major pathway for IGF-1 processing is proposed to involve post-translational modification, whereas that for IGF-2 occurs via transcription and RNA splicing. This results in a precise developmental pattern of IGF biosynthesis which appears to be species specific. The mechanisms which regulate fetal IGF biosynthesis have yet to be completely defined. Early on in development, substrate availability to the cells is believed to regulate paracrine IGF biosynthesis. Nutrition may continue to be the primary regulator throughout life. With maturation however, higher neural control, via GH, develops as a secondary regulating system.

Acknowledgements. We are grateful to Solveig Johansson for typing this manuscript. These studies have been supported by the Swedish Medical Research Council, Expressen Prenatal Fund, Sävstaholmsföreningen and the Nordic Insulin Fund.

References

Adams SO, Nissley SP, Handwerger S, Rechler MM (1983) Developmental patterns of insulin-like growth factor-I and -II synthesis and regulation in rat fibroblasts. Nature 302: 150-153

Ashton IK, Spencer EM (1983) Effect of partially purified human somatomedin on human fetal and postnatal cartilage in vitro. Early Hum Dev 8:135-140

Bell GI, Gerhard DS, Fong NM, Sanchez-Pescador R, Rall LB (1985) Isolation of the human insulin-like growth factor genes: insulin-like growth factor II and insulin genes are contiguous. Proc Natl Acad Sci USA 82: 6450-6454

Brown AL, Graham DE, Nissley SP, Hill DJ, Strain AJ, Rechler MM (1986) Developmental regulation of insulin-like growth factor II mRNA in different rat tissues. J Biol Chem 261: 13144-13150

Canalis E (1980) Effect of insulin-like growth factor I on DNA and protein synthesis in cultured rat calvaria. J Clin Invest 66: 709-719

Carlsson-Skwirut C, Jörnvall H, Holmgren A, Andersson C, Bergman T, Lundquist G, Sjögren B, Sara VR (1986) Isolation and characterization of variant IGF-1 as well as IGF-2 from adult human brain. FEBS Lett 201: 46-50

Carlsson-Skwirut C, Andersson C, Sara VR (1987) Circulating forms of human foetal somatomedin. Acta Endocrinol (Copenh) 114: 37-40

Clemmons DR (1984) Multiple hormones stimulate the production of somatomedin by cultured human fibroblasts. J Clin Endocrinol Metab 58: 850-856

Clemmons DR, Shaw DS (1983) Variables controlling somatomedin production by cultured human fibroblasts. J Cell Physiol 115: 137-142

Clemmons DR, Elgin RG, Han VKM, Casella SJ, D'Ercole AJ, Van Wyk JJ (1986) Cultured fibroblast monolayers secrete a protein that alters the cellular binding of somatomedin-C/insulin-like growth factor I. J Clin Invest 77: 1548-1556

Conover CA, Rosenfeld RG, Hintz RL (1986) Hormonal control of the replication of human fetal fibroblasts: role of somatomedin C/insulin-like growth factor I. J Cell Physiol 128: 47-54

Daughaday WH, Yanow CE, Kapadia M (1986) Insulin-like growth factors I and II in maternal and fetal guinea pig serum. Endocrinology 119: 490-494

D'Ercole AJ, Applewhite GT, Underwood LE (1980) Evidence that somatomedin is synthesized by multiple tissues in the fetus. Dev Biol 75: 315-328

D'Ercole AJ, Hill DJ, Strain AJ, Underwood LE (1986) Tissue and plasma somatomedin-C/insulin-like growth factor I concentrations in the human fetus during the first half of gestation. Pediatr Res 20: 253-255

de Pagter-Holthuizen P, van Schaik FMA, Verduijn GM, van Ommen GJB, Bouma BN, Jansen M, Sussenbach JS (1986) Organization of the human genes for insulin-like growth factors I and II. FEBS Lett 195: 179-184

de Pagter-Holthuizen, Jansen M, van Schaik FMA, van der Kammen R, Oosterwijk O, Van den Brande JL, Sussenbach JS (1987) The human insulin-like growth factor II gene contains two development-specific promoters. FEBS Lett 214: 259-264

de Vroede M, Tseng LY-H, Katsoyannis PG, Nissley SP, Rechler MM (1986) Modulation of insulin-like growth factor I binding to human fibroblast monolayer cultures by insulin-like growth factor carrier proteins released to the incubation media. J Clin Invest 77: 602-613

Enberg G, Carlquist M, Jörnvall H, Hall K (1984) The characterization of somatomedin A, isolated by microcomputer-controlled chromatography, reveals an apparent identity to insulin-like growth factor 1. Eur J Biochem 143: 117-124

Ewton DZ, Florini JR (1980) Relative effects of the somatomedins, multiplication-stimulating activity, and growth hormone on myoblasts and myotubes in culture. Endocrinology 106: 577-583

Ewton DZ, Florini JR (1981) Effects of somatomedins and insulin on myoblast differentiation in vitro. Dev Biol 56: 31-39

Florini JR, Ewton DZ, Falen SL, Van Wyk J (1986) Biphasic concentration dependency of stimulation of myoblast differentiation by somatomedins. Am J Physiol 250: 771-778

Francis GL, Read LC, Ballard FJ, Bagley CJ, Upton FM, Gravestock PM, Wallace JC (1986) Purification and partial sequence analysis of insulin-like growth factor-1 from bovine colostrum. Biochem J 233: 207-213

Freemark M (1986) Epidermal growth factor stimulates glycogen synthesis in fetal rat hepatocytes: comparison with the glycogenic effects of insulin-like growth factor I and insulin. Endocrinology 119: 522-526

Froesch ER, Zapf J, Audhya TK, Ben-Porath E, Segen BJ, Gibson KD (1976) Nonsuppressible insulin-like activity and thyroid hormones: major pituitary-dependent sulfation factors for chick embryo cartilage. Proc Natl Acad Sci USA 73: 2904-2908

Gluckman PD, Butler JH (1985) Circulating insulin-like growth factor-I and -II concentrations are not dependent on pituitary influences in the midgestation fetal sheep. J Dev Physiol 7: 405-409

Gluckman PD, Butler JH, Comline R, Fowden A (1987) The effects of pancreatectomy on the plasma concentrations of insulin-like growth factors 1 and 2 in the sheep fetus. J Dev Physiol 9: 79-88

Graham DE, Rechler MM, Brown AL, Frunzio R, Romanus JA, Bruni CB, Whitfield HJ, Nissley SP, Seelig S, Berry S (1986) Coordinate developmental regulation of high and low molecular weight mRNAs for rat insulin-like growth factor II. Proc Natl Acad Sci USA 83: 4519-4523

Hall K (1972) Human somatomedin - determination, occurence, biological activity and purification. Acta Endocrinol [Suppl] (Copenh) 163: 1-52

Han VKM, D'Ercole AJ, Lund PK (1987) Cellular localization of somatomedin (insulin-like growth factor) messenger RNA in the human fetus. Science 236: 193-197

Hari J, Pierce SB, Morgan DO, Sara VR, Smith MC, Roth RA (1987) The receptor for insulin-like growth factor II mediates an insulin-like response. Embo J 6: 3367-3371

Haselbacher GK Andres RY, Humbel RE (1980) Evidence for the synthesis of a somatomedin similar to insulin-like growth factor I by chick embryo liver cells. Eur J Biochem 111: 245-250

Hill DJ, Milner RDG (1980) Increased somatomedin and cartilage metabolic activity in rabbit fetuses injected with insulin in utero. Diabetologia 19: 143-147

Hill DJ, Crace CJ, Strain AJ, Milner RDG (1986) Regulation of amino acid uptake and deoxyribonucleic acid synthesis in isolated human fetal fibroblasts and myoblasts: effect of human placental lactogen, somatomedin-C, multiplication-stimulating activity, and insulin. J Clin Endocrinol Metab 62: 753-760

Honegger A, Humbel RE (1986) Insulin-like growth factors I and II in fetal and adult bovine serum. J Biol Chem 251: 569-575

Jansen M, van Schaik FMA, Ricker AT, Bullock B, Woods DE, Gabbay KH, Nussbaum AL, Sussenbach JS, Van den Brande JL (1983) Sequence of cDNA encoding human insulin-like growth factor I precursor. Nature 306: 609-611

Lenoir D, Honegger P (1983) Insulin-like growth factor I (IGF-I) stimulates DNA synthesis in fetal rat brain cell cultures. Dev Brain Res 7: 205-213

Lund PK, Moats-Staats BM, Hynes MA, Simmons JG, Jansen M, D'Ercole AJ, Van Wyk JJ (1986) Somatomedin-C/insulin-like growth factor-I and insulin-like growth factor-II mRNAs in rat fetal and adult tissues. J Biol Chem 261: 14539-14544

Marquardt H, Todaro GJ, Henderson LE, Oroszlan S (1981) Purification and primary structure of a polypeptide with multiplication-stimulating activity from rat liver cell cultures. J Biol Chem 256: 6859-6865

McMorris FA, Smith TM, DeSalvo S, Furlanetto RW (1986) Insulin-like growth factor I/somatomedin C: a potent inducer of oligodendrocyte development. Proc Natl Acad Sci USA 83: 822-826

Moses AC, Nissley SP, Short PA, Rechler MM, White RM, Knight AB, Higa OZ (1980) Increased levels of multiplication-stimulating activity, an insulin-like growth factor, in fetal rat serum. Proc Natl Acad Sci USA 77: 3649-3653

Philipps AF, Persson B, Hall K, Lake M, Skottner A, Sanengen T, Sara VR (1988) The effects of biosynthetic insulin-like growth factor/supplementation on somatic growth, maturation and erythropoiesis on the neonatal rat. Pediatr Res (in press)

Póvoa G, Enberg G, Jörnvall H, Hall K (1984) Isolation and characterization of a somatomedin-binding protein from mid-term human amniotic fluid. Eur J Biochem 144: 199-204

Prewitt TE, D'Ercole AJ, Switzer BR, van Wyk JJ (1982) The relationship of immunoreactive somatomedin-C to dietary protein and energy on the growing rat. J Nutr 112: 144-150

Rechler MM, Nissley SP (1985) The nature and regulation of the receptors for insulin-like growth factors. Ann Rev Physiol 47: 425-442

Rechler MM, Fryklund L, Nissley SP, Hall K, Podskalny JM, Skottner A, Moses AC (1978) Purified human somatomedin A and rat multiplication stimulating activity; mitogens for cultured fibroblasts that cross-react with the same growth peptide receptors. Eur J Biochem 82: 5-12

Richman RA, Benedict MR, Florini JR, Toly BA (1985) Hormonal regulation of somatomedin secretion by fetal rat hepatocytes in primary culture. Endocrinology 116: 180-188

Rinderknecht E, Humbel RE (1978a) The amino acid sequence of human insulin-like growth factor I and its structural homology with proinsulin. J Biol Chem 253: 2769-2776

Rinderknecht E, Humbel RE (1978b) Primary structure of human insulin-like growth factor II. FEBS Lett 89: 283-286

Rotwein P (1986) Two insulin-like growth factor I messenger RNAs are expressed in human liver. Proc Natl Acad Sci USA 83: 77-81

Rotwein P, Pollock KM, Didier DK, Krivi GG (1986) Organization and sequence of the human insulin-like growth factor I gene. J Biol Chem 261: 4828-4832

Rubin JS, Mariz I, Jacobs JW, Daughaday WH, Bradshaw RA (1982) Isolation and partial sequence analysis of rat basic somatomedin. Endocrinology 110: 734-740

Sara VR (1987) The role of somatomedins in fetal growth. In: Lindblad BS (ed) Perinatal nutrition. Academic, New York

Sara VR, Hall K (1984) The biosynthesis and regulation of fetal somatomedin. In: Ellendorff F, GLuckman PD, Parvizi N (eds) Fetal neuroendocrinology. Perinatology Press, Ithaca, pp 213-229

Sara VR, Hall K, Sjögren B, Finnson K, Wetterberg L (1979a) The influence of early nutrition on growth and the circulating levels of immunoreactive somatomedin. Am J Dev Physiol 1: 343-350

Sara VR, Hall K, Wetterberg L, Fryklund L, Sjögren B, Skottner A (1979b) Fetal brain growth: the role of the somatomedins and other growth-promoting peptides. In: Giordans G et al. (eds) Somatomedins and growth. Academic, New York, pp 225-230

Sara VR, Hall K, Lins PE, Fryklund L (1980a) Serum levels of immunoreactive somatomedin A in the rat: some developmental aspects. Endocrinology 107: 622-625

Sara VR, Hall K, Ottosson-Seeberger A, Wetterberg L (1980b) The role of the somatomedins in fetal growth. Endocrinology 80: 453-456

Sara VR, Hall K, Misaki M, Fryklund L (1983) Ontogenesis of somatomedin and insulin receptors in the human fetus. J Clin Invest 71: 1084-1094

Sara VR, Carlsson-Skwirut C, Andersson C, Hall E, Sjögren B, Holmgren A, Jörnvall H (1986) Characterization of somatomedins from human fetal brain: identification of a variant form of insulin-like growth factor I. Proc Natl Acad Sci USA 83: 4904-4907

Schmid C, Steiner T, Froesch ER (1983 a) Insulin-like growth factors stimulate synthesis of nucleic acids and glycogen in cultured calvaria cells. Calcif Tissue Int 35: 578-585

Schmid C, Steiner T, Froesch ER (1983 b) Preferential enhancement of myoblast differentiation by insulin-like growth factors (IGF I and IGF II) in primary cultures of chicken embryonic cells. FEBS Lett 161: 117-121

Schmid C, Steiner T, Froesch ER (1984) Insulin-like growth factor I supports differentiation of cultured osteoblast-like cells. FEBS Lett 173: 48-52

Scott J, Cowell J, Robertson ME, Priestley LM, Wadey R, Hopkins B, Pritchard J, Bell GI, Rall LB, Graham CF, Knott TJ (1985) Insulin-like growth factor-II gene expression in Wilms' tumour and embryonic tissues. Nature 317: 260-262

Soares MB, Ishii DN, Efstratiadis A (1985) Developmental and tissue-specific expression of a family of transcripts related to rat insulin-like growth factor II mRNA. Nucleic Acids Res 13: 1119-1134

Soares MB, Turken A, Ishii D, Mills L, Episkopou V, Cotter S, Zeitlin S, Efstratiadis A (1986) Rat insulin-like growth factor II gene - a single gene with two promoters expressing a multitranscript family. J Mol Biol 192: 737-752

Soliman AT, Hadi HABD, Aref MK, Hintz RL, Rosenfeld RG, Rogol AD (1986) Serum insulin-like growth factors I and II concentrations and growth hormone and insulin responses to arginine infusion in children with protein-energy malnutrition before and after nutritional rehabilitation. Pediatr Res 10: 1122-1130

Svoboda ME, Van Wyk JJ, Klapper DG, Fellows RE, Grissom FE, Schlueter RJ (1980) Purification of somatomedin-C from human plasma: chemical and biological properties, partial sequence analysis, and relationship to other somatomedins. Biochemistry 19: 790-797

Weidman ER, Bala RM (1980) Direct mitogenic effects of human somatomedin on human embryonic lung fibroblasts. Biochem Biophys Res Commun 92: 577-585

White RM, Nissley SP, Short PA, Rechler MM, Fennoy I (1982) Developmental pattern of a serum binding protein for multiplication stimulating activity in the rat. J Clin Invest 69: 1239-1252

Yang YW-H, Rechler MM, Nissley SP, Coligan JE (1985) Biosynthesis of rat insulin-like growth factor II. J Biol Chem 260: 2578-2582

Zumstein PP, Lüthi C, Humbel RE (1985) Amino acid sequence of a variant pro-form of insulin-like growth factor II. Proc Natl Acad Sci USA 82: 3169-3172

Endocrine Control of Lung Development*

G. C. Liggins[1] and J.-C. Schellenberg

Introduction

Thorough reviews of the role of hormones in lung maturation are available (Ballard 1984; Smith 1984; Liggins and Schellenberg 1986) and will not be attempted in the present paper which has the purpose of examining recent information regarding the complex relationships of a number of hormones that interact either inside or outside the lungs to promote lung maturation. Many of the studies of hormonal effects on lung maturation in the past 15 years have been made in small laboratory animals which have certain disadvantages limiting their usefulness in dissecting hormonal interrelationships. In the first place, it is not possible to determine the concentrations of administered hormones or to ascertain the changes in other hormones that may occur as a result of treatment. Furthermore, the hormone must be administered either to the mother, in which case it is difficult to determine whether the observed effects are direct or indirect, or by a single injection into the fetus, in which case interpretation is confounded by responses to surgical stress. For these reasons, fetal sheep are being used more extensively, since in these animals the concentration of hormones can be measured serially and hormone treatments can be widely separated in time from operative stress. An added advantage of the larger, monotocous species is the ease of performing selective ablation of endocrine organs. This review will consider primarily work in sheep and will refer to similarities and differences in rats and rabbits only as appropriate.

Hormones in the Ovine Fetus in Late Gestation

The hormone-dependent phase of fetal lung maturation (120–147 days) occurs in an ever-changing hormonal environment that requires definition if maturational changes are to be placed in context. All of the hormones reported to have an effect of one sort or another on lung maturation – in particular cortisol, triiodothyronine

---

* This work was supported by the New Zealand Medical Research Council.
[1] Postgraduate School of Obstetrics and Gynaecology, National Women's Hospital, Claude Road, Auckland, New Zealand

The Endocrine Control of the Fetus
Ed. by W. Künzel and A. Jensen
© Springer-Verlag Berlin Heidelberg 1988

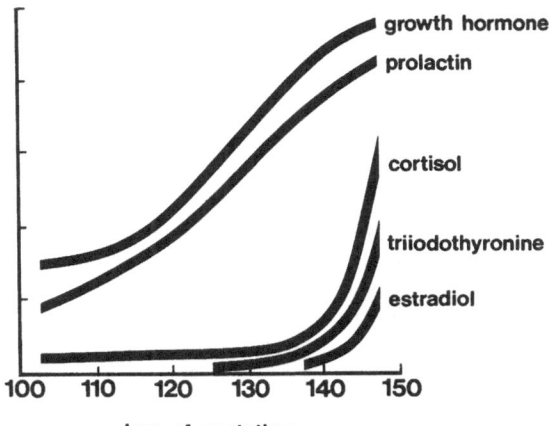

Fig. 1. Patterns of hormone concentrations in plasma of fetal sheep in the last third of pregnancy

(T_3), oestradiol-17β, catecholamines, prolactin and prostaglandins – rise markedly in concentration at variable times before birth (Fig. 1). Cortisol levels begin to rise at 125 days from values of approximately 0.5 µg/dl to 2 µg/dl at 140 days and thereafter to 10–20 g/dl at term (Klein et al. 1978). Oestrogen concentrations follow a similar temporal pattern, total oestradiol-17β levels rising from approximately 0.05 µg/dl at 125 days to 0.2–0.5 µg/dl at term (Findlay and Searmark 1973). Both T_3 and prolactin are barely measurable at 125 days and rise steadily to reach concentrations of about 30 ng/dl (Klein et al. 1978) and 100 ng/dl (Mueller et al. 1979) respectively at term. Prostaglandin E_2 levels are relatively high in late pregnancy and incresae threefold shortly before term (Challis et al. 1976). Adrenaline concentrations remain low (<4 ng/dl) until the onset of parturition, when there is a several-fold but variable increase (Eliot et al. 1981). Growth hormone levels are high throughout the second half of pregnancy and, although they rise in late pregnancy, the percentage increase is relatively small (Gluckman et al. 1979).

Hormonal Responses to Glucocorticoid Treatment

The assumption is sometimes made that maturational responses to corticosteroid treatment of rat or rabbit fetuses reflect solely responses to corticosteroid, whereas the treatment is probably associated with a number of hormonal changes which have not been defined. Indeed, injection of saline into rabbit fetuses is associated with accelerated lung maturation thought to be due to increased levels of corticosteroids (Ballard et al. 1978; Robert et al. 1975) but possibly also reflecting increased levels of other hormones. In vitro experiments in which fetal calf serum is incorporated into the incubation medium may also have problems of interpretation on account of hormones contained in the serum (Gross and Wilson 1982) and it is necessary to use serum-free media.

Although treatment with glucocorticoids stimulates a wide variety of maturational responses both in vivo and in vitro, evidence obtained from experiments in hypophysectomized fetal sheep suggests that cortisol in physiological concentrations has little effect on lung maturation in the absence of 'normal' concentrations of certain other hormones. Liggins et al. (1981) hypophysectomized fetal sheep at 100–122 days' gestation and studied lung development at term. Hypophysectomized, untreated animals had retarded lung development as determined by pressure-volume relationships and by the content of saturated phosphatidylcholine in lung tissue and lavage fluid. Treatment with cortisol for 72 h in doses that increased plasma concentrations at least to those normally observed at parturition had no effect on any of the indices of maturation. In contrast to the lack of response to cortisol, treatment with corticotrophin (ACTH) significantly increased all the indices of maturation even though the concentration of plasma cortisol was less than half than in cortisol-treated fetuses. The authors concluded that the action of cortisol on the fetal lung is absolutely dependent on the presence of one or more pituitary-dependent hormones. Furthermore, since ACTH partially restored retarded lung maturation, whereas cortisol did not, they concluded also that ACTH has effects on the lung that are not mediated solely by cortisol. The unlikely possibility that ACTH has direct effects on the lung was excluded in subsequent experiments in which adrenalectomized fetal sheep at term treated with ACTH showed no reversal of the retarded development of their lungs (Liggins et al. 1985).

Given appropriate support by other hormones that will be considered in detail below, cortisol has a wide variety of actions not only directly on the lung but also on extrapulmonary sites that mediate indirect actions.

Pulmonary Actions of Cortisol

The actions of cortisol on the lung may be broadly divided into effects on structure and effects on surfactant synthesis and secretion. Structural effects may be further subdivided into those in epithelial cells and those in connective tissues. While surfactant synthesis has attracted the interest of many workers since the maturational response to corticosteroid was first described (Liggins 1969), relatively little work has been done on details of epithelial changes and even less on connective tissues, although these components make major contrictions to pulmonary mechanics.

Bunton and Plopper (1984) studied the structural alterations in fetal rhesus monkeys treated with triamcinolone during the mid-pseudoglandular or the mid-canalicular stage of lung development. Triamcinolone accelerated maturation of interstitial and epithelial cells and increased airspace size and the relative contribution of airspace to total lung volume at both ages in a dose-dependent manner. In addition, growth of airspace septa was inhibited by treatment during the pseudoglandular phase, whereas treatment during the canalicular phase accelerated alveolization.

The actions of glucocorticoids on surfactant synthesis are mediated by specific receptors (Ballard 1977) and have two main directions. First, synthesis of phosphatidylcholine by choline incorporation is enhanced. Agreement is lacking on the rate-limiting biochemical step but it is likely to be either choline phosphate cytidyltransferase (Rooney et al. 1979a) or choline phosphotransferase (Van Golde 1982). Second, corticosteroids accelerate glycogenolysis, which releases glucose and triose phosphate intermediates to serve as substrates and sources of reduced nicotinamide adenine dinucleotide phosphate (NADPH) in the synthesis of phosphatidylcholine. In the fetal mouse, glucocorticoids have a biphasic action in glycogen synthesis, promoting deposition in early pregnancy and breakdown in late pregnancy (Brehier and Rooney 1981). Alveolar surfactant increases markedly in response to fetal treatment with corticosteroids or ACTH in various experimental animals including sheep (Avery 1975), and there is a close temporal relationship between the concentration of plasma cortisol and both alveolar and tissue surfactant in fetal sheep (Kitterman et al. 1981), but whether this is a direct effect on type II cells or is an indirect effect – possibly mediated by an increase in adrenaline or adrenergic receptors – is unknown. Studies of human fetal lung in organ culture demonstrated secretion of lamellar bodies in response to a mixture of cortisol and prolactin but not to cortisol alone (Mendelson et al. 1981), which is consistent with the finding referred to earlier that cortisol fails to stimulate surfactant secretion in hypophysectomized fetal sheep (Liggins et al. 1981).

Extrapulmonary Actions of Cortisol

Cortisol has a number of extrapulmonary actions that lead to hormonal changes that may in turn promote lung maturation either directly or by interacting with cortisol. Probably the most important of these is its action at physiological concentrations on monodeiodination of thyroxine (T_4). In sheep, cortisol and T_3 concentrations rise together before parturition as a result of increased conversion of T_4 to T_3 (Thomas et al. 1978) and reduced metabolic clearance of T_3 (M. Fraser, 1987, unpublished observations) and these changes can be reproduced by administering cortisol. In the adrenal medulla of fetal rats, corticosteroids induce phenylethanolamine-N-methyltransferase (PNMT; Margolis et al. 1966), an enzyme catalysing the conversion of noradrenaline to adrenaline. The action of cortisol probably contributes to the sharp rise in adrenaline concentrations at term in sheep. The observation that ACTH but not cortisol increases alveolar surfactant in hypophysectomized sheep (Liggins et al. 1981) is consistent with the possibility that high intramedullary concentrations of cortisol associated with ACTH infusion induce PNMT activity, whereas the lower intramedullary concentrations associated with cortisol infusion fail to do so.

In the ovine placenta, cortisol induces activity of 17α-hydroxylase which promotes synthesis of oestrogens and is responsible for the prepartal surge of oestrogen synthesis which might take part in lung maturation (see below). The placenta is thought to be the source of prostaglandin E_2 released in increased amounts before parturition, probably in response to a fall in the rate of progesterone secretion consequent on activation of 17α-hydroxylase.

Interactions of Cortisol with Other Hormones

In the absence of support by other hormones, cortisol in physiological concentrations appears to be incapable of accelerating lung maturation in the fetuses of sheep and possibly othe species. This is suggested by the absence of response not only in hypophysectomized fetuses but also in immature fetuses (124–128 days' gestation) when the concentrations of hormones in general are low (Liggins and Schellenberg 1986). Although effects on lung maturation of a number of hormones apart from cortisol have been described, no evidence that any one of them interacts with another hormone other than cortisol has been forthcoming.

Thyroid Hormones

Interactions of glucocorticoids and T_3 have been investigated in vitro in both rats and rabbits (Smith and Torday 1974; Gross and Wilson 1982; Ballard et al. 1984). Each hormone alone stimulates choline incorporation into phosphatidylcholine in cultures of either explants or dispersed cells; when combined the effects are additive or supra-additive. Responsiveness to both hormones is dependent on synthesis of protein and RNA (Gross et al. 1983; Ballard et al. 1984) which is consistent with gene activation. Supra-additive responses have been observed also in explants of fetal rat lung after intrafetal injections of T_3 and dexamethasone (Torday and Dow 1984). No effect on pressure–volume relationships or alveolar surfactant content was observed in fetal sheep delivered at 128 days' gestation after continuous infusion of T_3 for 3.5 days at a dose rate that increased the concentration of T_3 in fetal plasma to levels several times greater than that of term animals (Liggins and Schellenberg 1986). Combined treatment with T_3 and cortisol failed to enhance the mechanical response compared to cortisol alone but alveolar surfactant content was increased.

The absence of synergism of cortisol and T_3 in fetal sheep, as determined by mechanical properties of the lung, and its presence in fetal rats, as determined by choline incorporation, could have many explanations including differences between species, corticosteroids (dexamethasone was used in rats) and doses. Plasma levels of T_3 after intrafetal injections in rats are not known but Gross et al. (1984) noted a

28-fold increase in fetal T_3 after maternal treatment. It is possible that small increases in the rate of choline incorporation were present in fetal sheep but were not reflected in the measured indices.

Prolactin

Mendelson et al. (1981) observed that treatment of human fetal lung explants with cortisol and prolactin increased choline incorporation and stimulated secretion of lamellar material into the prealveolar ducts, whereas treatment with the hormones separate was without effect. Liggins and Schellenberg (1986) found no evidence of synergism of cortisol and prolactin in fetal sheep.

Oestrogen

Maternal administration of oestradiol-17β in pregnant rabbits causes both morphological and biochemical maturation of the fetal lungs (Khosla et al. 1982), and intraamniotic injection in fetal rats is associated with differentiation of type II cells (Thuresson-Klein et al. 1985). Effects of combined treatment with cortisol and oestrogen have not been reported. It is difficult to assess the response to combined treatment in fetal sheep because treatment with cortisol alone is associated with a marked increase in the concentration of oestrogen in fetal plasma. Treatment with oestrogen alone has no effect; Andujo et al. (1987) infused fetal sheep from 111 to 127 days with oestradiol-17β in a dose that increased the concentration of unconjugated hormone to levels at least equal to those at parturition and found no differences compared to controls in the flux of phosphatidylcholine in tracheal fluid or in morphological development. However, effects attributed to cortisol when infused alone in fetal sheep in late gestation could represent a synergistic response to cortisol and oestrogen. This possibility can be excluded at least at 128 days because infusions of cortisol from 124 to 128 days have no effect on maturational indices, although they are accompanied by an increase in oestradiol-17β to high levels (Liggins and Schellenberg 1986).

Adrenaline

Since glucocorticoids increase the concentrations of β-adrenergic receptors (Cheng et al. 1980), interaction of cortisol and adrenaline in promoting surfactant secretion is to be expected. In vitro, an alveolar type II cell line exposed to glucocorticoid for 24 h showed an enhanced response to β-adrenergic agents (Smith 1977). In the fetal sheep, simultaneous infusion of cortisol and adrenaline for 84 h was associated with a greater content of saturated phosphatidylcholine in alveolar lavage fluid than when the hormones were infused separately, but distensibility and stability of the lungs were no greater than when cortisol was infused alone (Schellenberg and Liggins 1987). The paradox that maturation in hypophysectomized fetal sheep is accelerated by ACTH but not by cortisol infusion (Liggins et al. 1981) could be explained by postulating that ACTH stimulates secretion of adrenaline by the adrenal medulla (see above).

T_3 and Prolactin

Although neither T_3 nor prolactin singly enhances distensibility or stability of the lungs of fetal sheep infused with cortisol (Schellenberg and Liggins 1987), a striking synergism is observed when the two hormones together are combined with cortisol. Both distensibility and stability of the lungs of fetuses delivered at 128 days are increased to near-term values and alveolar surfactant is sharply increased although not to term levels. The interaction of glucocorticoid, T_3 and prolactin has not been investigated in other species.

Thyrotrophin Releasing Hormone

Administration of thyrotrophin releasing hormone (TRH) to pregnant rabbits for 48 h is associated with a 150% increase in alveolar surfactant in the fetuses but no change in tissue concentration (Rooney et al. 1979b). The response to TRH has been attributed to increased secretion of thyroxine by the fetal thyroid, stimulated by transplacental passage of TRH which is known to cross the placental barrier readily in humans (Roti et al. 1981). However, the actions of TRH are not simply thyrotrophic; TRH stimulates secretion not only of thyroxine but also of prolactin, and in addition has effects on the central nervous system, where it stimulates sympathetic activity (Jackson 1982). In fetal sheep, infusion of TRH for 6.5 days from 123–128 days' gestation is associated with increased concentrations of T_3 and prolactin, but neither the distensibility nor the stability of the lungs is greater than in saline-treated controls (Liggins and Schellenberg 1986), as would be anticipated from the experiments described above in which infusion of T_3 and prolactin failed to change these indices of maturation. However, the combination of infusion of cortisol for 3.5 days with TRH not only increased distensibility and stability of the lungs to near-term values but also increased alveolar surfactant.

The hormonal responses to treatment with cortisol and TRH are complex. In the doses used in these experiments, the concentration of cortisol was approximately twice that normally found at term, and prolactin and T_3 were at approximately term values. In addition, the concentration of oestradiol-17β also increased to term levels. Thus, the interpretation of the response is difficult. What is clear, however, in these experiments is that at a gestational age of 128 days, neither a combination of cortisol and oestradiol-17β (the latter caused by the former) nor a combination of cortisol, T_3 and oestradiol, a combination of cortisol, oestradiol and prolactin, or a combination of T_3 and prolactin (resulting from infusion of TRH) had any effect on the mechanical properties of the lungs. Yet a remarkable synergism of the hormones was found when all the hormones were simultaneously increased to approximately term levels. Since a similar synergism was noted when cortisol, T_3 and prolactin were infused together as single hormones, the response does not appear to depend on actions of TRH other than those mediated by the secretion of thyroid hormones and prolactin. The nature of the interaction of these hormones remains uncertain. It may depend in part on induction of receptors; although cortisol receptors are present at 128 days in a similar concentration to that at term (Flint and Burton 1984), receptors for thyroxine, prolactin and oestrogen could be low and be induced by cortisol or any one of the other hormones. Furthermore, since a specific β_2-adrenergic antagonist substantially impairs the responses to cortisol and TRH (G. C. Liggins, 1986, unpublished observations), it is likely that the effects are mediated in part through adrenergic mechanisms.

Conclusions

The various maturational responses that can be induced in vivo by treatment with glucocorticoids in late pregnancy almost certainly depend on the presence in relatively high concentrations of other hormones. Earlier in gestation, when hormone levels are low, responses to glucocorticoids are minimal or absent but they can be restored by fetal infusion of TRH which recreates the hormonal environment normally found at term, including raised levels of prolactin, T_3 and oestrogen. Prolactin and T_3 are absolute requirements for the synergistic response but the role of oestrogen is uncertain.

Since TRH readily crosses the placenta, maternal treatment with glucocorticoid and TRH in women has the potential to accelerate fetal lung maturation at a gestational age when the response to corticosteroid is often attenuated.

References

Andujo O, Rosenfeld CR, Nielsen HC, Parker CR, Snyder JM (1987) Failure to detect a stimulatory effect of estradiol-17β on ovine fetal lung maturation. Pediatr Res 22: 145-149

Avery ME (1975) Pharmacological approaches to the acceleration of fetal lung maturation. Br Med Bull 31: 13-17

Ballard PL (1977) Glucocorticoid receptors in the lung. Fed Proc 36: 2660-2665

Ballard PL, Gluckman PD, Brehier A, Kitterman JA, Kaplan SL, Rudolph AM, Grumbach MM (1978) Failure to detect an effect of prolactin on pulmonary surfactant and adrenal steroids in fetal sheep and rabbits. J Clin Invest 62: 879-882

Ballard PL (1984) Combined hormonal treatment and lung maturation. Semin Perinatol 8: 283-291

Ballard PL, Hovey ML, Gonzales LH (1984) Thyroid hormone stimulation of phosphatidyl-choline synthesis in cultured fetal rabbit lung. J Clin Invest 74: 848–905

Brehier A, Rooney SA (1981) Phosphatidylcholine synthesis and glycogen depletion in fetal mouse lung. Exp Lung Res 2: 273–287

Bunton TE, Plopper GC (1984) Triamcinolone-induced structural alterations in the development of the lung of the fetal rhesus macaque. Am J Obstet Gynecol 148: 203–215

Challis JRG, Dilley SR, Robinson JS (1976) Prostaglandins in the circulation of the fetal lamb. Prostaglandins 11: 1041–1052

Cheng JB, Goldfien A, Ballard PL, Roberts JM (1980) Glucocorticoids increase pulmonary β-adrenergic receptors in fetal rabbit. Endocrinology 107: 1646–1648

Eliot RJ, Klein AH, Glatz TH, Nathanielsz PW (1981) Plasma norepinephrine, epinephrine and dopamine concentrations in maternal and fetal sheep during spontaneous parturition and in premature sheep during cortisol induced parturition. Endocrinology 108: 1678–1682

Findlay JK, Seamark RF (1973) The occurrence and metabolism of oestrogens in the sheep foetus and placenta. In: Pierrepoint CG (ed) The endocrinology of pregnancy and parturition. Alpha Omega Alpha, Cardiff, p 54

Flint APF, Burton RD (1984) Properties and ontogeny of the glucocorticoid receptor in the placenta and fetal lung of the sheep. J Endocrinol 103: 31–42

Gluckman PD, Mueller PL, Kaplan SL, Rudolph AM, Grumbach MM (1979) Hormone ontogeny in the ovine fetus. 1. Circulating growth hormone and mid and late gestation. Endocrinology 104: 162–168

Gross I, Wilson CM (1982) Fetal lung in organ culture. IV. Supra-additive hormone interactions. J Appl Physiol 52: 1420–1425

Gross I, Ballard PL, Ballard RA, Jones CT, Wilson CM (1983) Corticosteroid stimulation of phosphatidylcholine synthesis in cultured fetal rabbit lung: evidence for de novo protein synthesis mediated by glucocorticoid receptors. Endocrinology 112: 829–837

Gross I, Dynia DW, Wilson CM, Ingleson LD, Gewolb IH, Rooney SA (1984) Glucocorticoid-thyroid hormone interactions in fetal rat lung. Pediatr Res 18: 191–196

Jackson IMD (1982) Thyrotropin-releasing hormone. N Engl J Med 306: 145–155

Khosla SS, Gobran LI, Rooney SA (1982) Stimulation of phosphatidylcholine synthesis by 17β-estradiol in fetal rabbit lung. Biochim Biophys Acta 617: 282–290

Kitterman JA, Liggins GC, Campos GA, Clements JA, Forster CS, Lee CH, Creasy RK (1981) Prepartum maturation of the lung in fetal sheep: relation to cortisol. J Appl Physiol 51: 384–390

Klein AH, Oddie TM, Fisher DA (1978) Effect of parturition on serum iodothyronine concentrations in fetal sheep. Endocrinology 103: 1453–1457

Liggins GC (1969) Premature delivery of foetal lambs infused with glucocorticoids. J Endocrinol 45: 515–523

Liggins GC, Schellenberg J-C (1986) Hormones and lung maturation. In: Johnston BM, Gluckman PD (eds) Respiratory control and lung development in the fetus and newborn. Perinatology Press, Ithaca, p 107

Liggins GC, Kitterman JA, Campos GA, Clements JA, Forster CS, Lee CH, Creasy RK (1981) Pulmonary maturation in the hypophysectomized ovine fetus. Differential responses to adrenocorticotrophin and cortisol. J Dev Physiol 3: 1–14

Liggins GC, Schellenberg J-C, Finberg K, Kitterman JA, Lee CH (1985) The effects of $ACTH_{1-24}$ or cortisol on pulmonary maturation in the adrenalectomized fetus. J Dev Physiol 7: 105–111

Margolis FL, Roffi J, Jost A (1966) Norepinephrine methylation in fetal rat adrenals. Science 154: 275–276

Mendelson CR, Johnston JM, MacDonald PC, Snyder JM (1981) Multihormonal regulation of surfactant synthesis by human fetal lung in vitro. J Clin Endocrinol Metab 53: 307–317

Mueller PL, Gluckman PD, Kaplan SL, Rudolph AM, Grumbach MM (1979) Hormone ontogeny in the ovine fetus. V. Circulating prolactin in mid- and late gestation and in the newborn. Endocrinology 105: 129–134

Robert MF, Bator AT, Taeusch HW (1975) Pulmonary pressure-volume relationships after corticotropin (ACTH) and saline injections in fetal rabbits. Pediatr Res 9: 760-762

Rooney SA, Gobran LI, Marino PA, Maniscalco WM, Gross I (1979a) Effects of betamethasone on phospholipid content, composition and biosynthesis in fetal rabbit lung. Biochim Biophys Acta 572: 64-76

Rooney SA, Marino PA, Gobran LI, Gross I, Warshaw JB (1979b) Thyrotropin-releasing hormone increases the amount of surfactant in lung lavage from fetal rabbits. Pediatr Res 12: 623-625

Roti E, Gnudi A, Braverman LE, Robuschi G, Emanuele R, Bandini P, Benassi L, Pagliani A, Emerson CH (1981) Human cord blood concentrations of thyrotropin, thyroglobulin and iodothyronine after maternal administration of thyrotropin releasing hormone. J Clin Endocrinol Metab 53: 813-817

Schellenberg J-C, Liggins GC, Manzai M, Kitterman JA, Lee CH (1987) Synergistic hormonal effects on lung maturation in fetal sheep. J Appl Physiol (in press)

Schellenberg J-C, Liggins GC (1987) Triiodothyronine, prolactin and oestradiol levels during pulsatile administration of thyrotropin releasing hormone and continuous infusion of cortisol and triiodothyronine in fetal sheep. Proceedings of the NZ Society of Endocrinology, May 9187, abstr No 10

Smith BT (1977) Cell line A459. A model system for the study of alveolar type II cell function. Am Rev Resp Dis 115: 285-293

Smith BT (1984) Pulmonary surfactant during fetal development and neonatal adaptation: hormonal control. In: Robertson B, Van Golde LMG, Batenburg JJ (eds) Pulmonary surfactant. Elsevier, Amsterdam, p 357

Smith BT, Torday JS (1974) Factors affecting lecithin synthesis in fetal lung cells in culture. Pediatr Res 8: 848-851

Thomas AL, Krane EJ, Nathanielsz PW (1978) Changes in the fetal thyroid axis after induction of premature parturition by low dose intravascular cortisol infusion to fetal sheep at 130 days gestation. Endocrinology 103: 17-23

Thuresson-Klein A, Moawad AH, Hedqvist P (1985) Estrogen stimulates formation of lamellar bodies and release of surfactant in the rat fetal lung. Am J Obstet Gynecol 151: 506-514

Torday JS, Dow KE (1984) Synergistic effects of triiodothyronine and dexamethasone on male and female fetal rat lung surfactant synthesis. Dev Pharmacol Ther 7: 133-139

Van Golde LMG (1982) The CDP choline pathway: cholinephosphotransferase. In: Farrell PM (ed) Lung development: biological and clinical perspectives. Academic, New York, p 337

Fetal Endocrine and Metabolic Response to Placental Insufficiency

J. F. CLAPP III[1]

Introduction

The purpose of this chapter is two-fold: first, to review the data which describes the fetoplacental endocrine and metabolic response to placental insufficiency; second, to use this information to support the hypothesis that these responses represent critical adaptations which insure survival in the face of a progressive reduction in nutrient supply. Initially the data, which has been obtained in several species, will be reviewed. Next, the interactions between the endocrine, metabolic and hemodynamic responses will be examined in an attempt to identify a central regulatory mechanism. Finally, the survival value of the responses will be assessed. Throughout, discussion will be limited to the dysmature, asymmetrically growth-retarded fetus. The more general problem of the 'small-for-gestational-age' fetus without clear documentation of the proportionality of growth will not be addressed.

Endocrine and Metabolic Response to Placental Insufficiency

Human Studies

Descriptive studies have provided us with information on changes in maternal blood levels, cord blood values and some indices of placental function in human pregnancies complicated by placental insufficiency with resultant asymmetric intrauterine growth retardation. Positive findings have included evidence of decreased estriol production along with low levels of human placental lactogen in the maternal compartment (Chard et al. 1985; Klapper 1984; Langer et al. 1986; Villar and Belzian 1986; Westergaard 1985). The latter appears to be more specific for the dysmature, asymmetrically growth retarded fetus (Chard et al. 1985; Langer et al. 1986; Westergaard et al. 1985) but, as pointed out by Villar and Belzian (1986), the predictive value of either is probably no better than that obtained with routine clinical measurements in an unselected populace. However, as discussed by Klopper

[1] Department of Obstetrics and Gynecology, C217 Health Science Complex, University of Vermont, College of Medicine, Burlington, VT 05405, USA

The Endocrine Control of the Fetus
Ed. by W. Künzel and A. Jensen
© Springer-Verlag Berlin Heidelberg 1988

(1984), they appear quite worthwhile when used longitudinally in 'at risk' pregnancies. The major value of the cord blood data is that they allow comparison with the findings from animal models. To date, the similarity of the values indicates that the mechanisms identified through animal experimentation appear to be applicable to the human condition. These data have recently been reviewed (Robinson et al. 1985) and demonstrate that the dysmature, asymmetrically growth-retarded human newborn has evidence of increased erythrocyte production, hypoglycemia, hypoinsulinemia, and high levels of lactate, alanine, and triglycerides, reflecting a relative deficiency of nutrient supply in utero.

The value of the clearance of dehydroepiandrosterone sulfate, initially proposed by Gant et al. (1971) as a test of placental function, in the diagnosis of placental insufficiency with secondary fetal growth retardation has been the subject of continued controversy (Klopper 1984; Rabe et al 1986; Villar and Belzian 1986). Finally, use of a variety of placental incubation techniques has produced provocative data which support the concept that the enzymology of the placenta of the dysmature, asymmetrically growth-retarded fetus adapts to maintain substrate delivery by accelerating alternate mechanisms of nutrient supply (Biale 1985; Reynolds et al. 1986).

Data from Animal Models

Fortunately, as diversity strengthens mechanistic interpretation, several different approaches have been used to produce animal models of placental insufficiency with resultant asymmetric intrauterine growth retardation. The primary perturbations employed include a reduction in uterine blood flow either by vascular ligation (Jones 1985; Vileisis and D'Ercole 1986; Wigglesworth 1964) or constriction (Clark and Mack 1986), placental destruction by repetitive embolization (Charlton and Johengen 1986; Clapp et al. 1980; Creasy et al. 1972), limitation of placental growth through carunculectomy (Alexander 1964; Harding et al. 1985; Owens et al. 1987) or maternal heat stress (Alexander et al. 1987; Bell et al. 1987) and a decrease in maternal substrate levels (Charlton and Johengen 1985; Koritnik et al. 1981; Mellor 1983).

A chronic, sustained reduction in uterine blood flow in the sheep, rat, and guinea pig during the latter portion of gestation results in a consistent, marked reduction in placental weight with an asymmetric fetal morphometric outcome (Clark and Mack 1986; Jones 1985; Vileisis and D'Ercole 1986; Wigglesworth 1964). A decrease in the placental transfer of glucose and amino acids has been demonstrated (Nitzan et al. 1979; Saintonge and Rosso 1981) along with a decrease in fetal levels of glucose, insulin, thyroxine and somatomedin C and increased levels of glucagon (Clark and Mack 1986; Jones 1985; Vileisis and D'Ercole 1986; Wigglesworth 1964). In addition, there is evidence that hepatic DNA, RNA and protein synthesis are suppressed while phosphoenolpyruvate carboxykinase activity is induced. The latter findings can be reproduced by the addition of serum from the umbilical vein of growth-retarded fetuses to normal hepatocytes, suggesting that these changes may be regulated by a placentally produced substance (Jones 1985). To date, in vivo assessment of substrate turnover or fetal metabolic rate has not been carried out in any of these models. In addition, serial data during the evolution of placental insufficiency are limited (Jones 1985).

Repetitive embolization of the uterine circulation produces evidence of asymmetric growth retardation in the fetal lamb that is directly related in magnitude to the reduction in placental mass (Charlton and Johengen 1986; Clapp et al. 1982a, b) and the concomitant reduction in uteroplacental blood flow (Clapp et al. 1982a). In this model serial studies before, during, and after the progressive placental damage indicate that the fetoplacental unit responds to the repetitive insult with an early increase in fetal cortisol levels without evidence of fetal hypoxia (Clapp et al. 1982c). When the rise is progressive and exceeds 30 ng/ml premature labor ensues. Although adrenocorticotropic hormone (ACTH) levels have not been measured, the rise in fetal cortisol levels with the onset of embolization occurred at a time in gestation when the fetal adrenal should have been unresponsive to ACTH and may have been placental in origin (Clapp et al. 1982d). In addition, the rate of release of progesterone into the maternal circulation in the postembolization period is increased, with a marked increase in day-to-day variability. This change in hormonal pattern was associated with a change in uterine and umbilical flow distribution (Clapp et al. 1980, 1982a) and placental perfusion balance (Clapp et al. 1982e) that optimized fetal substrate availability (Clapp et al. 1981). Serial studies by all investigators (Charlton and Johengen 1986; Clapp et al. 1980, 1981; Creasy et al. 1972) note that fetal pH, pO_2, PCO_2, hematocrit, glucose, and lactate levels following the onset of embolization are initially stable but that evidence of decreased oxygen and substrate availability appears at some point in the course and is then progressive. The timing of the change in blood levels and substrate differences across the umbilical circulation appears to be delayed because of an early slowing and/or cessation of fetal growth in response to an as yet unknown signal which does not appear to be hypoxia (Charlton and Johengen 1986; Clapp et al. 1980, 1981). Thus, although absolute uptake of substrate is reduced it appears to be matched to the fetal needs which are also reduced due to the cessation of growth (Charlton and Johengen 1986; Clapp et al. 1981). In this regard it is of interest that fetal intravenous nutritional supplementation begun at the onset of embolization completely prevents the metabolic, hemodynamic, and morphometric consequences of the progressive placental damage (Charlton and Johengen 1986). From a mechanistic point of view the important difference in these animals appears to be continued regulatory role for the placenta and strongly reinforcing the morphometric conclusions reached in earlier studies (Clapp et al. 1982b). Unfortunately, additional measures of substrate turnover, placental function, and endocrine profile have not been obtained in the embolization model.

Both carunculectomy and chronic heat stress (Alexander 1964; Alexander et al. 1987; Bell et al. 1987; Harding et al. 1985; Owens et al. 1987; Robinson et al. 1985) restrict placental growth, with subsequent asymmetric fetal growth retardation. In the carunculectomy model it is uncertain whether the primary event is a restriction of flow or a restriction of growth (Owens et al. 1986). However, the recent findings of Alexander et al. (1987) suggest that a decrease in flow is the primary event in response to heat stress. In both models there are direct relationships between fetal weight, placental weight, and uterine and umbilical blood flow. Most of these relationships appear linear while a few appear to oscillate between linear and curvilinear best fits at different time points in gestation. It is clear that in both models placental transfer of flow-limited substances is reduced (Bell et al. 1987; Owens et al. 1986). Fetal hormonal profiles are available only in the carunculectomy model (Robinson et al. 1985). As might be anticipated with the chronic stress of placental insufficiency, the fetal levels of catecholamines, cortisol, glucagon, and β-endorphins are increased, while the levels of most growth-promoting factors (T_3, rT_3, T_4, somatomedins, insulin, and prolactin) are decreased. ACTH, insulin-like growth-factor 2 (IGF_2), ovine placental lactogen (OPL), and growth hormone levels are unchanged. Evidence of substrate restriction is present in both models late in gestation, with a decrease in oxygen tension, glucose level, and placental glucose transfer capacity as well as elevated levels of hemoglobin, lactate, and alanine in the fetal circulation. Uptakes of oxygen and glucose are decreased but at a level commensurate with the reduction in fetal size (Alexander et al. 1987; Harding et al. 1985; Owens et al. 1987; Robinson et al. 1985).

Nutritional deprivation in late pregnancy restricts placental growth and produces a similar fetal morphometric picture save that there is a greater reduction in fetal crown-rump length (Charlton and Johengen 1985; Koritnik et al. 1981; Mellor 1983). As in the earlier models, the decrease in fetal weight is commensurate with the reduction in placental weight. As with embolization, there is an abrupt decrease in fetal growth rate within several days of the onset of nutritional restriction which is not reversed by resuming feeding if the period of nutritional deprivation has been prolonged (Mellor 1983). Uterine blood flow has not been measured in this model, but umbilical blood flow and substrate uptake are both reduced late in the course (Charlton and Johengen 1985). In contrast to earlier models, in this model oxygenation remains normal, indicating that transfer function is appropriate for the decreased growth rate. Only limited hormonal data are available and these suggest that there are minimal changes in fetal levels of growth hormone and no change in prolactin.

A Proposed Central Regulatory Mechanism

All four of the perturbations are associated with a decrease in placental size and/or transfer function that is directly related to the magnitude of the retardation in fetal growth. In the models in which serial early information is available, the primary response to the stimulus which produces placental insufficiency appears to be an abrupt reduction in fetal growth rate (Clapp et al. 1981; Mellor 1983) and/or a reduction in maternal and fetal placental blood flow (Alexander et al. 1987; Clapp et al. 1980, 1982a). Both sets of observations suggest that a placental mechanism signals the early suppression of fetoplacental growth before placental function actually becomes limiting. This response appears to be induced early by multiple stimuli including changes in maternal substrate levels, disrupted regional placental perfusion, placental damage, or an anatomic restriction of placentation. The recent description of a growth-suppressing substance in the umbilical venous effluent from the placentas of growth-retarded fetuses (Jones 1985) is consistent with this hypothesis, as are the observations on the role of initial placentation and functional placental mass on fetal growth rate in normal, embolized, and nutritionally deprived ovine pregnancy (Clapp et al. 1982a; Koritnik et al. 1981). Likewise, the value of concomitant fetal nutritional supplementation in maintaining placental and fetal growth despite repetitive embolic damage indicates that the mechanism can be altered and/or suppressed by a change in the local environment.

If indeed this or other placental mechanisms are the primary regulators of fetoplacental growth, then the additional endocrine, metabolic, and hemodynamic responses should be viewed as secondary adaptations to, and/or a reflection of, the decrease in placental and fetal growth rate. Clearly, the changes in maternal and fetal hormonal profiles and placental flows, fetal substrate levels, and fetal oxygen and substrate utilization could well be secondary to a simple restriction of placental surface area that is matched to the demands of fetal growth. Indeed, the progressive hemodynamic and endocrine changes associated with repetitive embolic placental damage (Clapp et al. 1982a, b, e), the early changes in uterine and the late changes in umbilical velocity wave forms in human intrauterine growth retardation (Giles et al. 1986; Trudinger et al. 1983, 1985), and the extreme fragility of the fetus late in gestation after carunculectomy (Harding et al. 1985; Owens et al. 1986, 1987) indicate that this is indeed the case. Thus, our current state of knowledge suggests that the process of placentation and the placental response to a variety of negative stimuli initiate a restriction of fetal growth which is mediated by an as yet incompletely understood mechanism or mechanisms.

Survival Value of the Endocrine, Hemodynamic, and Metabolic Responses

Clearly the reduction and/or cessation of fetal growth rate as the primary response to noxious maternal and placental stimuli that ultimately limit placental growth and transfer function has survival value as it immediately decreases absolute utilization of oxygen and metabolic substrates (Bell et al. 1987; Charlton and Johengen 1986; Clapp et al. 1980; Nitzan et al. 1979; Owens et al. 1987). Likewise, the changes in the distribution of fetal cardiac output (Creasy et al. 1972) and the associated relative sparing of brain, heart, and adrenal growth which occurs later in the course have distinct survival value. The late and progressive fall in substrate levels and the changes in hormonal levels can be viewed simply as additional mechanisms which merely reinforce and magnify the reduction in growth rate in the face of a progressive compromise between placental function and fetal demands. As noted in several models (Charlton and Johengen 1986; Clapp et al. 1980; Creasy et al. 1972; Harding et al. 1985; Owens et al. 1986, 1987) these clearly are late changes, as they only appear at a point in time when progressive fetal acidosis, death, and/or fetal escape are the only remaining responses to superimposed surgical or experimental stress. As such they have survival value. Likewise, both the acute and chronic responses in the maternal and placental blood flows and their distribution act to decrease shunting, maximize exposure time and optimize transfer function (Bell et al. 1987; Charlton and Johengen 1985; Clapp et al. 1980, 1982a, e; Owens et al. 1986). Again, this adaptation has its limitations and, when it reaches its limit, forms a clear indication for delivery in human pregnancy (Giles et al. 1986; Trudinger et al. 1983, 1985). Thus, all facets of the fetoplacental response to noxious stimuli have survival value. Each shows itself to be needed in order to cope with progression of the underlying process. Indeed, the day-to-day fluctuations in fetal growth rate observed by Mellor (1983) may simply represent intermittent suppression due to transient decreases in placental perfusion or maternal substrate levels. These fluctuations may simply reflect the primary way the fetus maintains reserve mechanisms to cope with superimposed stress for a protracted period of time. When this primary and all secondary adaptations are exhausted, or when the rate of progression of the process is too rapid for sufficient adaptation to occur, progressive fetal deterioration and/or premature labor result.

Acknowledgement. This work was supported by NIH grants HD 11122 and HD 21268.

References

Alexander G (1964) Studies on the placenta of the sheep: effect of surgical reduction in the number of caruncles. J Reprod Fertil 7: 307–322

Alexander G, Hales JRS, Stevens D, Donnelly JB (1987) Effects of acute and prolonged exposure to heat on regional blood flows in pregnant sheep. J Dev Physiol 9: 1–15

Bell AW, Wilkening RB, Meschia G (1987) Some aspects of placental funciton in chronically heat-stressed ewes. J Dev Physiol 9: 17–29

Biale Y (1985) Lipolytic activity in the placentas of chronically deprived fetuses. Acta Obstet Gynecol Scand 64: 111–114

Chard T, Sturdee J, Cockrill B, Obiekwe BC (1985) Which is the best placental function test? A comparison of placental lactogen and unconjugated estriol in the prediction of intrauterine growth retardation. Eur J Obstet Gynecol Reprod Biol 19: 13–17

Charlton V, Johengen M (1985) Effects of intrauterine nutrional supplementation on fetal growth retardation. Biol Neonate 48: 125–142

Charlton V, Johengen J (1987) Fetal intravenous nutritional supplementation ameliorates the development of embolization-induced growth retardation in sheep. Pediatr Res 22: 55–61

Clapp JF, Szeto HH, Larrow RW, Hewitt J, Mann LI (1980) Umbilical blood flow response to embolization of the uterine circulation. Am J Obstet Gynecol 138: 60–67

Clapp JF, Szeto HH, Larrow R, Hewitt J, Mann LI (1981) Fetal metabolic response to experimental placental vascular damage. Am J Obstet Gynecol 140: 446–451

Clapp JF, McLaughlin MK, Larrow R, Farnham J, Mann LI (1982a) The uterine hemodynamic response to repetitive unilateral vascular embolization in the pregnant ewe. Am J Obstet Gynecol 82: 309–318

Clapp JF, Larrow R, Hewitt J, Mann LI (1982b) Fetoplacental morphometric correlates in intrauterine growth retardation. Soc Gynecol Invest Abstr 159

Clapp JF, Auletta FJ, Farnham J, Larrow R, Mann LI (1982c) The ovine fetoplacental endocrine response to placental damage. Am J Obstet Gynecol 144: 47–54

Clapp JF, Thabault NC, Hubel CA, McLaughlin MK, Auletta FJ (1982d) Ovine placental cortisol production. Endocrinology 111: 1728–1730

Clapp JF, McLaughlin MK, Larrow R, Farnham J, Mann LI (1982e) Perfusion balance in the ovine placenta following embolic damage. Soc Gynecol Invest Abstr 305

Clark KE, Mack CE (1986) A new model of growth retardation-fetal growth is blood flow dependent. Soc Gynecol Invest Abstr 137P

Creasy RK, Barrett CT, De Swiet M, Kahanpaa KV, Rudolph AM (1972) Experimental intrauterine growth retardation in the sheep. Am J Obstet Gynecol 112: 566–573

Gant N, Hutchinson H, Siteri P, MacDonald P (1971) Study of the metabolic clearance rate of dehydroisoandrosterone sulfate in pregnancy. Am J Obstet Gynecol 111: 555–561

Giles WB, Lingman G, Marsal K, Trudinger BJ (1986) Fetal volume blood flow and umbilical artery flow velocity waveform analysis: a comparison. Br J Obstet Gynaecol 93: 461–467

Harding JE, Jones CT, Robinson JS (1985) Studies on experimental growth retardation in sheep. The effects of a small placenta in restricting transport to and growth of the fetus. J Dev Physiol 7: 427–442

Jones CT (1985) Reprogramming of metabolic development by restriction of fetal growth. Biochem Soc Trans 13: 89–91

Klopper A (1984) Diagnosis of growth retardation by biochemical methods. Clin Obstet Gynecol 11: 437–456

Koritnik DR, Humphrey WD, Kaltenbach CC, Dunn TG (1981) Effects of maternal undernutrition on the development of the ovine fetus and the associated changes in growth hormone and prolactin. Biol Reprod 24: 125–137

Langer O, Damus K, Maiman M, Divon M, Levy J, Bauman W (1986) A link between relative hypoglycemia-hypoinsulinemia during oral glucose tolerance tests and intrauterine growth retardation. Am J Obstet Gynecol 155: 711–716

Mellor D (1983) Nutritional and placental determinants of foetal growth rate in sheep and consequences for the newborn lamb. Br Vet J 139: 307–324

Nitzan M, Orloff S, Shulman JD (1979) Placental transfer of analogs of glucose and amino acids in experimental intrauterine growth retardation. Pediatr Res 13: 100–103

Owens JA, Falconer J, Robinson JS (1986) Effect of restriction of placental growth on umbilical and uterine blood flows. Am J Physiol 250: R427–R434

Owens JA, Falconer J, Robinson JS (1987) Effect of restriction of placental growth on fetal and utero-placental metabolism. J Dev Physiol 9: 225–238

Rabe T, Hosch R, Kiesel L, Runnebaum B, Keller PJ, Kubli F (1986) Diagnosis on intrauterine fetal growth retardation by DHAS half-life. Eur J Obstet Gynecol Reprod Biol 22: 41–51

Reynolds JW, Barnhart BJ, Carlson CV (1986) Feto-placental steroid metabolism in growth retarded human fetuses. Pediatr Res 20: 166–168

Robinson JS, Falconer J, Owens JA (1985) Intrauterine growth retardation: clinical and experimental. Acta Paediatr Scand [Suppl] 319: 135–142

Saintonge J, Rosso P (1981) Placental blood flow and the transfer of nutrient analogs in large, average and small guinea pig litter mates. Pediatr Res 15: 152–156

Trudinger BJ, Giles WB, Cook CM (1983) Feto-placental blood flow resistance and placental microvascular anatomy: a doppler ultrasound-pathological correlation. J Ultrasound Med 2: 59–78

Trudinger BJ, Giles WB, Cook CM (1985) Uteroplacental blood flow velocity-time wave forms in normal and complicated pregnancies. Br J Obstet Gynaecol 92: 39–44

Vileisis RA, D'Ercole AJ (1986) Tissue and somatomedin-C/insulin-like growth factor I in fetal rats made growth retarded by uterine artery ligation. Pediatr Res 20: 126–130

Villar J, Belzian JM (1986) The evaluation of methods used in the diagnosis of intrauterine growth retardation. Obstet Gynecol Surv 41: 187–199

Westergaard JG, Teisner B, Hau J, Grudzinskas JG, Chard T (1985) Placental function studies in low birthweight infants with and without dysmaturity. Obstet Gynecol 65: 316–318

Wigglesworth JS (1964) Experimental growth retardation in the foetal rat. J Pathol Bacteriol 88: 1–13

Endocrine Aspects
of Fetal Behaviour

Behavioural States and Blood Flow Changes in the Human Fetus

J. W. Wladimiroff[1], J. van Eyck[1], J. A. G. W. van der Wijngaard[1],
M. J. Noordam[1], K. L. Cheung[1], and H. F. R. Prechtl[2]

Introduction

In the human fetus behavioural states were studied from fetal heart rate recordings alone (de Haan 1979; van Geyn et al. 1980), and in conjunction with body movements (Timor-Tritsch et al. 1978; Natale 1985), resulting in the description of quiet and activity phases. The observations did not provide conclusive proof of the presence of true behavioural states in the human fetus since both fetal heart rate and fetal heart rate variability are affected by fetal motility (Wheeler and Guerard 1974). Nijhuis et al. (1982) stated that the presence of behavioural states can only be established following fulfillment of a number of criteria:

(a) particular conditions of several variables must recur in specific, fixed combinations;
(b) these combinations must be temporally stable;
(c) there should be clear transitions between states.

It was therefore necessary to study several independent variables with respect to the consistency and stability of their association and the simultaneity of their transition from one condition into another. Nijhuis et al. (1982) clearly identified fetal behavioural states on the basis of eye and body movements and fetal heart rate patterns in low risk multigravidae at 36–38 weeks of gestation. Four distinct behavioural states could be detected, named 1F to 4F, states 1F and 2F being most prevalent.

State 1F is a state of quiescence, which may be regularly interrupted by brief gross body movements, mostly startles. Eye movements are absent. Heart rate is stable, with a small oscillation bandwidth. Isolated acceleration occur, strictly related to movements. This fetal heart rate pattern is called FHRP-A.

State F2 frequent and periodic gross body movements – mainly stretches and retroflexions – and movements of the extremities occur. Eye movements are continually present (REMs and SEMs). Heart rate (called FHRP-B) has a wider oscillation bandwidth than FHRP-A and there are frequent accelerations during movements.

[1] Department of Obstetrics and Gynaecology, Erasmus University, Rotterdam, The Netherlands
[2] Department of Developmental Neurology, Academic Hospital Groningen, Groningen, The Netherlands

The Endocrine Control of the Fetus
Ed. by W. Künzel and A. Jensen
© Springer-Verlag Berlin Heidelberg 1988

The first objective of the present study was to establish in normal term pregnancies a possible relationship between the flow velocity waveform in the fetal descending aorta, internal carotid artery and umbilical artery and fetal behavioural states, in particular states 1F and 2F according to the classification of Nijhuis et al. (1982). The second objective was to elucidate a possible role of fetal behavioural states with respect to blood flow in the same vessels in intrauterine growth retardation.

Methods

Using a combined real-time scanner and pulsed Doppler system as described by Eik-Nes et al. (1980), the mean blood flow velocity at the lower thoracic level of the fetal descending aorta was recorded. The Doppler probe (Pedoff) was attached to the linear array real-time transducer so that the Doppler beam intersected the fetal descending aorta at a fixed angle of 45°. The beam direction with the sample gate position (electronic marker) could be displayed on the real-time screen. The blood flow velocity waveform was recorded during fetal apnoea. A combined mechanical sector scanner and pulsed Doppler system (Diasonics CV 400) was used for recording the maximum flow velocity waveform in the fetal internal carotid artery at the level of the bifurcation into the middle and anterior cerebral artery (Wladimiroff et al. 1986) and for recording the maximum flow velocity waveform in the umbilical artery. Only recordings depicting simultaneous arterial and venous flow velocity patterns were accepted as originating from the umbilical cord. Venous blood flow had to be constant to ensure the presence of fetal apnoea. Maximum flow velocity waveforms were recorded during fetal behavioural states 1F and 2F according to Nijhuis et al. (1982). In order to establish fetal behavioural states, the following parameters were simultaneously recorded:

1. Fetal heart rate (FHR), which was obtained from a Doppler ultrasound cardiotocograph (Hewlett Packard 8040A, carrier frequently 1 MHz);
2. Fetal eye movements, which were studied from a transverse view of the fetal face using the Diasonics CV 400 (carrier frequency 3.5 MHz);
3. Fetal body movements from a two-dimensional real-time linear array scanner (Toshiba Sal 20A, carrier frequency 3.5 MHz) for a sagittal view of the fetal trunk.

The three transducers were placed in such way that there was minimal interference between the three ultrasound modes. Flow velocity recordings were only performed when a clear fetal behavioural state was identified and when this state had been present over a period of at least 3 min. All recordings were performed during fetal apnoea. The maximum amount of time for the completion of a flow velocity recording following a state determination was 3 min.

The blood flow velocity waveforms were recorded on video tape over a 15-s period which included an average of 30 consecutive cardiac cycles. In each subject a minimum of three flow velocity waveform recordings in each vessel in each behavioural state (1F and 2F) was obtained. The degree of pulsatility of the waveform was quantified by calculating the pulsatility index (PI) according to Gosling and King (1975) using a microcomputer (Apple III). The PI changes predominantly reflect changes in peripheral vascular resistance. Since at normal fetal heart rates, changes in cardiac output are mainly regulated through changes in FHR (Kirkpatrick et al. 1976; Marsal et al. 1984), it was decided to standardize cardiac output by dividing the PI values in each subject and for each fetal behavioural state into groups, each representing an FHR range of 5 beats/min. This standardization will also rule out the effect of cardiac cycle length dependency of the PI when comparing PI values between behavioural states 1F and 2F. Changes in PI values with respect to the fetal behavioural state were tested using the paired Student's t test. The relationship between PI and FHR was tested by analysing the slopes of the individual regression lines using the Student's t test.

Subjects

Blood flow velocity waveforms in the lower thoracic part of the fetal descending aorta, in the internal carotid artery and in the umbilical artery were related to fetal behavioural states in 40 normal subjects (van Eyck et al. 1985, 1987) and in 20 cases of intra-uterine growth retardation (IUGR) (van Eyck et al. 1986, 1988). In the normal subjects, fetal birth weight was between the 10th and 90th percentile for gestational age according to Kloosterman's tables corrected for maternal parity and fetal sex (Kloosterman 1970). All mothers were non smokers, no medications were prescribed. IUGR was defined as follows:

1. A progressive asymmetric slow-down in increase of fetal head and upper abdominal circumference resulting in values below the 5th percentile of the nomograms by Campbell and Wilkin (1975)
2. Fetal birth weight below the 5th percentile for gestational age according to Kloosterman's tables, corrections being made for maternal parity and fetal sex (Kloosterman 1970)

Results

Normal Pregnancy

Fetal Descending Aorta ($n = 13$)

The mean number of cardiac cycles studied for all 13 patients was 55 (min. 26, max. 101) in state 1F and 58 (min. 26, max. 95) in state 2F, a total of 1480 cycles. FHR in state 1F ranged between 103 and 171 beats/min and in state 2F between 94 and 185 beats/min.

Paired flow velocity data in state 1F and 2F were observed in the FHR range between 106 and 165 beats/min. Statistical analysis of the paired differences was feasible in the FHR range between 121 and 150 beats/min, resulting in six groups, i.e. 121–125, 126–130, . . ., 146–150 beats/min, a total number of 1320 cardiac cycles. A higher PI ($p < 0.01$) was observed in state 1F than in state 2F. A significant reduction ($p < 0.001$) in PI with increasing FHR was established in state 2F; this was mainly determined by a significant rise ($p < 0.02$) in end-diastolic flow velocity.

Fetal Internal Carotid Artery ($= 12$)

The mean number of cardiac cycles studies in the fetal internal carotid artery for all 12 subjects was 56 (min. 47, max. 68) in state 1F and 59 (min. 26, max. 76) in state 2F, a total of 1381 cycles. FHR in state 1F ranges between 109 and 164 beats/min and in state 2F between 109 and 185 beats/min. Paired analysis of the PI data in state 1F and 2F was feasible in the FHR range between 121 and 145 beats/min, resulting in five groups, i.e. 121–125, 126–130, . . ., 141–145 beats/min, a total number of 1021 cardiac cycles. There was a statistically significant difference in mean PI between states 1F and 2F for all of the FHR ranges studied ($p < 0.001$).

A significant reduction in PI with increasing FHR was established in state 1F ($p < 0.005$) and 2F ($p < 0.001$).

Umbilical Artery ($n = 15$)

The mean number of cardiac cycles studied in the umbilical artery for all 15 subjects was 59 (min. 38, max. 82) in state 1F and 58 (min. 33, max. 76) in state 2F, a total of 1752 cycles. FHR in state 1F ranged between 105 and 185 beats/min and in state 2F between 102 and 188 beats/min. Paired analysis of the flow velocity data in state 1F and 2F was statistically feasible in the FHR range between 121 and 150 beats/min, resulting in six groups, i.e. 121–125, 126–130, . . ., 146–150 beats/min, a total of 1449 cardiac cycles. the 95% confidence interval of the paired difference in mean PI between states 1F and 2F displayed a narrow distribution around zero, reflecting a virtually complete overlap of PI values originating from states 1F and 2F. An inverse relationship between PI and FHR both in behavioural state 1F ($p < 0.05$) and 2F ($p < 0.001$) was established.

Intra-uterine Growth Retardation

Fetal Descending Aorta ($n = 12$)

The mean number of cardiac cycles studied for all 12 patients was 81 (min. 32, max. 127) in state 1F and 86 (min. 52, max. 153) in state 2F, a total of 2004 cycles. FHR in state 1F ranged between 111 and 169 beats/min and in state 2F between 115 and 176 beats/min. Paired analysis of the flow velocity data in states 1F and 2F was statistically feasible in the FHR range between 121 and 165 beats/min, resulting in nine groups, i.e. 121-125, 126-130, ..., 161-165 beats/min, a total of 1888 cardiac cycles.

The 95% confidence interval of the paired difference in mean PI between state 1F and 2F displayed for each FHR range in narrow distribution around zero, reflecting a virtually complete overlap of PI values originating from states 1F and 2F. Using Student's t test a significant reduction ($p < 0.0001$) in PI with increasing FHR was established in both states 1F and 2F.

Fetal Internal Carotid and Umbilical Artery ($n = 8$)

The mean number of cardiac cycles studied in the fetal internal carotid artery for all eight subjects was 50 (min. 31, max. 72) in state 1F and 48 (min. 29, max. 63) in state 2F, a total of 782 cycles. FHR in state 1F ranged between 98 and 158 beats/min and in state 2F between 115 and 167 beats/min. Calculating the mean PI independent of FHR and behavioural state, all eight values were below -1SD and in two patients below -2SD of the nomogram according to Wladimiroff et al. (1987).

The mean number of cardiac cycles studied in the umbilical artery was 47 (min. 15, max. 66) in state 1F and 41 (min. 18, max. 62) in state 2F, a total of 699 cycles. FHR in state 1F ranged between 99 and 171 beats/min and in state 2F between 107 and 162 beats/min. The mean PI irrespective of FHR and behavioural state was situated above $+1$SD in all eight patients and above $+2$SD in five patients according to the nomogram by Wladimiroff et al. (1987).

Paired analysis of the PI data in states 1F and 2F was feasible for the fetal internal carotid artery in the FHR range between 126 and 145 beats/min (561 cardiac cycles) and for the umbilical artery also in the FHR range between 126 and 145 beats/min (529 cardiac cycles). There was no statistically significant difference in mean PI between states 1F and 2F for all FHR ranges studied in the fetal internal carotid artery and umbilical artery. Moreover, the 95% confidence interval of the paired differences in mean PI between state 1F and 2F displayed for each FHR range a narrow distribution around zero, reflecting a virtually complete overlap of PI values originating from states 1F and 2F. A significant inverse relationship between PI and FHR was established for both behavioural states (1F: $p < 0.001$; 2F: $p < 0.01$) in the umbilical artery and for behavioural state 1F ($p < 0.001$) in the fetal internal carotid artery.

Discussion

The marked difference observed between PI in state 1F and in state 2F in our study is almost entirely determined by changes in end-diastolic blood flow velocity (EDV). The elevated EDV in the fetal descending aorta, suggesting reduced fetal peripheral vascular resistance in behavioural state 2F, may be explained by the need for an increased perfusion of the fetal skeletal musculature (trunk and lower extremities) to meet the energy demands during raised muscular activity in this particular behavioural state. This is further supported by our data from umbilical artery velocity waveforms, in which no such state dependency was observed. The decrease in PI in the internal carotid artery during behavioural state 2F, as demonstrated in our study, suggests a reduction in peripheral vascular resistance in the fetal brain. The possible clinical importance of flow velocity waveform recordings in the human fetal internal carotid artery is mainly based on reduction in peripheral vascular resistance in the fetal cerebrum, i. e. a brain sparing effect in intrauterine growth retardation (Wladimiroff et al. 1986, 1987).

While in normal pregnancies and under standardized FHR conditions, the PI in the lower thoracic part of the fetal descending aorta (van Eyck et al. 1985) and fetal internal carotid artery shows significant changes with respect to the fetal behavioural state, this is not so for the umbilical artery. This independency of fetal behavioural state suggests the existence of a fetal regulatory mechanism for the state-dependent changes in the fetal descending aorta and internal carotid artery. The decrease in PI with rising fetal heart rate observed in all three vessels studied is mainly determined by the definition presented by Gosling and King (1975) for PI calculations, i. e. at a lower FHR a more gradual end-diastolic slow-down of the blood flow velocity takes place. In IUGR, the PI in the descending aorta showed a marked increase compared to normal pregnancy. In the normal fetus, increased peripheral perfusion documented by a reduction in PI and rise in EDV was established in state 2F, but this change did not occur in IUGR. This may be explained by the fact that the chronic hypoxia present in IUGR stimulates the peripheral arterial chemoreceptors (Dawes et al. 1968; Itskovitz et al. 1982) and subsequent release of vasoconstrictive agents, such as vasopressin and catecholamines (Iwamoto et al. 1979; Oosterbaan 1985; Mott 1985). This peripheral vasoconstriction seems to overrule state-dependent PI fluctuations. Consequently the increased energy demand, needed for raised muscular activity during state 2F, may not be adequately met. Flow measurements in the fetal internal carotid artery of normal pregnancies demonstrated reduced PI values during state 2F, reflecting a decrease in cerebral vascular resistance during this behavioural state, but this state dependency seems to be absent during IUGR. In contrast to the marked PI increase in the fetal descending aorta, state independency in the fetal internal carotid artery was associated with only moderate reduction in PI, suggesting the onset of circulatory redistribution with the aim of favouring cerebral blood flow (brain sparing effect). The degree of PI reduction at this stage seems, however, to be sufficient to overrule behavioural state dependency. The inverse relationship between PI and fetal heart rate for both states 1F and 2F in the fetal descending aorta and umbilical artery and for state 1F in the fetal internal carotid artery is mainly determined by the cycle length dependency of the formula from which the PI is calculated.

Conclusion

In the normal growing human fetus at term, blood flow velocity waveforms obtained from the lower thoracic part of the fetal descending aorta and the internal carotid artery show fetal behavioural state dependency. The fetal origin of the behavioural state-dependent PI fluctuations is suggested by the absence of behavioural state-dependent PI fluctuations in the umbilical artery. In intra-uterine growth retardation, the PI in the lower thoracic part of the fetal descending aorta and in the fetal internal carotid artery shows no fetal behavioural state dependency. In all studies there is a significant inverse relationship between PI and fetal heart rate.

References

Campbell S, Wilkin D (1975) Ultrasonic measurement of fetal abdomen circumference in the estimation of fetal weight. Br J Obstet Gynaecol 82: 689–697

Dawes GS. Lewis BV, Milligan IE, Roach MR, Talner NS (1968) Vasomotor responses in the hind limbs of foetal and new-born lambs to asphyxia and aortic chemoreceptor stimulation. J Physiol 195: 55–81

de Haan R, Patrick J, Chess JF, Jaco MT (1979) Definition of sleep state in the new-born infant by heart rate analysis. Am J Obstet Gynecol 127: 753–758

Eik-Nes SH, Brubakk AO, Ulstein MK (1980) Measurement of human fetal blood flow. Br Med J 1: 283–284

Gosling RG, King DH (1975) Ultrasound angiology. In: Marcus AW, Adamson L (eds) Arteries and veins. Churchill Livingstone, Edinburgh, pp 61–98

Itskovitz J, Goetzmann BW, Rudolph AM (1982) The mechanism of late decerlation of the heart rate and its relationship to oxygenation in normoxemic and chronically hypoxemic fetal lambs. Am J Obstet Gynecol 142: 66–73

Iwamoto HS, Rudolph AM, Keil LC, Heymann MA (1979) Hemodynamic responses of the sheep to vasopressin infusion. Clin Res 44: 430–436

Kirkpatrick SE, Pitlick PT, Naliboff J, Friedman WF (1976) Frank-Starling relationship as an important determination of fetal cardiac output. Am J Physiol 231: 495–502

Kloosterman GJ (1970) On intrauterine growth. Int J Gynecol Obstet 8: 895–912

Marsal K, Eik-Nes SH, Lindblad A, Lingman G (1984) Blood flow in the fetal descending aorta. Intrinsic factors affecting fetal blood flow. Ultrasound Med Biol 10: 339–348

Mott JC (1985) Humoral control of the fetal circulation. In: Jones CT, Nathanielsz PW (eds) Physiological Development of the fetus and newborn. Academic, London, pp 113–121

Natale R, Nasello-Paterson C, Turliuk R (1985) Longitudinal measurement of fetal breathing, body movements, heart rate, and heart rate accelerations and decelerations at 24 to 32 weeks of gestation. Am J Obstet Gynecol 151: 256–263

Nijhuis JG, Prechtl HFR, Martin CB Jr, Bots RSGM (1982) Are there behavioural states in the human fetus? Early Hum Dev 6: 177–195

Timor-Tritsch IE, Dierker LJ, Zador I, Hertz RH, Rosen MG (1978) Fetal movements associated with fetal heart rate accelerations and decelerations. Am J Obstet Gynecol 131: 276–280

van Eyck J, Wladimiroff JW, Noordam MJ, Tonge HM, Prechtl HFR (1985) The blood flow velocity waveform in the fetal descending aorta; its relationship to fetal behavioural states in normal pregnancy at 37–38 weeks. Early Hum Dev 12: 137–143

van Eyck J, Wladimiroff JW, Noordam MJ, Tonge HM, Prechtl HFR (1986) The blood flow velocity waveform in the fetal descending aorta; its relationship to behavioural states in the growth-retarded fetus at 37–38 weeks of gestation. Early Hum Dev 14: 99–107

Oosterbahn HP (1985) Amniotic oxytocin and vasopression in the human and the rat. Thesis, University of Amsterdam

van Eyck J, Wladimiroff JF, van den Wijngaard JAGW, Noordam MJ, Prechtl HFR (1987) The blood flow velocity waveform in the fetal internal carotid and umbilical artery; its relationship to fetal behavioural states in normal pregnancy at 37–38 weeks of gestation. Br J Obstet Gynecol 94: 736–741

van Eyck J, Wladimiroff JW, Noordam MJ, van den Wijngaard JAGW, Prechtl HFR (1988) The blood flow velocity waveform in the fetal internal carotid and umbilical artery; its relationship to fetal behavioural states in the growth retarded fetus at 37–38 weeks of gestation. Br J Obstet Gynecol (In press)

van Geyn HP, Jongsma HW, de Haan J, Eskes TKAB, Prechtl HFR (1980) Heart rate as an indication of the behavioural state: studies in the new-born infant and prospects for fetal heart rate monitoring. Am J Obstet Gynecol 136: 1061–1066

Wheeler T, Guerard P (1974) Fetal heart rate during late pregnancy. J Obstet Gynecol Br Commonw 64: 348–356

Wladimiroff JW, Tonge HM, Stewart PA (1986) Doppler ultrasound assessment of cerebral blood flow in the human fetus. Br J Obstet Gynecol 93: 471–474

Wladimiroff JW, van den Wijngaard JAGW, Degani S, Noordam MJ, van Eyck J, Tonge HM (1987) Cerebral and umbilical arterial blood flow velocity waveforms in normal and growth-retarded pregnancies; a comparative study. Obstet Gynecol 69: 705–709

Electrophysiological and Neurochemical Response to Asphyxia in the Ovine Fetus

P. D. Gluckman[1], C. J. Cook, C. E. Williams, A. J. Gunn, and B. M. Johnston

Asphyxial neural insults are the major cause of encephalopathy in the neonate and it is generally considered that the origin of the hypoxic/ischaemic episode is prenatal in at least 90% of cases (Hill and Volpe 1982). It is therefore of considerable potential importance to gain a better understanding of the fetal neural responses to asphyxial insults (Freeman 1985). The early studies particularly of Myers et al. (1984) demonstrated that the magnitude of insult is not directly related to the degree of hypoxia and that other factors than hypoxaemia contribute to the development of neural demage. Such factors included the concurrent development of acidosis and/or hypotension, reducing cerebral perfusion. Recent studies of the biochemistry of neuronal death suggest a number of potential metabolic pathways by which ischaemia may cause neuronal death. One favoured hypothesis is that asphyxia leads to intracellular calcium accumulation, with a variety of biochemical consequences including intracellular hydroxyapatite deposition, disruption of intracellular organelles, activation of kinases, ecosanoid release and free radical formation (Schanne et al. 1979; Farber et al. 1981; Siesjo 1981; Raichle 1983). Alternative hypotheses to explain neural death have focused on the role of excitatory amino acids (Rothman 1984; Meldrum 1985; Wieloch 1985), although recent evidence has linked their receptors directly to the calcium channel (MacDermott et al. 1986), as well as to models based on depolarization followed by influx of chloride ions (Rothman 1985).

Recent advances in methodology have expanded the range of technique than can be used to study neural function in utero. The studies here described were performed in the chronically instrumented late-gestation fetal lamb (term 147 days). For some years assessment of neural function in utero has largely been restricted to monitoring of the electrocorticogram(ECoG) for the definition of fetal behavioural state or to postmortem histology. In general the ECoG has been recorded over the parietal dura and two major states have been defined. The high voltage (HV) state is considered analogous to quiet sleep, during which there is high voltage activity on the electrocorticogram, shown by power spectral analysis to be largely in the range 1–5 Hz. It is associated with some electromyographic activity in postural muscles and an absence of fetal breathing movements. The alternating state is that of low voltage (LV) ECoG activity, in which there is a loss of power particularly below 5 Hz. This is associated with a loss of postural muscle activity, while rapid eye

---

[1] Developmental Physiology Laboratory, Department of Paediatrics, University of Auckland, Private Bag, Auckland, New Zealand

The Endocrine Control of the Fetus
Ed. by W. Künzel and A. Jensen
© Springer-Verlag Berlin Heidelberg 1988

movements (REM) and fetal breathing movements are generally present if the fetus is healthy. This state has generally been considered analogous to the postnatal REM state (Clewlow et al. 1983). Transitional states are recognized but there is a lack of consensus regarding the extent to which these may represent periods of arousal.

Advances in microcomputers have meant that real-time power spectral analysis of the ECoG is now possible. In our laboratory we have used a signal-processing laboratory software language, ASYST (Macmillan Publishing Co. NY), on IBM-AT equipment for this purpose. While some maturational changes in the power spectrum have been reported (Szeto et al. 1985; Umans et al. 1985), the simple ECoG is a rather naive approach to the monitoring of fetal neural function. The imposition of hypoxia on the fetus, usually by means of altering the maternal FIO_2, generally leads to the cessation of fetal breathing and loss of REM. Some workers have regarded the latter as sufficient evidence that the apnoea of hypoxia is dependent on a state change (Koos et al. 1987). Certainly HV usually follows the onset of hypoxia but LV activity may be seen during such a challenge. During such challenges there is a reduction in power at high frequency, and if the asphyxia is sufficiently severe and prolonged there is a loss of power at all frequencies and a flattening of the ECoG. More immediate flattening can be observed if cerebral perfusion is reduced concurrently by means of inflatable cuffs around both carotid arteries. If the hypoxic-ischaemic insult is maintained for a sufficiently long period (30–60 min), residual abnormality in the ECoG may be seen. This may be apparent in the form of convulsive activity or changes in the power spectrum. The metabolically compromised cortex shows loss of power in the higher frequencies. At a technical level we have found the use of driven electrodes greatly reduces the significance of artefacts on the signal.

With the development of techniques for stereotactic neurosurgery on the fetal ovine brain under sterile conditions, (Gluckman and Parsons 1983), we have considered the recording of the ECoG at other sites. The hippocampus is one of the most sensitive regions for asphyxial damage. We have chronically implanted electrodes in the fetal hippocampal cortex and recorded ECoG activity at this site. During maternal inhalational asphyxia with or without concurrent carotid clamping there is flattening of the hippocampal ECoG with post-insult evidence of hyperactivity in the lower frequency of the power spectrum. We not infrequently have recorded convulsive activity at this site, but we have not recorded any concurrent activity from the parietal cortex.

The ECoG can only monitor aspects of intrinsic activity in the neocortex and archicortex. The evoked potential is a useful technique for examining the neural response to a specific stimulus. Several groups have studied the somatosensory and auditory evoked potentials in the exteriorized fetus and the somatosensory evoked potential has been studied in relationship to asphyxia in such a preparation (Hrbek et al. 1974). However, such an approach must of necessity be limited, particularly as the most important aspect of an investigation of cerebral asphyxia is the recorvery period.

We and others (Plessinger et al. 1986; Cook et al. 1987a) have used an implantable earphone in the fetal ear to stimulate the auditory evoked potential. Recording electrodes are placed extradurally in a 120-day fetus 12 mm lateral and 10 mm anterior to the bregma. We have demonstrated that the brainstem potentials are first recorded between 115 and 188 days of gestation. It seems probable that the limiting factor is in the middle or inner ear, as all components of the response are present soon after it appears. The five classical waves can be demonstrated. In studies using an array of indwelling electrodes we were able to demonstrate that the level of origin of the waveforms was similar to that in postnatal animals. Waveform latencies decrease with advancing gestation and are longer in the LV state. Of particular value has been to demonstrate that the maximum tolerated frequency of stimulation without loss of waveform or prolongation of latency increases markedly with maturation from 1 Hz at 120 days to 40 Hz at 140 days.

While the auditory brainstem potential is valuable for investigation of brainstem function and of lesser value for cortical components (although we have identified and studied later latency cortical components), the somatosensory evoked potentials is particularly valuable in the study of cortical function. Although we have used forelimb stimulation in some studies, we have generally used stimulating electrodes implanted on an upper lip, because of the relatively large representation of the snout on the somatosensory cortex. Even so, placement of the extradural recording electrodes is more important than for the auditory potential. At 125 + days the snout potential is recorded with an array of electrodes centred 22 mm anterior and 12 mm lateral to the bregma. Some potentials can be recorded prior to 100 days, but full maturation of the waveforms is not seen until after 125 days of gestation. Maturational reduction in latencies and increases in following frequencies are observed, and again the latencies are slightly prolonged in LV states (Cook et al. 1987b). It is important to exteriorize stimulating and recording electrodes on opposite sides of the ewe to reduce the stimulus artefact which otherwise obscures the early potentials.

The early studies of Myers et al. (1984), supported by studies in the exteriorized fetus by Hrbek et al. (1974), suggested the importance of acidosis in the generation of hypoxia-induced neural damage. We have used the evoked potential methodology to examine this question in the chronically instrumented fetus (Cook et al. 1987c). Either isocapnic hypoxia or hypercapnic hypoxia was induced in fetal lambs between 125 and 130 days by altering maternal inspired gases. In the isocapnic insult the fetal PO_2 fell from 18.4 ± 0.6 Torr to 9.9 ± 1.3 Torr but the fetal arterial pH remained at 7.37 ± 0.07. In the hypercapnic insult the fetal PO_2 fell similarly to 6.8 ± 1.3 Torr but the pH fell to 7.25 ± 0.03. During the isocapnic insult the amplitude of both the somatosensory and auditory evoked waveforms fell and the latencies were prolonged. However, there was rapid and complete recovery within 1 h after the termination of the 1-h insult. During the hypercapnic insult there was a similar change in amplitude and latency of both potentials. However – in marked contrast to the isocapnic insult – there was no evidence of recovery in either potential up to 72 h following the insult, the longest period studied. In more limited studies performed in older fetuses (135 days) similar results were obtained with respect to the somatosensory potential, but full recovery was seen with respect to the auditory potential. These results demonstrate strikingly that a slight fall in arterial pH is associated with a marked loss of the ability of fetal brain to recover from an hypoxic episode. The arterial pH only fell to 7.25, a level most clinicians would accept in a fetus during labour.

These results suggest that the acceptable level of acidosis may be considerably lower than previously thought. No histological correlation was performed in this study but it seems probable that the persistence of abnormality in the evoked responses is a consequence of neuronal loss. Acidosis is likely to aggravate neuronal sensitivity to hypoxia by several mechanisms. First, acidosis, by promoting depolarization of the neuron, favours intracellular calcium accumulation, strongly suggested to be a final common pathway of anoxic cellular death (Farber et al. 1981; Siesjö 1981). Further, lactate itself may be directly toxic, possibly further promoting calcium entry (Raichle 1983; Myers et al. 1984).

Not all regions of the neural axis are equally susceptible to hypoxia. Myers et al. (1984) correlated sensitivity to regions with high glucose and lactate concentrations and suggested that these areas were both more metabolically active, thus having a higher demand for oxygen, and more likely to develop acidosis. Maturational changes in neuronal activity might explain why the auditory pathway was more sensitive to asphyxial damage at younger gestational ages (Cook et al. 1988c). Other reasons for local variation in sensitivity include the vascular anatomy (Volpe et al. 1985) and the degree of innervation by excitatory amino-acid terminals (Weiloch 1985).

It is well recognized that growth-retarded fetuses have a raised susceptibility to asphyxial neural damage. The experimentally growth-retarded fetus is frequently acidotic and this could be one basis for greater susceptibility. It has been a clinical observation that growth-retarded neonates appear on clinical examination to be neurologically advanced for gestational age. We have examined this in fetuses with experimental growth retardation induced by preconception reduction in the number of caruncles (implantation sites) in the uterus. We have shown that the greater the degree of growth retardation the shorter the latency of waveforms with both potentials and the greater the maximum stimulation frequency tolerated, indices compatible with advanced neural maturation. The significance of this observation in mechanistic terms remains to be elucidated. One feature of an immature neural system is plasticity, and it would be of considerable interest if the growth-retarded fetus showed premature reduction on the capacity for a plastic response to neural deficit. This would have the consequence that any insult late in gestation or during labour might have greater residua in the growth-retarded than in the normal fetus. The mechanism by which growth retardation leads to premature neural development is a matter for speculation. It is clearly a relatively general phenomenon – we have documented changes in two separate sensory systems and the clinical data relates primarily to motor systems. Potential considerations include the chronic changes in plasma catecholamines and cortisol reported in the growth retarded fetus (Jones and Robinson 1983; Robinson et al. 1980). The recent observations both in the sheep from our laboratory (Basset and Gluckman 1987) and in the guinea pig (Jones et al. 1987) that fetal growth retardation is associated with marked elevations in insulin-like growth factor 2 (IGf-2) also merits consideration in view of recent evidence that the latter has a role in neural development.

Microdialysis tubing implanted into neural tissue forms the basis of recently developed method for continuous monitoring of neurochemical events. We have adapted this method for the chronically instrumented fetal lamb. A small loop of microdialsis tubing (approximately 1-2 mm in length) is implanted stereotactically through a stainless steel cannula into the fetal hippocampus. It is fixed by dental cement to the fetal skull and impermeable microtubing is used to exteriorize the afferent and efferent connections to the maternal flank. By infusing artifical cerebrospinal fluid and collecting the dialysate, changes in extracellular concentrations of ions and neurotransmitters can be followed. We have followed dialysate calcium concentrations during and following hypercapnic hypoxic challenges. During the insult extracellular calcium concentrations fall considerably, consistent with the hypothesis that there is a marked increase in intracellular calcium during asphyxial insults. During the initial recovery period dialysate calcium levels rise, consistent with recovering neurons pumping calcium out. We have also shown that later in the recovery period there is an increase in glutamate in the dialysate, consistent with an increase in excitatory amino-acid neurons firing following the insult. This technique offers the exciting potential of neurochemical monitoring of specific regions of the fetal brain during an following the insult.

Both calcium and the excitatory amino-acids are implicated in the most generally held models of brain damage. Asphyxia leads to depolarization particularly of the excitatory aspartate/glutaminergic neurons, with the consequence of increased metabolic demand. This may explain the high sensitivity of regions of the brain, such as the hippocampus, with considerable glutaminergic innervation (Weiloch 1985). Recently it has been shown that the activation of the glutaminergic receptor leads to opening of calcium channels, and indeed that the receptor is part of the channel protein complex (MacDermott et al. 1986). Normally intracellular calcium concentrations are maintained at very low levels and asphyxia leads to an increase in both calcium entry into neurons by several mechanisms and to a reduction in energy-dependent removal of calcium from the cell. It is suggested that intracellular calcium accumulation is at least one mechanism which can cause neuronal death (Schanne et al. 1979; Farber et al. 1981; Siesjö 1981; Raichle 1983).

As the biochemistry of asphyxial death has been reduced to one of several models, approaches to pharmacological protection against neural damage become more rational. Previous attempts based on use of barbiturates have been disappointing, probably because they do not reflect interruption to pathways central to the induction of damage (Trauner 1986; Weiloch 1985). Most attempts at neuroprotection have been based on post-insult administration and thus the likelihood of success is not likely to be high. The perinatal period remains the one situation where prophylactic neuroprotection may be possible, because of the narrow window when the fetus is at high risk and because the highest risk population can be identified (Hill and Volpe 1982; Rosen 1985).

Some studies in the adult have suggested that, because of the potential central role of calcium in neuronal death, calcium channel antagonists may be neuroprotective (Deshpande and Wieloch 1986). Recently it was reported that in the neonatal rat, a calcium channel antagonist given prophylactically was neuroprotective as judged by gross hemispheric weight, histological loss of volume in the corpus striatum and cortex, and the ratio of homovanillic acid to dopamine in striatum (Silverstein et al. 1986).

We have extended the latter study (Gluckman et al. 1987). Twenty-one-day-old rats were subjected to unilateral carotid ligation followed by 2 h inhalational asphyxia. In vehicle-treated animals this was associated with subsequent gross motor abnormalities and on histological examination with reduced hemispheric size, reduced neuronal density and reduced muscarinic receptor density in the corpus striatum on the ligated side. In rats pretreated with flunarizine (30 mg/kg body wt.), a calcium channel blocker that crosses the blood-brain barrier easily and thus can reach relatively high neural concentrations without reducing cardiac output, there was no evidence of behavioural abnormality or hemispheric damage. Thus it appears that in this model pertreatment with flunarizine is neuroprotective. While the easiest explanation is that it directly prevented intracellular calcium accumulation, other mechanisms such as alterations in capillary blood flow, in blood viscosity and its potential anticonvulsant effects must be considered. Our studies are now extending to studies in the fetal lamb using the various approaches described above to monitor the insult as well as subsequent histological examination.

Acknowledgements. This research was funded by grants from the Medical Research Foundation of New Zealand, the Wellcome Trust, the Neurological Foundation of New Zealand and The Foundation for the Newborn.

References

Bassett NS, Gluckman PD (1987) Insulin-like growth factors (IGF's) in experimental fetal growth retardation in the sheep. Endocrinology 120: 234A

Clewlow F, Dawes GS, Johnson BM, Walker DW (1983) Changes in breathing, electrocortical and muscle activity in unanaesthetized fetal lambs with age. J Physiol 341: 463–467

Cook CJ, Williams CE, Gluckman PD (1987a) Brainstem auditory evoked potentials in the fetal lamb, in utero. J Dev Physiol 9: 429–440

Cook CJ, Gluckman PD, Johnston BM, Williams CE (1987b) The development of the somatosensory evoked potential in the unanaesthetized fetal lamb. J Dev Physiol 9: 441–456

Cook CJ, Gluckman PD, Williams CE, Johnston BM (1988) Hypercapnic but not isocapnic hypoxia is associated with persistent changes in neural function in utero. Am J Ob Gyn (submitted)

Deshpande JK, Wieloch T (1986) Flunarizine, a calcium entry blocker, ameliorates ischaemic brain damage in the rat. Anesthesiology 64: 215–224

Farber JL, Chien KR, Mittnacht S (1981) The pathogenesis of irreversible cell injury in ischaemia. Am J Pathol 102: 271–281

Freeman JM (1985) Preface to: Freeman JM (ed) Prenatal and perinatal factors associated with brain disorders. National Institutes of Health, Bethesda MD, pp iii–iv (NIH Publication no 85-1149)

Gluckman PD, Parsons BY (1983) Stereotaxic method and atlas for the ovine fetal forebrain. J Dev Physiol 5: 101–128

Gluckmanm PD, Mydlar T, Bennet L, Cook CJ, Faull R, Gorter S, Gunn AJ, Johnston BM (1987) A calcium channel antagonist ameliorates hypoxia induced damage in young rats with unilateral carotid ligation. In: Jones CT, Nathanielsz PW (eds) Fetal and Neonatal Development. Perinatology Press, Ithaca (in press)

Hill A, Volpe JJ (1982) Hypoxic-Ischaemic Brain Injury in the Newborn. Seminars in Perinatology 6 (1): 25–41

Hrbek A, Karlsson K, Kjellmer I, Olsson T, Riha M (1974) Cerebral reactions during intrauterine asphyxia in the sheep. II. Evoked electroencephalogram responses. Pediatr Res 8: 56–63

Jones CT, Robinson JS (1983) Studies on experimental growth retardation in sheep. Plasma catecholamines in fetuses with small placenta. J Dev Physiol 5: 77–87

Jones CT, Lafeber HN, Price DA, Parer JT (1987) Studies on the growth of the fetal guinea pig. Effects of reduction in uterine blood flow on the plasma sulphation-promoting activity and on the concentration of insulin-like growth factors I and II. J Dev Physiol 9: 181–201

Koos BJ, Sameshima H, Power GG (1987) Fetal breathing, sleep state, and cardiovascular responses to graded hypoxia in sheep. J Appl Physiol 62 (3): 1033–1039

MacDermott AB, Mayer ML, Westbrook GL, Smith SJ, Barker JL (1986) NMDA-receptor activation increases cytoplasmic calcium concentration in cultured spinal cord neurones. Nature 321: 519–522

Meldrum B (1985) Possible therapeutic applications of antagonists of excitatory amino acid neurotransmitters. Clin Sci 68: 113–122

Myers RE, de Courten-Myers GM, Wagner KR (1984) Effect of hypoxia on fetal brain. In: Beard RW, Nathanielsz PW (eds) Fetal physiology and medicine, 2nd edn. Dekker, New York, pp 419–458

Plessinger MA, Woods JR (1986) Fetal auditory brainstem response: external and intrauterine auditory stimulation. Am J Physiol 250: 137–141

Raichle ME (1983) The pathophysiology of brain ischaemia. Ann Neurol 13: 2–10

Robinson JS, Hart IC, Kingston EJ, Jones CT, Thorburn GD (1980) Studies on the growth of the fetal sheep. The effects of reduction of placental size on hormone concentration in fetal plasma. J Dev Physiol 2: 239–248

Rosen MG (1985) Factors during labor and delivery that influence brain disorders. In: Freeman JM (ed) Prenatal and perinatal factors associated with brain disorders. National Institutes of Health, Bethesda MD, pp 237–262 (NIH Publication no 85-1149)

Rothman SM (1984) Synaptic release of excitatory amino acid neurotransmitter mediates anoxic neuronal death. J Neurosci 4 (7): 1884–1891

Rothman SM (1985) The neurotoxicity of excitatory amino acids is produced by passive chloride influx. J Neurosci 5 (6): 1483–1489

Schanne FA, Kane AB, Young EE, Farber JL (1979) Calcium dependence of toxic cell death: a final common pathway. Science 206: 700–702

Siesjö BK (1981) Cell damage in the brain: a speculatice synthesis. J Cereb Blood Flow Metabol 1 (2): 155–185

Silverstein FS, Buchanan K, Hudson C, Johnston MV (1986) Flunarizine limits hypoxic-ischaemia induced morphologic injury in immature rat brain. Stroke 17 (3): 477–482

Szeto HH, Vo TDH, Dwyer G, Dogramajian ME, Cox MJ, Senger G (1985) The ontogeny of fetal lamb electrocortical activity: a power spectral analysis. Am J Obstet Gynecol 153: 462–466

Trauner DA (1986) Barbiturate therapy in acute brain injury. J Pediatr 109: 742–746

Umans JG, Cox MJ, Hinman DJ, Dogramajian JE, Senger G, Szeto HH (1985) The development of electrocortical activity in the fetal and neonatal guinea pig. Am J Obstet Gynecol
153: 467–471
Volpe JJ, Herscovitch P, Perlman JM, Kreussere KL, Raichle EM (1985) Positron emission to-
mography in the asphyxiated term newborn: parasagittal impairment of cerebral blood
flow. Ann Neurol 17: 287–296
Wieloch T (1985) Neurochemical correlates to selective neuronal vulnerability. Prog Brain Res
63: 69–85

Endocrine Control of Thermoregulation

Central Noradrenergic and Serotonergic Mechanisms in Temperature Regulation and Adaptation

K. Brück[1]

Stability of body temperature results from the actions of several effector systems integrated in a feedback control system. Notably, the immediate control of the effector systems is neuronal not hormonal (Fig. 1); blood-borne hormones may, however, modulate the actions of the thermoregulatory system. All controlling efferent nervous pathways originate in the hypothalamus, which is thought to represent the central regulator or integration centre of the temperature control system (Brück and Zeisberger 1987; Simon et al. 1986).

A 'central shivering pathway' mediating shivering originates in the hypothalamus. Vasomotor control is exerted through the central sympathetic control system via the medial forebrain bundle. Nonshivering thermogenesis (NST) has been

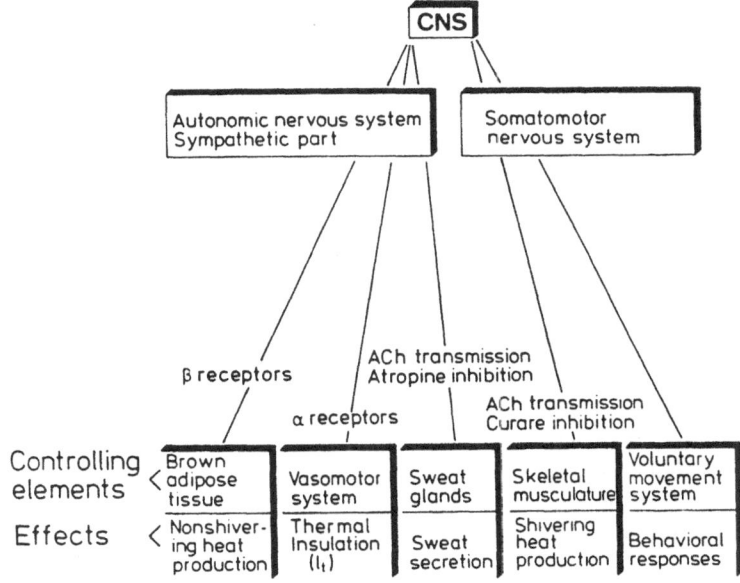

Fig. 1. Neural control of the thermoregulatory 'conrolling elements' (effectors) and their effects (actions)

[1] Department of Physiology, Justus Liebig University, Aulweg 129, 6300 Giessen, FRG

The Endocrine Control of the Fetus
Ed. by W. Künzel and A. Jensen
© Springer-Verlag Berlin Heidelberg 1988

shown to be under the control of the hypothalamic nucleus ventromedialis (Perkins et al. 1981; for further references see Brück and Hinckel 1982; Brück and Zeisberger 1987). In contrast to technical regulatory systems, the organismal thermoregulatory system undergoes certain long-term alterations when challenged by repeated or prolonged exposures to thermal stimuli. Moreover, a number of nonthermal factors, e.g. changing progesterone levels during the menstrual cycle (Hessemer and Brück 1985), as well as immunological processes resulting in fever and the production of antipyretic hormones like vasopressin (when acting on the central nervous system; Zeisberger 1985), may change its properties. Thus, this system possesses a high degree of adaptibility.

The adaptive modifications may concern

(a) the capacity of the final control elements and
(b) the functional characeristics of the control system.

One example of the first type is the increasing capacity of NST, which is based on the proliferation of brown adipose tissue (BAT) during cold acclimation of small animals; another example is the increasing sweat rate capacity, which is a most important factor in heat acclimation. NST is also the prevailing mechanism of thermoregulatory heat production in newborn mammals and in the human neonate. As for the second type of adaptive modification, it has been shown that the threshold temperature and the gains of the effector responses may change during acclimation. For instance, repeated cold exposure may lead to a shift of the shivering threshold to a lower body temperature, as will be demonstrated in the next paragraph; even a single cold exposure has been shown to decrease shivering threshold temperature. Such changes could be partly related to alterations in neuronal brain stem interconnections including noradrenergic and serotonergic pathways; to argue for this is the main objective of the present review.

Adaptive Modifications in the Thermoregulatory System

Figure 2 shows the results of a cold test in a human subject exposed to a low environmental temperature. With decreasing ambient temperature shivering occurs, indicated by the increased electrical muscle activity (EMA) and by an increase in the metabolic rate (calculated from oxygen uptake, $\dot{V}O_2$). These reations occur before any substantial decrease in body core temperature is to be seen; they are obviously actuated by a drop in skin temperature. However, this does not mean that they are independent of deep body temperature. As these reactions do not correlate with the temperature measured either at the skin or in the body core, threshold response and gain are related to a weighted mean body temperature, $\bar{T}_b$, calculated from mean skin temperature, T_{sk}, and deep body temperature, T_{core}, using the equation

$$\bar{T}_b = 0.87 \cdot T_{core} + 0.13 \cdot \bar{T}_{sk}.$$

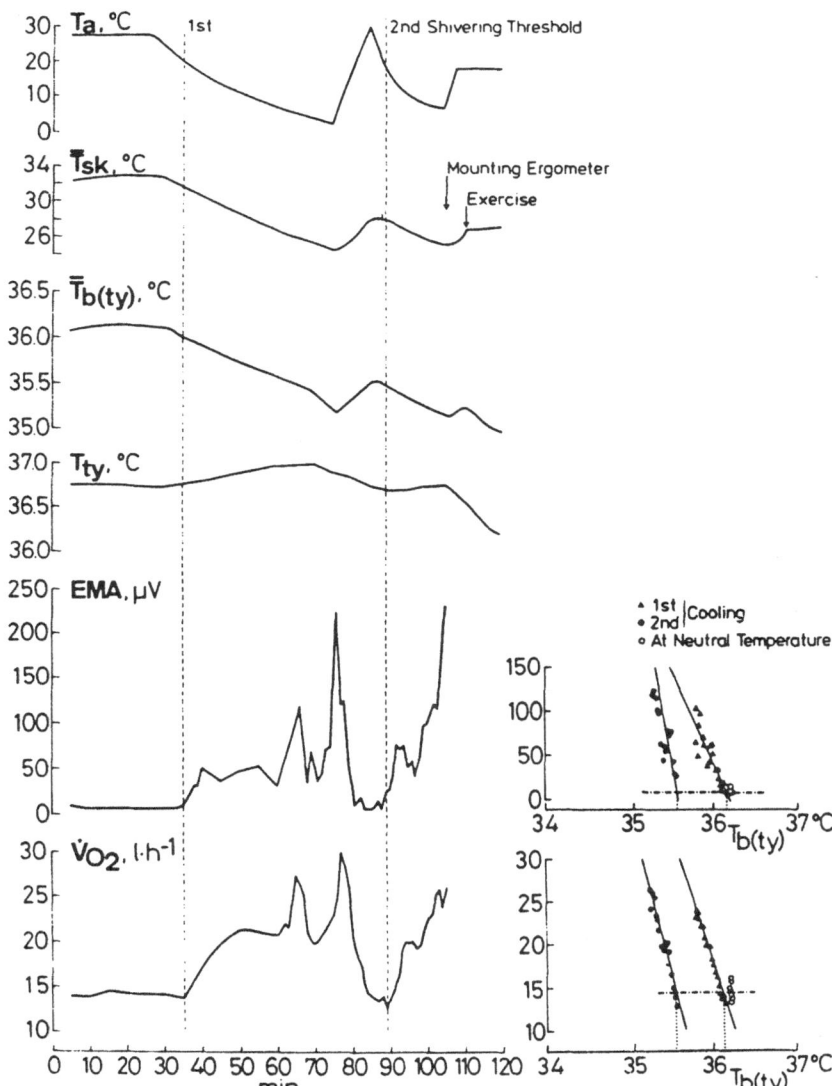

Fig. 2. Cooling test in adult human subject. Shivering threshold is shifted to a lower level of mean body temperature at the second cooling phase after short rewarming (inset, right). T_a, ambient temperature; $\bar{T}_{sk}$, mean skin temperature; T_{ty}, tympanic temperature; $\bar{T}_{b(ty)}$, mean body temperature; EMA, electrical muscle activity from m. latissimus dorsi; $\dot{V}O_2$, oxygen uptake. (Based on V. Schmidt and K. Brück, unpublished data)

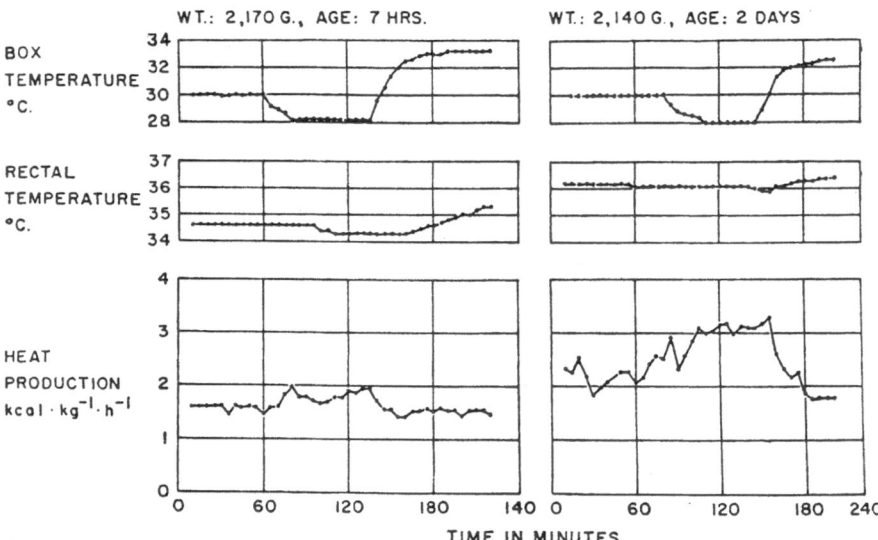

Fig. 3. Thermoregulatory heat production in a premature infant on the 1st and 2nd days of life. Note that on the 1st day it was only when rectal temperature dropped below 35 °C that heat production increased; this can be ascribed to a shivering threshold deviation caused by postnatal cooling ('short-term adaptation'). (Reproduced from Brück 1978)

Two regression lines are calculated from sub- and suprathreshold data points; the point of intersection of the two lines represents the threshold temperature. After a short rewarming period, repetition of cold exposure again evoked a cold defence reaction, but the shivering threshold was shifted to a lower mean body temperature (see inset, Fig. 2). This phenomenon has been called 'central short-term acclimation' (Brück and Hinckel 1984). Previous observations in premature infants, who could be maintained at ambient temperatures so low as to yield steady-state body core temperatures of near 35 °C without arousing any cold defence reaction (Fig. 3), have been ascribed to such short-term acclimation (Brück 1978). Recent observations in the newborn rabbit (Hull et al. 1986) may also be explained on the basis of short-term acclimation: after a 1-h preperiod at an ambient temperature of 28 °C, animals kept in a thermal gradient (tube with increasing temperature from one end to the other) moved to an ambient temperature at which body temperature was maintained at about 35 °C; by contrast, another group, kept at 40 °C for 1 h, preferred a higher ambient temperature at which their body temperatures attained values around 40 °C.

As shown in guinea pigs, the normal shivering threshold can be restored by exposing the animals to a warm environment (above 32 °C) for more than 30 min. After daily cold exposures, however, the shivering threshold remains decreased; this has been described as *long-term hypothermic or tolerance adaptation* and has been found in man as well as in various mammalian species (Brück 1986).

Neuronal Correlate for the Thermoregulatory and Adaptive Mechanisms

As for the possible neurophysiological correlate for the temperature–effector relationship and its thermoadaptive modifications, the first question that arises is: Where do the thermal signals governing the effector system come from?

Thermoreception

First we must mention the thermal receptors at the skin which serve not only for temperature sensation but also temperature regulation. This is strongly suggested by experiments like those demonstrated in Fig. 2. Here, shivering was evoked before there was a substantial decrease in body core temperature, but it occurred simultaneously with a decrease in skin temperature. There is ample experimental evidence, however, for the existence of internal thermosensitive structures. By means of fine, chronically implanted thermodes it was possible to selectively cool or heat small areas of the brain stem in the goat (Jessen 1985) and in the guinea pig (for references see Brück and Zeisberger 1987). The thermal sensitivity of the respective areas was estimated from the magnitude of the thermoregulatory responses, i.e. the metabolic rate or respiratory evaporative heat loss. The greatest thermosensitivity was found in the pre-optic area and the anterior hypothalamus, slight thermosensitivity in the midbrain, but nearly none in the posterior hypothalamus. These results agree with the results of electrophysiological studies in several species. More recently, further insight into internal receptors was obtained by single unit recording from hypothalamic slices and from cell cultures obtained from embryonic hypothalamus tissue (Boulant and Dean 1986).

By extending the selective cooling or warming studies it has been shown that areas other than the skin and hypothalamus possess thermosensitive properties, namely, the spinal cord (Simon 1974), the skeletal musculature and the abdominal wall (for references see Jessen 1985). We must therefore speak of a *multiple input system*.

Thermointegrative Structures and the Thermoafferent System

The Thermoafferent System

There is now evidence that thermal information from the various sites of the body is fed into the hypothalamus, which therefore displays thermointegrative properties in addition to its thermosensitivity. Thus, it could be shown that the activity of single units recorded from the posterior hypothalamus increased during heating of both the pre-optic area and the spinal cord (Wünnenberg and Hardy 1972). On the basis of these and other studies it is possible to distinguish between the anterior, pre-optic *thermoreceptive region* and the posterior *thermointegrative region* of the hypothalamus.

How does thermal information reach the hypothalamus? From electro- and thermophysiological studies, it appears that a pathway must be postulated connecting the thermosensors in the pre-optic area and in the anterior hypothalamus with the effector neurons in the posterior hypothalamus (see the very simplified diagram in Fig. 4). It must be inferred that the warm units of the body core, which are more

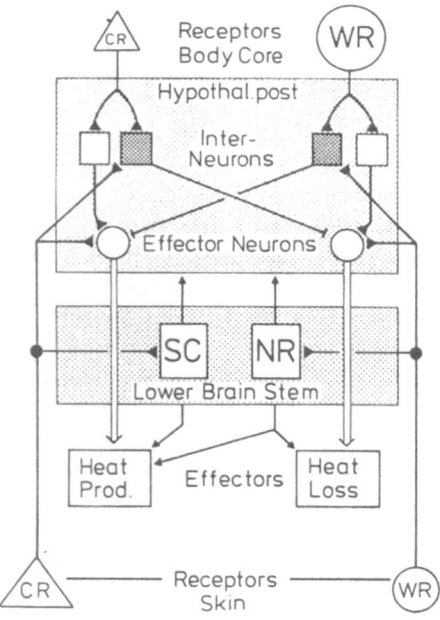

Fig. 4. Simplified model of neuronal connections related to temperature control. *NR*, raphe nuclei; *SC*, subceruleus area; *CR*, cold receptors; *WR*, warm receptors. Size of symbols indicates roughly respective density of receptors. ◁ excitatory, ⊣ inhibitory synaptic connections. The imperfectly known connections between subceruleus area, raphe nuclei and hypothalamus are omitted (see Fig. 8)

numerous, stimulate the effector neurons for heat dissipation directly or via interneurons; in addition, through inhibitory interneurons, they inhibit the effector neurons that govern extra heat production.

Noradrenergic and Serotonergic Interconnections

The thermal afferents from the skin have been postulated to diverge from the lemniscal spinothalamic tract, forming a direct projection to the hypothalamic thermoregulatory structures. More recently, evidence has been obtained that a part of the cutaneous thermal input is conveyed via the spinoreticular pathway (the so-called unspecific system) to the reticular formation, and from there to the hypothalamus via the raphe nuclei and the ventral noradrenergic system, passing the subceruleus region (Brück and Hinckel 1982; Fig. 4).

Evidence for these projection systems has been obtained in several steps

1. Micro-injection of noradrenaline (NA) into the integrative part of the hypothalamus evoked a metabolic rise, whereas micro-injection of serotonin (5-HT) inhibited the metabolic rate (Behr et al. 1983), when it had been increased by exposure to a cold environment.
2. Horseradish peroxidase injections into the same area resulted in retrograde labelling of a number of neuronal perikarya in the lower brain stem.
3. Electrical stimulation in this area of the lower brain stem caused a rise in the metabolic rate, which was prevented, however, by micro-injection of an α-adrenergic blocking agent (phentolamine) into the hypothalamus. This was taken as evidence that the metabolic effect was indeed mediated by an ascending catecholaminergic pathway (ACP; Szelenyi et al. 1977).
4. Acute aminergic denervation of the hypothalamus (by 6-hydroxydopamine) caused effects opposite to hypothalamic NA injection, i.e. a downward displacement of the shivering threshold (Behr et al. 1983).

With smaller stimulation electrodes it was possible to separate functionally two lower brain stem areas (Fig. 4). The metabolic response could be reproduced by stimulation of a relatively small region located below the locus ceruleus (subceruleus or SC area). As shown in Fig. 5, both shivering and NST increased (the latter indicated by the local temperature rise in the BAT, the former by the increase in EMA) following electrical SC stimulation (for references see Brück and Hinckel 1982). By contrast, electrical stimulation of a more ventromedial area coinciding with the nucleus raphe magnus (NRM) *inhibited* shivering thermogenesis when it had been evoked by external cooling, as shown in Fig. 6. Furthermore, both the SC area and the NRM have been shown to receive afferents from the skin of the trunk and extremities. This proves that the proposal of a second afferent pathway conveying cutaneous thermal information is correct (for references see Brück and Hinckel 1982).

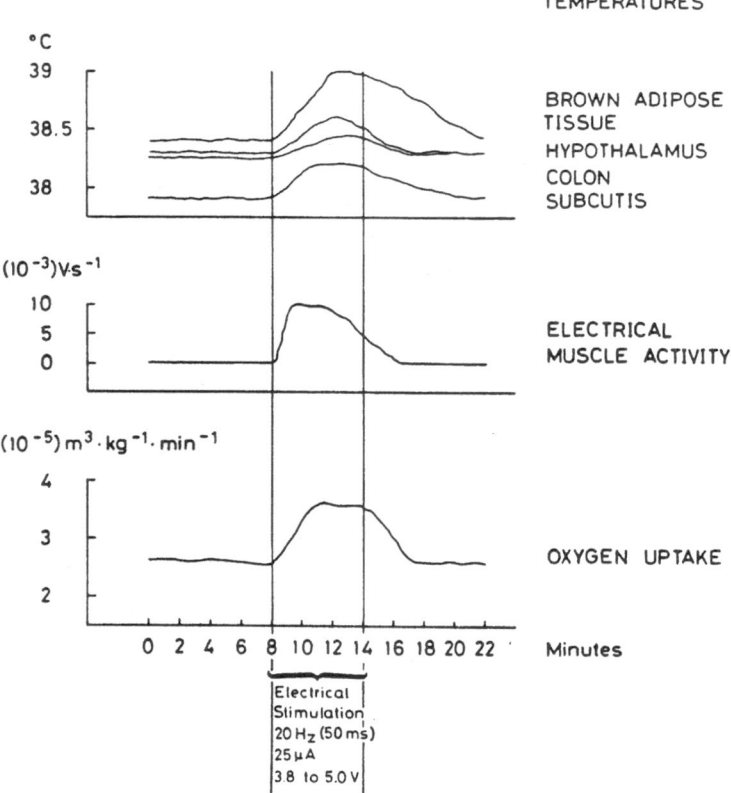

Fig. 5. Thermogenetic response evoked by electrical stimulation of the subceruleus area in a guinea pig. Increase in electrical muscle activity (EMA) and increase of the temperature within the interscapular adipose tissue (brown fat) indicate that both shivering and nonshivering thermogenesis were evoked. (From Brück and Hinckel 1980)

By use of single unit recordings various cell types could be identified in the hypothalamus which responded to electrical stimulation of the two lower brain stem areas, SC and NRM (Brück and Hinckel 1980, 1982). One type of interneuron (IN_C, Fig. 8) responded to both NRM and SC stimulation. Whereas SC stimulation had an inhibitory effect, NRM stimulation had an excitatory effect on this cell type, which was characterized by warmth responsiveness with regard to skin temperature (Fig. 7). Furthermore, the effects of iontophoretically applied NA and 5-HT could be studied in the same cell (Ishikawa and Hinckel 1985): NA showed an inhibitory effect, 5-HT an excitatory effect. This cell type was activated by NRM stimulation and skin warming, but inhibited by SC stimulation. There were also cell types which were activated by skin cooling as well as by SC stimulation and by iontophoretically applied NA.

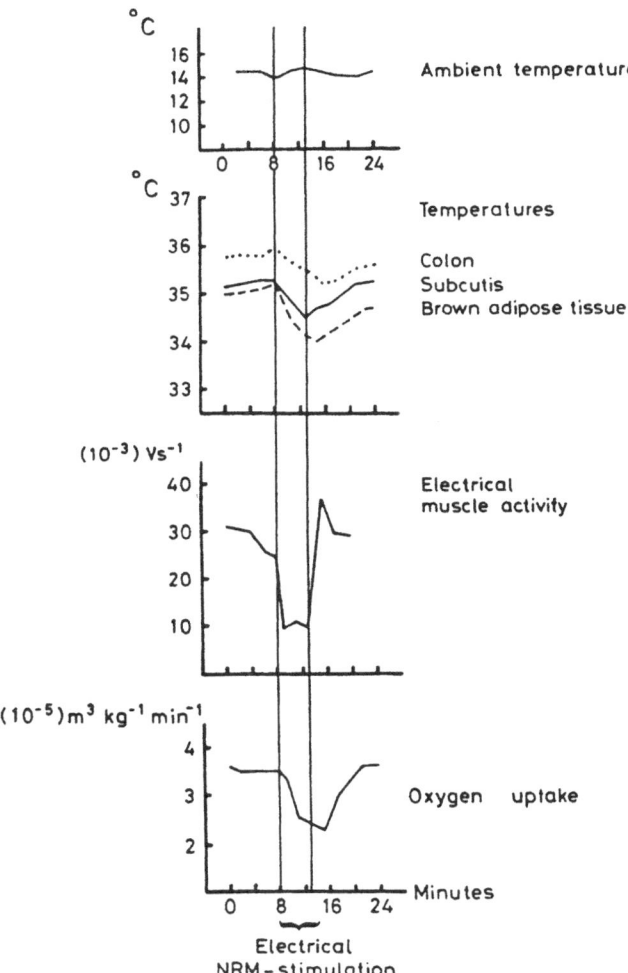

Fig. 6. Electrical stimulation of nucleus raphe magnus (NRM) in a guinea pig exposed to a cool environment (15 °C) arousing cold defence reactions (shivering, indicated by increased EMA and oxygen uptake). These regulatory responses were inhibited by stimulation of the ventral part of the NRM. (From Hinckel et al. 1983)

In sum, there is now ample avidence that afferent projections from the skin to the hypothalamus traverse the raphe system and the SC area. The latter is certainly noradrenergic; as for the former there is, as shown above, evidence for serotonergic transmission, but other transmitters may be involved.

In addition, these two brain stem areas give rise to descending pathways (Figs. 4, 8). The pathway descending from the NRM is serotonergic and projects to both spi-

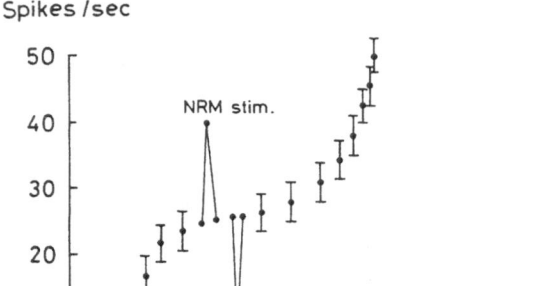

Fig. 7. Response of a single neuronal unit located in the transition area between the anterior and posterior hypothalamus to changes in trunk skin temperature and to electrical stimulation of the SC area and the NRM. (From Brück and Hinckel 1980)

nal motoneuron pools and to the dorsal horn. The pathways descending from the SC area are noradrenergic and project mainly to the spinal motoneuron pools and to the dorsal horn (for references see Brück and Hinckel 1982; Hinckel and Perschel 1987).

Neuronal Changes Related to Adaptive Modifications

The system described so far shows some adaptive changes, which may be part of the neurophysiological correlate for the adaptive modifications described, i.e. the threshold temperature deviations of the effector systems. Thus, in cold-acclimated guinea pigs (CA), the maximum static firing rate of the SC units was markedly reduced when compared with controls kept at normal room temperature. On average, the static maximum spike rate for the units in CA guinea pigs was 11 spikes/s and that for the controls 24 spikes/s. In the CA guinea pigs no units were found with maximum frequency rates above 15 spikes/s (Brück and Hinckel 1982, 1984). On the other hand, the NRM base and peak activity of warm-responsive units was larger in CA guinea pigs (Hinckel and Perschel 1987).

The diagram in Fig. 8 may be useful in the interpretation of the functional significance of the adaptive neuronal changes described. Reduced average activity of the SC units would decrease the shivering threshold by different actions. First, reduction of the inhibitory input to an inhibitory interneuron would diminish the heat production at any given ambient temperature. To restore the original metabolic rate,

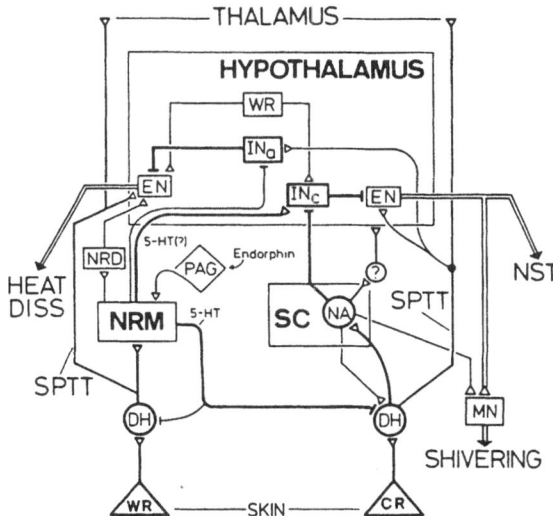

Fig. 8. Neuronal connectivity model demonstrating functional interconnections between the thermointegrative hypothalamus, the lower brain stem and the afferent projections from the skin and the preoptic area. *NRM*, nucleus raphe magnus; *NRD*, nucleus raphe dorsalis; *SC*, subceruleus area; *CR*, cold receptors; *WR*, warm receptors; *DH*, dorsal horn neurons; *MN*, motoneurons in lower brain stem and spinal cord; *IN*$_A$ and *IN*$_C$, interneurons; *PAG*, periaqueductal and periventricular grey matter; *EN*, effector neurons; *SPTT*, spinothalamic tract; *NST*, nonshivering thermogenesis; *NA*, noradrenaline; *5-HT*, serotonin; ◁ excitatory pathways; ⊣ inhibitory pathways. (Modified from Brück and Hinckel 1982, 1984)

the activity of cold receptors in the skin would have to be raised by lowered skin temperature. Consequently, the threshold temperature for eliciting the metabolic response would be shifted to a lower level. Second, reduced SC activity in cold-adapted individuals would diminish the peripheral drive for shivering via the postulated ascending stimulatory pathway and by the well-documented descending pathway projecting to the motoneurons.

The tendency to a lower shivering threshold would be supported by the activation of the inhibitory interneurons (IN$_C$, Fig. 8) due to the increased NRM activity in cold-adapted individuals. The increased NRM activity might also contribute to a decrease of the shivering threshold, in that it depresses the cold input from the skin via the serotonergic inhibitory pathway descending to the dorsal horn (this would be analogous to the well-known descending inhibition of the nociceptive input).

Finally, the question arises how the neuronal adaptive changes demonstrated in the NRM and SC area are affected during the acclimation process. According to anatomical studies, the reticular formation possesses reciprocal connections with the limbic system. One may assume that internal signals accompanying thermal discomfort may be integrated over time in the limbic system and this may result in a long-term modification of the lower brain stem units. Humoral transmission may be involved in the sequence of events. A long-lasting shift of shivering threshold to a lower temperature level can be caused by peripheral NA infusion. This suggests the existence of a kind of peripheral-central feedback mechanism with respect to the NA action (Zeisberger 1982; Roth et al. 1986). Prolonged cold exposure has been shown, in fact, to increase the plasma NA level to a magnitude comparable to that after NA infusion (see Zeisberger, this volume). The central effect of the peripheral NA level may be mediated by NA on its own which might penetrate the blood-brain barrier at special sites of the circumventricular system, or by indirect action, conceivably baroreceptor stimulation or some other unknown mechanism.

References

Behr R, Zeisberger E, Merker G (1983) Response of the guinea-pig *(Cavia aperea porcellus)* to external cooling after aminergic denervation of the anterior hypothalamus. J Therm Biol 8: 125–128

Boulant JA, Dean JB (1986) Temperature receptors in the central nervous system. Ann Rev Physiol 48: 639–654

Brück K (1978) Heat production and temperature regulation. In: Stave U (ed) Perinatal Physiology. Plenum, New York, pp 455–498

Brück K (1986) Basic mechanisms in thermal long-term and short-term adaptation. J Therm Biol 11: 73–77

Brück K, Hinckel P (1980) Thermoregulatory noradrenergic and serotonergic pathways to hypothalamic units. J Physiol 304: 193–202

Brück K, Hinckel P (1982) Thermoafferent systems and their adaptive modifications. Pharmacol Ther 17: 357–381

Brück K, Hinckel P (1984) Thermal afferents to the hypothalamus and thermal adaptation. J Therm Biol 9: 7–10

Brück K, Zeisberger E (1987) Adaptive changes in thermoregulation. Pharmacol Ther 35: 163–215

Hessemer V, Brück K (1985) Influence of menstrual cycle on shivering, skin blood flow, and sweating responses measured at night. J Appl Physiol 59: 1902–1910

Hinckel P, Cristante L, Brück K (1983) Inhibitory effects of the lower brain stem on shivering. J Therm Biol 8: 129–131

Hinckel P, Perschel WT (1987) Influence of cold and warm acclimation on neuronal responses in the lower brain stem. Can J Physiol Pharmacol 65: 1281–1289

Hull D, Vinter J (1986) The preferred environmental temperature of newborn rabbits. Biol Neonate 50: 323–330

Ishikawa Y, Hinckel P (1985) The role of serotonin and noradrenaline in thermoafferent pathways to the hypothalamus. Pflügers Arch 405 [Suppl] (2): R 70

Jessen C (1985) Thermal afferents in the control of body temperature. Pharmacol Ther 28: 107–134

Perkins MN, Rothwell NJ, Stock MJ, Stone TW (1981) Activation of brown adipose tissue thermogenesis by the ventromedial hypothalamus. Nature 289: 401–402

Roth J, Schwandt HJ, Zeisberger E (1986) Amounts of catecholamines excreted in guinea-pigs during long-term acclimation to 5 °C and 28 °C. Pflügers Arch 406 [Suppl]: R 22

Simon E (1974) Temperature regulation: the spinal cord as a site of extrahypothalamic thermoregulatory functions. Rev Physiol Biochem Pharmacol 71: 1–76

Simon E, Pierau FK, Taylor DCM (1986) Central and peripheral thermal control of effectors in homeothermic temperature regulation. Physiol Rev 66: 235–300

Szelényi Z, Zeisberger E, Brück K (1977) A hypothalamic alpha-adrenergic mechanism mediating the thermogenic response to electrical stimulation of the lower brainstem in the guinea pig. Pflügers Arch 370: 19–23

Wünnenberg W, Hardy JD (1972) Response of single units of the posterior hypothalamus to thermal stimulation. J Appl Physiol 33: 547–552

Zeisberger E (1982) The role of noradrenergic systems in thermal adaptation. In: Hildebrandt G, Hensel H (eds) Biological adaptation. Thieme, Stuttgart, pp 140–147

Zeisberger E (1985) Role of vasopressin in fever regulation and suppression. Trends Pharmacol Sci 6: 428–430

Role of Catecholamines in Thermoregulation of Cold-Adapted and Newborn Guinea Pigs

E. Zeisberger[1] and J. Roth

Introduction

The strain of birth – particularly that caused by hypoxia and pressure on the head – produce in the fetus an unusually high release in catecholamines. The levels of these 'stress' hormones in the human fetus are probably increased as early as several days before delivery. Catecholamine concentrations in human fetal scalp samples taken at the beginning of normal delivery, when the mother's cervix was barely dilated (2–3 cm), were about 5 times as high as the concentrations in a resting adult. After birth the catecholamine levels were found to have doubled or tripled again; thus, levels in umbilical samples of neonates were found to be about 15–20 times higher on average than levels in the venous blood of resting adults. Neonates thus had catecholamine levels more than 5–8 times higher than stressed adults such as women during delivery or men during heavy exercise. Their levels were even higher than in patients with pheochromocytoma.

It seems that the increased release of catecholamines at term serves not only as a highly effective protection system in emergencies such as asphyxia during delivery, but helps to prepare the neonate for life and survival in the new environment outside the uterus. The adaptational effects of a catecholamine surge during delivery include alteration of blood flow to protect heart and brain against potential asphyxia, promotion of normal breathing of the neonate, immediate mobilization of fuel for energy and, possibly, enhancement of maternal-infant bonding (for review see Lagercrantz and Slotkin 1986). After delivery, catecholamines continue to be important for the responses of neonates to stressful situations such as mobilization of fuels from body stores during starvation, and the activation of nonshivering heat production in the multilocular brown adipose tissue, which is unique to neonates, in the cold. According to our investigations in guinea pigs, they also influence the central nervous system (CNS) and are responsible for the reduction of the shivering threshold to a lower body temperature.

In the present paper we review our earlier results concerning the magnitude of NST and the shivering threshold temperature in newborn and adult warm- or cold-adapted guinea pigs, and compare these values with the peripheral release of catecholamines in these animals as determined in our most recent experiments.

---

[1] Physiological Institute, University of Giessen, Aulweg 129, D-6300 Giessen, FRG

The Endocrine Control of the Fetus
Ed. by W. Künzel and A. Jensen
© Springer-Verlag Berlin Heidelberg 1988

Nonshivering Thermogenesis

Neonates of many mammalian species of low body weight (under 10 kg), among them also the human infant, are born equipped with an additional, very efficient mechanism of heat production. Besides involuntary tonic or rhythmic muscle activity, which may at higher intensity be accompanied by a visible tremor (shivering), they possess a means of regulative chemical heat production called 'nonshivering thermogenesis' (NST). It is now generally accepted that this heat production takes place in brown adipose tissue (BAT) distributed at various strategically important sites of the body. In most mammals, BAT is located around the kidneys, heart and aorta, along the intercostal muscles and sternum, in the subcutaneous inter- and subscapular regions and between the muscles of the neck, particularly around the carotid arteries and jugular veins. This tissue has a remarkable aerobic capacity, able to produce heat at a rate equivalent to 500 W/kg. This is an order of magnitude greater than the aerobic power of muscle during maximal exercise, which is about 60 W/kg. The biochemical mechanism of conversion of substrate energy into heat, which is based on uncoupling of oxidative phosphorylation in the mitochondria of BAT, has been described in several recent reviews (Cannon and Nedergaard 1984; Nicholls and Locke 1984; Rothwell and Stock 1984).

In order to meet the high oxygen requirement for thermogenesis, BAT has a rich vasculature, estimated to be 4–6 times as dense as that of white fat. This can support a blood flow in excess of $20 \ ml \cdot min^{-1} \cdot g^{-1}$ in cold-adapted rats infused with noradrenaline (NA), which means a 25-fold increase. Considerable evidence now exists to show that NA released from sympathetic nerves is the primary stimulus to heat production in BAT and that these effects are mediated mostly by interaction with a β_1-receptor. For further details on NST and BAT and their integration in the thermoregulatory system, see reviews by Brück 1970, 1978; Jansky 1973; Rothwell and Stock 1984.

Magnitude at Birth and Postnatal Development of NST in Guinea Pig and Rat

The magnitude of NST, and the distinction between NST and shivering, were assessed at thermoneutral ambient temperature from the maximal thermogenic response to peripheral injection of exogenous NA. This test is now widely used for determination of animals' capacity for NST. The magnitude at birth and postnatal development of NST in guinea pig and rat are compared in Fig. 1, which shows considerable interspecies variation. In the guinea pig the highest magnitude of NST is found at birth; it then declines with age, dependent on ambient temperature. In the rat, however, the capacity for NST is low at birth, and it increases until the age of 3 weeks, when it achieves its maximum magnitude and then declines again, somewhat more slowly than in the guinea pig but in essentially the same manner.

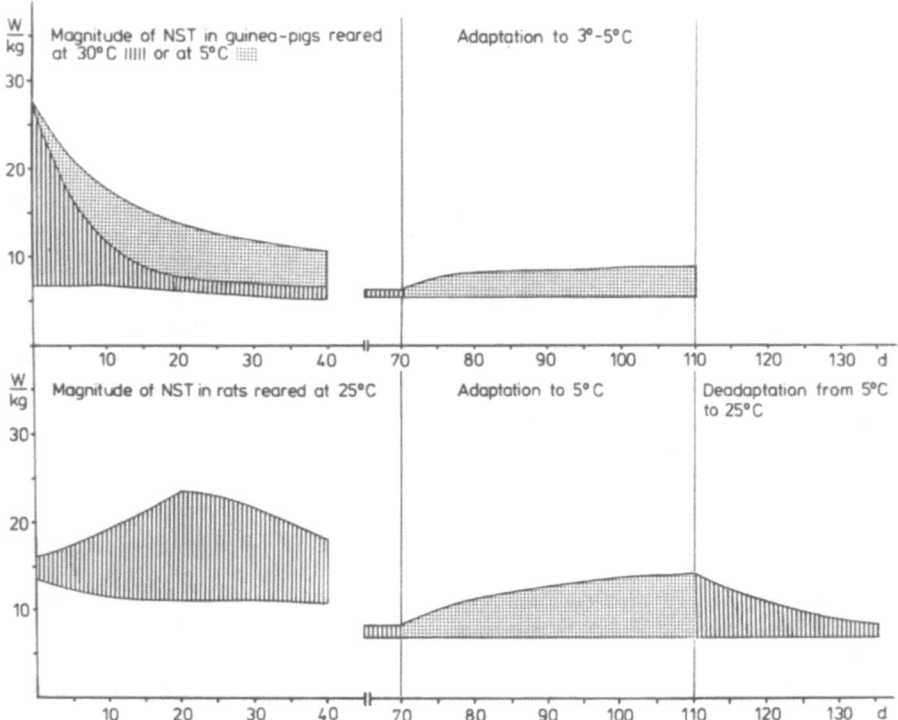

Fig. 1. Comparison of the postnatal development of nonshivering thermogenesis (NST) in guinea pig (data from Zeisberger et al. 1967) and rat (data from Jansky 1973) and of the time course of adaptation and deadaptation to cold (data from Jansky et al. 1967 and Zeisberger et al. 1967). The lower limit of the *shaded areas* denotes basal metabolic rate; the *height of shading* gives the magnitude of NST at different ages

This can be explained by the different maturity states at which these two species are born. While the guinea pig is born mature, the rat does not achieve this maturity until the age of 21 days. In other words, the increase in NST occurs much earlier in guinea pig, at the fetal stage.

It seems that the magnitude of NST correlates with the activity and mass of BAT as a proportion of body weight, which declines with age in most mammals (Brück 1970). In mouse, rat and hamster, BAT develops mainly after birth, while in the guinea pig it is fully active at birth. The reason for the rapid postnatal involution of BAT is not known, but it may reflect reduced sympathetic stimulation. The decline in activity of brown fat can be prevented or reversed by a number of factors stimulating sympathetic activity, the most potent of which is cold exposure. Chronic exposure to cold causes large increases in the mass and protein content of BAT, which result initially from hypertrophy but are then associated with hyperplasia of

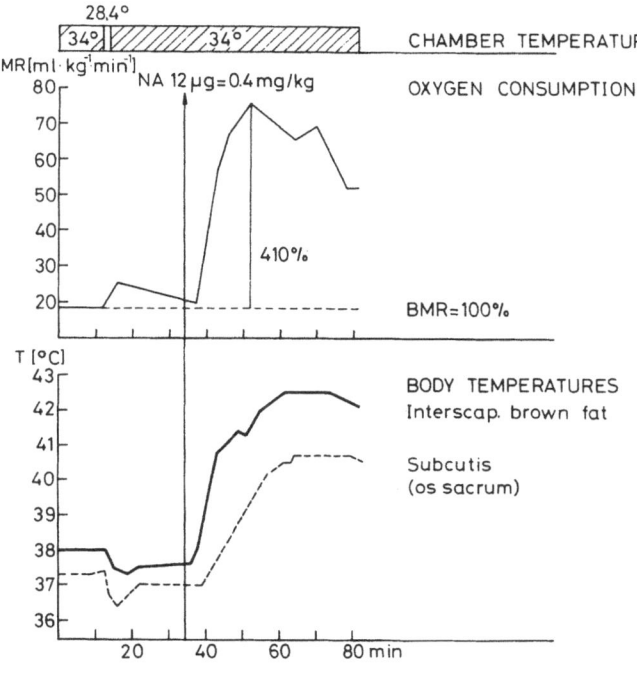

Fig. 2. Test of thermoregulatory and thermogenetic abilities of a premature guinea pig. Personal observation (Zeisberger)

brown adipocytes. Increases in food intake also produce hypertrophy and hyperplasia of BAT which may form the basis of diet-induced thermogenesis (Rothwell and Stock 1984). The shaded areas in the middle panel in Fig. 1 denote the extent to which NST can be reactivated in guinea pig and rat by means of chronic exposure to cold. The time course af adaptation and deadaptation in both species is very similar but the absolute magnitude of NST is different, being higher in the rat.

Evidence for Existence of NST in the Fetal Guinea Pig

Experimental results supporting the hypothesis that BAT and NST are already developed in guinea pig at the late fetal stage may be seen in Fig. 2. This records a single experiment in a small premature guinea pig. The animal was born sparsely haired, with closed eyes and low motility, and with a body weight of 34 g (normal weight at birth: 90–110 g). Immediately after birth it was transferred into a warm respiratory chamber (34 °C) and supplied with thermocouples in BAT and the subcutis (over os sacrum). One hour later the experiment was started. Body temperature stabilized at 38 °C, the subcutaneous temperature was slightly lower, and the resting metabolic rate corresponded to 6,3 W/kg body wt. The animal responded to a short experimental cooling with an immediate increase in oxygen uptake, demonstrating its sensory and thermoregulatory abilities. After peripheral injection of NA (0.4 mg/kg body wt.) considerable thermogenesis was initiated in BAT, leading to massive hyperthermia. The peak oxygen uptake corresponds to a metabolic rate of 25.5 W/kg, nearly equal to neonates' capacity for NST.

Peripheral Catecholamine Levels

From the foregoing it can be deduced that the activity of BAT and the magnitude of NST must correlate with the activity of the sympathetic system. This activity may be assessed from the release of catecholamines or from their turnover. Since in the past chemical methods for detection of catecholamines were not sensitive enough, catecholamine release could be measured only in the urine of adult animals during cold adaptation (Leduc 1961). Measurements in small urine samples from newborn animals and in blood plasma have been made possible only recently by the invention of new, highly sensitive methods for detection of catecholamines and their metabolites.

For these and other reasons which will be apparent later, we determined in the recent studies the peripheral release of catecholamines and their metabolites in urine samples from newborn and adult guinea pigs before and in the course of thermal adaptation. Figure 3 shows the levels of the main peripheral metabolite of NA and adrenaline (A) in the guinea pig, 3-methoxy-4-hydroxyphenylglycol (MHPG), in relation to excreted creatinine in four groups of newborn animals (10 animals in total) reared at 23 °C in comparison to those found in a group of 12 adult animals. The release of MHPG during the first 4 days after birth (mean of 14 determinations) was about six times higher than the release in adult animals (12 determinations). The determinations in the same animals at postnatal days 5 and 7 indicate a continuous decline of the MHPG level, corresponding to the rapid diminution of NST in the warm-reared guinea pigs.

Figure 4 compares the levels of MHPG with those of NA and A in newborn and adult guinea pigs. Data from human infant urine samples (Abeling et al. 1984) are shown for comparison. This comparison indicates that in newborns of both guinea pigs and humans the catecholamine turnover is several times higher than in adults. In the guinea pig, however, it is higher absolutely. Here it must be noted that the value for the human infant represents an average from urine samples taken on postnatal days 1–90, whereas in the guinea pig we used samples from the first 4 days of life. These data show that the catecholamine levels remain increased over a longer period after birth, which again correlates with prolonged appearance of NST. The comparison also shows that both NA and A levels are raised after birth.

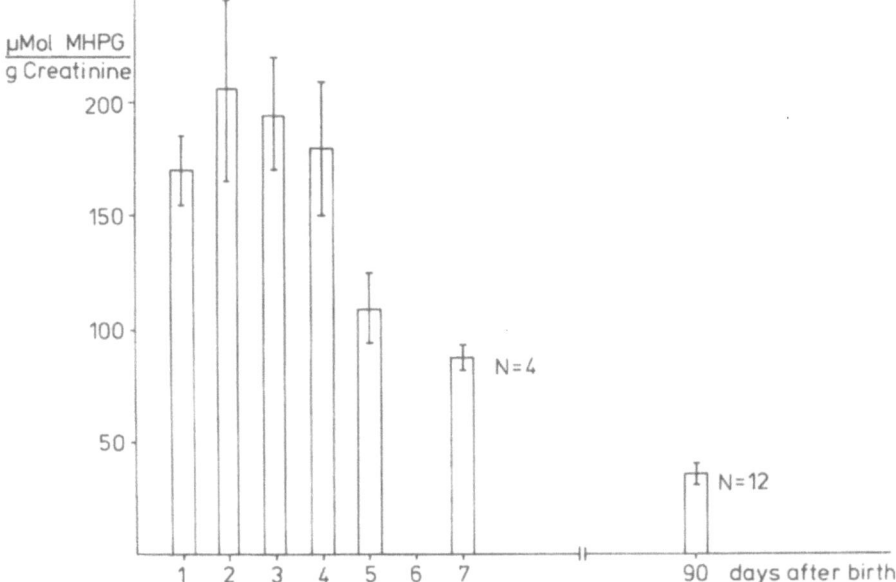

Fig. 3. Comparison of peripheral catecholamine metabolism in newborn and adult guinea pigs based on 3-methoxy-4-hydroxyphenylglycol (MHPG) recovered in urine. Personal observation (for details on methods see Roth et al. 1987). Bars indicate SEM

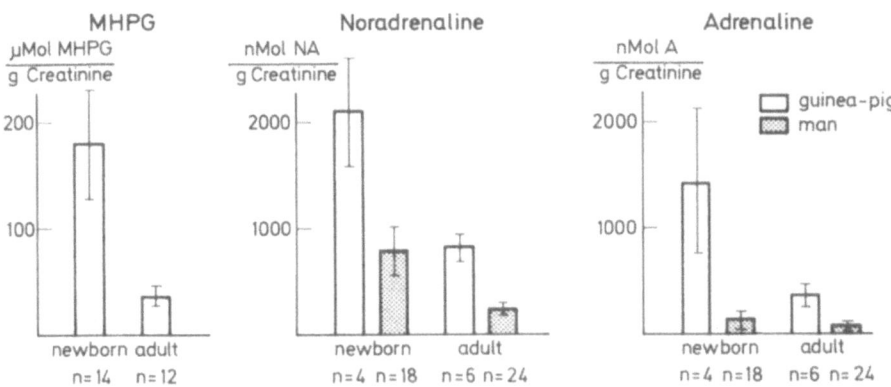

Fig. 4. Levels of MHPG, noradrenaline (NA) and adrenaline (A) in urine samples from newborn and adult guinea pigs (data from Roth et al. 1987) and humans (*shaded columns;* data from Abeling et al. 1984) determined by means of high-pressure liquid chromatography. Bars indicate SEM

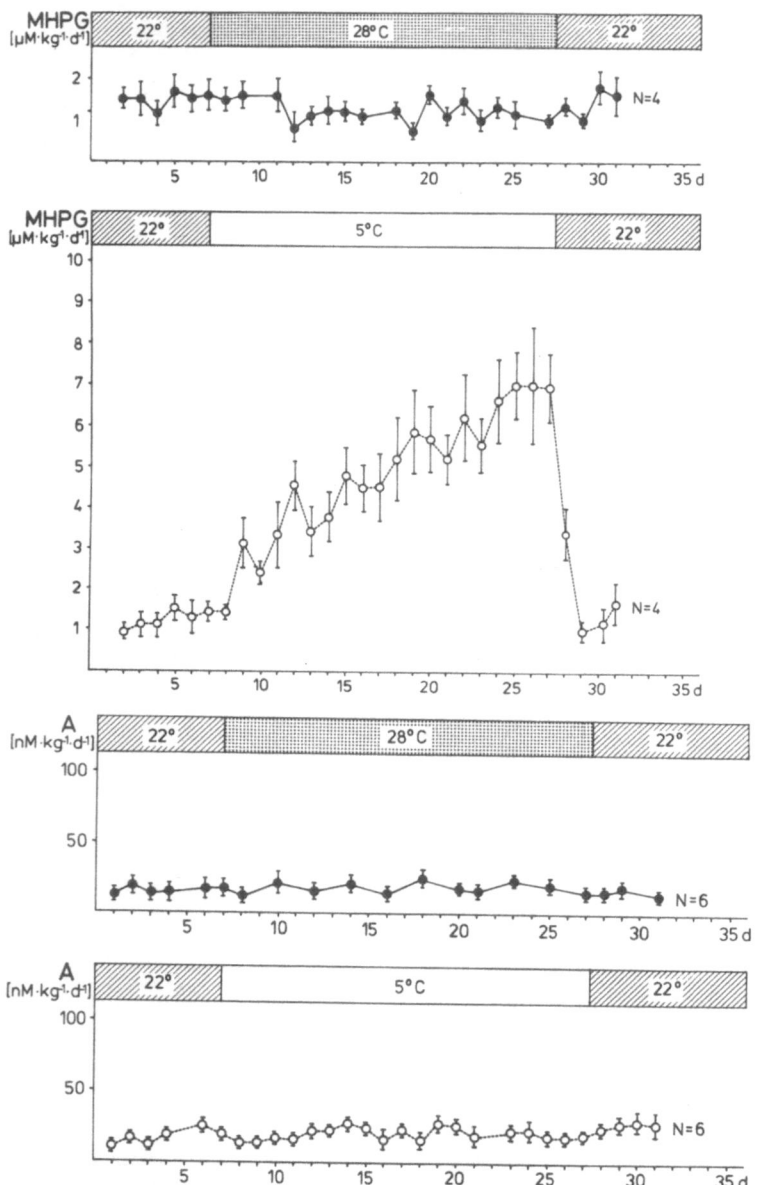

Fig. 5. The daily excretion of MHPG *(top two panels)* and of adrenaline *(bottom two panels)* in urine of guinea pigs before, during and after adaptation to an ambient temperature of 5 °C or 28 °C. Data from experiments described by Roth et al. 1987

Thus the measurement of peripheral catecholamine levels in newborns indicates high activity of the sympathetic system and of adrenal glands, which may explain the prolonged activity of BAT. Furthermore, the reactivation of involuted BAT during cold adaptation is accompanied by strong activation of the sympathetic system as can be seen from Fig. 5. This shows the results of a systematic study in 12 adult guinea pigs reared for 3 months at a room temperature of 22 °C until they reached a weight of 300–400 g and lost their capacity for NST. Thereafter they were put into individual metabolic cages. One week after adaptation to confinement in the metabolic cages, catecholamine levels were determined in urine samples taken daily under the following regime: All animals were kept at 22 °C during the first experimental week, then six animals were exposed to an ambient temperature of 5 °C for 3 weeks and another six to an ambient temperature of 28 °C; in the 4th week all animals were kept again at 22 °C.

Since there were no changes in the excretion of A during thermal adaptation, it must be concluded that in guinea pigs adrenal glands are not activated by exposure to a low temperature (5 °C), and that the six-fold increased levels of MHPG during cold adaptation can be ascribed solely to the activation of the sympathetic nervous system. In fact, levels of free NA excreted in urine increased from 45 nmol·kg^{-1}· d^{-1} at 22 °C to 300 nmol·kg^{-1}·d^{-1} in the 3rd week of cold adaptation (Roth et al. 1987). Corresponding changes in catecholamine levels were also found in blood plasma samples of these animals (Roth et al. 1988).

Thus, during cold adaptation the activity of the sympathetic system increases to an extent similar to that found in the early postnatal stage. This activation correlates with the time course of reactivation of NST and with the increase of the content of 32000-dalton uncoupling protein in the mitochondria from BAT in cold-adapted guinea pigs (Rial and Nicholls 1984).

Catecholamine Effects on the CNS

In addition to acting peripherally, affecting the vasomotor system, thermogenesis in BAT, and mobilization of metabolic substrates, catecholamines seem to influence the CNS. This can be deduced from our investigations on shivering thresholds (Zeisberger and Brück 1976). Shivering is activated in the guinea pig if stimulatory signals from peripheral cold receptors exceed the inhibitory input from central warm receptors into the central thermointegrative structures located in the hypothalamus. This integrative area also receives modulatory aminergic inputs from the lower brain stem (Brück and Zeisberger 1978, 1987).

The effects of manipulation of the central noradrenergic input on shivering thresholds in several experimental studies are summarized in the Fig. 6, upper panel. In newborn and cold-adapted guinea pigs, which normally start to shiver at lower body temperature than the warm-adapted animals, it was possible to shift the onset of shivering to a higher body temperature by an intrahypothalamic injection of

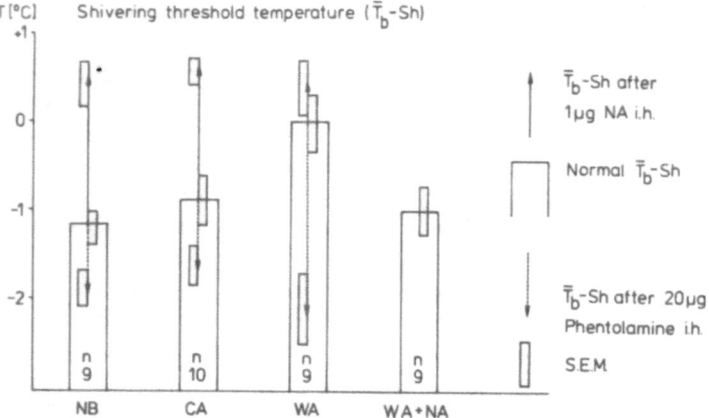

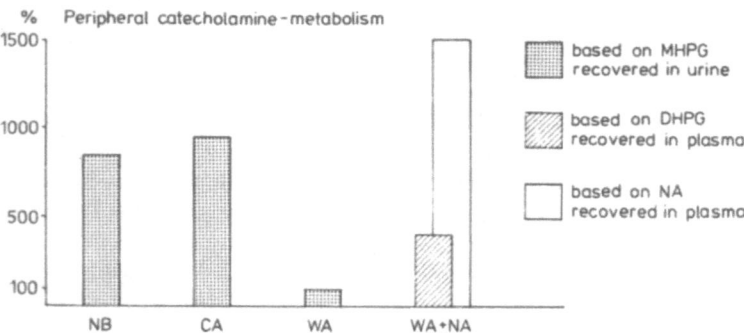

Fig. 6. *Top:* Columns show changes in mean body temperature (including subcutaneous as well as body core temperature) at which groups of newborn *(NB)* and cold-adapted *(CA)* animals started to shiver in comparison to warm-adapted *(WA)* guinea pigs. *Arrows* indicate changes in shivering threshold temperature after intrahypothalamic microinjections of exogenous NA or phentolamine. The *right column (WA + NA)* shows the shivering threshold temperature in warm-adapted animals after peripheral infusion of NA. Data from Zeisberger 1982 and Roth et al. 1988. *Bottom:* Relative magnitude of peripheral metabolism of catecholamines in the same groups of animals, based on measurements of MHPG in urine (Roth et al. 1987) or of dihydroxyphenylglycol (DHPG) and NA in blood plasma (Roth et al. 1988). These data indicate that high peripheral metabolism of NA is correlated with reduced release of NA in the hypothalamus

NA. The same dose of NA was less effective in the warm-adapted animals. By contrast, the warm-adapted animals responded to a blockade of adrenergic receptive sites in the hypothalamic area, induced by micro-injection of phentolamine, with a greater shift of shivering threshold than cold-adapted and newborn animals.

These results suggest that the activity of the noradrenergic brain stem system, and, consequently, the concentration of the endogenous transmitter at postsynaptic sites in the hypothalamus is normally higher in warm-adapted than in cold-adapted or newborn animals. Surprisingly, this indicates that the hypothalamic release of NA is reduced in cold-adapted guinea pigs, in which an increased peripheral release of NA could be expected, similarly to that found in the rat during cold adaptation (Leduc 1961).

In order to find out whether a reciprocal relationship exists between peripheral and central release of catecholamines, the peripheral catecholamine metabolism was assessed on the basis of MHPG recovered in urine of newborn and differently adapted adult animals. This comparison is shown in the lower part of Fig. 6. It shows that peripheral catecholamine metabolism in newborn and cold-adapted guinea pigs is between eight and nine times higher than that in warm-adapted animals, which supports the idea that peripheral and central catecholamine release are reciprocally interrelated. It is therefore possible that the activity in the modulatory noradrenergic afferents into the thermointegrative area of the hypothalamus is inhibited by feedback signals arriving at the brain stem after extensive stimulation of the peripheral sympathetic system. In an attempt to verify this hypothesis, a strong elevation of the sympathetic activity was simulated in warm-adapted guinea pigs at neutral ambient temperature by means of a peripheral infusion of exogenous NA $(0.25 \text{ mg} \cdot \text{kg}^{-1} \cdot \text{h}^{-1})$ lasting 2 h. In the frist experimental study (Zeisberger 1982) the shivering threshold was shifted downwards by more than 1 °C by the end of the NA infusion, and it remained at the lower temperature for as long as 6 h after the infusion was discontinued. Similar infusions of A had the reverse effect and raised the shivering threshold. Recently, these experiments were repeated in animals implanted with indwelling intra-arterial catheters so that blood samples could be taken before and after the peripheral NA infusion and catecholamines in plasma determined (Roth et al. 1988). The columns on the right in Fig. 6 show the main results of the study. At the end of peripheral NA infusion the shivering threshold was lowered by about 1 °C, agreeing with the previous study. Thus, with the increased peripheral NA levels, the warm-adapted animals began to shiver at body temperatures as low as those at which newborn or cold-adapted animals start shivering under normal circumstances.

At the end of peripheral NA infusion the concentration of NA in the blood plasma increased 15-fold. Since the peripheral metabolism of catecholamines increased only four times, as can be inferred from plasma levels of DHPG, it can be concluded that most of the NA present in plasma passed into the urine without being metabolized. The calculations of peripheral metabolism of catecholamines have been based on measurements of DHPG, the precursor of MHPG, in blood plasma samples.

This study demonstrates that an artificial increase of peripheral catecholamine levels, in amounts comparable to those found in newborns and in adults after severe cold exposure, may influence parts of the central structures controlling the thermoregulatory set point. The details of these complicated interrelations between peripheral and central catecholaminergic systems have yet to be revealed. They appear to play a role in adjustment of autonomic responses to different stressor stimuli. By such feedback mechanisms the organism may correct inappropriate and exaggerated response to first stimulus confrontation. They enable the organism to improve and optimize responses in subsequent exposures to the stressor and thus have considerable adaptive value. They also form the basis of short- and long-term shifts of shivering threshold, which are the only thermoadaptive mechanisms left in adult humans.

Summary

The extremely high release of catecholamines at birth serves not only for protection against possible emergencies, such as asphyxia, but helps to prepare the neonate for life and survival in the new environment outside the uterus. The catecholamine levels are already increased in the fetus, which may be important for developement of brown adipose tissue. After delivery the catecholamine release remains increased for long periods of time, mostly dependent on the level of cold stress. We have found a good correlation between the magnitude of nonshivering thermogenesis and the peripheral release of catecholamines during postnatal development and thermal adaptation in the guinea pig. In addition to acting peripherally, affecting the vasomotor system, thermogenesis in brown adipose tissue, and mobilization of metabolic substrates, catecholamines seem to influence the central nervous system. This can be deduced from changes in shivering threshold temperature, which depends on release of noradrenaline in the hypothalamus. Our investigations indicate that high peripheral metabolism of noradrenaline correlates with reduced release of noradrenaline in the hypothalamus. The surge of catecholamines during delivery, and their release in the early postnatal period, might have a determining influence on the development of these interrelations.

References

Abeling NG, v Gennip AH, Overmars H, Voute PA (1984) Simultaneous determination of catecholamines and metanephrines in urine by HPLC with fluorometric detection. Clin Chim Acta 137: 211-226

Brück K (1970) Nonshivering thermogenesis and brown adipose tissue in relation to age, and their integration in the thermoregulatory system. In: Lindberg O (ed) Brown adipose tissue. Elsevier, New York, pp 117-154

Brück K (1978) Heat production and temperature regulation. In: Stave U (ed) Perinatal physiology. Plenum, New York, pp 455-498

Brück K, Zeisberger E (1978) Significance and possible central mechanisms of thermoregulatory threshold deviations in thermal adaptation. In: Wang LCH, Hudson JW (eds) Strategies in cold: natural torpidity and thermogenesis. Academic, New York, pp 654-694

Brück K, Zeisberger E (1987) Adaptive changes in thermoregulation and their neuropharmacological basis. Pharmac Ther 35: 163-215

Cannon B, Nedergaard J (1984) The biochemistry of an inefficient tissue. Essays Biochem 20: 110-164

Jansky L (1973) Nonshivering thermogenesis and its thermoregulatory significance. Biol Rev 48: 85-132

Jansky L, Bartunkova R, Zeisberger E (1967) Acclimation of the white rat to cold: noradrenaline thermogenesis. Physiol Bohemoslov 16: 366-372

Lagercrantz H, Slotkin TA (1986) The "stress" of being born. Sci Am 54: 92-102

Leduc J (1961) Catecholamine production and release in exposure and acclimation to cold. Acta Physiol Scand 53: 1-101

Nicholls DG, Locke R (1984) Thermogenic mechanisms in brown fat. Physiol Rev 64: 1-64

Rial E, Nicholls DG (1984) The mitochondrial uncoupling protein from guinea-pig brown adipose tissue. Biochem J 222: 685-693

Roth J, Zeisberger E, Schwandt HJ (1987) Changes in peripheral metabolism of catecholamines in guinea-pig during thermal adaptation. J Therm Biol 12: 39-44

Roth J, Zeisberger E, Schwandt HJ (1988) Influence of increased catecholamine levels in blood plasma during cold-adaptation and intramuscular infusion on thresholds of thermoregulatory reactions in guinea-pigs. J Comp Physiol B 157: 855-863

Rothwell NJ, Stock MJ (1984) Brown adipose tissue. In: Baker PF (ed) Recent advances in physiology, vol 11. Churchill Livingstone, Edinburgh, pp 349-384

Zeisberger E (1982) The role of noradrenergic systems in thermal adaptation. In: Hildebrandt G, Hensel H (eds) Biological adaptation. Thieme/Stratton, Stuttgart-New York, pp 140-147

Zeisberger E (1987) The roles of monoaminergic neurotransmitters in thermoregulation. Can J Physiol Pharmacol 65: 1395-1401

Zeisberger E, Brück K (1976) Alterations of shivering threshold in cold- and warm-adapted guinea-pigs following intrahypothalamic injections of noradrenaline and of an adrenergic alpha-receptor blocking agent. Pflügers Arch 362: 113-119

Zeisberger E, Brück K, Wünnenberg W, Wietasch C (1967) Das Ausmaß der zitterfreien Thermogenese des Meerschweinchens in Abhängigkeit vom Lebensalter, Pflügers Arch 296: 276-288

Maturation of Thermoregulatory and Thermogenic Mechanisms in Fetal Sheep

P. D. Gluckman[1], T. R. Gunn, B. M. Johnston, G. C. Power, and K. T. Ball

Survival of the neonate depends in part on the establishment of adequate homoeostatic control of body temperature. In utero the fetus is in a relatively stable thermal environment with its body temperature maintained approximately 0.5 °C above maternal temperature (Gunn and Gluckman 1983). The fetus maintains this temperature by the balance between metabolic heat generation and loss of heat across the placenta and a small amount lost across the fetal skin (about 15%; Gilbert et al. 1985). As a result, whenever the maternal temperature rises the fetal temperature must passively follow. As the fetal environmental temperature is not likely to fall below 37 °C, thermogenic mechanisms cannot be anticipated to be important to fetal survival. However, at birth there is an immediate fall in environmental temperature, and active thermoregulation becomes essential to survival. Our research has addressed three related questions: firstly, have thermoregulatory mechanisms differentiated in utero, secondly, has the fetus the ability to initiate thermogenesis in utero and, thirdly, what factors determine the immediate onset of thermogenesis at birth?

To study these questions we developed an experimental approach in the chronically instrumented fetal lamb (Gunn and Gluckman 1983). By means of a coil of tubing placed around the fetal trunk we could cool the fetus by running tap water through the coil. A standard protocol was devised by which the fetal temperature was reduced by 2 °C for a period of 60 or 120 min. There was only a small fall in maternal temperature, so the fetal temperature fell below that of maternal temperature. In our early studies we were impressed by the observation that relatively little heat had to be drawn from the uterus for the fetal temperature to fall quite rapidly even in late gestation. By contrast, in the postnatal lamb, even if prematurely delivered following glucocorticoid induction, thermogenic capacity seemed quite well developed (Alexander et al. 1983). This suggested that either central thermoregulatory and/or peripheral thermogenic mechanisms were relatively immature in utero.

Postnatally the hypothalamus integrates a number of thermogenic responses and is considered to be the site of thermoregulatory control. The major mechanisms include shivering, thyroid hormone and catecholamine release to increase metabolic rate, and peripheral vasoconstriction to reduce heat loss. In the neonate the uncoupled oxidation of brown fat to produce heat is the critical mechanism. This non-

---

[1] Developmental Physiology Laboratory, Department of Paediatrics, University of Auckland, Private Bag, Auckland, New Zealand

The Endocrine Control of the Fetus
Ed. by W. Künzel and A. Jensen
© Springer-Verlag Berlin Heidelberg 1988

shivering thermogenesis (NST) within brown fat contributes at least 50% of the heat generated in the cooled neonatal lamb (Alexander et al. 1983). It is measured as a rise in plasma free fatty acids (FFA) and glycerol and by a rise in the temperature of brown fat relative to body temperature. NST is generally considered to be initiated by sympathetic innervation, although systemic catecholamines may also enhance it. Thyroid hormones are synergistic with catecholamines to promote NST (Fregly et al. 1979). There is increasing evidence that brown fat converts thyroxine into triiodothyronine (Silva 1986). Blood flow to brown fat is increased markedly during cooling.

When the fetus was cooled using the protocol described above the fetal PO_2 fell markedly (Gunn and Gluckman 1983). Measurement of umbilical blood flow showed that blood flow had increased (Gunn et al. 1985) – not decreased due to cold-induced vasoconstriction of the cord, which was the alternative possibility; thus we concluded that fetal oxygen consumption did increase with fetal cooling. Using electromyographic electrodes on limb muscles we have been able to show that the fetus responds immediately to surface cooling with the onset of enhanced muscle activity suggestive of shivering (Gluckman et al. 1983). Interestingly, the application of cold stimulus to the skin was associated with immediate onset of continuous fetal breathing movements (Gluckman et al. 1983), although these were not sustained indefinitely (Johnston et al. 1988). This suggested that stimulation of cutaneous thermoreceptors is one mechanism by which breathing is initiated at birth. When the fetus was cooled by means of a loop of tubing placed in the fetal stomach, breathing was not stimulated. Further, as the sleep state pattern was not disturbed by this approach it was possible to demonstrate that shivering was only demonstrable in quiet (high voltage) sleep and was suspended in low voltage (REM) sleep. Thus the control of shivering seems analogous to that seen in the postnatal lamb and has differentiated by 110 days, gestation. However, clearly, shivering cannot in itself generate sufficient heat to maintain fetal temperature in face of a mild cold stress. In paralysed (by gallamine), cooled fetuses the thermal profile is not dissimilar to that in fetuses capable of shivering (Gluckman et al. 1983). Presumably the relatively small muscle mass of the fetus limits the effectiveness of shivering. However, it is likely that it is the increase in muscle activity (both respiratory and postural) that leads to the increased oxygen consumption, and, indeed, we have recent direct measurements in ventilated fetuses following umbilical cord snare to confirm this.

We have shown that with cooling the fetal hypothalamus responds appropriately in terms of endocrine responses. There is a prompt rise in plasma thyrotropin (TSH) concentrations (Fraser et al. 1985). However, this is not accompanied by either a rise in plasma thyroxine or triiodothyonine. Presumably the presence of the placenta means that any increase in thyroid hormone secretion is associated with immediate inner ring deiodination to the inactive reverse triiodothyronine and to other iodothyronines. Similarly, plasma adrenaline and noradrenaline levels rise and this is associated with a marked rise in fetal plasma glucose and a decrease in plasma insulin concentrations. Plasma cortisol also rises (Gunn et al. 1986).

Appropriate cardiovascular responses can also be demonstrated. Amniotic cooling is associated with a rise in heart rate and blood pressure. These rises can be abolished by catecholaminergic antagonists, suggesting that appropriate hypothalamic autonomic responses have been initiated (Gunn et al. 1985). Kawamura et al. (1986) have shown, using microspheres, that there is appropriate peripheral vasoconstriction with a reduction in skin blood flow and that there is a marked increase in blood flow to brown fat.

Thus the major components of the central hypothalamic response to thermoreceptor stimulation can be demonstrated in utero: shivering is initiated, autonomic activation leads to catecholamine release and cardiovascular changes, and the release of TSH is stimulated. Thus we can conclude that thermoregulatory mechanisms differentiate well before birth. However, cooling of the fetus does not led to any significant rise in FFA or glycerol concentrations (Gunn et al. 1986; Power et al. 1987) or to a relative rise in brown fat temperature (recorded in the perinephric region by an implanted thermistor). Marginal changes are seen in FFA and glycerol but they do not reach statistical significance. By contrast, at birth there is an immediate and marked rise in all indices of NST. We therefore conclude that the fetal environment in some way restricts the ability of the fetus to respond to cooling by the initiation of NST even though central thermoregulatory mechanisms have differentiated and premature birth is associated with the onset of NST.

What possible explanations need to be considered? Firstly, we have considered the possibility that the failure of thyroid hormone to increase might prevent NST. Breall et al. (1984) have shown that an intact thyroid gland is needed prenatally for the neonatal onset of NST. However, the infusion of triiodothyronine (T3) for periods of 7-10 days to raise T3 levels well in excess of normal fetal levels failed to have any effect on the response to cooling. The possibility that the innervation of and/or metabolic pathways within brown fat may be immature in the fetus must also be considered. However, fetal sheep brown fat responds appropriately to endocrine stimuli in vitro (Klein et al. 1984). Brown fat is innervated well before 130 days of gestation (Alexander and Stevens 1980). Experiments to be described in detail below exclude immaturity of brown fat metabolic pathways as an explanation.

Hypoxia inhibits NST in the neonate and we have considered the possibility that the rise in PO_2 at birth may be the limiting factor signalling the induction of NST. The possibility of a placental factor that inhibits NST and which is withdrawn at birth must also be considered. These latter two potential mechanisms, together with a more detailed study of which mechanisms initiate NST at birth, have been considered in a recent series of experiments (Power et al. 1987). Late-gestation fetuses were studied 24 h after surgery. In the first study the fetus was first cooled by means of the tubing coil placed around the thorax. After 1 h the fetal PO_2 was elevated to postnatal levels (100-120 Torr) by means of ventilation with oxygen via an exteriorised tracheostomy tube. This led to a significant although relatively small rise in FFA release and in brown fat temperature. Ventilation with nitrogen had no effect, suggesting that any effect was due to oxygen rather than to mechanical ventilation of the lungs. However, as the effect of oxygenation was relatively small, it seems unlikely that oxygen delivery to brown fat is the primary factor preventing the initiation of NST in utero, particularly as cooling is associated with an increase in brown fat blood flow in utero. It does remain possible that, although PO_2 is not limiting, a rise in PO_2 is part of the signal for the initiation of NST at birth.

The fetus was then separated from placental influences by snaring of the umbilical cord. There was an immediate rise in FFA and glycerol concentrations and in brown fat temperature (Power et al. 1987). Cord snare was not associated with any change in the metabolic clearance of glycerol, showing that the marked rise following cord snare did reflect the onset of NST.

The dominant factor in the initiation of NST at birth would thus appear to be the separation from the placenta. What is the mechanism of this effect? It has been suggested that the neonatal surge in T3 at birth might be a major signal for the initiation of NST (Fisher and Klein 1980). Certainly an intact thyroid gland is necessary for neonatal thermogenesis (Breall et al. 1984). Following cord snare, T3 levels rise by 40%–100% over the next hour (Sack et al. 1976). To investigate this possibility we infused cooled, ventilated fetuses with a high dose of T3 0.5 h before cord snare. T3 infusion had no effect on the indices of NST, although NST is induced within 15 min of cord snare (irrespective of whether T3 was previously infused or not). Thus, the failure of T3 to induce NST is not due to inadequate length of exposure. We can conclude that, although T3 is necessary for neonatal NST, the initiation of NST is not dependent on a rise in circulating T3 concentrations. It has also been suggested that the rise in catecholamines at birth may be the signal for the initiation of NST (Padbury et al. 1981). However, measurement of plasma catecholamine levels during these studies did not show any marked change at the time of cord snare, although there was a gradual rise during the study. Thus, catecholamines do not appear to be the trigger for the initiation of NST. Further studies, in which the β-adrenergic agonist isoprotenerol was infused into the fetus prior to cord snare and cooling, suggested that placental separation and cutaneous thermal stimulation were the critical factors in the initiation of NST. In this last experiment the order of manipulation was reversed: the fetus was first ventilated, then subjected to cord snare, then cooled, all during an isoprotenerol infusion. NST was only markedly stimulated after the onset of cooling. This would suggest that, although placental separation is necessary, direct sympathetic neural stimulation of brown fat in response to cold, rather than an elevation in circulating catecholamine levels, is central to the initiation of NST.

The signal for the initiation of NST at birth thus appears to be dependent on stimulation of cutaneous cold receptors and removal of a placental influence. In the presence of the placenta NST cannot be demonstrated, but immediately on separation, provided there is a cold stimulus, it is rapidly initiated. This excludes immaturity of brown fat as an adequate explanation of the failure of NST in utero. We therefore postulate that the placenta is likely to be secreting into the fetal circulation a factor which inhibits the ability of brown fat to respond to either hormonal or neural stimulation. This factor must have a very short half-life as NST is quickly initiated after cord snare. If this factor was an ecosanoid or similar substance which is cleared by the lung, then the variable effect of ventilation in the cooled fetus to allow a degree of NST could be explained by a reduction in plasma concentration of this inhibitor due to increased pulmonary blood flow secondary to ventilation. Indeed, it has been reported that PGE_2 is antilipolytic (Steinberg et al. 1964). Similarly, it has been suggested that withdrawal of a hypothetical placental inhibitor plays a role in the initiation of breathing at birth (Adamson et al. 1987) and in the rapid fall in plasma growth hormone concentrations at birth (see Gluckman et al., this volume, pp. 201-209).

In summary, our studies allow us to conclude that the central mechanisms for responding to a cold stress have differentiated well before birth in the sheep fetus. However, thermogenic responses are defective and in particular marked non-shivering thermogenesis cannot be demonstrated unless placental separation has occurred – there being a less although consistent and significant effect of oxygenation alone. The signal for the initiation of NST is not thyroid hormone or circulating catecholamines; the most plausible hypothesis is that the initiation of NST depends on removal of a placental inhibitor allowing autonomic neural stimulation of brown fat metabolism. The increase in oxygen delivery to brown fat may also play a role.

Acknowledgments. This work was funded by a programme grant of the Medical Research Council of New Zealand.

References

Adamson SL, Richardson BS, Homen J (1987) Initiation of pulmonary gas exchange by fetal sheep in utero. J Appl Physiol 62: 989-998

Alexander G, Stevens D (1980) Sympathetic innervation and the development of structure and function of brown adipose tissue: studies on lambs chemically sympathectomised in utero with 6-hydroxydopamine. J Dev Physiol 2: 119-137

Alexander G, Nicol D, Thorburn G (1983) Thermogenesis in prematurely delivered lambs. In: Comline R, Cross K, Dawes G, Nathanielsz P (eds) Fetal and neonatal physiology. Sir Joseph Barcroft Centenary Symposium. Cambridge University Press, Cambridge, pp 410-417

Breall JA, Rudolph AM, Heymann MA (1984) Role of thyroid hormone in postnatal and metabolic adjustments. J Clin Invest 73: 1418-1824

Fisher DA, Klein AH (1980) The ontogenesis of thyroid function and its relationship to neonatal thermogenesis. In: Tulchinsky D, Ryan KJ (eds) Maternal fetal endocrinology. Saunders, Philadelphia, pp 281-293

Fraser M, Gunn TR, Butler JH, Johnston BM, Gluckman PD (1985) Circulating thyrotropin (TSH) in the ovine fetus: evidence for pulsatile release and the effect of hypothermia in utero. Ped Res 19: 208-212

Fregly MJ, Field FP, Katovich MJ, Barney CC (1979) Catecholamine-thyroid hormone interaction in cold acclimatized rats. Fed Proc 38: 2162-2169

Gilbert RD, Schroder H, Kawamura T, Dale PS, Power GG (1985) Heat transfer pathways between the fetal lamb and ewe. J Appl Physiol 59: 634-638

Gluckman PD, Gunn TR, Johnston BM (1983) The effect of cooling in breathing and shivering in unanaesthetized fetal lambs in utero. J Physiol 343: 495-506

Gunn TR, Gluckman PD (1983) The development of temperature regulation in the fetal lamb. J Dev Physiol 5: 167-179

Gunn TR, Johnston BM, Iwamoto HS, Fraser M, Nicholls MG, Gluckman PD (1985) Hemodynamic and catecholamine responses to hypothermia in the fetal lamb in utero. J Dev Physiol 7: 241-249

Gunn TR, Butler J, Gluckman PD (1986) Metabolic and hormonal responses to cooling the fetal lamb in utero. J Dev Physiol 8: 55-56

Johnston BM, Gunn TR, Gluckman PD (1988) Surface cooling rapidly induces coordinated activity in upper and lower airway muscles of the fetal lamb in utero. Pediatr Res 23: 257-261

Kawamura T, Gilbert RD, Power GC (1986) Effect of cooling and heating on the regional distribution of blood flow in fetal sheep. J Dev Physiol 8: 11-22

Klein AH, Reviczky A, Padbury JF (1984) Thyroid hormones augment catecholamine-stimulated brown adipose tissue thermogenesis in the ovine fetus. Endocrinology 114: 1065-1069

Padbury JF, Diakomanolis ES, Hobel CJ, Perelman A, Fisher DA (1981) Neonatal adaptation: sympatho-adrenal response to umbilical cord cutting. Pediatr Res 15: 483-487

Power GC, Gunn TR, Johnston BM, Gluckman PD (1987) Oxygen supply and the placenta limit thermogenic responses in fetal sheep. J Appl Physiol 63: 1896-1901

Sack J, Beaudry M, DeLamater PV, Oh W, Fisher DA (1976) Umbilical cord cutting triggers hypertriiodothyroninemia and nonshivering thermogenesis in the newborn lamb. Pediatr Res 10: 169-175

Silva JE (1986) Brown adipose tissue an extrathyroidal source of triiodothyronine. News Psych Sci 1: 119-122

Steinberg D, Vaughan M, Nestel P, Strand O, Bergstrom S (1964) Effects of the prostaglandins on hormone-induced mobilization of free fatty acids. J Clin Invest 43: 1533-1537

Perinatal Activation of Brown Adipose Tissue

B. Cannon[1], E. Connolly[1], M.-J. Obregon[1, 2], and J. Nedergaard[1]

Introduction

Discussions concerning the significance of brown adipose tissue for the newborn mammal have become facilitated within the last few years by a general agreement on the following basic facts:

- That in all mammals, newborns as well as adults, facultative non-shivering thermogenesis *is* brown-fat thermogenesis
- That the molecular background to the unique ability of brown adipose tissue to function as a heat-producing organ is the existence in this tissue of a large abundance of mitochondria endowed with the uncoupling protein thermogenin, and
- That the amount of thermogenin in a mammal (under most circumstances) is the rate-limiting factor for thermogenesis.

In the following we shall first briefly introduce the molecular background for thermogenesis and examine how, in different types of newborns, the correlation between thermogenin amount and thermogenic capacity holds. We shall especially point to the problems involved in understanding the ability of certain mammals to recruit their brown fat already during intra-uterine life. Finally we shall examine brown fat and thermogenesis in the human fetus and newborn, both under normal and pathological conditions.

Thermogenin and Thermogenesis

The specialization of brown adipose tissue for heat production is expressed in several properties of the tissue, such as the high degree of vascularization (to bring oxygen to the tissue and to lead the heat produced away to the rest of the body), the presence of stored triglycerides (serving as an immediately available source of substrate for combustion when heat production is acutely necessary), the large capacity

[1] The Wenner-Gren Institute, Biologihus F3, University of Stockholm, S-106 91 Stockholm, Sweden
[2] Departemento de Endocrinología Experimental, Instituto de Investigaciones Biomédicas, Facultad Autónoma de Medicina (U. A. M), Arzobispo Morcillo 4, 20829 Madrid, Spain

The Endocrine Control of the Fetus
Ed. by W. Künzel and A. Jensen
© Springer-Verlag Berlin Heidelberg 1988

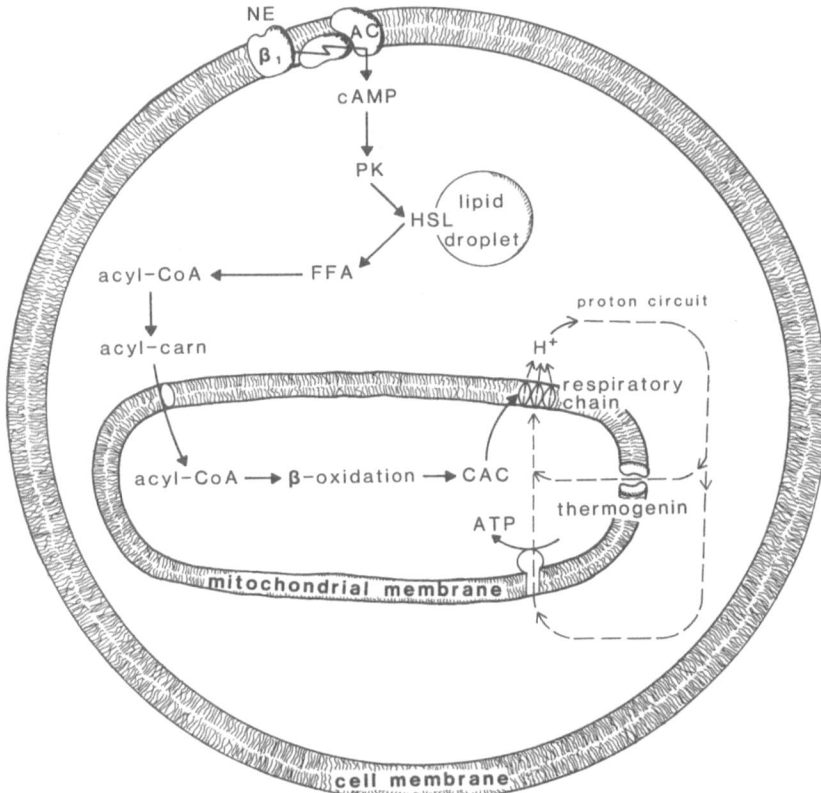

Fig. 1. *The mechanism of brown adipose tissue thermogenesis.* Norepinephrine *(NE)* released by the sympathetic nervous system binds to the β-adrenergic receptor, inducing an elevation in intracellular cyclic adenosine monophosphate *(cAMP)* via adenylate cyclase *(AC)* activation. Via protein kinases *(PK)*, cAMP activates hormone-sensitive lipase *(HSL)*, which then liberates free fatty acids *(FFA)* from the intracellular lipid depots. The FFA are transported into the mitochondria where they are oxidized by β-oxidation, the citric acid cycle *(CAC)*, and, finally, by the respiratory chain. The opening of the uncoupling protein thermogenin after β-stimulation allows protons to enter the mitochondria freely, thus uncoupling respiration from ATP synthesis and releasing substrate energy as heat. (For detailed reviews of these mechanisms see Cannon and Nedergaard 1982, 1985; Nedergaard and Cannon 1984)

of lipid-degrading enzymes, and the abundance of mitochondria richly endowed with respiratory chain enzymes (for the biochemical combustion process). These specializations are, however, only enhanced expressions of properties which are also found in several other tissues of the body.

The property of brown adipose tissue which makes it unique is the presence in the mitochondrial membrane of a protein which is only found in this tissue. This protein, which in different contexts is known as the (GDP-)guanosine diphosphate binding protein, the uncoupling protein (UCP) or the 32000-dalton protein, and which here will be referred to as *thermogenin*, gives the mitochondria the ability to dissipate the chemical energy of combusted substrates in the form of heat, without having to first transform the energy into ATP for subsequent hydrolysis. This thermogenin-mediated energy dissipation is, however, regulated so that it only occurs under physiological conditions where it is necessary, and this regulation is mediated via sympathetic innervation of the tissue.

We shall not here further review the function of the tissue as such since this has recently been performed in several comprehensive review articles (Nedergaard and Lindberg 1982; Nicholls and Locke 1984; Cannon and Nedergaard 1985), as well as in both the classical (Lindberg 1970) and the recent (Trayhurn and Nicholls 1986) books with the title *Brown Adipose Tissue*. The mechanism of brown adipose tissue thermogenesis is shown schematically in Fig. 1.

Brown Adipose Tissue in the Fetus and Neonate

Patterns of Perinatal Development

The perinatal development of brown adipose tissue in mammals has been reviewed in detail recently (Nedergaard et al. 1986), and we shall here summarize some pertinent points.

In the perinatal development of mammalian species, three different patterns can be distinguished, representing different perinatal strategies. These are general developmental patterns and are, therefore, also of great importance when brown-fat development is discussed.

In species with *precocial* newborns (which include most of the larger mammals, e.g. sheep and cattle, but also smaller species such as the guinea pig), the newborns are, in most respects, miniature copies of the adult and each litter consists of one or a few members. They are born with full pelts and fully developed eyes, and they are generally able to walk within hours of birth.

In species with *altricial* newborns (which include the common laboratory animals such as the rat and the mouse), the newborns have not at all reached this degree of development. The litters are generally rather large, and the newborns huddle together in the first days of life. These newborns are naked, blind, can only crawl and are helpless without their mother.

Finally, in species with *immature* newborns (including only a few true mammals such as hamsters, and notably the kangaroos) the animal at birth can to a large extent be considered merely a fetus, and a large part of what might normally be understood as fetal development occurs postnatally.

Thermogenic Demands in Newborns of Different Species

There are large differences between these three perinatal groups with regard to the thermogenic demands upon the newborn.

In the *precocial* newborns, the demand for thermogenesis is already at its maximum at birth. The newborn has to function from the start as a self-regulating homeothermic unit, and in many cases the demand for thermogenesis is greatest just at the moment of birth, when the newborn meets the world with a fur wet from amniotic fluids and is perhaps faced with strong, cooling winds.

In the *altricial* newborns, the situation is less drastic. Although, of course, the newborns are no longer insulated by the womb, the mother may still, by arranging a kind of 'nest', make sure that a stable and comparatively warm environment is created. Further, huddling behaviour allows the newborns to warm each other. In these newborns one can imagine a thermoregulatory demand which progressively arises as they get older and develop the means to leave the comfort of the nest.

Finally, investigations of temperature regulation in the *'immature'* group indicate that at the moment of birth these newborns lack the thermoregulatory function altogether. These mammals are apparently born poikilothermic, changing their body temperature in parallel with that of the surroundings. Only with time, during postnatal development, will a thermoregulatory capacity begin to appear.

Non-shivering Thermogenesis in Newborns of Different Species

By injection of norepinephrine into newborn mammals, it is possible to obtain a measure of the capacity for non-shivering thermogenesis during postnatal development. Results of such experiments using typical examples from the newborn groups are shown in Fig. 2 D–F.

The *precocial* newborns (e.g. the guinea pig) give a very large metabolic response to norepinephrine already on the day of their birth (Fig. 2 D), i.e. they are born with a fully developed capacity for non-shivering thermogenesis which can accommodate their immediate demands. In the *altricial* newborns, the response to norepinephrine progressively develops with postnatal age (Fig. 2 E), whereas in the *'immature'* newborns the capacity for non-shivering thermogenesis suddenly starts to develop at a defined time after birth (Fig. 2 F).

We have compiled data (Fig. 2 A–C) on the total amount of thermogenin in the different groups of newborns [measured by mitochondrial GDP-binding (for review see Cannon and Nedergaard 1985) and expressed per gram body weight to the power 0.67, i.e. in proportion to the body surface area, where the major heat loss occurs]. The marked parallelism between this thermogenic index and the capacity for non-shivering thermogenesis in the different groups is very clear. This not only confirms that regulatory non-shivering thermogenesis in the newborn mammal is located in the brown adipose tissue but is also fully in agreement with the view ex-

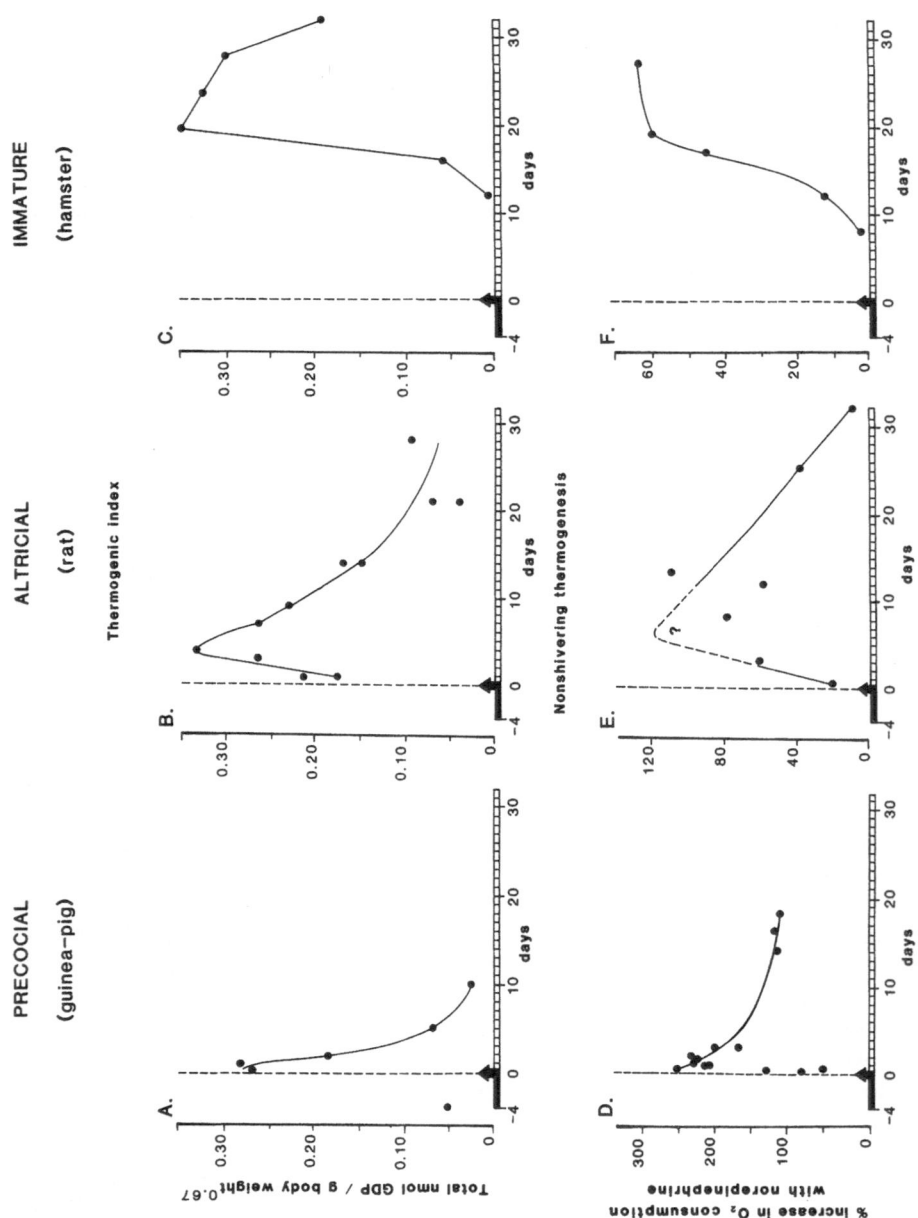

Fig. 2. *The development of the thermogenic index* (**A-C**) *with age in the three newborn types and its relationship to non-shivering thermogenic capacity* (**D-F**). GDP-binding is a measure of thermogenin content (for review see Cannon and Nedergaard 1985). (Adapted from Nedergaard et al. 1986, in which references to the original studies used to compile the curves can be found)

pressed above that the amount of thermogenin is the rate-limiting factor for thermogenesis (Cannon et al. 1981; Cannon and Nedergaard 1985).

What Regulates the Recruitment Process?

When an animal is in a physiological condition in which its need for thermogenesis is increased, a concerted growth and activation (recruitment) of the brown adipose tissue occurs, which is the product of a series of events on different organizational levels. The tissue becomes more vascularized, the number of cells in the tissue is increased, the amount of mitochondria both in the tissue and apparently also within each cell is increased, and the mitochondria contain more thermogenin per milligram of mitochondrial mass (for reviews see Trayhurn and Nicholls 1986).

Although in no way proven, there is good reason to think that it is the same physiological, neurohormonal signal which initiates all these events. It may therefore be possible to follow the recruitment process simply as the amount of mRNA coding for thermogenin present in the tissue, since the amount of thermogenin is the rate-limiting step in thermogenesis.

The regulation of thermogenin mRNA synthesis has already been studied in some detail in adult mammals. The amount of thermogenin mRNA is increased in a very rapid and dramatic way when the adult mammal is exposed to cold (Jacobsson et al. 1985, 1986, 1987; Ricquier et al. 1986). Evidence suggests that this cold-induced increase in the amount of thermogenin mRNA is under adrenergic control (Ricquier et al. 1986; Jacobsson et al. 1986), and in vivo experiments indicate that both α- and β-adrenergic components are involved (Jacobsson et al. 1986).

Regulation of Thermogenin mRNA Synthesis in the Mammalian Fetus and Newborn

The simplest hypothesis concerning the regulation of thermogenin synthesis in the newborn is, of course, that the increase in the expression of the thermogenin gene is also under adrenergic control. According to this hypothesis, the newborn, just like the adult, experiences a cold stress which leads to an increased sympathetic activity to brown adipose tissue, and the norepinephrine which is consequently released is responsible for gene activation.

In order to investigate this hypothesis, the amount of thermogenin mRNA in rat pups experiencing different environmental temperatures was examined. Rat pups were taken at birth and placed at 28 °C or at 36 °C (in the absence of the dam) for the times shown in Fig. 3. It can be clearly seen that the severe cold stress of the pups at 28 °C induced a large increase in the level of thermogenin mRNA; the level

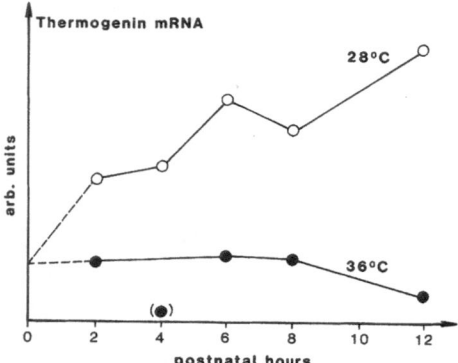

Fig. 3. *Effect of ambient temperature on postnatal level of thermogenin mRNA in the brown adipose tissue of the rat. p.p,* post-partum. Thermogenin mRNA was measured as described in Obregon et al. (1987)

continued to rise for at least 12 hours. Exposure of neonatal pups to 36 °C, however, completely abolished the postnatal elevation in thermogenin mRNA. It seems, therefore, that ambient temperature and thermal stress are of major importance to the postnatal expression of the thermogenin gene.

Thus, the most simple theory – that thermogenin gene activation in the newborn is due to postnatal thermal stress – is supported by the experimental data available, at least in the altricial newborns.

Thyroid Influence on Thermogenin Expression

Another neurohormonal factor involved in the regulation of thermogenin gene expression could be thyroid hormones. Thyroid hormones have earlier been discussed in relation to perinatal function of brown adipose tissue, since newborn hypothyroid rats are less cold-resistant than their euthyroid counterparts (Steele and Wekstein 1972) and the lipolytic and respiratory responses are markedly reduced in the brown adipose tissue of the hypothyroid neonate (Hemon 1976; Klein et al. 1984). These defects have been associated with an impairment of the function of the β-adrenergic receptor in the tissue, since dibutyryl-cAMP was able to produce similar lipolytic and respiratory responses in hypo- and euthyroid newborn brown fat (Hemon 1976; Klein et al. 1984).

Pregnant rats were therefore made hypothyroid by treatment with methimazole and the amount of thermogenin mRNA in the fetuses and the newborn was monitored (this treatment did not influence the normal growth curve of the offspring).

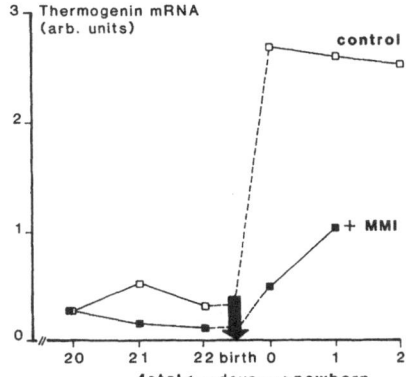

Fig. 4. *The effect of methimazole* (MMI-) *induced hypothyroidism on the perinatal level of thermogenin mRNA in rat brown adipose tissue.* Pregnant rats were given 0.02 vol% methimazole in the drinking water from day 14 of gestational age. (For further details see Obregon et al. 1987)

Several features are evident from the results of this study (Fig. 4). Firstly, the large postnatal increase in thermogenin mRNA in normal rats is again conspicuous, and it is clear that the increase in expression occurs as a single step between a low 'fetal' value (which proceeds into the first hours after birth, cf. Fig. 3) and a high postnatal value which is unaltered for several days after birth (i.e. the postnatal period when thermogenin synthesis proceeds, cf. Fig. 2B). Secondly, the postnatal increase in thermogenin mRNA is markedly blunted in hypothyroid animals, indicating that the euthyroid state is essential for the normal postnatal recruitment of brown adipose tissue. Whether this is a direct effect of thyroid hormone, or whether it is perhaps related to thyroid hormone being necessary for β-adrenergic receptor function (Seydoux et al. 1982) (and β-adrenergic receptor function being essential for thermogenin recruitment), is not presently known. Finally, even in the fetus the amount of thermogenin mRNA in the tissue is not at its lowest constitutive level, but must be under positive regulation, as hypothyroidism is able to further diminish it. This points to the problem of the fetal recruitment of brown adipose tissue and of thermogenin mRNA.

Problem of Fetal Recruitment of Brown Adipose Tissue

The problem of the regulation of perinatal recruitment of thermogenin can be summarized as follows. There is, especially in the precocial species (but apparently also in the altricial, cf. Fig. 4), good reason to conclude that the brown adipose tissue of the fetus is already recruited in several respects, one of which is an increased level of thermogenin mRNA (Freeman and Patel 1984) despite the fact that the fetus is experiencing no thermal stress. This prenatal recruitment is necessary to acquire a tissue which is maximally active already *at* birth, and not a considerable time after. Thus it must be suggested that:

Either

1. This recruitment is induced by norepinephrine, as seems to be the case in the adult. This idea introduces the following implications:
    a)   That there must be an alternative, fetal way of regulating sympathetic activity to brown adipose tissue, which cannot be due to activation of thermoregulatory centres, and
    b 1) That the fetal tissue, due to this chronic sympathetic stimulation, must be producing heat in utero, a somewhat inefficient and unlikely situation, or alternatively
    b 2) That an inhibitory factor for thermogenesis (but not for recruitment) must exist in utero.

Or

2. Chronic norepinephrine stimulation is not responsible for brown adipose tissue recruitment in utero. This too has implications:
    a 1) That the generally accepted view – that chronic norepinephrine stimulation is causative of recruitment – is wrong, or else
    a 2) That a completely different and so far uncharacterized pathway for regulation of recruitment also exists, but only in the fetus. After birth, the control of brown adipose tissue recruitment is then switched to the adrenergic system of the adult.

The problem of the fetal recruitment of brown adipose tissue is thus far from solved and provides an important area for future research.

Brown Fat in the Human Infant

The infant represents the only stage in the life of the human where it is generally accepted that brown fat-mediated non-shivering thermogenesis is present and physiologically important for the defence of the individual against low environmental temperature. Unlike in the experimental animal, the unequivocal demonstration of brown fat thermogenesis in infants cannot be performed for both practical and ethical reasons. There is, however, a great deal of work concerned with the response of the infant to changes in environmental temperature and with its ability to thermoregulate. From this knowledge, together with other observations which might individually appear circumstantial, one can infer the presence of active brown adipose tissue in the human newborn. The significance of the tissue decreases with the age of the individual, but man may nevertheless retain a functional amount in adult life.

Distribution and Characteristics

The infant has at birth a distribution of brown fat which is similar to that of other mammalian species. It includes interscapular, cervical, axillary, thoracic, periaortic and perirenal deposits (Aherne and Hull 1966; Heaton 1972). These deposits display the characteristic brown adipose tissue structure, i.e. multilocular cells, abundant mitochondria and central nucleus, etc. (Aherne and Hull 1966). Though clearly more prominent in the newborn, these cells can be observed throughout adult life, tending towards a white-fat appearance in later life (Pawlikowski 1955; Heaton 1972; Tanuma et al. 1976; Naeye 1974, 1976; Valdes-Dapena et al. 1976).

Brown fat development begins in the mid-term fetus (Heim et al. 1968; Merklin 1974; Hull 1983) and the tissue can be distinguished as early as 20 weeks after conception (Moragas and Torán 1983). There is an increase in the cytoplasmic volume of the cells up until and after birth (Aherne and Hull 1966), indicating a postnatal recruitment similar to that seen in the altricial newborns. This is consistent with a retention of the tissue fat stores during the first 24 h of life (Moragas and Torán 1983), after which time depletion occurs (Aherne and Hull 1966).

Recently, the presence of thermogenin has been demonstrated in the mitochondria of the brown fat of infants, and human thermogenin is found to have immunocrossreactivity with the protein found in rat brown-fat mitochondria (Lean and James 1983). Thus, the newborn human tissue contains the specific apparatus required for non-shivering thermogenic activity.

Physiological Observations

Do physiological observations confirm the altricial nature of the human infant inferred from morphological observations? Immediately following birth there is a rapid fall in *body temperature,* which may be interpreted as a period of insufficient thermogenesis. However, the resistance of the newborn to cold develops rapidly, and by day 1 the rectal temperature can be maintained under conditions of mild cold stress (Pribylová and Znamenacek 1964; Rylander 1972).

The interpretation of the studies on *oxygen consumption in the cold* in the newborn human are complicated by differing conditions of cold exposure being used. Despite these problems, however, it seems that immediately after birth a response to cold is present but not maximal, and during the 1st day there is a rapid rise in the ability of the infant to respond (Fig. 5 A). This postnatal development of the cold response was also pronounced in the study of Brück (1961), where the altricial nature of the human newborn was clear.

There is some evidence for the development of brown fat-mediated *non-shivering thermogenesis* in the human neonate, since infusion of norepinephrine into newborn babies leads to an elevation in the rate of oxygen consumption which is independent of physical activity (Karlberg et al. 1962, 1965, see Fig. 6). This norepinephrine-stimulated rise in oxygen consumption develops after birth, apparently

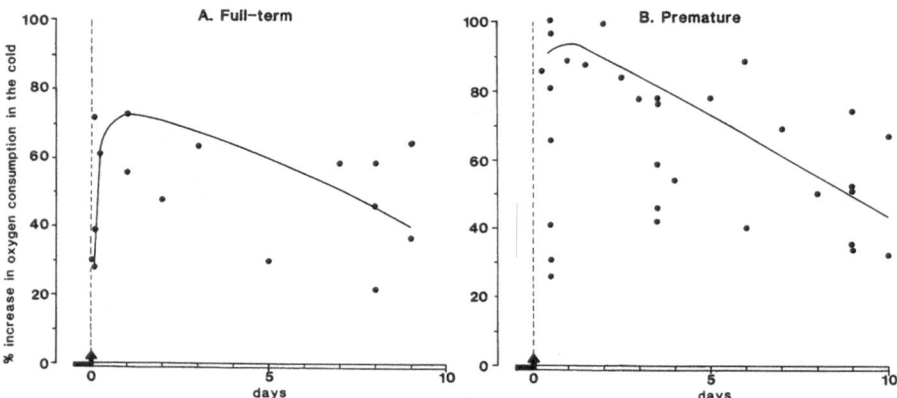

Fig. 5. *Response of full-term* (**A**) *and premature* (**B**) *newborn infants to cold stress.* Results are from studies using mild cold stress in which no shivering was observed. **A** From Hill and Rahimtulla 1965; Pribylová and Znamenacek 1964; Smales and Kime 1978. **B** From Adams et al. 1964; Hey and Katz 1969; Mestyán et al. 1968

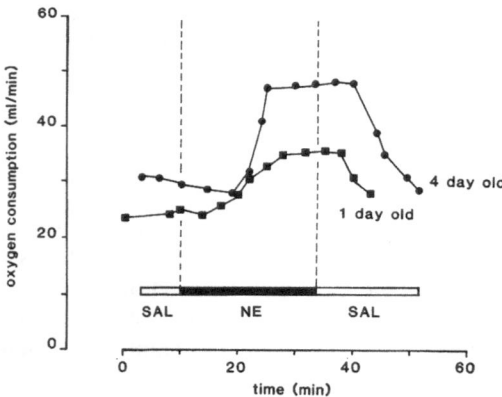

Fig. 6. *Non-shivering thermogenesis in the newborn infant.* The curve shows the whole-body oxygen consumption of two newborn infants aged 1 day (3.39 kg) and 4 days (3.84 kg) in response to the infusion of norepinephrine *(NE)* as compared to saline *(SAL)*. (Data from Karlberg et al. 1965)

being greater at 4 days than at 1 day (Fig. 6), thus illustrating the human newborn's altricial character. This conclusion is indirectly confirmed by an elevation of the urinary excretion of norepinephrine from infants exposed to cool environments, compared to their warm counterparts (Stern et al. 1965). Using this as an index, it is possible to say that brown fat thermogenesis in the infant increases in intensity from day 3 to day 11 of life.

The available evidence would, therefore, tend to indicate that the newborn human displays an altricial development of its brown fat.

Disorders and Physiological Influences in the Human Neonate

Although *premature* babies are more prone to cold stress than their full-term counterparts, this is probably a consequence of their disadvantageous body weight/surface area ratio and not due to a deficiency in heat production. In fact, premature babies probably require more heat production because of their small size, and this is apparent in the faster depletion of the brown adipose tissue fat stores of the premature compared to the full-term baby (Aherne and Hull 1966). When exposed to a cool environment, the premature infant responds by elevating his oxygen consumption, and the development of this response with age is remarkably similar to that of the full-term infant, showing a wide variability in the first hours of life, indicating a rapid activation of thermogenesis, followed by a decreasing response with age (Fig. 5 B).

Maintenance of the infant at *thermoneutral* temperatures leads to a retention of the brown adipose tissue fat deposits, even if the child has been undernourished (Heim et al. 1968). Thus human neonatal brown fat defends its fat stores for use primarily in thermogenesis – exactly like, for example, the newborn rabbit (Heim and Kellermayer 1966). Infants that have been reared in *cool temperatures* have depleted brown adipose tissue fat stores compared to those at thermoneutrality (Heim et al. 1968) and prolonged exposure leads to increased cold resistance in the infant (Glass et al. 1968). As might be expected, the brown adipose tissue fat stores of a 4-month old hypothermic baby were found to be depleted, compared, for example, to a similar infant who suffered cot death (Heaton 1973).

Infants that develop chronic congenital heart disease and various other syndromes (Naeye 1974) are characterized by a *chronic hypoxic state;* on post-mortem examination, these victims have been reported to possess a higher proportion of the multilocular brown adipocytes than accidental death controls (Naeye 1974, 1976; Valdes-Dapena et al. 1976). This pattern of retarded brown fat involution has also been observed in victims of the *sudden infant death* syndrome and is associated with hypoxia due to a defective neuronal control of breathing (Naeye 1974, 1976; Valdes-Dapena et al. 1976). Hypoxia probably produces these effects by increasing the sympathetic drive to the tissue.

Fetal malnutrition *(hypotrophy)* in 'small for gestational age' infants does not lead to a loss of the acute oxygen consumption response in the cold (Heim 1981). However, severe *malnutrition* of newborn infants leads to a nearly complete impairment of their response to cold. Such malnourished infants have lipid-depleted brown fat and this is probably the reason for the absence of thermogenesis (Brooke et al. 1973).

Conclusions

The perinatal development of brown adipose tissue and non-shivering thermogenesis in the newborn mammal can follow one of three general patterns: precocial, altricial or immature. These patterns are consistent with the thermogenic demands made upon each of the newborn types. The regulation of brown adipose tissue recruitment in the fetus remains unclear but the euthyroid state appears to be essential for normal postnatal tissue development, at least in the altricial newborn. The human infant appears to fit into the altricial pattern of development of neonatal brown adipose tissue and non-shivering thermogenesis.

References

Adams FH, Fujiwara T, Spears R, Hodgman J (1964) Temperature regulation in premature infants. Pediatrics 33: 487–495

Aherne W, Hull D (1966) Brown adipose tissue and heat production in the newborn infant. J Pathol Bacteriol 91: 223–234

Brooke OG, Harris M, Salvosa CB (1973) The response of malnourished babies to cold. J Physiol 233: 75–91

Brück K (1961) Temperature regulation in the newborn infant. Biol Neonate 3: 65–119

Cannon B, Nedergaard J (1982) The function and properties of brown adipose tissue in the newborn. In: Jones CT (ed) Biochemical development of the fetus and the neonate. Elsevier, Amsterdam, pp 697–730

Cannon B, Nedergaard J (1985) The biochemistry of an inefficient tissue: brown adipose tissue. Essays Biochem 20: 110–164

Cannon B, Nedergaard J, Sundin U (1981) Thermogenesis, brown fat and thermogenin. In: Musacchia XJ, Jansky L (ed) Survival in cold. Elsevier-North Holland, Amsterdam, pp 99–120

Freeman KB, Patel HV (1984) Biosynthesis of the 32-kdalton uncoupling protein in brown adipose-tissue of developing rabbits. Can J Biochem 62: 479–485

Glass L, Silverman WA, Sinclair JC (1968) Effect of the thermal environment on cold resistance and growth of small infants after the first week of life. Pediatrics 41: 1033–1046

Heaton JM (1972) The distribution of brown adipose tissue in the human. J Anat 112: 35–39

Heaton JM (1973) A study of brown adipose tissue in hypothermia. J Pathol 110: 105–108

Heim T (1981) Energy requirements of thermoregulatory heat production in newly born. In: Monset-Couchard M, Minkowski A (ed) Physiological and biochemical basis for perinatal medicine. Karger, Basel, pp 158–174

Heim T, Kellermayer M (1966) Effect of starvation on brown adipose tissue in the newborn rabbit. Acta Physiol Acad Sci Hung 30: 107–109

Heim T, Kellermeyer M, Dani M (1968) Thermal conditions and the mobilization of lipids from brown and white adipose tissue in the human neonate. Acta Paediatr Acad Sci Hung 9: 109–120

Hemon P (1976) Some aspects of rat metabolism in the brown adipose tissue of normal and hypothyroid rats during early postnatal development. Biol Neonate 28: 241–255

Hey EN, Katz G (1969) Temporary loss of a metabolic response to cold stress in infants of low birthweight. Arch Dis Child 44: 323–330

Hill JR, Rahimtulla KA (1965) Heat balance and the metabolic rate of new-born babies in relation to environmental temperature; and the effect of age and of weight on basal metabolic rate. J Physiol 180: 239-265

Hull D (1983) Brown adipose tissue in the newborn human infant. Int J Obes 7: 503

Jacobsson A, Stadler U, Glotzer MA, Kozak LP (1985) Mitochondrial uncoupling protein from mouse brown fat: molecular cloning, genetic mapping and mRNA expression. J Biol Chem 260: 16250-16254

Jacobsson A, Nedergaard J, Cannon B (1986) Alpha- and beta-adrenergic control of thermogenin mRNA expression in brown adipose tissue. Biosci Rep 6: 621-631

Jacobsson A, Cannon B, Nedergaard J (1987) Increased turnover rate of thermogenin mRNA in physiologically active brown adipose tissue. FEBS Lett 224: 353-356

Karlberg P, Moore RE, Oliver TK Jr (1962) The thermogenic response of the newborn infant to noradrenaline. Acta Paediatr 51: 284-292

Karlberg P, Moore RE, Oliver TK (1965) Thermogenic and cardiovascular responses of the newborn baby to noradrenaline. Acta Paediatr Scand 54: 225-238

Klein AH, Reviczky A, Padbury JF (1984) Thyroid hormones augment catecholamine-stimulated brown adipose tissue thermogenesis in the ovine fetus. Endocrinology 114: 1065-1069

Lean MEJ, James WPT (1983) Uncoupling protein in human brown adipose tissue mitochondria. Isolation and detection by specific antiserum. FEBS Lett 163: 235-240

Lindberg O (ed) (1970) Brown adipose tissue. Elsevier, New York

Merklin RJ (1974) Growth and distribution of human fetal brown fat. Anat Rec 178: 637-646

Mestyán J, Jarai I, Fekete M (1968) The total energy expenditure and its components in premature infants maintained under different nursing and environmental conditions. Pediatr Res 2: 161-171

Moragas A, Torán N (1983) Prenatal development of brown adipose tissue in man. A morphometric and biomathematical study. Biol Neonate 43: 80-85

Naeye RL (1974) Hypoxemia and the sudden infant death syndrome. Science 186: 837-838

Naeye RL (1976) Brain-stem and adrenal abnormalities in the sudden infant death syndrome. Am J Clin Pathol 66: 526-530

Nedergaard J, Cannon B (1984) Thermogenic mitochondria. In: Ernster L (ed) New comprehensive biochemistry (Bioenergetics). Elsevier, Amsterdam, pp 291-314

Nedergaard J, Lindberg O (1982) The brown fat cell. Int Rev Cytol 74: 187-286

Nedergaard J, Connolly E, Cannon B (1986) Brown adipose tissue in the mammalian neonate. In: Trayhurn P, Nicholls DG (eds) Brown adipose tissue. Arnold, London, pp 152-213

Nicholls DG, Locke RM (1984) Thermogenic mechanisms in brown fat. Physiol Rev 64: 1-64

Obregon MJ, Pitamber R, Jacobsson A, Nedergaard J, Cannon B (1987) Euthyroid status is essential for the perinatal increase in thermogenin mRNA in brown adipose tissue of rat pups. Biochem Biophys Res Comm 148: 9-14

Pawlikowski T (1955) The human brown adipose tissue. Folia Morphol 6: 209-216

Pribylová H, Znamenacek K (1964) Some aspects of thermoregulatory reactions in newborn infants during the first hours of life. Biol Neonate 6: 324-339

Ricquier D, Bouillaud F, Toumelin P, Mory G, Bazin R, Arch J, Pénicaud L (1986) Expression of uncoupling protein mRNA in thermogenic or weakly thermogenic brown adipose tissue. Evidence for a rapid beta-adrenoceptor-mediated and transcriptionally regulated step during activation of thermogenesis. J Biol Chem 261: 13905-13910

Rylander E (1972) Age dependent reactions of rectal and skin temperatures of infants during exposure to cold. Acta Paediatr Scand 61: 597-605

Seydoux J, Giacobino JP, Girardier L (1982) Impaired metabolic response to nerve stimulation in brown adipose tissue of hypothyroid rats. Mol Cell Endocrinol 25: 213-226

Smales ORC, Kime R (1978) Thermoregulation in babies immediately after birth. Arch Dis Child 53: 58-61

Steele RE, Wekstein DR (1972) Influence of thyroid hormone on homeothermic development of the rat. Am J Physiol 222: 1528-1533

Stern L, Lees MH, Leduc J (1965) Environmental temperature, oxygen consumption, and catecholamine excretion in newborn infants. Pediatrics 36: 367-373

Tanuma Y, Ohata M, Ito T, Yokochi C (1976) Possible function of human brown adipose tissue as suggested by observation on perirenal brown fats from necropsy cases of variable age groups. Arch Histol Jpn 39: 117–145
Trayhurn P, Nicholls DG (eds) (1986) Brown adipose tissue. Arnold, London
Valdes-Dapena MA, Gillant MM, Catherman R (1976) Brown fat retention in sudden infant death syndrome. Arch Pathol Lab Med 100: 547–549

Endocrine Control of
the Fetal Carbohydrate Metabolism

Carbohydrate Metabolism During Fetal Development

J. GIRARD[1]

During the last part of pregnancy a decline in maternal blood glucose concentration is observed in most species in the postabsorptive period or after an overnight fast (for a review, see Girard et al. 1984). Since the absolute rate of maternal glucose production and utilization is increased by 15%–70% in late pregnancy (for a review, see Leturque et al. 1987a), this suggests that the relative maternal hypoglycemia that appears in late pregnancy is related to an increased glucose utilization by the growing conceptus and a concomitant decrease in glucose utilization by maternal non-uterine tissues.

Impact of Pregnancy on Maternal Metabolism

To quantify the impact of the growing conceptus on the rate of glucose utilization by the mother, several investigators have measured maternal glucose utilization by tracer methodology (Kalhan et al. 1979; Leturque et al. 1981; Gilbert et al. 1982; Hay et al. 1984a) and uterine glucose utilization by the Fick principle (uterine blood flow × venous arterial glucose concentration difference across the uterus; Hay et al. 1983a; Peeters et al. 1984; Gilbert et al. 1984; Block et al. 1985; Johnson et al. 1986) or the radioactive 2-deoxyglucose technique (Leturque et al. 1986, 1987b; Hauguel et al. 1988). In singleton pregnant sheep at midgestation (71–81 days), uterine glucose utilization is 15 mg/min (Bell et al. 1986), i. e., 10% of maternal glucose utilization (Hay et al. 1983a). At term, approximately one-third of the maternal glucose utilization is accounted for by the uterus (Hay et al. 1983a; Fig. 1). In twin pregnancy in the sheep, uterine glucose utilization can represent 60% of maternal glucose utilization (Bergman et al. 1974). In other species studied, uterine glucose utilization near term represents 30%–50% of maternal glucose utilization (Fig. 1). In addition, the rate of glucose utilization by nonuterine maternal tissues is decreased in late pregnancy (Fig. 1). Thus, the pregnant animals supply glucose to their uterus at the expense of their own tissues. A shift towards an increased utilization of free fatty acids to replace glucose in nonuterine tissues (mainly skeletal muscles) is strongly supported by recent studies in pregnant sheep (Pethick et al. 1983).

[1] Centre de Recherches sur la Nutrition, 9 rue Jules Hetzel, 92190 Meudon-Bellevue, France

The Endocrine Control of the Fetus
Ed. by W. Künzel and A. Jensen
© Springer-Verlag Berlin Heidelberg 1988

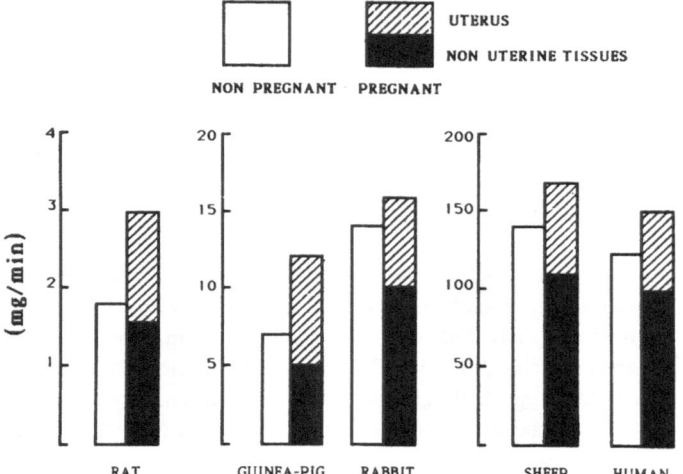

Fig. 1. Glucose utilization rates in non-pregnant and late pregnant females from several species (Leturque et al. 1987a)

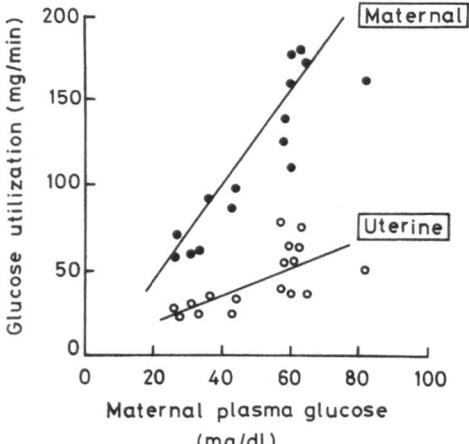

Fig. 2. Glucose utilization by maternal and uterine tissues as a function of maternal arterial plasma glucose concentration in pregnant sheep (adapted from Hay et al. 1983a)

Glucose utilization by the gravid uterus and by nonuterine maternal tissues is directly related to maternal arterial plasma glucose concentration in the sheep in late pregnancy (Hay et al. 1983a; Fig. 2). As maternal plasma insulin concentration varies with plasma glucose concentration, and as the placenta is endowed with a large number of insulin receptors (Posner 1974), it has been suggested that uterine glucose uptake could be regulated by maternal plasma insulin levels. Experiments in which insulin was infused into the mother to reach high levels ($> 300 \,\mu U/ml$)

while the maternal plasma glucose concentration was kept at normal levels by a variable infusion rate of glucose (glucose clamp technique) have shown that insulin has no short-term effect on uterine glucose uptake (Hay et al. 1984b). No studies have been performed addressing the question of chronic regulation of uterine glucose uptake by insulin.

Placental Glucose Metabolism and Transfer

Experiments performed in the sheep have shown that the placenta metabolizes a large fraction of the glucose that the mother delivers to the pregnant uterus. At midgestation, uteroplacental tissues use more than 80% of the glucose taken up by the uterus (Bell et al. 1986) and about 70% at term (Meschia et al. 1980). At midgestation, the rate of placental lactate production (3.5 mg/min) is much lower than at term (12 mg/min) despite a larger placental weight (490 g versus 300 g; Bell et al. 1986; Meschia et al. 1980). In addition, lactate is released almost entirely into the maternal circulation at midpregnancy (Bell et al. 1986), whereas in late gestation 60% of the lactate produced by the placenta is transferred to the fetus (Meschia et al. 1980). Lactate production by the placenta is also a characteristic found in all other species studied (Comline and Silver 1976; Peeters et al. 1984; Gilbert et al. 1984; Block et al. 1985; Hauguel et al. 1986; Johnson et al. 1986; Duée et al. 1987). Since the placenta also consumes large amounts of oxygen (Meschia et al. 1980; Bell et al. 1986), placental lactate production represents an intense aerobic metabolism designed to produce energy for the active transport of amino acids and micronutrients and for the synthesis and secretion of specific placental hormones.

The major determinant of the rate of placental glucose uptake is the maternal arterial plasma glucose concentration (Sparks et al. 1983; Fig.3). Glucose uptake by the uteroplacenta is directly related to maternal arterial plasma glucose concentration in the sheep in late pregnancy (Sparks et al. 1983).

The rate of glucose transfer from placenta to fetus depends primarily on the concentration of glucose in maternal and fetal arterial plasma and on the properties of the placental membrane (i.e., the number of glucose carriers) rather than on the rate of placental blood flow. Restriction of uterine blood flow has a negligible effect on placental glucose transfer as long as the mother is normoglycemic and the decrease in blood flow does not cause severe fetal hypoxia (Simmons et al. 1979).

In vivo experiments using the euglycemic-hyperinsulinemic clamp technique have shown that insulin does not affect placental glucose uptake, metabolism, and transfer in the sheep (Hay et al. 1984b; Rankin et al. 1986). In vitro studies using perfused term human placenta have also shown that glucose uptake and metabolism are not affected by insulin (Challier et al. 1986). By contrast, high levels of insulin (400 µU/ml) increase glucose utilization in the rat placenta by 30% on day 19 of gestation but are without effect on term placenta (21 days of gestation; Leturque et al. 1986, 1987c). Thus, short-term variations in maternal plasma insulin levels do

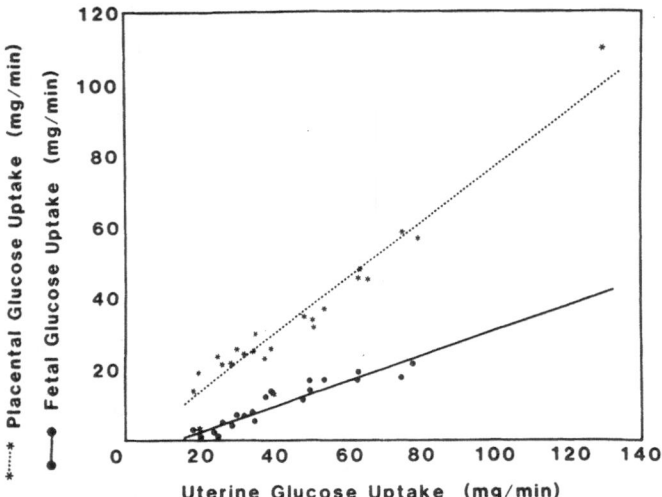

Fig.3. Glucose uptake by the uteroplacental tissues and the fetus as a function of uterine glucose uptake (Sparks et al. 1983)

not seem to increase placental glucose uptake, metabolism or transfer to a large extent at term, despite the presence of a large number of insulin receptors in the placenta. These receptors may be involved in the regulation of placental growth rather than in the short-term control of placental glucose metabolism.

Fetal Glucose Metabolism

Fetal Glucose Uptake

In large animals (cow, horse, sheep), net fetal glucose uptake from the placenta has been measured by the Fick principle method (Comline and Silver 1976; Silver and Comline 1976; Hay et al. 1981). The values average $4-6 \text{ mg} \cdot \text{min}^{-1} \cdot \text{kg}^{-1}$ fetal weight. More recently, the rate of fetal glucose utilization has been measured in the lamb by infusing [^{14}C]glucose or [^{3}H]glucose into the fetal circulation and using a three-pool model (Hay et al. 1981, 1984c). The rates of fetal glucose utilization were not statistically different from the rates of umbilical glucose uptake measured simultaneously in well-fed ewes. This suggests that the rates of glucose production by fetal liver and kidney are negligible in fetuses from well-fed mothers (for a review, see Girard 1986). In small animals (rat, rabbit) fetal glucose utilization has been estimated using radioactive 2-deoxyglucose (Leturque et al. 1986; Hauguel et al. 1988). The values obtained in the fetal rabbit are slightly higher ($9 \text{ mg} \cdot \text{min}^{-1} \cdot \text{kg}^{-1}$)

Table 1. Glucose utilization rates (mg·min^{-1}·kg^{-1}) in the placenta and individual tissues of near-term fetuses of several species

	Sheep	Rabbit[a]	Rat[a]
Whole fetus	5	9	20
Brain	53	9	30
Heart	63	10	60
Hind-limb muscles	4	4	20
Intestine	8	–	–
Liver	0	2	10
Adipose tissue	–	6	–
Placenta	100	10	30

[a] For rabbit and rat, the glucose metabolic index of individual tissues measured with radioactive 2-deoxy glucose was corrected using the lumped constant of adult tissue, i.e., 0.5 for the brain, 1 for muscles, 0.7 for adipose tissue, 0.8 for the whole fetus, 0.8 for the placenta (Leturque et al. 1986)

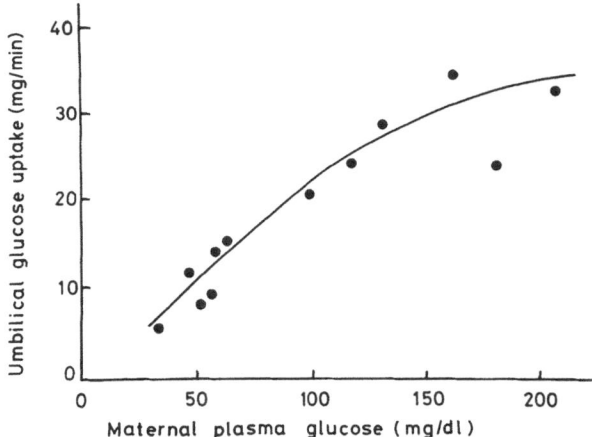

Fig. 4. Umbilical glucose uptake as a function of maternal arterial plasma glucose concentration (adapted from Simmons et al. 1979)

than in larger mammals, while the values obtained in the rat fetus are much higher (20 mg·min^{-1}·kg^{-1}; Table 1).

In sheep, fetal glucose uptake is directly related to the maternal arterial plasma glucose concentration (Hay et al. 1984d) but tends to plateau when glucose concentration is higher than 140 mg/dl (Simmons et al. 1979; Fig. 4), suggesting saturation of transplacental glucose carriers and/or pathways of fetal glucose utilization. Fetal glucose uptake depends primarily on maternal glucose concentration rather than on the rate of uterine blood flow. Fetal glucose uptake and arterial glucose concentra-

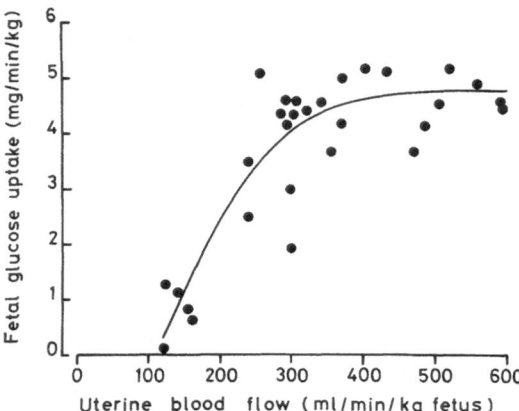

Fig. 5. Fetal glucose uptake as a function of placental blood flow (adapted from Wilkening et al. 1985)

tion remain normal as the uterine blood flow rate decreases from 600 to 300 ml· min^{-1}·kg^{-1} of fetus, if maternal blood glucose concentration is maintained constant by a glucose clamp (Wilkening et al. 1985; Fig. 5). At uterine blood flow rates lower than 300 ml·min^{-1}·kg^{-1} of fetus, the fetal glucose uptake decreases (Wilkening et al. 1985) and hypoxia causes fetal hyperglycemia due to an initial decrease in fetal glucose utilization followed by increased fetal glucose production (Jones et al. 1983). Studies in the fetal lamb have shown that the rate of glucose oxidation and the fraction of glucose oxidized are directly related to fetal plasma glucose concentration (Hay et al. 1983 a, b). The fraction of glucose oxidized averages 60%, suggesting that 40% of glucose taken up by the fetus is involved in glycogen and fatty acid synthesis and carbon accretion in fetal body. The rate of glucose utilization by individual organs can be estimated in the near-term fetal sheep, taking into account the difference in arteriovenous glucose concentration across those organs (Singh et al. 1984; Makowski et al. 1972; Fisher et al. 1980; Charlton et al. 1979; Wilkening et al. 1987) and fetal organ blood flows (Rudolph and Heymann 1970; Table 1). As fetal limb muscles represent 70% of fetal weight they used a large amount of the glucose (12 mg/min) supplied by the mother (18 mg/min). The brain (3 mg/min), the heart (1 mg/min), and the intestines (1 mg/min) constitute a smaller percentage of fetal glucose utilization at term in the lamb. More recently, the rates of glucose utilization by the placenta and individual organs of fetal rat (Leturque et al. 1987 b, c) and fetal rabbit (Hauguel et al. 1988) have been determined using radioactive 2-deoxyglucose (Table 1). In general, the rates of glucose utilization in the organs of fetal rabbit are lower than those measured in fetal sheep (except in muscles and liver), but the rates of glucose utilization in fetal rat tissues are much higher than in fetal sheep (except in brain and heart).

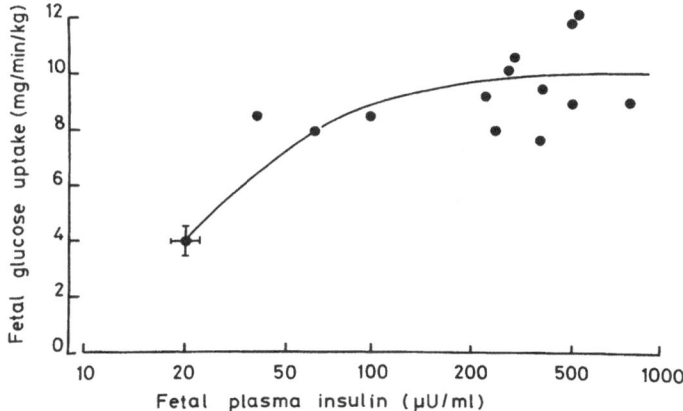

Fig. 6. Fetal glucose uptake as a function of fetal arterial plasma insulin concentration (adapted from Hay et al. 1985)

Effects of Insulin on Fetal Glucose Uptake

The response of fetal insulin secretion to changes in fetal glucose concentration has been studied in several species (Girard et al. 1974; Bassett and Madill 1974; Philipps et al. 1978; Fowden 1985). The plasma insulin concentration is directly related to fetal glycemia. Thus, insulin could be a prime regulator of fetal glucose utilization.

Insulin infusion into fetal lamb induces hypoglycemia, increased umbilical glucose uptake, and increased glucose utilization (Simmons et al. 1978; Bloch et al. 1986; Milley et al. 1986). However, as glucose concentration was lowered in these experiments it was not possible to make accurate estimates of the specific effects of insulin on fetal glucose utilization. To study the specific effects of insulin on fetal glucose metabolism, Hay et al. (1985) infused varying amounts of insulin into fetal lambs and maintained fetal glucose concentration at a constant level by a glucose infusion (glucose clamp). At constant glycemia, fetal glucose utilization and oxidation rates are directly related to plasma insulin concentration, reaching maximal rate at insulinemia greater than 100 µU/ml (Hay et al. 1985; Fig. 6). In addition, insulin increases fetal glucose oxidation rates by 80% (Hay and Meznarich 1986) and substitutes glucose oxidation for that of other substrates (mainly amino acids).

Pancreatectomy of the fetal lamb between 115 and 121 days of gestation leads to chronic insulin deficiency and hyperglycemia in utero and reduces fetal glucose uptake (Fowden et al. 1986). By contrast, fetal pancreatectomy has marginal effect on glucose uptake by the uterus and uteroplacental tissues (Fowden et al. 1986). In the rat fetus, hyperinsulinemia (180–260 µU/ml) increases glucose utilization by the heart and, to a lesser extent, skeletal muscles and liver, whereas the rate of glucose utilization by the brain remains unaffected (Leturque et al. 1987b).

Fetal Lactate Uptake and Utilization

Fetal lactate utilization has been determined recently by the infusion of [^{14}C]lactate into the fetal circulation (Sparks et al. 1982). Lactate utilization in the term fetal lamb (6 mg·min^{-1}·kg^{-1} fetal weight) is three times higher than the umbilical lactate uptake (2 mg·min^{-1}·kg^{-1} fetal weight). Thus, some fetal organs are lactate producers while some are net lactate consumers. Studies of net lactate flow across several organs have established that fetal hind limb (Singh et al. 1984; Wilkening et al. 1987) and to a smaller extent fetal brain (Jones et al. 1975) produce lactate, whereas fetal heart (Fisher et al. 1980), intestine (Charlton et al. 1979), and liver (Gleason et al. 1985) consume lactate. Of the lactate used by the fetus, 70% is oxidized (Hay et al. 1983 b), and the amount of lactate oxidized is directly related to fetal lactate concentration. The fraction of lactate that is not oxidized participates in carbon accretion in the fetal body and in liver glycogen and fatty acid synthesis.

References

Bassett JM, Madill D (1974) The influence of maternal nutrition on plasma hormone and metabolite concentrations of foetal lambs. J Endocrinol 61: 465–477

Bell AW, Kennaugh JM, Battaglia FC, Makowski EL, Meschia G (1986) Metabolic and circulatory studies of fetal lamb at midgestation. Am J Physiol 250: E538–E544

Bergman EN, Brockman RP, Kaufman CF (1974) Glucose metabolism in ruminants: comparison of whole-body turnover with production by gut, liver and kidneys. Fed Proc 33: 1849–1854

Bloch CA, Banach W, Landt K, Devaskar S, Sperling MA (1986) Effect of fetal insulin infusion on glucose kinetics in pregnant sheep: a compartmental analysis. Am J Physiol 251: E448–E456

Block SM, Sparks JW, Johnson RL, Battaglia FC (1985) Metabolic quotients of the gravid uterus of the chronically catheterized guinea-pig. Pediat Res 19: 840–845

Challier JC, Hauguel S, Desmaizières V (1986) Effect of insulin on glucose uptake and metabolism in the human placenta. J Clin Endocrinol Metab 62: 803–807

Charlton VE, Reis BL, Lofgren DJ (1979) Consumption of carbohydrates, amino acids and oxygen across the intestinal circulation in the fetal sheep. J Dev Physiol 1: 329–336

Comline RS, Silver M (1976) Some aspects of foetal and utero-placental metabolism in cows with indwelling umbilical and uterine vascular catheters. J Physiol 260: 571–586

Duée PH, Simoes-Nunes C, Pégorier JP, Gilbert M, Girard J (1987) Uterine metabolism of the conscious gilt during late pregnancy. Pediatr Res 22: 587–590

Fisher DJ, Heymann MA, Rudolph AM (1980) Myocardial oxygen and carbohydrate consumption in fetal lambs in utero and in adult sheep. Am J Physiol 238: H399–H405

Fowden AL (1985) Pancreatic endocrine function and carbohydrate metabolism in the fetus. In: Albrecht ED, Pepe GJ (eds) Research in perinatal medicine. Perinatology Press, Ithaca, pp 71–90

Fowden AL, Silver M, Comline RS (1986) The effect of pancreatectomy on the uptake of metabolites by the sheep fetus. Q J Exp Physiol 71: 67–78

Gilbert M, Sparks JW, Girard J, Battaglia FC (1982) Glucose turnover rate during pregnancy in the conscious guinea-pig. Pediatr Res 16: 310–313

Gilbert M, Hauguel S, Bouisset M (1984) Uterine blood flow and substrate uptake in conscious rabbit during late gestation. Am J Physiol 247: E574–E580

Girard J (1986) Gluconeogenesis in late fetal and early neonatal life. Biol Neonate 50: 237-258

Girard JR, Kervran A, Soufflet E, Assan R (1974) Factors affecting the secretion of insulin and glucagon by the rat fetus. Diabetes 23: 310-317

Girard J, Leturque A, Burnol A-F, Ferré P, Satabin P, Gilbert M (1984) Glucose homeostasis in the rat. In: Shafrir E, Renold AE (eds) Lessons from animal diabetes. Libbey, London, pp 667-675

Gleason CA, Roman C, Rudolph AM (1985) Hepatic oxygen consumption, lactate uptake, and glucose production in neonatal lambs. Pediatr Res 19: 1235-1239

Hauguel S, Desmaizières V, Challier JC (1986) Glucose uptake, utilization and transfer by the human placenta as function of maternal glucose concentration. Pediatr Res 20: 269-273

Hauguel S, Leturque A, Gilbert M, Kandé J, Girard J (1987) Glucose utilization by the placenta and fetal tissue in fed and fasted pregnant rabbits. Pediatr Res (in press)

Hay WW Jr, Meznarich HK (1986) The effect of hyperinsulinaemia on glucose utilization and oxidation and on oxygen consumption in the fetal lamb. Q J Exp Physiol 71: 689-698

Hay WW Jr, Sparks JW, Quissell BJ, Battaglia FC, Meschia G (1981) Simultaneous measurements of umbilical uptake, fetal utilization rate and fetal turnover rate of glucose. Am J Physiol 240: E662-E668

Hay WW Jr, Sparks JW, Wilkening RB, Battaglia FC, Meschia G (1983a) Partition of maternal glucose production between conceptus and maternal tissues in sheep. Am J Physiol 245: E347-E350

Hay WW Jr, Myers SA, Sparks JW, Wilkening RB, Meschia G, Battaglia FC (1983b) Glucose and lactate oxidation rates in the fetal lamb. Proc Soc Exp Biol Med 173: 553-563

Hay WW Jr, Gilbert M, Johnson RL, Battaglia FC (1984a) Glucose turnover rates in chronically catheterized non-pregnant and pregnant rabbits. Pediatr Res 18: 276-280

Hay WW Jr, Sparks JW, Gilbert M, Battaglia FC, Meschia G (1984b) Effect of insulin on glucose uptake by the maternal hindlimb and uterus, and by the fetus in conscious pregnant sheep. J Endocrinol 100: 119-124

Hay WW Jr, Sparks JW, Battaglia FC, Meschia G (1984c) Maternal-fetal glucose exchange: necessity of a three-pool model. Am J Physiol 246: E526-E534

Hay WW Jr, Sparks JW, Wilkening RB, Battaglia FC, Meschia G (1984d) Fetal glucose uptake and utilization as functions of maternal glucose concentration. Am J Physiol 246: E237-E242

Hay WW Jr, Meznarich HK, Sparks JW, Battaglia FC, Meschia G (1985) Effect of insulin on glucose uptake in near-term fetal lambs. Proc Soc Exp Biol Med 178: 557-564

Johnson RL, Gilbert M, Block SM, Battaglia FC (1986) Uterine metabolism of the pregnant rabbit under chronic steady-state conditions. Am J Obstet Gynecol 154: 1146-1151

Jones CT, Ritchie JWK, Walker D (1983) The effects of hypoxia on glucose turnover in the fetal sheep. J Dev Physiol 5: 223-235

Jones MD Jr, Burd LI, Makowski EL, Meschia G, Battaglia FC (1975) Cerebral metabolism in sheep: a comparative study of the adult, the lamb, and the fetus. Am J Physiol 229: 235-239

Kalhan SC, D'Angelo LJ, Savin SM, Adam PAJ (1979) Glucose production in pregnant women at term gestation. Sources of glucose for human fetus. J Clin Invest 63: 388-394

Leturque A, Gilbert M, Girard J (1981) Glucose turnover during pregnancy in anesthetized post-absorptive rats. Biochem J 196: 633-636

Leturque A, Ferré P, Burnol A-F, Kandé J, Maulard P, Girard J (1986) Glucose utilization rates and insulin sensitivity in vivo in tissues of virgin and pregnant rats. Diabetes 35: 172-177

Leturque A, Hauguel S, Ferré P, Girard J (1987a) Glucose metabolism in pregnancy. Biol Neonate 51: 64-69

Leturque A, Revelli JP, Hauguel S, Kandé J, Girard J (1987b) Hyperglycemia and hyperinsulinemia increase glucose utilization in fetal rat tissues. Am J Physiol 253: E616-E620

Leturque A, Hauguel S, Kande J, Girard J (1987c) Glucose utilization by the placenta of anaesthetized rats: effects of insulin, glucose and ketone bodies. Pediatr Res 22: 483–487

Makowski EL, Schneider JM, Tsoulos NG, Colwill JR, Battaglia FC, Meschia G (1972) Cerebral blood flow, oxygen consumption and glucose utilization of fetal lambs in utero. Am J Obstet Gynecol 114: 292–303

Meschia G, Battaglia FC, Hay WW, Sparks JW (1980) Utilization of substrates by the ovine placenta in vivo. Fed Proc 39: 245–249

Milley JR, Papacostas JS, Tabata BK (1986) Effect of insulin on uptake of metabolic substrates by the sheep fetus. Am J Physiol 251: E349–E356

Peeters LLH, Martenson L, Van Kreel BK, Wallenburg HCS (1984) Uterine arterial and venous concentrations of glucose, lactate, ketones, free fatty acids and oxygen in the awake pregnant guinea pig. Pediatr Res 18: 1172–1175

Pethick DW, Lindsay DW, Barker PJ, Northrop AJ (1983) The metabolism of circulating nonesterified fatty acids by the whole animal, hind-limb muscle and uterus of pregnant ewes. Br J Nutr 49: 129–143

Philipps AF, Carson BS, Meschia G, Battaglia FC (1978) Insulin secretion in fetal and newborn sheep. Am J Physiol 235: E34–E38

Posner BI (1974) Insulin receptors in human and animal placental tissues. Diabetes 23: 209–217

Rankin JHG, Jodarski G, Shanahan MR (1986) Maternal insulin and placental 3-O-methyl-glucose transport. J Dev Physiol 8: 247–253

Rudolph AM, Heymann MA (1970) Circulatory changes during growth in the fetal lamb. Circ Res 26: 289–299

Silver M, Comline RS (1976) Fetal and placental O_2 consumption and the uptake of different metabolites in the ruminant and horse during late gestation. In: Reneau DD, Grote J (eds) Oxygen transport to tissues. Plenum, New York, pp 731–736

Simmons MA, Jones MD Jr, Battaglia FC, Meschia G (1978) Insulin effect on fetal glucose utilization. Pediatr Res 12: 90–92

Simmons MA, Battaglia FC, Meschia G (1979) Placental transfer of glucose. J Dev Physiol 1: 227–243

Singh S, Sparks JW, Meschia G, Battaglia FC, Makowski EL (1984) Comparison of fetal and maternal hind limb metabolic quotients in sheep. Am J Obstet Gynecol 149: 441–449

Sparks JW, Hay WW Jr, Bonds D, Meschia G, Battaglia FC (1982) Simultaneous measurements of lactate turnover and umbilical lactate uptake in the fetal lamb. J Clin Invest 70: 179–192

Sparks JW, Hay WW Jr, Meschia G, Battaglia FC (1983) Partition of maternal nutrients to the placenta and fetus in the sheep. Eur J Gynec Reprod Biol 14: 331–340

Wilkening RB, Battaglia FC, Meschia G (1985) The relationship of umbilical glucose uptake to uterine blood flow. J Dev Physiol 7: 313–319

Wilkening RB, Molina RD, Battaglia FC, Meschia G (1987) Effect of insulin on glucose/oxygen and lactate/oxygen quotients across the hind limb of fetal lambs. Biol Neonate 51: 18–23

Relationship Between Alterations in Uterine Blood Flow and the Handling of Glucose by Fetus and Placenta

C. T. JONES[1]

Introduction

The transport of glucose across the placenta involves a carrier exhibiting the kinetic characteristics of facilitated diffusion (Widdas 1952). It ensures a close relationship between fetal and maternal plasma glucose concentration, although in species such as the sheep fetal values are much below those in the maternal circulation (Tsoulos et al. 1971; Simmons et al. 1979). This has led to the suggestion that the placenta in the sheep, at least, is relatively impermeable to glucose. However, such a view may be considered to be simplistic as it has become increasingly obvious that a large proportion of the glucose taken up by the placenta is for its own use and not for transport to the fetus (Hay et al. 1984; Sparks et al. 1983; Jones et al. 1987). Moreover, a substantive proportion of that used by the placenta is derived from the fetal rather than the maternal circulation. Hence a potentially important factor in the regulation of glucose supply to the fetus is partition in the placenta of glucose metabolism between transport and consumption.

Transport of glucose exhibits Michaelis-Menton kinetics with an apparent K_m of about 4 mM. At this concentration glucose transport has been reported to be about 1 mmol/min (Simmons et al. 1979; Jones et al. 1987), i.e. around 10%–20% of the normal uterine glucose supply to the placenta. At this level of extraction a fall in uterine blood flow and hence in glucose supply could increase fractional extraction to the extent that glucose concentration on passage through the uteroplacental circulation is reduced and hence uptake capacity falls. In theory there are two possible consequences of such a fall in supply. It might depress placental utilisation, although this is unlikely as its metabolism is particularly dependent on glucose supply. Alternatively, the availability of glucose to the fetus could be reduced.

Therefore the object of the studies reported here is to evaluate the effects of changes in uterine blood flow on the partition of substrate utilisation between fetus and placenta.

[1] Laboratory of Cellular and Developmental Physiology, The Institute for Molecular Medicine, University of Oxford, Oxford OX3 9DS, UK

The Endocrine Control of the Fetus
Ed. by W. Künzel and A. Jensen
© Springer-Verlag Berlin Heidelberg 1988

Materials and Method

In the present experiments sheep 128–137 days pregnant and with chronically implanted fetal and maternal vascular catheters were used. The techniques for measurement of glucose turnover using [^{14}C]glucose have been described elsewhere (Jones et al. 1987). Uterine blood flow was reduced either by application of a 'cuff catheter' around the common uterine artery (Gu et al. 1985) or by infusion of adrenaline at 0.5 µg/min per kilogram of maternal body weight into the maternal arterial circulation (Gu and Jones 1986).

Results

Effects of Normal Fluctuations in Uterine Blood Flow

One of the striking features of glucose balance across the normally perfused sheep placenta is that from time to time net output of glucose from the placenta and fetus is observed. Such apparently random and negative arteriovenous differences across both sides of the placenta can be explained if related to natural variations in uterine blood flow or, more particularly, to supply of glucose to the uterus. Hence there is a remarkably close relationship between the net uteroplacental consumption of glucose and the uterine supply (Fig. 1) such that when supply to the placenta falls be-

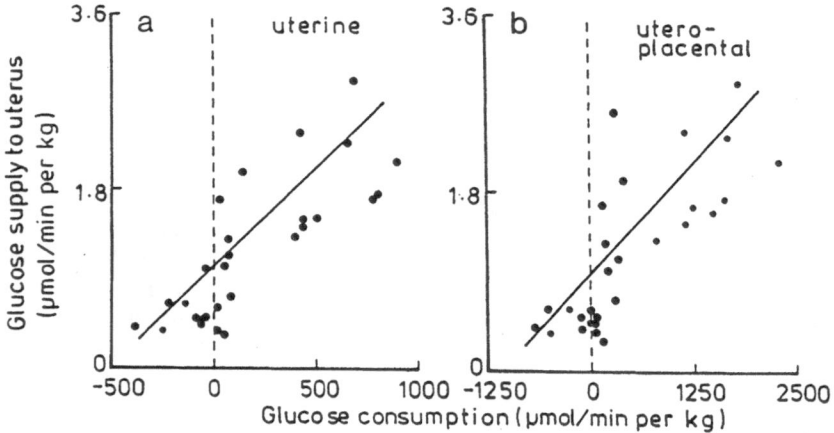

Fig. 1a, b. The relationship between net glucose uptake across the uterine circulation and glucose supply to the uterus in sheep 125–135 days pregnant. **a** Relationship between glucose supply to the uterus (blood flow × arterial plasma concentration) and net consumption of glucose by the uterus and its contents ($y = 1.67x + 1207$; $r = 0.793$; $p < 0.001$). **b** Relationship between glucose supply to the uterus and uteroplacental net consumption (excluding fetal uptake) of glucose ($y = 0.78x + 1257$; $r = 0.786$; $p < 0.001$). (From Jones et al. 1987)

low 5-10 mmol/min net output into the maternal circulation from the uteroplacenta is seen (Fig. 1).

This net output, although arising to a limited extent from the small rate of *de novo* glucose synthesis by the placenta and from mobilization of the limited pools of placental glucose, is derived mostly from the fetus. Hence studies investigating the fate of [14C]glucose infused into the maternal or fetal circulation demonstrate a substantial placental metabolism of glucose derived from the umbilical circulation (Fig. 2). Moreover, the extent to which the fetus supplies glucose to the placenta relates closely to the uterine supply (Fig. 2). Hence as the latter falls the fetus increases its contribution to placental glucose metabolism from 5%-10% to at least 40%. It is notable that the point at which the fetus starts making a major contribution to placental glucose consumption is when net extraction of glucose from the uterine circulation is at least about 20% and hence significant diminution of glucose concentration in the uteroplacental vascular bed would be expected.

When glucose supply to the uteroplacenta is not at the low end of the normal range the fetal supply of glucose is insufficient to maintain placental oxygen consumption, the latter being largely sustained by glucose (Jones et al. 1987). This pla-

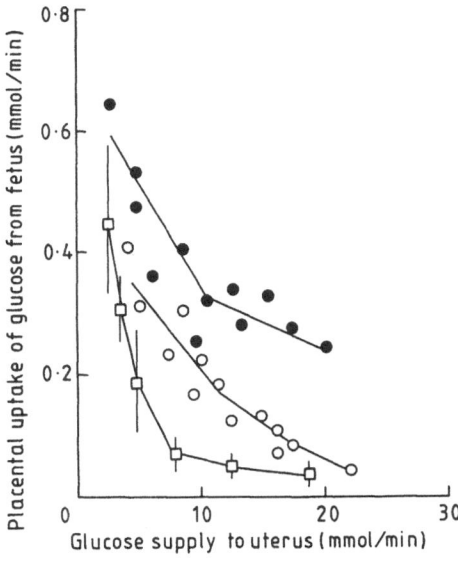

Fig. 2. The relationship in fetal sheep at 127-135 days' gestation between glucose supply to the uterus and placental uptake of glucose from the fetus in ewes with normal uterine blood flow (□); ewes in which uterine blood flow was reduced to 35%-44% of normal by application of an occlusion catheter around the common uterine artery for 60 min (O); and in ewes in which uterine blood flow was reduced to 60%-67% of control by infusion of adrenaline at 0.5 μg/min per for 60 min (●). Uptake of glucose was calculated from umbilical arteriovenous differences for [14C]glucose (Jones et al. 1987). Values are means ± SD for 4-8 separate experiments

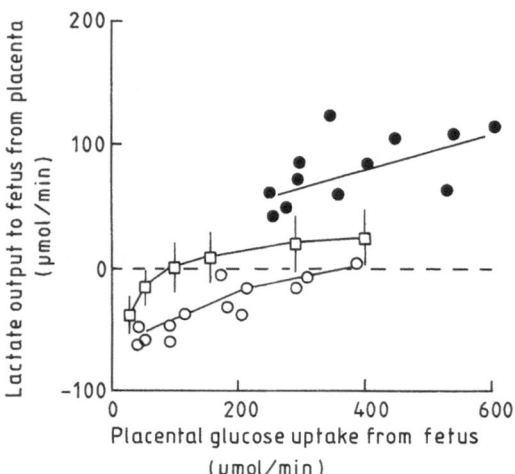

Fig. 3. Relationship between the uptake of glucose by the placenta from the fetus and placental lactate production to the fetus from fetally derived glucose. Glucose uptake was calculated from umbilical arteriovenous difference for [U-^{14}C]glucose after a constant intrafetal infusion for 2 h (Jones et al. 1987). Lactate production was calculated from [^{14}C]lactate balance across the umbilical circulation on the assumption that the pool of placental glucose from which lactate was derived had the same specific radioactivity as that in the fetal artery (Jones et al. 1987). Other details as for Fig. 2

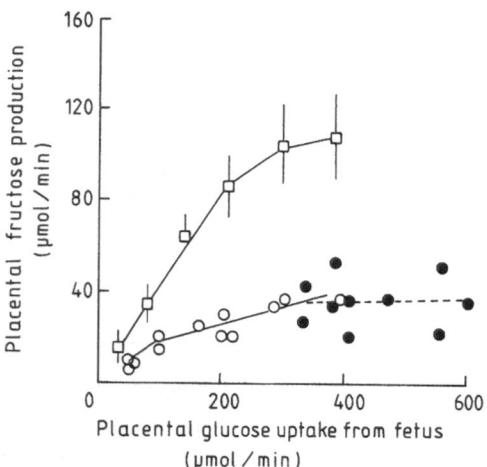

Fig. 4. Relationship between the output of fructose to the fetus from the placenta and the uptake of glucose by the placenta from the fetus. The substrate was [U-^{14}C]glucose infused into the fetal circulation (Jones et al. 1987). Other details are as for Figs. 2 and 3

cental glucose must therefore have another fate. Previous studies have shown that at normal uterine blood flow rates over 90% of the fetally derived glucose taken up by the placenta is converted to lactate and fructose (Jones et al. 1987; Gu and Jones 1986; Sparks et al. 1982). Moreover, the lactate and fructose output to the fetus are coupled closely to the rate of placental consumption of glucose from the fetus and hence to its supply to the uteroplacenta (Figs. 3, 4). As glucose supply to the placenta from the uterine circulation falls, and uptake from the fetus rises, there is a progressive increase in placental output of lactate and fructose. A high rate of uteroplacental supply of glucose is associated with lactate extraction by the placenta from the fetus and low rates of placental lactate production (Figs. 3, 4).

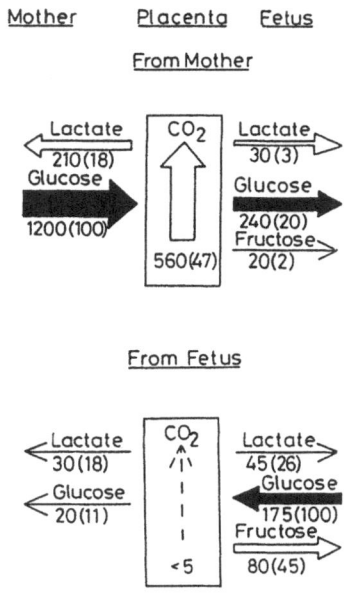

Flux = μmol glucose/min
(% utilization)

Fig. 5. Flux and fate of glucose taken up by the placenta from the fetus and from the ewe of sheep 135–137 days pregnant. The values are for a uterine glucose supply of 3–10 mmol/min. Glucose uptake was calculated from umbilical arteriovenous difference for [U-14C]glucose; lactate, fructose and glucose output was calculated on the assumption that this pool of placental glucose had the same specific radioactivity as that in the fetal arterial plasma. Net glucose uptake from the maternal circulation was calculated from uterine arteriovenous difference; lactate output to the ewe from this pool was estimated from net output less that for lactate output from fetal glucose; a similar calculation was made for lactate output to the fetus from maternal glucose. Glucose and fructose output to the fetus from the maternal glucose pool was estimated from umbilical arteriovenous difference, after allowing for the contribution that uptake or output from fetal glucose makes to these. Placental conversion of maternal glucose to CO_2 was estimated from uteroplacental oxygen comsumption on the assumption that this is sustained entirely by glucose. (From Jones et al. 1987)

What therefore is the logic of glucose being supplied to the placenta from the fetal circulation and then being converted largely to fructose and lactate for transport back to the fetus? This can be deduced by studying the gross fluxes for glucose transport across and utilization by the placenta (Fig. 5). With uteroplacental glucose supply in the middle of the normal range high net fluxes are maintained across both sides of the placenta, with only about 25% of the glucose taken up from the maternal circulation finding its way to the fetus and the remainder being oxidized in the placenta or converted to lactate. The placental uptake of glucose from the fetus is not much less than that transported to it from the mother, but surprisingly little is transported back into the maternal circulation, as most is metabolized to lactate and fructose. Hence, net glucose flux from both sides of the placenta is kept to a small proportion of total flux because of high rates of placental metabolism. For the fetus this means that glucose, which readily crosses the placenta, can be more effectively constrained to the fetal compartment for its own metabolism (Jones et al. 1987). For the placenta, which needs to maintain a fairly constant and high rate of glucose consumption, it means that during periods of depressed uterine blood flow the placenta can supplement its glucose needs by calling on the fetus; to meet this need the fetus maintains a constant and rapid high rate of glucose turnover and recycling.

What therefore happens when uterine blood flow is reduced?

Effects of Reduced Uterine Blood Flow on Uteroplacental Glucose Balance

When uterine blood flow is reduced by compression of the common uterine artery, uteroplacental oxygen consumption is unaltered whilst net consumption of glucose is reduced to 25% of normal (Gu et al. 1985). Under these circumstances the proportion of instances in which net output of glucose to the ewe was apparent did not increase (Gu et al. 1985). The reduced uterine blood flow was associated with an increased placental consumption of glucose from the fetus, so that even at normal rates of glucose supply to the uteroplacenta uptake from the fetus was higher than normal (Fig. 2). The enhanced extraction of glucose from the fetus is not reflected in an elevation of fructose production, which remains low (Fig. 4). Furthermore, there is little evidence of significant output of lactate to the fetus, as the placenta now becomes a substantial consumer of lactate produced by the fetus (Fig. 3; Table 1) in response to the associated hypoxaemia and reduced fetal oxygen consumption (Gu et al. 1985).

Net fluxes in this situation are summarized in Fig. 6. The sharp reduction in net glucose flux to the placenta is unable to sustain placental oxygen consumption, but gross maternal-to-fetal glucose flux is sustained if not increased. Gross glucose flux from fetus to placenta is also increased, but as placental lactate and fructose production from this glucose is much depressed, loss of this glucose to the maternal circulation rises. This leads to a substantial shortfall in substrate to sustain placental oxygen consumption. It is made up by the large increase in lactate production by the fetus and uptake by the placenta. The lactate output is sufficient, and required,

Table 1. Production of lactate by the placenta during reduced uterine blood flow in sheep 127–135 days pregnant

	Placental lactate production to fetus (µmol/min)		
	From fetal glucose	Net umb. AV balance	Gross output to fetus
Control	23.9 ± 8.4	103 ± 29	103
Compression of uterine artery	−29.6 ± 8.9	−914 ± 420	−914
Maternal adrenaline	74.3 ± 29.6	−310 ± 146.2	−385

Results are means ± SD from 8–12 experiments. Lactate production from fetal glucose was determined by constant infusion of [U-^{14}C]glucose into the fetus and measurement of conversion rates to lactate. Net output was determined by measurement of umbilical arteriovenous difference for lactate. Gross output to the fetus was calculated from net output and rates of [^{14}C]lactate production. Adrenaline was infused into the maternal circulation at 0.5 µg/min per kg body wt. to reduce uterine blood flow to 63.6 ± 5.3% of control. Alternatively the common uterine artery was compressed to reduce blood flow to 37.3 ± 5.1% of control

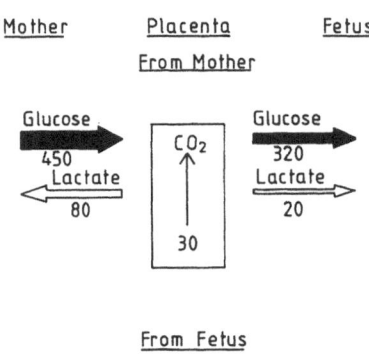

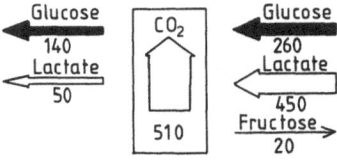

Fig. 6. Flux and fate of glucose taken up by the placenta from the fetus and ewe in sheep 127–137 days pregnant when uterine blood flow is reduced to 35%–44% of control by application of a compression catheter around the common uterine artery (Gu et al. 1985). Other details as for Fig. 5

to maintain most of the placental oxygen consumption when uterine blood flow is reduced sharply by compression of the uterine artery.

Uterine blood flow can also be reduced by infusion of adrenaline into the maternal circulation at about 0.5 µg/min per kg body wt. (Gu et al. 1986; Rosenfeld et al. 1976). When this was used the depression of flow to 60%–67% of normal was

Mother   Placenta   Fetus

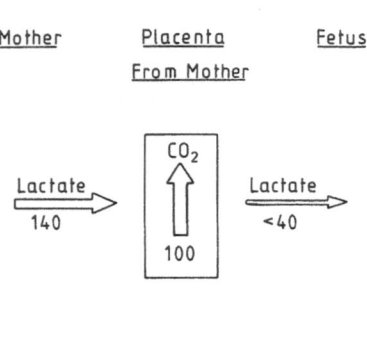

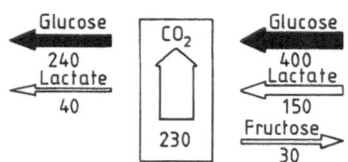

Flux = μmol glucose equiv./min

Fig. 7. Flux and fate of glucose taken up by the placenta from the fetus and ewe of sheep 125–136 days pregnant during reduction of uterine blood flow to 60%–67% of control through maternal intravenous infusion of adrenaline at 0.5 μg/min per kg body wt. (Gu and Jones 1986). Other details as for Fig. 5

less than that achieved by compression of the common uterine artery at 35%–44% of control values (Gu and Jones 1986; Gu et al. 1985). Despite this the uteroplacental effects of reducing flow with maternal adrenaline infusion were much more pronounced.

In contrast to the effects of uterine artery compression, maternal adrenaline infusion reduced uteroplacental oxygen consumption by 40%, whilst fetal consumption was unaffected (Gu and Jones 1986; Gu et al. 1985). In spite of this there was little if any net consumption of uterine glucose by the uteroplacenta, but substantial output to the ewe (Gu and Jones 1986). The associated elevation of fetal plasma glucose concentration occurred with a simultaneous marked rise in placental extraction of glucose from the fetus, such that the rates of uptake were high over a wide range of glucose supply rates to the uterus (Fig. 2). This was associated with enhanced placental conversion of fetal glucose to lactate but not to fructose (Figs. 3, 4). Despite the increase in placental conversion of fetal glucose to lactate, the rise in fetal plasma lactate concentration led to enhanced gross uptake of lactate from the fetus by the placenta (Table 1).

Summarizing the fluxes at the placenta when uterine blood flow is reduced by maternal adrenaline infusion (Fig. 7), there is little evidence of significant net uptake of glucose from the uterine circulation. This is supported by the fact that despite the placenta being hypoxic, it becomes a net consumer and not producer of lactate at the uterine circulation. The only site of net glucose supply is uptake from

Table 1. Production of lactate by the placenta during reduced uterine blood flow in sheep 127-135 days pregnant

	Placental lactate production to fetus (μmol/min)		
	From fetal glucose	Net umb. AV balance	Gross output to fetus
Control	23.9± 8.4	103± 29	103
Compression of uterine artery	−29.6± 8.9	−914±420	−914
Maternal adrenaline	74.3±29.6	−310±146.2	−385

Results are means ± SD from 8-12 experiments. Lactate production from fetal glucose was determined by constant infusion of [U-^{14}C]glucose into the fetus and measurement of conversion rates to lactate. Net output was determined by measurement of umbilical arteriovenous difference for lactate. Gross output to the fetus was calculated from net output and rates of [^{14}C]lactate production. Adrenaline was infused into the maternal circulation at 0.5 μg/min per kg body wt. to reduce uterine blood flow to 63.6±5.3% of control. Alternatively the common uterine artery was compressed to reduce blood flow to 37.3±5.1% of control

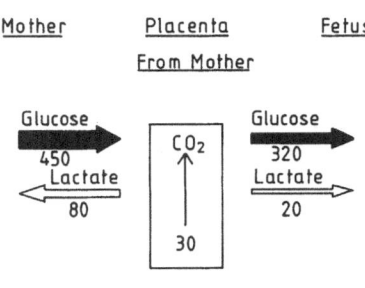

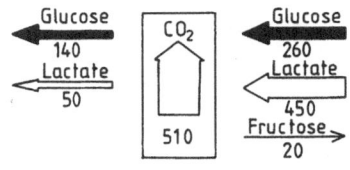

Fig. 6. Flux and fate of glucose taken up by the placenta from the fetus and ewe in sheep 127-137 days pregnant when uterine blood flow is reduced to 35%-44% of control by application of a compression catheter around the common uterine artery (Gu et al. 1985). Other details as for Fig. 5

Flux = μmol glucose equiv./min

to maintain most of the placental oxygen consumption when uterine blood flow is reduced sharply by compression of the uterine artery.

Uterine blood flow can also be reduced by infusion of adrenaline into the maternal circulation at about 0.5 μg/min per kg body wt. (Gu et al. 1986; Rosenfeld et al. 1976). When this was used the depression of flow to 60%-67% of normal was

this does not provide a mechanism for sensing changes in glucose supply and transmitting that information to the fetus, the latter being essential to intiate hepatic output of glucose in the fetus. Therefore we must search for a local endocrine system that is capable of modulating intraplacental glucose utilization and inducing changes in the intraplacental compartmentation of metabolism, and which intiates metabolic changes in the fetus mobilizing glycogen stores, enhancing lactate output and reducing fetal glucose utilization.

In conclusion, it is unwise to consider transport of glucose to the fetus independent of placental consumption, which clearly has profound effects on fetal metabolism, particularly when uterine blood flow is altered.

References

Gu W, Jones CT, Parer JT (1985) Metabolic and cardiovascular effects of sustained reduction of uterine blood flow. J Physiol 368: 100-129

Gu W, Jones CT (1986) The effect of elevation of maternal plasma catecholamines on fetus and placenta of pregnant sheep. J Dev Physiol 8: 173-186

Hay WW, Myers SA, Sparks JW, Wilkening RB, Meschia G, Battaglia FC (1983) Glucose and lactate oxidation rates in the fetal lamb. Proc Soc Exper Biol Med 173: 553-563

Hay WW, Sparks JW, Battaglia FC, Meschia G (1984) Maternal-fetal glucose exchange: necessity of a 3 pool model. Am J Physiol 246: E528-E534

Jones CT, Gu W, Harding JE (1987) Metabolism of glucose by fetus and placenta in sheep. The effects of normal fluctuations in uterine blood flow. J Dev Physiol 9: 369-389

Rosenfeld CR, Barton MD, Meschia G (1976) Effects of epinephrine on the distribution of blood flow in the ewe. Am J Obstet Gynecol 124: 156-163

Simmons MA, Battaglia FC, Meschia G (1979) Placental transfer of glucose. J Dev Physiol 1: 227-243

Sparks JW, Hay WW, Bonds DR, Meschia G, Battaglia FC (1982) Simultaneous measurements of lactate turnover rate and umbilical lactate uptake in the fetal lamb. J Clin Invest 70: 179-192

Sparks JW, Hay WW, Meschia G, Battaglia FC (1983) Partition of maternal nutrients to the placenta and fetus in sheep. Eur J Obstet Gynecol Reprod Biol 14: 331-340

Tsoulos NG, Colwill JR, Battaglia FC, Makowski EL, Meschia G (1971) Comparison of glucose, fructose and O_2 uptakes by fetuses of fed and starved ewes. Am J Physiol 221: 234-237

Widdas WF (1952) Inability of diffusion to account for placental glucose transfer in the sheep and consideration of the kinetics of a possible carrier transfer. J Physiol 118: 23-39

Wilkening RB, Anderson S, Martensson L, Meschia G (1982) Placental transfer as a function of uterine blood flow. Am J Physiol 242: H429-H436

Wilkening RB, Battaglia FC, Meschia G (1985) The relationship between umbilical glucose uptake to uterine blood flow. J Dev Physiol 7: 313-319

Pancreatectomy and Fetal Carbohydrate Metabolism

A. L. FOWDEN[1]

Introduction

In adult animals the pancreas has an important role in regulating carbohydrate metabolism. The pancreatic hormones insulin and glucagon maintain the circulating glucose concentration by controlling hepatic glucogenesis and peripheral glucose uptake. Removal of the pancreas in adult animals disturbs the balance between glucose supply and utilization and leads to hyperglycaemia and severe diabetes if the animals are not treated with insulin.

The fetal pancreas is active in utero and secretes insulin and glucagon into the circulation from early in gestation (Fowden 1985). However, the role of the pancreas in the regulation of fetal carbohydrate metabolism is not so immediately obvious, as glucose is supplied to the fetus at a relatively constant rate under normal conditions. Administration of exogenous insulin and glucagon to chronically catheterized fetuses has shown that these hormones can alter fetal glucose levels and influence the rates of glucose utilization and production in utero (Fowden 1985). However, because these studies used doses of hormone outside the physiological range, they provide little information about the role of the endogenous pancreatic hormones.

Fetal pancreatectomy is a feasible means of removing the endogenous hormones without the problems associated with the administration of antisera or diabetogenic drugs (Fowden 1987). The full extent of the endocrine deficiency produced by fetal pancreatectomy has still not been determined, but hypoinsulinaemia develops within 24 h of surgery and is maintained throughout gestation without any evidence of pancreatic regrowth (Table 1; Fowden and Comline 1984). The aims of this review are twofold; first, to consider the range of effects that pancreatic ablation has on fetal carbohydrate metabolism and, secondly, to discuss the consequences that these changes have on fetal growth and development.

[1] The Physiological Laboratory, Downing Street, Cambridge, CB2 3EG, UK

The Endocrine Control of the Fetus
Ed. by W. Künzel and A. Jensen
© Springer-Verlag Berlin Heidelberg 1988

Table 1. Mean ($\pm$ SE) arterial concentrations of plasma insulin, glucose, lactate and fructose in pancreatectomized and sham-operated fetuses, and mean weights of pancreas removed at surgery and at post mortem after delivery. (Data from Fowden and Comline 1984; Fowden et al. 1986a; Fowden 1987)

	Wt pancreas removed (mg) at		Plasma concentration			
	Surgery	Delivery	Insulin μU/ml	Glucose mmol/l	Lactate mmol/l	Fructose mmol/l
Pancreatectomized	884±39 (24)	53±15* (24)	6.4±0.6* (24)	0.95±0.04* (24)	2.62±0.12* (24)	7.01±0.40* (16)
Sham-operated	0 (10)	2297±225 (8)	17.2±1.0 (10)	0.59±0.04 (10)	1.56±0.12 (10)	3.70±0.23 (7)

Number of animals in parentheses
* Significantly different from sham-operated fetuses; $p < 0.05$

Fetal Plasma Glucose Concentrations

Pancreatectomy of the fetus leads to fetal hyperglycaemia (Table 1), which can be abolished by insulin infusion for 24 h (Fowden et al. 1986a). Fetal glucose levels rise within 24–48 h of surgery and then remain virtually unchanged until just before delivery (Fowden and Comline 1984). Normally, fetal glucose levels are 20%–30% of the maternal concentrations, but after pancreatectomy this value rises to 35%–50%. When the data from the sham-operated and pancreatectomized fetuses are combined, an inverse relationship is observed between this percentage and the fetal plasma insulin concentration (Fig. 1). Hence, the ratio of fetal to maternal glucose concentrations is useful in assessing on a daily basis the degree of insulin deficiency produced by pancreatectomy (Fowden 1987).

Hyperglycaemia is observed in pancreatectomized fetuses throughout the range of maternal glucose levels found under normal nutritional conditions (Fig. 2). Even when maternal hypoglycaemia is induced by fasting for 48 h, plasma glucose levels in the pancreatectomized fetus, although lower (0.76 ± 0.05 mmol/l, $n = 4$, Table 1), are still significantly higher than the values observed in fasted sham-operated animals (0.29 ± 0.04 mmol/l, $n = 4$, $p < 0.01$).

The degree of hyperglycaemia observed after pancreatectomy is dependent on the maternal glucose level (Fig. 2) and on the severity of the insulin deficiency. There is a significant inverse correlation between the plasma concentrations of insulin and glucose in individual fetuses when the data from all the animals are combined ($r = 0.647$, $n = 52$, $p < 0.01$). These observations show that the pancreas is essential for fetal glucose homoeostasis and indicate that insulin is the dominant pancreatic hormone involved in this process.

On the day before delivery, plasma glucose levels in the pancreatectomized fetuses are significantly higher than the mean value observed previously (Fig. 3). No

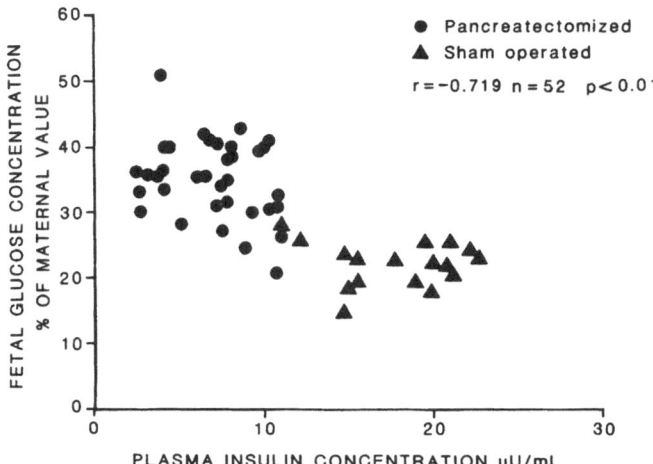

Fig. 1. Relationship between mean plasma insulin concentrations and mean ratio of fetal to maternal plasma glucose concentrations (expressed as a percentage) in sham-operated and pancreatectomized fetuses

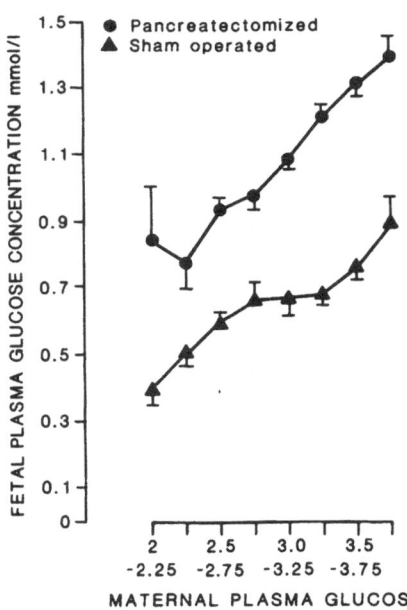

Fig. 2. Mean (± SE) plasma glucose concentrations in sham-operated (*n* = 6) and pancreatectomized (*n* = 11) sheep fetuses with respect to the maternal glucose level in fed ewes

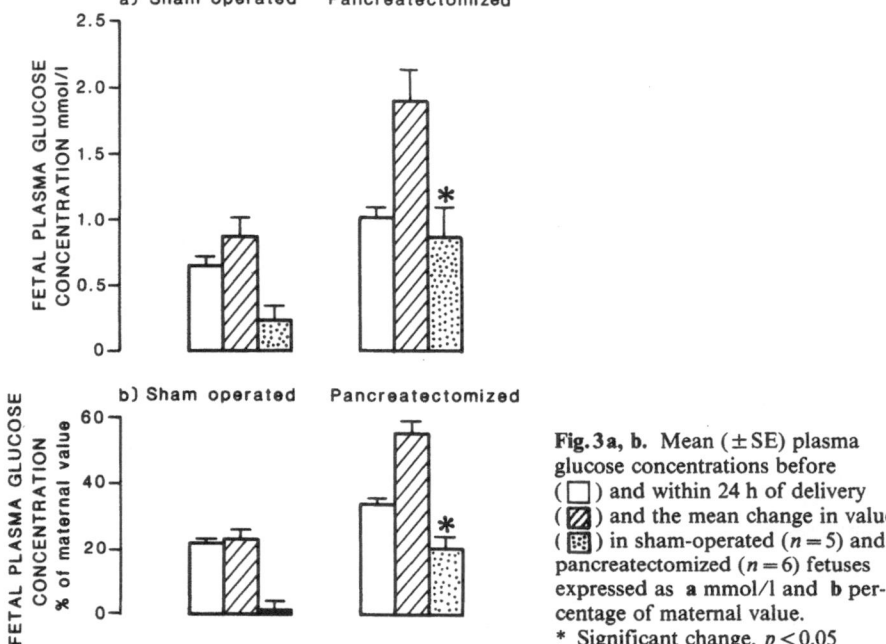

Fig. 3a, b. Mean (±SE) plasma glucose concentrations before (□) and within 24 h of delivery (▨) and the mean change in value (▦) in sham-operated ($n = 5$) and pancreatectomized ($n = 6$) fetuses expressed as **a** mmol/l and **b** percentage of maternal value.
* Significant change, $p < 0.05$

such change in plasma glucose is observed in the sham-operated animals (Fig. 3). The increase in the pancreatectomized fetuses is not due to a rise in maternal glucose levels and, hence, there is a significant increase in the ratio of fetal to maternal glucose levels in the pancreatectomized animals on the day before delivery (Fig. 3). The reasons for this rise in fetal plasma glucose remain obscure, but the observation suggests that there may be an increase in fetal glucose production before birth which the pancreatectomized fetus cannot use in the absence of insulin. Hyperglycaemic hormones such as cortisol and adrenaline rise in concentration in the fetus at this time, and hepatic glucose-6-phosphatase (G-6-Pase), which is essential for glucogenesis, also increases in activity before birth (Jones and Rolph 1985). Normal levels of cortisol and hepatic G-6-Pase are observed in the pancreatectomized fetus near term (unpublished observations), and it is therefore possible that changes in glucose turnover account for the prepartum increase in plasma glucose observed in these fetuses.

Other Plasma Carbohydrate Concentrations

The other major carbohydrates found in the circulation of the fetal sheep are lactate and fructose. The plasma concentrations of these substances also increase after pancreatectomy (Table 1) and are restored to normal values within 24 h of insulin infusion (Fowden et al. 1986a). When all the data from the intact and pancreatec- tomized fetuses are combined, there are significant inverse correlations between the mean plasma insulin concentrations and the mean plasma levels of fructose ($r = -0.645$, $n = 28$, $p < 0.01$) and lactate ($r = -0.717$, $n = 40$, $p < 0.01$). These obser- vations suggest that insulin is important in regulating not only glucose but also oth- er carbohydrate concentrations in utero. However, it is not clear from these studies whether the changes in lactate and fructose were directly or indirectly related to the insulin level. Infusion of exogenous glucose is known to raise the plasma concentra- tions of lactate and fructose in the sheep fetus and, hence, the changes in plasma lactate and fructose observed after pancreatectomy may be the result of hypergly- caemia rather than hypoinsulinaemia per se (see Fowden 1985). Certainly, plasma fructose levels in the intact and pancreatectomized fetuses are better correlated with the plasma glucose concentrations ($r = 0.730$, $n = 28$, $p < 0.01$) than with the plasma insulin levels ($r = -0.645$, $n = 28$, $p < 0.01$).

Fetal Uptake and Utilization of Carbohydrates

Pancreatectomy of the sheep fetus reduces the umbilical uptake of glucose and low- ers the fetal glucose/oxygen quotient by 50%–60% (Table 2) without any apparent changes in the uptake of glucose by the uteroplacental tissues or by the uterus as a whole (Fowden et al. 1986b). The rate of fetal glucose uptake and the actual glu- cose/oxygen quotient observed after pancreatectomy are similar to values found in intact fetuses made hypoinsulinaemic by fasting for 48 h (Fowden et al. 1986b). No significant changes are observed in the umbilical uptake of lactate or in the fetal lactate/oxygen quotient after pancreatectomy (Table 2).

The significant positive correlations observed between the plasma insulin con- centration and the glucose/oxygen quotient ($r = 0.730$, $n = 59$, $p < 0.01$) and be- tween plasma insulin and the umbilical uptake of glucose ($r = 0.918$, $n = 21$, $p < 0.01$) in the sham-operated and pancreatectomized fetuses show that the en- dogenous insulin level can influence both glucose uptake and consumption by the sheep fetus (Fowden et al. 1986b). Extrapolation of these relationships indicates that in the total absence of insulin, umbilical uptake of glucose would be approxi- mately 5.0 $\mu mol \cdot kg^{-1} \cdot min^{-1}$. Under normal conditions, there is no significant pro- duction of glucose by the sheep fetus in utero and the rate of fetal glucose utiliza- tion virtually matches the rate of umbilical glucose uptake (Hay et al. 1983). Hence the value of 5.0 $\mu mol \cdot kg^{-1} \cdot min^{-1}$ will represent the minimum rate of glucose utili-

Table 2. Mean ($\pm$ SE) umbilical uptake and substrate/oxygen quotients for glucose and lactate in pancreatectomized and sham-operated sheep fetuses during late gestation

	Umbilical uptake μmol/kg/min		Substrate/O_2 quotient	
	Glucose	Lactate	Glucose	Lactate
Pancreatectomized	$12.4 \pm 0.4*$	22.2 ± 2.9	$0.32 \pm 0.02*$	0.30 ± 0.04
	(7)	(4)	(9)	(9)
Sham-operated	27.5 ± 1.9	15.3 ± 1.2	0.63 ± 0.04	0.27 ± 0.05
	(8)	(4)	(8)	(5)

Number of animals in parentheses
* Significantly different from sham-operated animals; $p < 0.05$

zation by the insulin-insensitive tissues provided there is no endogenous glucose production in the pancreatectomized fetus. This rate of umbilical glucose uptake is insufficient to account for the normal glucose requirements of the brain without considering any of the other insulin-insensitive tissues in the fetus (Jones 1979; Fowden et al. 1986 a, b). These observations suggest that pancreatectomy not only reduces glucose consumption by the fetus as a whole but may also alter substrate utilization by individual fetal tissues.

The reduced glucose utilization of the pancreatectomized fetus explains in part its hyperglycaemia and may also contribute to the other metabolic changes that occur after pancreatectomy (Fowden and Comline 1984). The fetal hyperglycaemia after pancreatectomy would tend to reduce the transplacental glucose concentration gradient, which is known to be important in controlling umbilical glucose uptake in intact sheep fetuses (see Battaglia and Meschia 1978). However, the tendency towards a smaller transplacental glucose gradient after pancreatectomy does not appear to be the sole explanation for the reduced umbilical glucose uptake, as no correlation is observed between the gradient and glucose uptake in the pancreatectomized fetuses (Fowden et al. 1986b). Although the mechanisms of placental glucose transport do not appear to be affected by short-term changes in the insulin level (Jodarski et al. 1985), they may be influenced by the more long-term hypoinsulinaemia produced by pancreatic ablation. Further experiments over a wider range of transplacental gradients are needed in the pancreatectomized fetus to resolve whether or not insulin directly influences the transport systems of the placenta.

Tissue Glycogen Concentrations

In adult animals, hepatic glycogen content depends on the balance of the pancreatic and adrenal hormones. Insulin and cortisol favour glycogen deposition, while glucagon and adrenaline stimulate glycogenolysis. In the fetus, less is known about

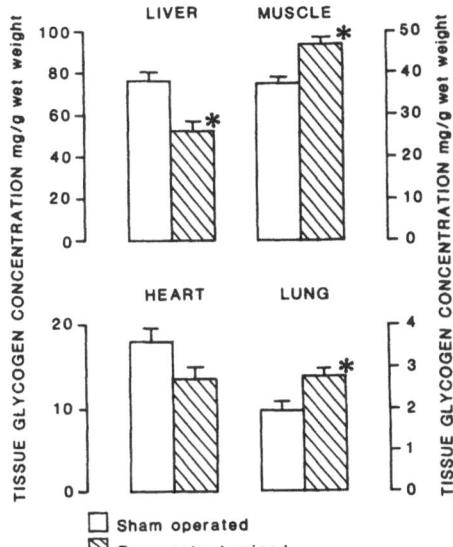

4. Mean (±SE) tissue glycogen concentrations in sham-operated (*n* = 9) and pancreatectomized (*n* = 11) fetuses between 136–141 days of gestation. * Significant difference, *p* < 0.05

the endocrine control of hepatic glycogen content, although cortisol is known to increase hepatic glycogen deposition in the fetus (Jost 1979; Barnes et al. 1978; Fowden et al. 1985). In order to determine the role of the pancreatic hormones in glycogen deposition in utero, tissue glycogen concentrations were measured in pancreatectomized fetuses near term (Fowden and Comline 1988).

Pancreatectomy of the fetus significantly lowers hepatic glycogen levels and significantly raises the glycogen content of the lungs and skeletal muscle compared with sham-operated fetuses of a similar gestational age (Fig. 4). No significant difference is observed in the levels of cardiac muscle glycogen between sham-operated and pancreatectomized fetuses. When the data from the sham-operated and pancreatectomized fetuses are combined, there is a significant, inverse correlation between the plasma insulin concentration in utero and the glycogen content of the lungs ($r = -0.605$, $n = 19$, $p < 0.01$), but no significant relationships are observed between plasma insulin and the glycogen content of the other tissues ($p > 0.05$ in all cases). As the mean plasma cortisol concentration is similar in sham-operated (35.4 ± 8.1 ng/ml, $n = 9$) and pancreatectomized fetuses (32.8 ± 5.8 ng/ml, $n = 11$, $p > 0.05$), these observations suggest that the pancreas is involved in regulating the pattern of glycogen deposition in the fetal lamb near term.

Fetal Growth and Development

Glucose is the principal metabolite of the sheep fetus and normally accounts for 25%-40% of the total daily requirement for carbon (Battaglia and Meschia 1978). It is used for growth and energy production, with the oxidative processes consuming 50%-60% of the glucose taken up by the fetus (Hay et al. 1983). The decrease in glucose uptake observed after pancreatectomy would reduce the availability of carbon, presumably both for carbon dioxide production and for accretion of new tissue. Other substrates may be able to substitute in part for glucose in the pancreatectomized fetus as there is no significant change in oxygen consumption per kilogram fetal weight after pancreatectomy (Fowden et al. 1986b). However, it is unlikely that the carbon requirement for growth can also be met from other sources, as there is no apparent change in umbilical lactate uptake and no evidence of increased tissue uptake of amino acids after pancreatectomy (Fowden and Comline 1984; Fowden et al. 1986b). Under these circumstances, it is not surprising to find that the pancreatectomized fetus is growth-retarded (Fowden 1987).

Pancreatectomy affects body weight more than crown-rump length (CRL) at term; the former is lowered by 30% while the latter is reduced by only 10% (Fowden et al. 1986a). Body weight increases more rapidly than CRL in normal fetuses during late gestation (Mellor 1984) and, hence, it might be anticipated that weight would be affected more severely than CRL by pancreatectomy. However, it is also possible that the soft tissues of the fetus are more sensitive than bone to changes in the availability of nutrients.

Individual organs also weigh less after pancreatectomy, but generally their reduction in weight is in proportion to the change in overall body weight (Fowden et al. 1986a). Only the spleen and thymus are disproportionately lighter in the pancreatectomized fetus. The explanation for the reduced growth rate of these specific organs remains obscure, although recent studies have shown that long-term insulin replacement in the pancreatectomized fetus restores these organs to their normal weight (A. L. Fowden and R. S. Comline, unpublished observations).

The reduced glucose uptake is unlikely to be the sole cause of the growth retardation in the pancreatectomized fetus. There are also changes in fat and amino acid metabolism and in the fetal concentrations of the somatomedins IGF_1 and IGF_2 after pancreatectomy (Fowden and Comline 1984; Gluckman et al. 1987). In part, the alterations in IGF appear to be related to the changes in plasma glucose, as IGF_2 levels are better correlated with the glucose level than with the insulin concentration after pancreatectomy (Gluckman et al. 1987). These observations suggest that glucose per se may be important in regulating IGF production in the fetal sheep.

The chronic hyperglycaemia of the pancreatectomized fetus may also have other effects on growth and development, particularly of individual organs and tissues. In adult animals, continual exposure to high glucose levels leads to microangiopathy and neural problems. Hyperglycaemia is known to cause a redistribution of the fetal cardiac output, with certain tissues receiving an increased blood flow at the expense of others (Crandell et al. 1985). Clearly, changes in perfusion of this kind could alter the relative growth rate of the various tissues involved by varying the supply of nutrients and growth factors. In addition, fetuses of diabetic mothers show abnormalities in cerebral development which have been attributed to hyperglycaemia in utero (Hill 1978). Although fetal glucose levels are low compared to adult concentrations even after fetal pancreatectomy, these observations indicate that the changes in metabolite levels induced by pancreatic ablation could have effects on the development of individual fetal tissues.

Conclusions and Implications

Pancreatectomy of the fetus has clearly shown that the pancreas has an important role in regulating fetal carbohydrate metabolism. In common with the adult, the fetal pancreas is involved in regulating the circulating glucose level and the rate of glucose uptake and utilization by the peripheral tissues. However, the extent to which the pancreatic hormones are involved in controlling hepatic glucogenesis in utero is still unclear. The pancreas does appear to influence the availability of glycogen for glucogenesis but simultaneous measurements of the rates of glucose utilization and umbilical glucose uptake are needed in the pancreatectomized fetus to make it possible to determine whether the pancreas has any direct effects on the endogenous production of glucose in the fetus.

Normal circulating carbohydrate concentrations were restored in the pancreatectomized fetus by insulin infusion, but the insulin concentration required to achieve this was twice the normal value (Fowden et al. 1986a). This suggests that insulin resistance develops during chronic hypoinsulinaemia and that there may be changes in receptor number or affinity under these conditions. The precise nature of these changes and of the role the pancreatic hormones have in the post-receptor events involved in fetal carbohydrate metabolism remains to be determined.

The effects that the pancreas has on fetal carbohydrate metabolism are predominantly anabolic and, hence, growth-promoting in utero. Thus, fetuses deficient in the pancreatic hormones are growth-retarded in a number of respects and may have difficulty surviving adverse conditions as a consequence. However, the extent to which the control of specific aspects of fetal carbohydrate metabolism and tissue growth can be attributed to individual metabolites and pancreatic hormones still requires further study.

Acknowledgements. The author would like to thank the many members of the Laboratory who helped with this work and the British Diabetic Association for their financial support.

References

Barnes RJ, Comline RS, Silver M (1978) Effect of cortisol on liver glycogen concentrations in hypophysectomized, adrenalectomized and normal foetal lambs during late or prolonged gestation. J Physiol 275: 567-579

Battaglia FC, Meschia G (1978) Principal substrates of fetal metabolism. Physiol Rev 58: 499-527

Crandell SS, Fisher DJ, Morriss FH (1985) Effects of ovine maternal hyperglycaemia on fetal regional blood flows and metabolism. Am J Physiol 249: E454-E460

Fowden AL (1985) Carbohydrate metabolism and pancreatic endocrine function in the fetus. In: Albrecht E, Pepe G (eds) Perinatal endocrinology. Perinatology Press, Ithaca, pp 71-90

Fowden AL (1987) Pancreatectomy of the fetus in utero: techniques and applications. In: Nathanielsz PW (ed) Animal models in fetal medicine VI. Perinatology Press, Ithaca, pp 216-240

Fowden AL, Comline RS (1984) The effects of pancreatectomy on the sheep fetus in utero. Q J Exp Physiol 69: 319-330

Fowden AL, Comline RS (1988) The effects of pancreatectomy on tissue glycogen concentrations in the fetal sheep. In: Jones CT (ed) Fetal and neonatal development. Perinatology Press, Ithaca (in press)

Fowden AL, Comline RS, Silver M (1985) The effects of cortisol on the concentration of glycogen in different tissues in the chronically catheterized fetal pig. Q J Exp Physiol 70: 23-32

Fowden AL, Mao XH, Comline RS (1986a) The effects of pancreatectomy on the growth and metabolite concentrations of the sheep fetus. J Endocrinol 110: 225-231

Fowden AL, Silver M, Comline RS (1986b) The effect of pancreatectomy on the uptake of metabolites by the sheep fetus. Q J Exp Physiol 71: 67-78

Gluckman PD, Butler J, Comline RS, Fowden AL (1987) The effects of pancreatectomy on the plasma concentrations of insulin-like growth factor 1 and 2 in the sheep fetus. J Dev Physiol 9: 79-88

Hay WW, Meyers SA, Sparks JW, Wilkering RB, Meschia G, Battaglia FC (1983) Glucose and lactate oxidation rates in the fetal lamb. Proc Soc Exp Biol Med 173: 553-563

Hill DE (1978) Effect of insulin on fetal growth. Semin Perinatol 2: 319-328

Jodarski GD, Shanahan MF, Rankin JHG (1985) Fetal insulin and placental 3-O-methyl glucose clearance in near-term sheep. J Dev Physiol 7: 251-258

Jones CT, Rolph TP (1985) Metabolism during fetal life: a functional assessment of metabolic development. Physiol Rev 65: 357-430

Jones MD (1979) Energy metabolism in the developing brain. Semin Perin 3: 121-129

Jost A (1979) Fetal hormones and fetal growth. Contrib Gynecol Obstet 5: 1-20

Mellor DJ (1984) Investigation of fetal growth in sheep. In: Nathanielsz PW (ed) Animal models in fetal medicine IV. Perinatology Press, Ithaca, pp 150-175

Validity of Amniotic Insulin Measurements in the Management of Pregnancy in Diabetic Women

P. A. M. Weiss[1]

Even before the 16th week of pregnancy, small amounts (1.3–2.5 μU/ml) of insulin are detectable in the amniotic fluid. During the 16th week the amniotic fluid insulin level surges to about 4 μU/ml, simultaneously with the onset of kidney function in the fetus. A further though slight increase occurs progressively towards the end of pregnancy, although in the lower percentiles of the distribution a mild decrease may occur (Weiss et al. 1984b; Table 1).

The ratio of the insulin levels in the cord blood, fetal urine, and amniotic fluid is roughly 100:90:80. In healthy women the amniotic insulin level is remarkably stable over periods of weeks (Weiss 1979). If the initial value measured is in the upper part of the normal range, this will also be true for subsequent readings. Similarly, should the initial value be low, subsequent readings will also be low.

Hourly measurements show that neither meals nor insulin administration exert a marked immediate influence on the amniotic fluid insulin level (Weiss et al. 1985; Fig. 1). The upper line in Fig. 1 shows the blood glucose level of an unstable overtly diabetic woman, the middle curve shows the corresponding amniotic glucose fluctuations, and the lower curve reflects her amniotic insulin level. A significantly reduced amniotic insulin concentration is found with intrauterine fetal death, placen-

Table 1. Insulin levels in amniotic fluid of healthy pregnant women between the 27th and 42nd weeks of gestation ($n = 458$)

Week of gestation	Case no.	Percentiles[a] (μU/ml)				
		3	10	50	90	97
27/28	45	2.0	3.2	6.0	10.4	11.2
29/30	36	2.0	3.2	6.1	10.8	11.7
31/32	63	2.0	3.2	6.2	11.3	12.2
33/34	86	1.9	3.0	6.3	11.7	12.8
35/36	80	1.8	2.9	6.4	11.8	13.5
37/38	70	1.5	2.8	6.6	12.2	14.9
39/40	60	1.2	2.8	6.8	13.4	17.2
41/42	18	1.1	3.4	7.5	15.3	18.0

[a] Rounded to nearest 0.1

[1] Graz University Clinic of Obstetrics and Gynecology, Auenbruggerplatz 14, 8036 Graz, Austria

The Endocrine Control of the Fetus
Ed. by W. Künzel and A. Jensen
© Springer-Verlag Berlin Heidelberg 1988

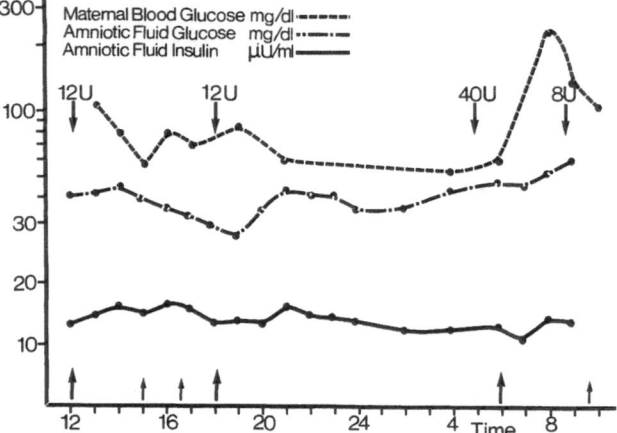

Fig. 1. Circadian course of the amniotic-fluid insulin level in an overtly diabetic woman (hourly measurements). *Top curve,* maternal blood glucose (mg/dl); *middle curve,* amniotic fluid glucose (mg/dl); *bottom curve,* amniotic fluid insulin (μU/ml). *Arrows pointing upwards,* meals; *arrows pointing downwards,* insulin injections. The scale applies to insulin as well as to glucose. Neither food intake nor insulin administration exert any immediate impact on the amniotic fluid insulin concentration (From Weiss 1988)

tal insufficiency, and major fetal malformations. On the other hand, an increase of amniotic insulin may be caused by moderate hemolytic disease, by glucocorticoid treatment to bring about fetal lung maturation, or by betamimetic therapy for the arrest of preterm labor (Weiss et al. 1984b). Figure 2 shows the course of the mean amniotic insulin level of 84 insulin-dependent diabetic women under conventional (i.e., not intensified) insulin therapy (Weiss and Hofmann 1985), together with the normal range and the mean amniotic insulin level in another group of 90 subjects undergoing tocolytic therapy.

The development of fetopathy in about 10% of gestational diabetes is strongly associated with elevated amniotic insulin values (Weiss et al. 1984a). Figure 3 shows insulin levels derived from pregnant women with diseased offspring and those from women with healthy offspring. In appropriately treated gestational diabetic women, and under intensive insulin therapy, amniotic insulin decreases to the normal range. The offspring are then healthy and the mean cord blood insulin level is within the normal range (Weiss et al. 1986; Fig. 4). However, in patients who either refuse insulin therapy or discontinue treatment themselves, amniotic fluid insulin levels remain elevated (Fig. 4). The offspring are severely affected. They suffer from obesity, hypoglycemia, hypocalcemia, hyperbilirubinemia, and respiratory distress syndrome. The vertical bar in Fig. 4 depicts the pronounced fetal hyperinsulinism of these infants.

Figure 5 shows the relation between amniotic insulin content and fetal outcome in insulin-dependent diabetic women (Weiss 1987). The lower curve corresponds to

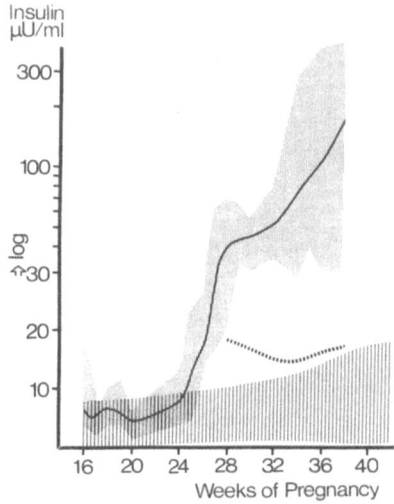

Fig. 2. Amniotic fluid insulin content in 84 diabetic women (White Class A-R) during conventional insulin therapy. *Continuous line,* mean; *stippled area,* range; *hatched area,* normal range of the amniotic fluid insulin content between the 3rd and 97th percentiles (Weiss et al. 1984); *dotted line,* mean amniotic fluid insulin content in 91 metabolically healthy women during tocolytic therapy (ritodrine) (From Weiss 1988)

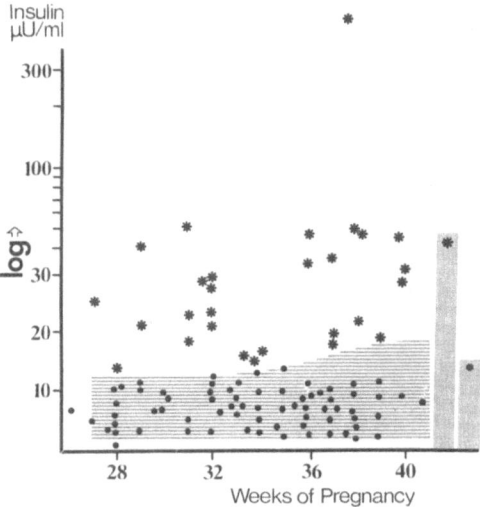

Fig. 3. 103 amniotic fluid insulin values from 75 gestational diabetic women. *Hatched area,* normal range (3rd to 97th percentile); *dots,* 75 values in 61 patients who subsequently gave birth to a healthy infant; *asterisks,* 28 values of 14 women who subsequently gave birth to a child with diabetogenic fetal disease; *bars,* the corresponding mean cord blood insulin concentration (From Weiss 1988)

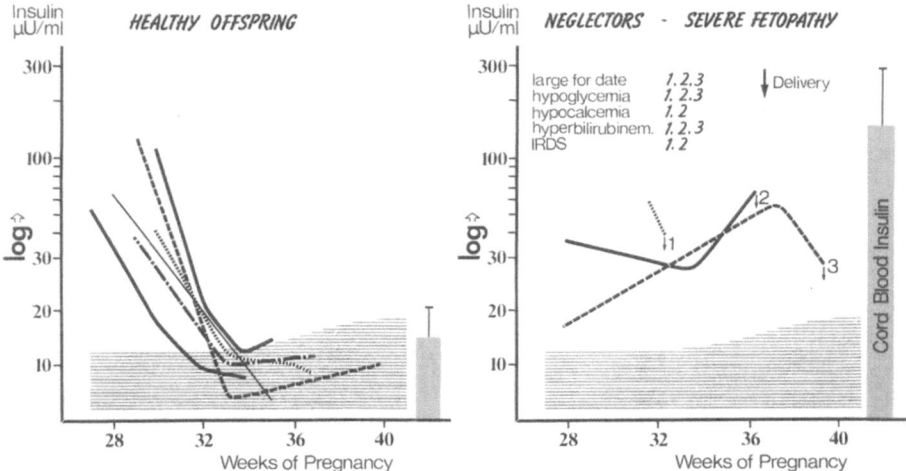

Fig. 4. Course of the amniotic fluid insulin concentration in six appropriately treated gestational diabetic women *(left)* and three "neglectors" who refused insulin therapy *(right)*. While the newborns of the neglectors had severe fetopathy (overly high birth weight, hypoglycemia, respiratory dystress syndrome - *IRDS*, hyperbilirubinemia, etc.), the others were completely healthy. *Hatched area,* normal range of the amniotic fluid insulin concentration (3rd to 97th percentile); *stippled bars,* cord blood insulin level (mean ± S. D.) (From Weiss 1988)

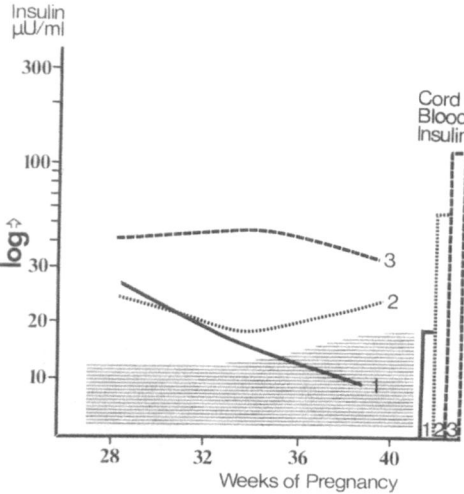

Fig. 5. Course of the mean amniotic fluid insulin level during pregnancy (45 diabetic women, 183 determinations). *1* healthy offspring; *2* biochemical fetopathy and *3* somatic fetopathy in the newborn. The bars at the *right* represent the corresponding mean cord blood insulin concentration

healthy newborns, and the middle curve gives the mean amniotic insulin concentration in women delivered of children with what we call *biochemical fetopathy*. This is a clinical entity of hyperinsulinism, hypoglycemia, and other biochemical disorders. The top curve, finally, shows the mean amniotic insulin content of fetuses with such biochemical fetopathy who were also overweight. We call this clinical picture with a birth weight above the 90th percentile *somatic fetopathy*. The vertical bars at the right represent the mean cord blood insulin values corresponding to the amniotic insulin content and the degree of diabetic impairment of the newborn.

We have up to now made about 2000 amniotic insulin determinations in diabetic and gestational diabetic women. In poorly adjusted diabetics, oversupply of glucose to the fetus, fetal hyperinsulinism, and diabetogenic fetopathy are most likely to result (except in cases of placental insufficiency). However, it is not really possible to predict accurately whether an individual fetus is injured, especially given the good degree of metabolic control which occurs in most of the patients who recieve intensified conventional or pump therapy, and in whom the mean blood glucose level ranges between 80 and 120 mg%. Persson and Lunell (1975) reported that there was no significant difference in metabolic adjustment between women pregnant with a healthy fetus and those with a diseased fetus. Burke et al. (1979) even reported on twins, one normal and the other with fetopathy. The cause appears to be differential sensitivity to glucose stimuli in the individual fetus. There may also be a different mechanism involving glucose transport through the placenta. Thus, we consider the amniotic insulin level to be the only specific parameter to distinguish between fetuses subject to diabetogenic compromise and those not at risk. Thus, in gestational diabetes we suggest insulin therapy in cases with increased amniotic insulin content.

One of the most compelling reasons for aggressive therapy in gestational diabetes is the damage to the fetal beta islet cells that ensues if treatment is insufficient. This damage is due to fuel oversupply and consequent fetal insulin hyperproduction during the course of pregnancy. Such afflicted offspring are significantly more prone to obesity in infancy and to type II diabetes later in life. This diabetic tendency has been termed fuel-mediated teratogenesis by Freinkel's group (1985). In treatment of insulin-dependent diabetic women, we increase the insulin doses when the amniotic insulin level is elevated despite satisfactory blood glucose profiles. In both gestational and insulin-dependent diabetes the enhanced insulin supply has no pronounced impact on maternal glycemia. In gestational diabetes, for instance, mean blood glucose concentration decreased under treatment with insulin (65 ± 29 U/24 h) from 98 ± 9 mg/dl to 82 ± 10 mg/dl (Weiss et al. 1986). The main effect appears to be a reduction of glucose deprivation of the mother because of decreased uptake by the fetus. Thus, fetal insulin production is reduced. In insulin-dependent diabetic women showing decreased amniotic insulin content and retardation of intrauterine growth, it is our policy to raise the mean maternal blood glucose level because there may be a lack of fetal glucose due to placental insufficiency, despite maternal hyperglycemia (Sarles and Adamsons 1978).

Finally, increased amniotic insulin values in late pregnancy influence our deci-

Table 2. Range of cord serum insulin levels (mean ± S. D.) in offspring born to women with normal glucose tolerance during pregnancy

Number of cases	764
Birthweight (g)	3334 ± 503
Week of pregnancy	40.0 ± 1.2
Cord serum glucose (mg/dl)	65 ± 29
Cord serum insulin (μU/ml)	6.9 ± 5.0
Per centiles (μU/ml) 3rd	2.0
10th	2.8
50th	5.7
90th	12.1
97th	17.4

sion about whether to induce labor since it is generally accepted that fetuses with hyperinsulinism are prone to silent intrauterine death (Salzberger and Liban 1975).

As to the newborn, the determination of the cord blood insulin level in infants of diabetic mothers and in babies who are large for their gestational age is routine at our institution. This acts as a control of the quality of treatment in diabetic women and helps to determine whether a high birth weight was due to fetal hyperinsulinism which must have been caused by an unrecognized maternal carbohydrate disturbance. Table 2 shows the normal range of the cord blood insulin concentration.

References

Burke BJ, Sheriff RJ, Savage PE, Dixon HG (1979) Diabetic twin pregnancy: an unequal result. Lancet 1: 1372–1373

Freinkel N, Metzger BE, Phelps RL, Dooley SL, Ogata ES, Radvany RM (1985) Heterogeneity of maternal age, weight, insulin secretion, HLA antigens, and islet cell antibodies and the impact of maternal metabolism on pancreatic B-cell and somatic development in the offspring. Diabetes 34 (suppl 2): 1–7

Persson B, Lunell NO (1975) Metabolic control in diabetic pregnancy. Variations in plasma concentrations of glucose, free fatty acids, glycerol, ketone bodies, insulin and human chorionic sommatomammotropin during the last trimester. Am J Obstet Gynecol 122: 737–745

Salzberger M, Liban E (1975) Diabetes and antenatal fetal death. Isr J Med Sci 11: 623–628

Sarles MD, Adamsons K (1978) Diabetes: New management concepts. Perinatal Care 2: 13–40

Weiss PAM (1979) Die Überwachung des Ungeborenen bei Diabetes mellitus an Hand von Fruchtwasserinsulinwerten. Wien Klin Wochenschr 91: 293–304

Weiss PAM (1988) Gestational diabetes. A survey and the Graz approach to diagnosis and therapy. In: Weiss PAM, Coustan DR (eds) Gestational diabetes. Springer, Vienna New York, pp 1–63

Weiss PAM, Hofmann H, Winter R, Pürstner P, Lichtenegger W (1984a) Gestational diabetes and screening during pregnancy. Obstet Gynecol 63: 776–780

Weiss PAM, Pürstner P, Winter R, Lichtenegger W (1984b) Insulin levels in amniotic fluid of normal and abnormal pregnancies. Obstet Gynecol 63: 371–375

Weiss PAM, Hofmann H (1985) Diabetes mellitus und Schwangerschaft. In: Burghardt E (ed) Spezielle Gynäkologie und Geburtshilfe. Springer, Vienna, pp 337–427

Weiss PAM, Winter R, Pürstner P, Lichtenegger W (1985) Amniotic fluid glucose values in normal and abnormal pregnancies. Obstet Gynecol 65: 333–339

Weiss PAM, Hofmann HMH, Winter R, Lichtenegger W, Pürstner P, Haas J (1986) Diagnosis and treatment of gestational diabetes according to amniotic fluid insulin levels. Arch Gynecol 239: 81–91

Mechanisms of Parturition

Initiation of Human Parturition[*]

G. C. Liggins[1] and T. Wilson

Introduction

Several thorough reviews of human parturition, each emphasising a particular viewpoint, have been published recently (Fuchs and Fuchs 1984; Liggins 1984; Thorburn 1985; Casey and MacDonald 1986) and it is not the present intention to attempt a further comprehensive review. Rather, consideration will be given to new developments that may bring together seemingly unrelated previous observations into a coherent hypothesis.

What has probably had the most unifying influence on current parturitional research is agreement that human parturition is controlled by a paracrine rather than an endocrine system. Whether oxytocin, prostaglandins (PGs), relaxin or a steroid is the favoured hormone of a particular worker, the hypothesis that is tested will propose that in one way or another the hormone will take part in a direct interaction between tissues of the conceptus (fetus, amnion, chorion) on the one hand and of the mother (decidua, myometrium, cervix) on the other.

Oxytocin

Source

The maternal and fetal neurohypophyses are probably the sole sources of oxytocin reaching the uterus, there being no evidence of a significant contribution from the maternal ovaries although they are known to have the capability of synthesis (Dawood and Khan-Dawood 1986). The need to resolve the conflicting reports on whether or not concentrations of oxytocin in the maternal circulation increase at the start of labour (see Fuchs and Fuchs 1984) has largely disappeared with the finding that the concentration of myometrial oxytocin receptors increases two-fold at that time (Fuchs et al. 1982a) but the question of the extent to which fetal oxytocin contributes to maternal levels is unresolved.

* Work supported by the New Zealand Medical Research Council
[1] Postgraduate School of Obstetrics and Gynaecology, University of Auckland, Auckland, New Zealand

The Endocrine Control of the Fetus
Ed. by W. Künzel and A. Jensen
© Springer-Verlag Berlin Heidelberg 1988

Umbilical arterial concentrations of oxytocin are higher than those in the maternal uterine vein in late pregnancy and increase in labour, raising the possibility of feto-maternal transfer which Dawood et al. (1979) have demonstrated in baboons. However, unless the transfer of fetal oxytocin to the maternal circulation leads to a rise in concentration, it seems unlikely that the fetal neurohypophysis can express its activity by this route. The alternative route to maternal uterine tissue provided by the amniotic fluid is possible, since there is a three- to four-fold increase in the concentration of oxytocin in amniotic fluid after the onset of labour, presumably as a result of increased urinary excretion of oxytocin. Loss of this route in fetuses with renal agenesis is not associated with disordered pregnancy length.

Action

Myometrium

Receptor sites for oxytocin have been characterized in human myometrium (see Soloff 1985). Binding to the receptor is specific and occupancy is associated with specific actions of oxytocin which are thought to be mediated by calcium, both by increasing influx through membrane channels and by reducing efflux by inhibiting $Ca^{2+}Mg^{2+}$ATPase. Calcium channel blockers are moderately effective in inhibiting contractions in preterm labour before cervical dilation has occurred (Ulmsten et al. 1980).

The progressive increase in the sensitivity of the gravid human uterus to oxytocin in the later weeks of pregnancy correlates well with increasing concentrations of myometrial oxytocin receptors (Fuchs et al. 1982a) and the relationship is probably causal although other factors could contribute. Of particular significance to the initiation of parturition is a two-fold increase in receptor concentrations found in myometrium taken at caesarean section in labour compared with similar samples taken from women before labour started (Fuchs et al. 1982a). As pointed out by Soloff (1985), circulating levels of oxytocin may be of limited significance if the response of the myometrium is dictated by the concentration of oxytocin receptors.

Decidua

A second uterine target for oxytocin action is the decidua which, like the myometrium, contains specific oxytocin receptors that increase in concentration throughout gestation (Fuchs et al. 1982a), reaching maximal values at term. Unlike in the myometrium, however, the concentration of receptors in the decidua is not significantly higher in labouring women. In the rabbit, a ten-fold increase in oxytocin binding sites occurs in both decidua and myometrium during the last day of pregnancy (Reimer et al. 1986).

The specific decidual response to oxytocin is increased synthesis and release of PGs, particularly $PGF_{2\alpha}$ and PGE_2. Fuchs et al. (1982a) compared the response to oxytocin of decidua taken before or after the start of labour and found no differences. This work needs to be repeated with improved methodology since these workers found that responses to oxytocin were achieved only when PG synthesis was inhibited with mefenamic acid during transport of the tissues. It is possible that recently described methods using perfused, dispersed cell preparations which are highly reproducible (Wilson et al. 1985) may yield different results.

The stimulus to PG synthesis is probably mediated at least in part by activation of phospholipase A_2 (PLA_2). Oxytocin stimulates the release of labelled arachidonic acid from perfused, dispersed human decidual cells in which tritiated arachidonic acid is incorporated into phospholipids (T. Wilson and G. C. Liggins, 1987, unpublished observations). Activation of PLA_2, a Ca^{2+}-dependent enzyme, may result from increased intracellular Ca^{2+} as described above. Alternatively or additionally, activation could occur by hydrolysis of phosphoinositide as described by Schrey et al. (1986) in isolated human decidual cells treated with 10^{-9}–10^{-6} M oxytocin.

Control of Action

A two-fold increase in the concentration of oxytocin receptors in the myometrium around the time of onset of labour as observed by Fuchs et al. (1982a) could initiate uterine activity if the sensitivity of the uterus to oxytocin were thereby raised sufficiently to become responsive to the prevailing levels of circulating oxytocin. That being the case, it becomes important to understand the factors that determine the concentrations of myometrial oxytocin receptors. Unfortunately, little is known of such factors in women. Although the progressive increase in receptors throughout pregnancy can be attributed to the similarly rising concentrations of oestrogen which are known to stimulate receptor formation in experimental animals (Soloff et al. 1983), the rapid increase with the onset of labour cannot be explained as an effect of oestrogen, plasma levels of which are unchanged at that time. Liggins (1984) suggested that receptor formation is induced by a prostanoid, supporting this speculation with evidence that intravascular infusion of subthreshold (oxytocic) doses of $PGF_{2\alpha}$ causes a marked increase in the myometrial sensitivity to oxytocin after a long latent period consistent with protein synthesis. In the rat, administration of 500 µg of $PGF_{2\alpha}$ subcutaneously on day 18 of gestation caused a rise in the concentration of myometrial oxytocin receptors as well as premature delivery (Alexandrova and Soloff 1980).

Prostaglandins

Source

Differing conclusions are reached when attempts are made to determine the major sources of the primary PGs, $PGF_{2\alpha}$ and PGE_2 from in vivo or in vitro studies. Measurements of their major metabolites, 13,14-dihydro-15-keto-$PGF_{2\alpha}$ (PGFM) and

11-deoxy-13,14-dihydro-15-keto-11,16-cyclo-prostaglandin E_2 (PGEM-II) in plasma
from women in labour reveal a marked rise in the concentration of PGFM but little
change in PGEM (Mitchell et al. 1982). Furthermore, the raised levels of PGFM
persist for 30 min or more after delivery and can be further increased by post-
partum injection of oxytocin, whereas levels of PGEM fall rapidly after delivery
(Fuchs et al. 1982). These observations strongly suggest that $PGF_{2\alpha}$ is the predominant PG in labour and that maternal tissues rather than fetal tissues are the major
source. In vitro studies, however, tell a different story. Although there are quantitative differences according to whether incubated tissues (Fuchs et al. 1982b), perfused tissues (Mitchell et al. 1978) or dispersed cells (Olson et al. 1983) are compared, the production of PGE_2 is at least equal to that of $PGF_{2\alpha}$. In particular,
amnion produces mainly PGE_2 (Okazaki et al. 1981) and the amount released is
greater from tissue taken at delivery than from tissue taken before labour starts (Olson et al. 1983; Manzai and Liggins 1984). Assuming plasma levels of metabolites
to be more indicative of production rates in vivo than results from incubated cells,
and since the major product of myometrium is prostacyclin (PGI_2; Barnford et al.
1980), the conclusion has been reached (Fuchs and Fuchs 1984; Liggins 1984) that
the decidua is the source of much of the $PGF_{2\alpha}$ released in labour. It is generally assumed that decidual PGs diffuse into the myometrium, and this is supported by the
effectiveness of small quantities of PGs instilled into the uterine cavity to activate
the uterus. The fate of PGE_2 formed in the amnion is not clear; some enters the amniotic fluid (Keirse et al. 1977) and some is probably metabolized to inactive products by enzymes in the chorion. Conversion to $PGF_{2\alpha}$ by 9-keto reductase in the decidua is unlikely because of the low specific activity of the enzyme (Niesert et al.
1987). A possible function of PGE_2 synthesized in amnion is discussed below.

Action

While the actions of oxytocin are limited to smooth muscle and decidual cells, the
actions of PGs probably extend to all cell types in the uterus although little is
known of most of them. In particular, prostaglandins are known to modulate contractility and promote gap junction formation in smooth muscle and to act on fibroblasts to stimulate collagenolysis and secretion of less highly charged glycosamino-
glycans (see Liggins 1984).

Transduction of signals from occupancy of PG receptors follows a number of
complex pathways which are only now beginning to be understood.

1. PG receptors may be coupled to calcium channels and their occupancy may allow influx of Ca^{2+}, since calcium channel blockers inhibit contractile responses
   in human muscle strips (Forman et al. 1979).
2. Receptor occupancy may activate phospholipase C, leading to hydrolysis of
   phosphoinositides and liberation of diacylglycerol and inositol phosphates. Further metabolism of these products yields phosphatidic acid (which may act as a

calcium ionophore) and 1,4,5-inositol triphosphate, which mobilizes intracellular stores of Ca^{2+} (Irvine 1982) and increases transmembrane Ca^{2+} influx (Kuno and Gardner 1987).

3. Diacylglycerol together with Ca^{2+} and phospholipid activates protein kinase C which may phosphorylate myosin, mimicking the action of myosin light chain kinase.

Control of Action

While control of oxytocin action, at least at the low concentrations prevailing at the onset of labour, is expressed mainly through receptor concentrations, control of PG action is probably exerted mainly through the rate of synthesis. Little is known of the natural variability of PG receptors. Infusion of $PGF_{2\alpha}$ intra-arterially in pregnant sheep or intravenously in pregnant women at a low rate that has no immediate oxytocic effects is followed after a latent period of 12 h or more by increasing responsiveness (Liggins 1977). Although these observations are consistent with induction of receptors by their agonist, this has yet to be demonstrated.

The overall rate of synthesis of PGs is subject to control in two ways: first, by the relative activities of the enzymes cycloxygenase, lipoxygenase and acyltransferase, which respectively determine whether arachidonic acid is further metabolized to PGs or leukotrienes or is reincorporated into phospholipids, and, second, by the rate of release of arachidonic acid from phospholipids. Although leukotrienes are synthesized in human fetal membranes and decidua (Saeed and Mitchell 1982) and a substantial proportion of arachidonic acid is rapidly reincorporated in non-uterine tissues (Irvine 1982), there is no evidence that either of these pathways is important in controlling PG synthesis in labour. Accordingly, phospholipases and cyclooxygenase are regarded as the likely points of control.

Cyclooxygenase

There is no convincing evidence that cyclooxygenase is rate-limiting. Although labour and abortion were induced by intra-amniotic injection of arachidonic acid in women (suggesting that it is not rate-limiting) (MacDonald et al. 1974), extra-amniotic injection in rhesus monkeys failed (Robinson et al. 1979). Okazaki et al. (1981) found increased activity of cyclooxygenase in amnion but not chorion or decidua after the onset of labour.

Phospholipases

Two phospholipases, both Ca^{2+}-dependent, are involved in the regulation of PG production. The first of these is phospholipase C (PLC), which catalyses the hydrolysis of phosphoinositides to form diacylglycerol and inositol phosphate. PLC is activated by occupancy of a variety of membrane receptors (including oxytocin and possibly PGs) and by agents such as ionophores that increase intracellular concentrations of Ca^{2+}. Activation of PLC enhances PG production by increasing the release of arachidonic acid in several ways:

(a) degradation of diacylglycerol by the successive action of diacylglycerol lipase and monoacylglycerol lipase, which liberates arachidonic acid (Okazaki et al. 1981);

(b) activation of PLA_2 by increased intracellular concentration of Ca^{2+} in response to 1,4,5-inositol triphosphate; and

(c) activation of PLA_2, by inactivating (by phosphorylation) an inhibitor of PLA_2, probably a lipocortin (see below).

The protein kinase involved in phosphorylating the lipocortin is likely to be protein kinase C (Khanna et al. 1986), which is activated by diacylglycerol.

The second phospholipase involved in regulating PG synthesis, and the one that is usually considered to be rate-limiting, is PLA_2, which catalyses the release of arachidonic acid from the sn-2 position of phospholipids. How much of the arachidonic acid serving as substrate for PG synthesis in labour derives from phosphoinositide turnover (hydrolysis of diacylglycerol) and how much from the hydrolysis of phospholipids catalysed by PLA_2 is uncertain, but it is generally considered that the massive synthesis occurring in labour depends to a substantial degree on the latter source. Thus, the factors determining the activity of PLA_2 are likely to have a key role in the initiation of labour.

Activation of PLA_2 by mobilization of Ca^{2+} is the subject of control a by large number of potential stimuli that fall broadly into two categories: those that stimulate phosphoinositide turnover and those like platelet activation factor (PAF) and phosphatidic acid, that act as natural ionophores. In addition, however, activity is determined by inhibitory proteins. This group of proteins, called lipocortins, may have an important role in initiating labour.

Protein with PLA_2-inhibitory activity induced by glucocorticoids was discovered independently by Blackwell et al. (1980) (macrocortin) and Hirata et al. (1980) (lipomodulin) and subsequent named lipocortin. Recently, it has been found to consist of two proteins, each of molecular weight (M_r) 35000–40000, which have been named lipocortin I and lipocortin II. Phosphorylation of the lipocortin molecule inactivates it (Hirata et al. 1984). Phosphorylation is catalysed by a protein kinase which is probably protein kinase C (PKC), since PLA_2 is activated by phorbol esters which mimic diacylglycerol in activating PKC. Thus phosphoinositide turnover is linked to activation of PLA_2 not only through mobilization of Ca^{2+} but also through PKC (Fig. 1).

The direct relevance of lipocortins to initiation of human parturition comes from a recent report by Wilson et al (1985) which describes the isolation from human amniotic fluid of a protein with inhibitory activity against PLA_2. An M_r of approximately 60000 distinguishes it from lipocortins I and II but in other respects the physical characteristics are similar. The protein is liberated during incubation of chorion, amnion and decidual tissue, particularly from chorion (Wilson et al. 1987).

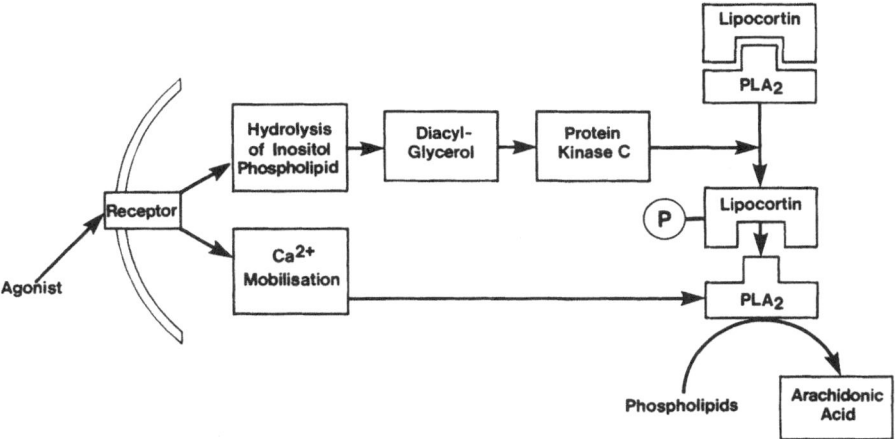

Fig. 1. Phosphorylation of lipocortin by protein kinase C inactivates it

The exciting feature of this novel lipocortin-like protein is the observation that the activity evident in medium from incubates of chorion taken before the start of labour disappears with labour although production of the protein continues. A structural modification, as indicated by a shift in isoelectric point, suggests that phosphorylation may occur with the onset of labour. It is not yet known whether the protein is glucocorticoid-dependent, but the high concentrations of cortisol $(10^{-5}\,M)$ required to inhibit synthesis of $PGF_{2\alpha}$ in endometrium (Skinner et al. 1984) and the absence of effects of large doses of corticosteroids on the initiation or progress of labour suggest that it is not corticosteroid-dependent (Casey et al. 1985). We speculate that synthesis of the protein may be progesterone-dependent. If this speculation is confirmed, it would provide a mechanism by which PG synthesis is inhibited through pregnancy and increases at term in those species (not human) in which progesterone levels fall before parturition. The agonist or agonists that may promote phosphorylation of the lipocortin, presumably via PKC, are unknown but clearly could be pivotal in initiating labour.

Working Model of Initiation of Parturition

Figure 2 outlines a proposed scheme describing the control of PG synthesis and the interactions with oxytocin. The initial event in the cascade is shown as two separate actions of PGE_2 diffusing from the amnion. Step 3 speculates first that PGE_2 induces formation of, or the response to, oxytocin receptors in decidua. This is supported by our unpublished observation that overnight preincubation of term decidua with PGE_2 enhances the release of PGF in response to oxytocin. Second, it

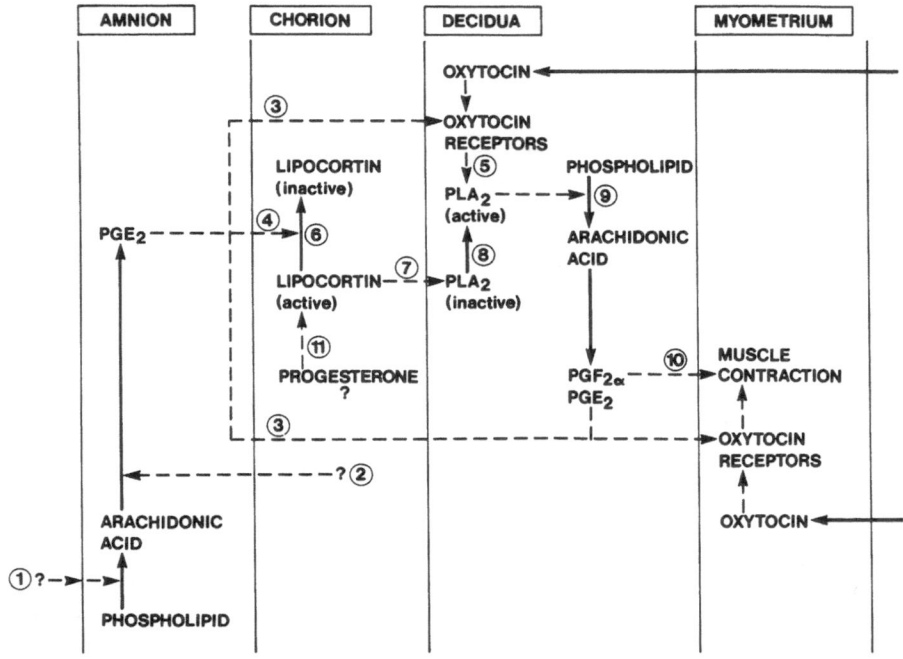

Fig. 2. Working model of initiation of human parturition showing the major interactions of oxytocin and PGs. The model postulates that increased synthesis of PGE$_2$ initiates a cascade of events in chorion, decidua and myometrium (see text for details). Relaxin and prolactin, both of which are synthesized by the decidua and have important functions and interactions with prostaglandins (see Liggins 1984), are omitted for clarity

speculates that PGE$_2$ stimulates inactivation of lipocortin, probably by phosphorylation. This is not yet supported by direct evidence. Factors responsible for initiating synthesis of PGE$_2$ by amnion remain uncertain but a number of substances in amniotic fluid have been proposed (step 1). They include platelet activation factor which is present in labour (Billah and Johnston 1983), an unidentified substance present in fetal urine (Casey et al. 1983), transforming growth factors (Mitchell and Casey 1984) and catecholamines which increase in concentration in late pregnancy and can stimulate release of PGE$_2$ in vitro (Di Renzo et al. 1984).

By whatever means it is achieved, inactivation of chorionic lipocortin (step 6) will activate PLA$_2$ (step 8) and stimulate the synthesis of prostaglandins (step 9). At the same time, increased concentrations of oxytocin receptors will also activate phospholipase C (not shown) and PLA$_2$ (step 5) and stimulate PG synthesis. Diffusion of PGF$_{2\alpha}$ from the decidua in large amounts to the myometrium causes increased contractility (step 10) while lesser amounts of PGE$_2$ may increase oxytocin receptors and the response to oxytocin.

We suggest that the synthesis of chorionic lipocortin is stimulated by progesterone (step 11). Progesterone inhibits release of arachidonic acid in a perifused dispersed cell system although not PLA_2 activity in a cell-free system (Wilson et al. 1986), an effect that could be mediated by activation of lipocortin.

In summary, it is proposed that the triggering of synthesis of PGE_2 in amnion by unidentified physiological stimuli or by physical, chemical or bacteriological insults sets in train a wave of events involving fetal and maternal tissues that culminates in the activities of smooth muscle cells and fibroblasts that we call labour.

References

Alexandrova M, Soloff MS (1980) Oxytocin receptors and parturition. III. Increases in oestrogen receptor and oxytocin receptor concentrations in the rat myometrium at term. Endocrinology 106: 739–743

Barnford DS, Jogea M, Williams KI (1980) Prostaglandin formation by the pregnant human myometrium. Br J Obstet Gynaecol 87: 215–218

Bazer FW, Vallet JL, Roberts RM, Sharp DC, Thatcher WW (1986) Role of conceptus secretory products in establishment of pregnancy. J Reprod Fertil 76: 841–840

Billah MM, Johnston JM (1983) Identification of phospholipid platelet-activating factor (1-0-alkyl-2-acetyl-sn-glycero-3-phosphocholine) in human amniotic fluid and urine. Biochem Biophys Res Commun 113: 51–58

Blackwell GJ, Carnuccio R, DiRosa M, Flower RJ, Parente L, Persico L (1980) Macrocortin: a polypeptide causing the antiphospholipase effect of glucocorticoids. Nature 287: 147–149

Casey ML, MacDonald PC (1986) Initiation of labor in women. In: Huszar G (ed) The physiology and biochemistry of the uterus in pregnancy and labor. CRC Press, Boca Raton, p 155

Casey ML, MacDonald PC, Mitchell MD (1983) Stimulation of prostaglandin E_2 production in amnion cells in culture by a substance(s) in human fetal and adult urine. Biochem Biophys Res Commun 114: 1056–1063

Casey ML, MacDonald PC, Mitchell MD (1985) Despite a massive increase in cortisol secretion in women during parturition, there is an equally massive increase in prostaglandin synthesis: a paradox? J Clin Invest 75: 1852–1857

Dawood MY, Khan-Dawood FS (1986) Human ovarian oxytocin: its source and relationship to steroid hormones. Am J Obstet Gynecol 154: 756–763

Dawood MY, Ylikorkala O, Trivedi D, Fuchs F (1979) Oxytocin in maternal circulation and amniotic fluid during pregnancy. J Clin Endocrinol Metab 49: 429–434

Di Renzo GC, Anceschi MM, Bleasdale JE (1984) Beta-adrenergic stimulation of prostaglandin production by human amnion tissue. Prostaglandins 27: 37–49

Forman A, Andersson KE, Persson CGA, Ulmsten U (1979) Relaxant effects of nifedipine on isolated human myometrium. Acta Pharmacol Toxicol 45: 81–87

Fuchs A-R, Fuchs F (1984) Endocrinology of human parturition: a review. Br J Obstet Gynaecol 91: 948–967

Fuchs A-R, Fuchs F, Husslein P, Soloff MS, Fernstrom MJ (1982a) Oxytocin receptors and human parturition: a dual role for oxytocin in the initiation of labor. Science 215: 1396–1398

Fuchs A-R, Husslein P, Sumulong L, Fuchs F (1982b) The origin of circulating 13,14-dihydro-15-keto-$PGF_{2\alpha}$ during delivery. Prostaglandins 24: 715–722

Fuchs A-R, Fuchs F, Husslein P, Fernstrom MJ, Soloff MS (1984) Oxytocin receptors in the human uterus during pregnancy and parturition. Am J Obstet Gynecol 150: 734–741

Hirata F, Schiffman E, Venkatasubramanian K, Salomon D, Axelrod J (1980) A phospholipase A_2 inhibitory protein in rabbit neurophils induced by glucocorticosteroids. Proc Natl Acad Sci USA 77: 2533–2536

Hirata F, Matsuda K, Notsu Y, Hattori T, Del Carmine R (1984) Phosphorylation of a tyrosine residue of lipomodulin in nitrogen stimulated murine thyrocytes. Proc Natl Acad Sci USA 81: 4717–4721

Irvine RF (1982) How is the level of free arachidonic acid controlled in mammalian cells? Biochem J 204: 3–16

Keirse MJNC, Hicks BR, Mitchell MD, Turnbull AC (1977) Increase of the prostaglandin precursor, arachidonic acid, in amniotic fluid during spontaneous labour. Br J Obstet Gynaecol 84: 937–940

Khanna NC, Masaaki T, Waisman DM (1986) Biochem Biophys Res Commun 547–554

Kuno M, Gardner P (1987) Ion channels activated by inositol 1,4,5-triphosphate in plasma membrane of human T-lymphocytes. Nature 326: 301–304

Liggins GC, Fairclough RJ, Grieves SA, Forster CS, Knox BS (1977) Parturition in the sheep. In: Knight J, O'Connor M (eds) The fetus and birth. Ciba Found Symp 47: 5

Liggins GC (1984) The paracrine system controlling human parturition. In: Jaffe RB, Dell'Aqua S (eds) The endocrine physiology of pregnancy and the peripartal period. Raven, New York, p 205

MacDonald PC, Parker JC, Schwarz BE, Johnston JM (1974) Initiation of parturition I. Mechanism of action of arachidonic acid. Obstet Gynecol 44: 629–636

Manzai M, Liggins GC (1984) Inhibitory effects of dispersed amnion cells on production rates of prostaglandin E and F by endometrial cells. Prostaglandins 28: 297–307

Mitchell MD, Casey ML (1984) Role of growth factors in PGE_2 formation in human amnion cells in monolayer culture. Proc Soc Gynecol Invest, p 33

Mitchell MD, Bibby JG, Hicks BR, Turnbull AC (1978) Specific production of prostaglandin E by human amnion in vitro. Prostaglandins 15: 377–382

Mitchell MD, Ebenhack K, Kraemer DL, Cox K, Cutter S, Strickland DM (1982) A sensitive radioimmunoassay for 11-deoxy-13,14-dihydro-15-keto-11,16-cyclo-prostaglandin E_2. Prostaglandins Leukotrienes Med 9: 549–557

Mitchell MD, MacDonald PC, Casey ML (1984) Stimulation of prostaglandin E_2 synthesis in human amnion cells maintained in monolayer culture by a substance(s) in amniotic fluid. Prostaglandins Leukotrienes Med 15: 399–407

Mitchell MD, Craig DA, Saeed SA, Strickland DM (1985) Endogenous stimulant of prostaglandin endoperoxide synthase activity in human amniotic fluid. Biochim Biophys Acta 833: 379–385

Niesert S, Christopherson W, Korte K (1987) Prostaglandin E_2 9-keto-reductase activity in human decidua vera tissue. Am J Obstet Gynecol (in press)

Okazaki T, Casey ML, Okita JR, MacDonald PC, Johnston JM (1981) Initiation of human parturition XII. Biosynthesis and metabolism of prostaglandin in human fetal membranes and uterine decidua. Am J Obstet Gynecol 139: 373–381

Okazaki T, Sagawa N, Okita JR, Bleasdale JE, MacDonald PC, Johnston JM (1981) Diacylglycerol metabolism and arachidonic acid release in human fetal membranes and decidua vera. J Biol Chem 256: 7316–7321

Olson DM, Skinner K, Challis JRG (1983) Prostaglandin production in relation to parturition by cells dispersed from human intrauterine tissues. J Clin Endocrinol Metab 57: 694–699

Reimer RK, Goldfien AC, Goldfien A, Roberts JM (1986) Rabbit uterine oxytocin receptors and in vitro contractile response: abrupt changes at term and the role of ecosanoids. Endocrinology 119: 699–709

Robinson JS, Mitchell MD, Challis JRG (1979) Parturition in the rhesus monkey. In: Keirse MJNC, Anderson ABM, Gravenhorst JB (eds) Human parturition. Nijhoff, The Hague, p 25

Saeed SA, Mitchell MD (1982) Formation of arachidonate lipoxygenasae metabolites by human fetal membranes, uterine decidua vera and placenta. Prostaglandins Leukotrienes Med 8: 635–640

Schrey MP, Read AM, Steer PJ (1986) Oxytocin and vasopressin stimulate inositol phosphate production in human gestational myometrium and decidua cells. Biosci Rep 6: 613–619

Skinner SJM, Liggins GC, Wilson T, Neale G (1984) Synthesis of prostaglandin F by cultured human endometrial cells. Prostaglandins 27: 821–838

Soloff MS, Fernstrom MA, Periyasamy S, Soloff S, Baldwin S, Wieder M (1983) Regulation of oxytocin receptor concentration in rat uterine explants by oestrogen and progesterone. Can J Biochem Cell Biol 61: 625–630

Soloff MS (1985) Oxytocin receptors and mechanisms of oxytocin action. In: Amico JA and Robinson AG (eds) Oxytocin: clinical and laboratory studies. Excerpta Medica, Amsterdam, p 259

Thorburn GD (1985) Prostaglandins and the regulation of myometrial activity: a working model. In: Jones CT and Nathanielsz PW (eds) The physiological development of the fetus and newborn. Academic, London, p 381

Ulmsten U, Andersson KE, Wingerup L (1980) Treatment of premature labor with the calcium antagonist nifedipine. Arch Gynecol 229: 1–6

Wilson T, Liggins GC, Aimer GP, Skinner SJM (1985) Partial purification and characterization of two compounds from amniotic fluid which inhibit phospholipase activity in human endometrial cells. Biochem Biophys Res Comm 131: 22–29

Wilson T, Liggins GC, Joe L (1987) Purification from incubated chorion of a phospholipase A_2 inhibitor active before but not after the onset of labor. Biochem Biophys Acta (submitted)

Wilson T, Liggins GC, Aimer GP, Watkins EJ (1986) The effect of progesterone on the release of arachidonic acid from human endometrial cells stimulated by histamine. Prostaglandins 31: 343–360

Activation of Fetal Hypothalamic-Pituitary Function and Birth*

J. R. G. Challis[1], A. N. Brooks, L. J. Fraher, L. J. Norman, A. D. Bocking, S. A. Jones, and L. Power

Introduction

Birth is initiated through activation of an endocrine organ communication system. In species such as the sheep it is clear that activation of the fetal hypothalamic-pituitary-adrenal axis plays a pivotal role in this process. Precocious activation of this axis leads to premature delivery, whereas ablation of the fetal pituitary or adrenal results in prolonged gestation. For these reasons, we have examined factors responsible for activation and maturation of hypothalamic-pituitary function in the fetal sheep during late pregnancy in association with spontaneous or induced parturition.

Normal Changes in Plasma Cortisol and Adrenocorticotrophic Hormone

It is well established that the fetal plasma cortisol concentration rises during late gestation in the ovine fetus. Before day 120 of gestation most of the cortisol in the fetal circulation is derived from transplacental transfer from the mother, whereas the prepartum increase is accounted for largely by an increase in fetal adrenal secretion of cortisol (Hennessy et al. 1982). The rise in cortisol is accompanied by an increase in the corticosteroid binding capacity (CBC) of the fetal circulation. The rise in CBC can be provoked by administration of adrenocorticotrophic hormone (ACTH) to the fetus (Challis et al. 1985) and would appear to be of hepatic or renal origin in response to the elevation in plasma cortisol. Thus changes in the bound and free cortisol concentration, which affect the feedback relationships of cortisol with the pituitary and hypothalamus, may be pronounced in the late gestation fetus.

* This work was supported by operating grants from the Medical Research Council of Canada (Group Grant in Reproductive Biology, JRGC; grant no. MA 9704, ADB), the Easter Seal Research Institute (JRGC) and the National Institute of Nutrition (LJF).
[1] The Lawson Research Institute, St. Joseph's Health Centre, University of Western Ontario Departments of Physiology and Obstetrics and Gynaecology, and Medical Research Council Group in Reproductive Biology, 268 Grosvenor Street, London, Ontario, N6A 4V2, Canada

The Endocrine Control of the Fetus
Ed. by W. Künzel and A. Jensen
© Springer-Verlag Berlin Heidelberg 1988

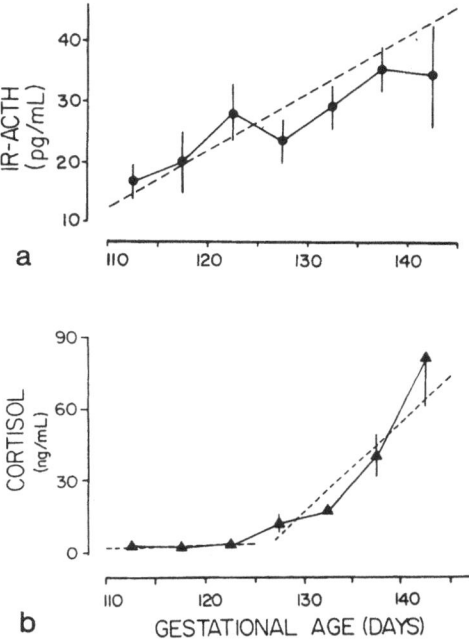

Fig. 1a, b. The concentration of **a** IR-ACTH and **b** cortisol in chronically catheterized fetal lambs from day 110 of gestation to within 2 days of spontaneous parturition. Each point is the mean ± SEM for up to 11 fetuses at successive 5-day intervals. Dashed lines in **a** regressioin analyses through individual animal data points, in **b** regression analyses through data points for individual sheep before day 125 and after day 125 of gestation. (From Norman et al. 1985)

Recent reports using specific radioimmunoassays have shown that there is a progressive rise in the fetal plasma ACTH concentration during the last 30 days of gestation (MacIsaac et al. 1985; Norman et al. 1985). This rise begins well before the exponential increase in plasma cortisol, which occurs mainly after days 120-125. One can calculate that immunoreactive ACTH (IR-ACTH) increases approximately 5 pg/ml for a 5-day period in fetal sheep between days 110 and 140, with a further rapid rise at the time of labour, perhaps in response to the stress of birth itself (Norman et al. 1985). Thus it may be concluded that there is a gradual increase in the trophic drive in fetal sheep from the pituitary to the adrenal gland during the last 30-40 days of pregnancy (Fig. 1).

We used high-pressure liquid chromatography (HPLC) to examine further the nature of the immunoreactive ACTH species released in relation to parturition. Samples of fetal and maternal plasma were extracted with C18 Sep-pak and eluted with 80% acetonitrile, 20% water (Bennett et al. 1981). These extracts were subjected to reverse phase HPLC on a Bondapak C18 column and eluted with a gradient of 20%-60% CH_3CN in 0.1% trifluoroacetic acid over 40 min at 1 ml/min. Concentra-

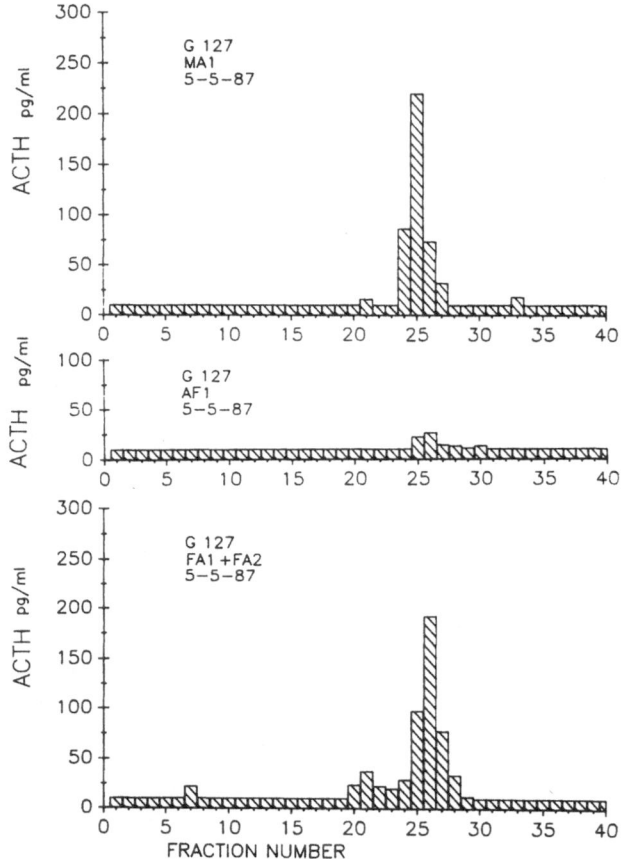

Fig. 2. Reverse phase HPLC profile of immunoreactive ACTH in samples of maternal arterial blood *(MA1)*, amniotic fluid *(AF1)* and fetal arterial blood *(FA1 + FA2)* collected from one sheep (G 127) within 30 min of spontaneous labour at term. Authentic ACTH$_{1-39}$ elutes in samples 25 and 26

tions of IR-ACTH in successive 1 ml fractions were determined using an antibody that recognizes primarily the mid-portion of ACTH$_{1-39}$. Samples of fetal and maternal plasma taken simultaneously within 30 min of delivery showed a single major peak of immunoreactivity corresponding in retention time to standard ACTH$_{1-39}$ (Fig. 2). Amniotic fluid collected at the time of delivery also contained immunoreactivity eluting at the position of standard ACTH$_{1-39}$.

These studies show that ACTH$_{1-39}$ is the major radioimmunoreactive form of ACTH measured in our assay system in fetal plasma at term. Similarly, ACTH$_{1-39}$ is the major form of ACTH present in the fetal circulation at earlier stages of gestation (see below). The design of our study, however, does not exclude the existence of other species of different molecular weight related to pro-opiomelanocortin (POMC) (Jones, 1983) that do not cross-react with our ACTH antibody.

Pituitary Responsiveness

We have examined the hypothesis that the progressive increase in ACTH concentration in the fetal circulation is due to progressive maturation of pituitary responsiveness to agents such as corticotrophin releasing factor (CRF). We found that exogenous CRF injected as a bolus into the fetal femoral vein provoked ACTH release in a dose-dependent fashion throughout late gestation (Norman et al. 1985). After CRF administration, ACTH rose between days 110 and 115 and between days 125 and 130 of pregnancy, but decreased again towards day 140. In contrast, the associated increment in plasma cortisol was small during early pregnancy, but basal and stimulated cortisol rose progressively between days 125–130 of gestation and day 140. These experiments provided evidence for maturation of pituitary adrenal responsiveness in the fetus during late pregnancy. The enhanced pituitary responsiveness could be due to an increase in the CRF receptor population or their coupling to adenylate cyclase, to the emergence of an adult population of corticotrophs in the pituitary, or to effects on post-translational processing of POMC. Maturation of fetal adrenal responsiveness is well known and is probably due to an increase in the number of ACTH receptors, enhanced receptor coupling to adenylate cyclase, and an increase in steroidogenic enzyme activity, including 17α-hydroxylase and 3β-hydroxysteroid dehydrogenase (Challis et al. 1984). From these studies it is apparent that pituitary responsiveness to CRF precedes the emergence of adrenal responsiveness to ACTH.

Of further interest in the above experiments was the sustained elevation of plasma IR-ACTH even at 4 h after bolus CRF injection. In part this could have been due to the continued presence of CRF in the fetal circulation. To examine this possibility fetal plasma CRF concentrations were measured after bolus injection of synthetic ovine CRF (1 µg). This study showed a rapid increase in the IR-CRF concentration to peak values within 5 min of the injection, with a gradual decline thereafter, and an initial half-life of approximately 60 min. However, IR-CRF was still present in the fetal circulation at 240 min after injection. The continued presence of CRF could explain the sustained stimulation of ACTH release after CRF injection. HPLC profiles of IR-ACTH showed that at both 15 min and 240 min after exogenous CRF administration the major peptide measured in the fetal circulation corresponded in retention time to authentic $ACTH_{1-39}$.

In contrast to the response to CRF, administration of arginine vasopressin (AVP) provoked a significant but short-lived elevation in the plasma ACTH concentration, with a rapid return to basal values by 30–60 min (Norman and Challis 1987a). The half-life of AVP is 2–10 min in the ovine fetus (Wiriyathian et al. 1983) and the ACTH response is consistent with the rapid disappearance of the peptide. At day 115 of gestation there was synergism between AVP and CRF in provoking ACTH release (Norman and Challis 1987a). This interaction was lost in older fetuses. Similar effects of AVP on ACTH output and synergism with CRF have been reported using dispersed pituitary cells in short-term culture (Durand et al. 1986). It is of interest that pharmacological doses of AVP do stimulate ACTH release throughout gestation (Norman and Challis 1987b) and that the synergism between AVP and CRF reappears in vivo, and with pituitary cells from adult sheep studied in vitro (S.A.Jones and J.R.G.Challis 1987, unpublished observations).

The apparent decrease in the ACTH response to CRF in late gestation is consistent with an emerging negative feedback effect of glucocorticoids at the pituitary. We found that exogenous dexamethasone blocked pituitary ACTH release in response to CRF, AVP, or AVP+CRF in fetal sheep (Norman and Challis 1987c). In the youngest fetuses (day 115) exogenous dexamethasone had no effect on the basal output of ACTH, although it did affect basal ACTH in fetuses after day 125. This time-frame would be consistent with the rise in plasma ACTH that occurs after days 120-125 of gestation following adrenalectomy of the fetal sheep (Wintour et al. 1980). It is also consistent with the changing ability of dexamethasone to inhibit basal CRF output by perifused hypothalamic tissue (see below).

We next examined whether exogenous CRF would activate pituitary adrenal function and cause premature parturition (Brooks et al. 1987). CRF was administered in pulses (1 µg over 12 min every 4 h) to fetal sheep for 7 days beginning at about day 120. This treatment provoked a significant rise in the basal ACTH and cortisol concentration. However, with successive days of treatment the pituitary ACTH response to CRF pulses decreased, while the adrenal cortisol response to endogenously released ACTH rose. The change in ACTH concentration after CRF was correlated negatively to the basal cortisol concentration at the time of the CRF pulse, suggesting that the blunted pituitary response was due to negative feedback of the elevated glucocorticoids. The blunted pituitary response could also be due to down-regulation of the CRF receptor population. However, when pituitary cells were prepared from fetuses that had been pretreated in vivo with CRF for 7 days and exposed to CRF in vitro, their output of cyclic AMP was greater than that of fetuses pretreated with saline. This observation indicates that exogenous CRF treatment in vivo had increased pituitary CRF receptor-adenylate cyclase coupling. These measurements after day 7 of treatment, however, do not allow one to draw conclusions concerning the possibility of desensitization during the in vivo treatment period. In these studies fetuses did not enter preterm labour. It is likely that a longer period of exogenous pulsed CRF treatment would be required to produce premature delivery, and others have found that exogenous CRF administered in a less controlled protocol will induce birth (Wintour et al. 1986).

We have also examined the pituitary and adrenal responses of fetal sheep measured as plasma ACTH and cortisol changes during prolonged (48 h) hypoxaemia. Hypoxaemia was provoked by adjusting a vascular clamp placed around the maternal common internal iliac artery and restricting uterine arterial blood flow. This

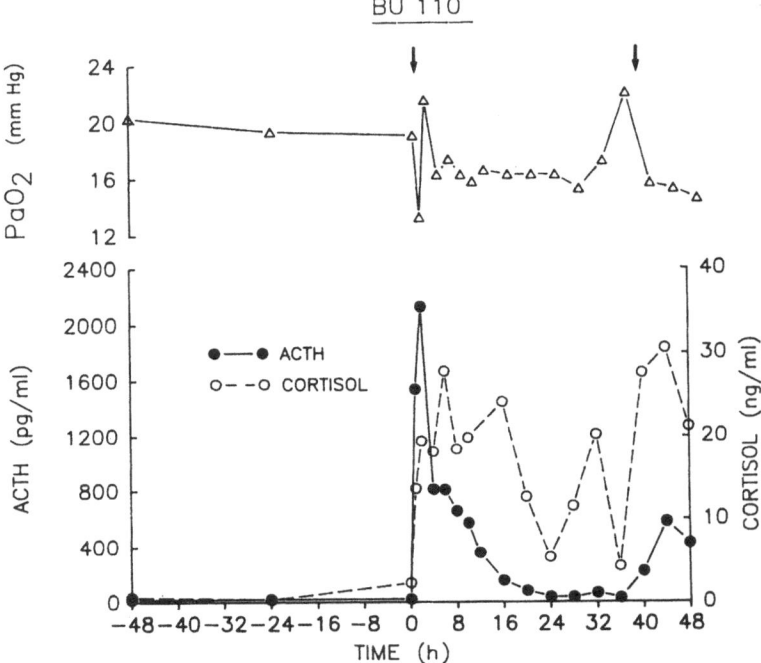

Fig. 3. Changes in PaO₂, ACTH (●——●) and cortisol (O--O) in arterial blood of a fetal sheep (BU 110) at day 125 of gestation after partial occlusion of the maternal common iliac artery with a vascular clamp applied at the times indicated (↓)

treatment caused a rapid decrease in fetal PaO_2, which remained depressed a 15–16 mmHg through the next 24–48 h. Plasma ACTH concentrations rose to a maximum after 1–2 h of hypoxaemia (Fig. 3). Using HPLC it was shown that this immunoreactivity co-chromatographed predominantly with authentic standard $ACTH_{1-39}$. However, the ACTH concentration then decreased towards baseline, even though the fetal PaO_2 remained depressed. In some animals the fetal PaO_2 was reduced further at $+24-+36$ h, and this was associated with a second increase in the plasma ACTH concentration. Thus, the profile of declining ACTH during the first 24 h of hypoxaemia was unlikely to be the result of depleted pituitary reserves of ACTH. This result suggests that the fetus may adapt to the low PaO_2 and adjust the threshold for ACTH release, although effects of elevated cortisol and negative feedback on ACTH cannot be excluded. The mean plasma cortisol concentration rose progressively with the initial ACTH response and remained elevated despite the subsequent decrease in ACTH. This presumably is due to an increase in fetal adrenal responsiveness in the presence of the slightly elevated ACTH concentrations, and if prolonged could result in preterm labour.

Controls of CRF

In view of the ability of exogenous CRF to elicit ACTH release we sought evidence
for the presence of endogenous CRF in fetuses, and we examined the controls of
CRF output from perifused hypothalami in vitro.

A radioimmunoassay to ovine CRF (oCRF) was developed using an antibody
raised in rabbits to CRF conjugated with thyroglobulin. The antibody does not
cross-react with a variety of other peptides studied, including AVP, LHRH, TRH,
enkephalins and ACTH. IR-CRF was present in acid extracts of medial basal hypothalami from fetuses between day 100 and term. The hypothalamic content of CRF
was similar at day 100 and at day 140 but varied appreciably between days 125 and
135, the time of the emerging negative and positive feedback controls of glucocorticoids and ACTH output. Hypothalamic tissue at days 100 and 140 was extracted
and the extract subjected to gel filtration on Sephadex G-75 columns. Succsessive
1-ml eluate fractions were collected, divided and assayed for IR-CRF content and
for biological activity by addition to anterior pituitary cells maintained in tissue culture. On Sephadex gel chromatography, hypothalami at both stages of gestation
contained peaks of IR-CRF that eluted in the same fractions as standard oCRF.
The output of IR-ACTH was measured after 2 h exposure of pituitary cells to successive fractions eluted from the column. It was established that the major biological activities corresponded to the major peaks of CRF immunoactivity.

In order to examine controls of CRF output, hypothalami from fetuses at
day 100 and at day 140 were then perifused in vitro. The basal output of immunoreactive CRF at day 140 was significantly greater than at day 100. At both times of
gestation there was significant release of IR-CRF in response to a bolus of 56 mM
KCl, and the change in IR-CRF after K^+ administration did not differ between the
two gestational ages. At day 100 dexamethasone had no effect on either the basal or
K^+-stimulated CRF release. However, at day 140 dexamethasone significantly decreased basal CRF release, although it had no effect on K^+-stimulated release.
These findings are consistent with an action of dexamethasone on nerve cell bodies,
since potassium provokes CRF release at the nerve terminals. We conclude that the
release mechanisms for IR-CRF are present by day 100, at least as assessed in vitro
(Fig. 4), and that negative feedback on basal, but not on stimulated, CRF output
was demonstrable only at day 140. These results are consistent with our earlier in vivo observations on the lack of dexamethasone negative feedback effect on basal release of ACTH at days 110–115 of gestation.

In concurrent work with human and ovine fetal membranes we have sought evidence for alternative sites of IR-CRF output. Amnion, chorion and decidual cells
from women at elective cesarean section were grown in monolayer culture and the
output of IR-CRF was compared with similar cultures of placental tissue. At day 2
of culture the output of IR-CRF by chorion and placenta was similar, and significantly greater than that from the amnion (Fig. 5). When media from these different
cultures was added to monolayers of adult ovine pituitary cells there was consistent
simulation of IR-ACTH output. Although it is well known that placental tissue pro-

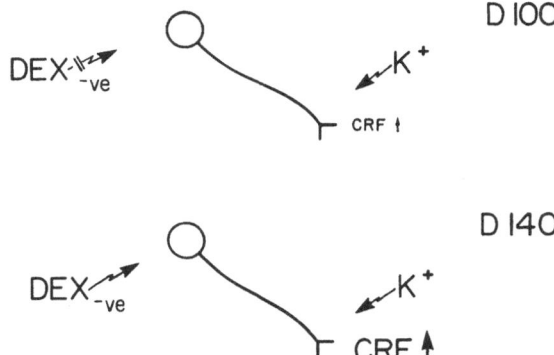

Fig. 4. Summary of dexamethasone and K⁺ effects on CRF relese from in vitro perifused hypothalamic tissue obtained from fetuses at days 100 and 140 of gestation

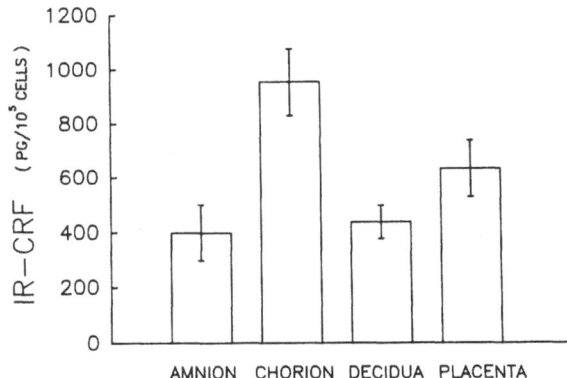

Fig. 5. Output of immunoreactive CRF (IR-CRF) from amnion, chorion, decidua and placenta obtained from nine patients at term elective cesarean section. Values are mean ± SEM for outputs on day 2 of culture

duces CRF, these experiments provide the first evidence for CRF production by the fetal membranes and for the biological activity of that material. CRF is present in umbilical cord and maternal blood in human pregnancy and in lower amounts in fetal blood in ovine pregnancy. The contribution of CRF produced by the fetal membranes and placenta to the controls of hypothalamic-pituitary-adrenal function in the fetus is unknown at the present, but provides an additional intriguing dimension to an already complicated story of endocrine ontogeny.

Acknowledgements. We are indebted to Nan Cumming, Susan White and Jessica Oosterhuis for their assistance with this work.

References

Bennett HPJ, Browne CA, Solomon S (1981) Biosynthesis of phosphorylated forms of cortico-tropin-related peptides. Proc Natl Acad Sci USA 78: 4713–4717

Brooks AN, Challis JRG, Norman LJ (1987) Pituitary and adrenal responses to pulsatile ovine corticotropin-releasing factor administered to fetal sheep. Endocrinology 120: 2383–2388

Challis JRG, Mitchell BF, Lye SJ (1984) Activation of fetal adrenal function. J Dev Physiol 6: 93–105

Challis JRG, Nancekievill EA, Lye SJ (1985) Possible role of cortisol in the stimulation of cortisol binding capacity (CBC) in the plasma of fetal sheep. Endocrinology 116: 1139–1144

Durand P, Cathiard A-M, Dacheux F, Naaman E, Saez JM (1986) In vitro stimulation and inhibition of adrenocorticotropin release by pituitary cells from ovine fetuses and lambs. Endocrinology 118: 1387–1394

Hennessy DP, Coghlan JP, Hardy KJ, Scoggins BA, Wintour EM (1982) The origin of cortisol in the blood of fetal sheep. J Endocrinol 95: 71–79

Jones CT (1983) The integration of adrenal and pituitary activity during development. In: MacDonald PC, Porter J (eds) Initiation of parturition: prevention of prematurity. 4th Ross Conference on Obstetric Research. Ross Laboratories, Columbus OH, p 17

MacIsaac RJ, Bell RJ, McDougall JG, Tregear GW, Wang X, Wintour EM (1985) Development of the hypothalamic-pituitary axis in the ovine fetus: ontogeny of action of ovine corticotropin-releasing factor. J Dev Physiol 7: 329–338

Norman LJ, Challis JRG (1987a) Synergism between systemic CRF and AVP on ACTH release in vivo varies as a function of gestational age in the ovine fetus. Endocrinology 120: 1052–1058

Norman LJ, Challis JRG (1987b) Dose-dependent effects of arginine vasopressin on endocrine and blood gas responses of fetal sheep during the last third of pregnancy. Can J Physiol Pharmacol 65: 2291–2296

Norman LJ, Challis JRG (1987c) Dexamethasone inhibits oCRF, AVP and oCRF+AVP stimulated release of ACTH during the last third of pregnancy in the sheep fetus. Can J Physiol Pharmacol 65: 1186–1192

Norman LJ, Lye SJ, Wlodek ME, Challis JRG (1985) Changes in pituitary responses to synthetic ovine corticotrophin-releasing factor in fetal sheep. Can J Physiol Pharmacol 63: 1398–1403

Wintour EM, Coghlan JP, Hardy KJ, Hennessy DP, Lingwood BE, Scoggins BA (1980) Adrenal corticosteroids and immunoreactive ACTH in chronically cannulated ovine fetuses with bilateral adrenalectomy. Acta Endocrinol 95: 546–552

Wintour EM, Bell RJ, Carson RS, MacIsaac RJ, Tregear GW, Vale W, Wang X-M (1986) Effect of long-term infusion of ovine corticotrophin-releasing factor in the immature ovine fetus. J Endocrinol 111: 469–475

Wiriyathian S, Porter JC, Naden RP, Rosenfeld CR (1983) Cardiovascular effects and clearance of arginine vasopressin in the fetal lamb. Am J Physiol 245: E24–E31

The Regulation and Effects of Myometrial Activity on the Fetus

P. W. Nathanielsz[1]

Introduction

In a review at the CIBA Foundation symposium meeting on The Fetus at Birth, in 1976, we described our early observations on the temporal relationships between myometrial contractures and fetal electrocorticogram (ECOG; Fig. 1). We observed long-term epochs of myometrial contractility as reflected by an increased intrauterine pressure (IUP) lasting 5–15 min which we designated *contractures* to distinguish them from labor and delivery contractions. We stated that in several instances there was a suggestion that high amplitude fetal ECOG activity may in some way be temporally related to these contractures. Our early hypothesis was that contractures in

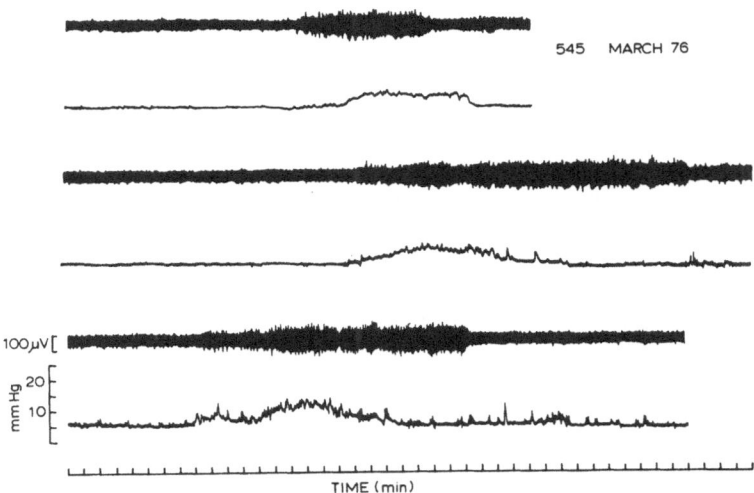

545   MARCH 76

100 μV

mm Hg 20 10

TIME (min)

Fig. 1. Simultaneous recording of uterine pressure (*lower trace* of each pair) and fetal electroencephalogram recording (*upper trace* of each pair) in a fetal sheep (reproduced with permission from Nathanielsz et al. 1976)

[1] Laboratory for Pregnancy and Newborn Research, New York State College of Veterinary Medicine, Cornell University, Ithaca, NY 14853, USA

The Endocrine Control of the Fetus
Ed. by W. Künzel and A. Jensen
© Springer-Verlag Berlin Heidelberg 1988

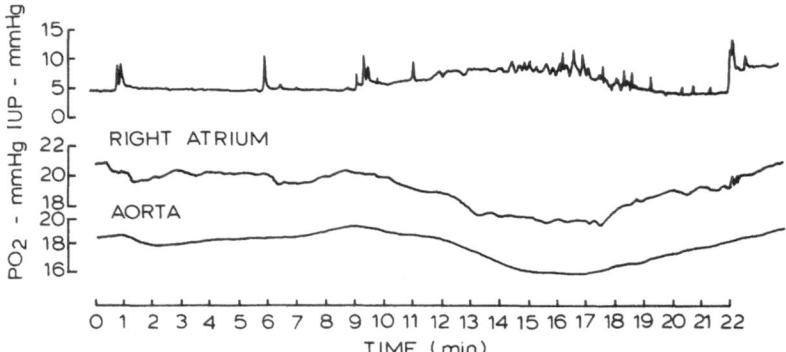

Fig. 2. Simultaneous measurement of IUP and right atrial and aortic arch PO_2 in a fetal sheep 113 days' gestation 5 days postoperatively. Fetus died during delivery following induction of labor with adrenocorticotropic hormone at 123 days' gestation (reproduced with permission from Jansen et al. 1979)

some way affected the fetus either by decreasing oxygenation or by a sensory stimulus (Nathanielsz et al. 1976).

Shortly after these studies were reported, we were fortunate to be able to use an indwelling PO_2 electrode designed by Dr. Dawood Parker to enable us to follow fetal vascular PO_2 continuously (Jansen et al. 1979). With this electrode we were able to show that contractures were accompanied by a fall in fetal PO_2 of 2 mm Hg or more (Fig. 2). To enable quantification of the PO_2 falls we arbitrarily determined that a fall of 2 mm Hg was necessary as a threshold to look for correlations. Several of the PO_2 falls associated with contractures are 4 mm Hg or more. To date, no investigators have carefully documented the profile of the PO_2 falls in a large group of animals under defined experimental conditions. This needs to be done. As a result of our initial studies with the PO_2 electrode we put forth the testable hypothesis that rhythmic changes in fetal PO_2 caused, at least in part, by alteration of uterine muscle tone, and/or sensory input to the fetus from the contracture, are able to influence fetal neural activity. Assessment of the quantitative importance of these changes and the possibility that they constitute a major pathway whereby maternal rhythms affect the fetus will require sophisticated analysis of continuous records of fetal PO_2 coupled with other measurements such as fetal and maternal blood flow measurements, myometrial electromyogram (EMG), IUP, changes in fetal shape, and significant fetal physiological variables.

The two potential input pathways which may act either together or separately to affect fetal function are shown in Fig. 3. Studies by several investigators up to 1984 which throw light on these systems have been reviewed elsewhere (Nathanielsz et al. 1984b). The purpose of this review is to discuss some of our most recent data on the regulation of contractures in the pregnant sheep. We will also discuss some recent data on the effect of increasing contracture frequency on fetal oxygenation.

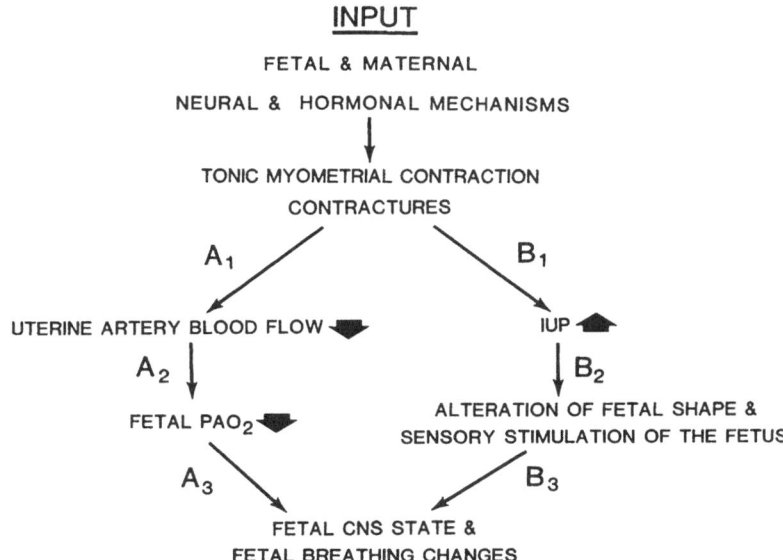

Fig. 3. Two potential pathways for production of changes in the fetal central nervous system *(CNS)* state and fetal breathing by contractures. *PaO$_2$*, arterial partial pressure of oxygen (reproduced with permission from Nathanielsz et al. 1984 c)

Methodology

The methods by which data are obtained will always determine the value of the data. Thus, it is necessary briefly to review our data acquisition and analytical techniques. We have developed long term recording techniques which permit us to record several electrophysiological signals for several days in succession, thereby obtaining information that is not distorted by circadian rhythms and in which there are adequate data points to begin to make some statistically firm conclusions. The relatively infrequent occurrence of contractures demands that conclusions drawn from small amounts of data be avoided. When only short periods of data acquisition are used, chance associations between contractures and other events will inevitably occur. Short term rhythms may also cloud the interpretation. Thus, careful, methodical hypothesis testing is required.

Our data acquisition system has been described at length (Nathanielsz et al. 1984 a, c). We now define a contracture on the basis of either EMG or IUP. On the basis of EMG, continuous activity for 3 min without quiescence longer than 16 s is considered to be the minimal duration of a contracture since more than 85% of contractures have this duration (Fig. 4). Our definition of the IUP changes accompanying a contracture requires that the change in IUP be at least 2 mm Hg above thresh-

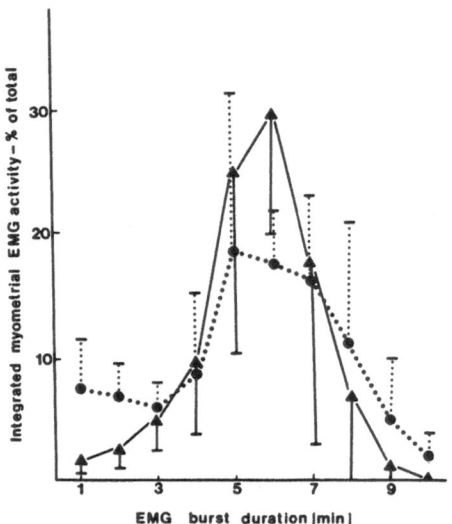

Fig. 4. Percentage of total activity (mean ± SD) in each 1-min group of myometrial electromyographic burst duration in four sheep during the first 8 days after operation at 60–132 days' gestation (group 1, ●·····●) and in three pregnant ewes at least 8 days after operation at 114–140 days' gestation (group , ▲———▲) (reproduced with permission from Figueroa et al. 1985 a)

old and last for 5 min (Jansen et al. 1979). The longer duration required when a contracture is monitored by the IUP change rather than the EMG reflects the fact that there is a finite relaxation time necessary before the IUP returns to baseline.

The question is often raised as to whether EMG or IUP is the better variable to monitor as a readout of myometrial activity. In essence, each variable will tell us different things. The uterine EMG is an index of the drive to the uterus and has the advantage of being easily recordable and easily analyzable. The IUP, on the other hand, is very position- and movement-sensitive and less easy to quantify when changes are small. In addition, our observations demonstrate quite clearly that, in the sheep at least, the IUP changes may not be uniform throughout the uterine cavity. This follows from the fact that in many parts of the uterus the fetus is in contact with the uterine wall. Thus, it is possible for a loculus of amniotic fluid to be isolated from the rest of the uterus.

Regulation of Myometrial Activity

The Effect of Starvation. Fowden and Silver (1983, 1985) have demonstrated that food withdrawal for 48 h in pregnant ewes after 137 days' gestation produces increases in IUP. However, monitoring IUP they noticed that prior to 137 days' gestation, although maternal plasma concentrations of 13,14-dihydro-15-keto prostaglandin $F_{2\alpha}$ (PGFM) rose, IUP did not increase. Since we are of the opinion that uterine EMG provides a better indication of drive to the pregnant myometrium, we thought it of value to investigate the effect of food withdrawal on uterine EMG (Milvae et al. 1986). We were able to demonstrate that during a 48-h period of food withdrawal at 122-127 days' gestation maternal plasma glucose concentration fell on the second day to $42.2 \pm 4.4\%$ (mean $\pm$ SEM) of resting. Maternal uterine venous PGFM concentration rose to $213.7 \pm 22.5\%$, estrone sulfate rose to $308 \pm 67\%$, and contracture frequency rose to $133 \pm 15\%$ of baseline. This experiment provided two pieces of information additional to those found by Fowden and Silver. Firstly, estrone sulfate was shown to rise during starvation; this is of interest in relation to potential roles for estrogen in regulating myometrial contractility. Secondly, it demonstrated that in certain situations EMG may be a more sensitive index of drive to the myometrium than IUP.

Diurnal Periodicity. Harbert (1977) and Novy et al. (1980) demonstrated circadian variation in myometrial activity by recording IUP in the pregnant monkey. Taylor et al. (1983) demonstrated similar circadian patterns using myometrial EMG. In this last study only contractions, i.e., short-lived activity, were analyzed and the major circadian variation was observed close to labor and delivery. We asked the question whether circadian patterns were present in either contractions or contractures in the pregnant sheep in late gestation. Five pregnant ewes were studied by recording uterine EMG continuously at 120-130 days' gestation and at least 5 days after surgery. Three animals had the day time reversed, being exposed to light from 1900-0900, and in two animals the normal lighting pattern occurred with lights on from 0700-2100. Myometrial EMG was analyzed using a 16-s delimiter. Cosinor analysis was used over at least 82 h in each animal and a significant periodicity was demonstrated in myometrial EMG events of less than 3 min duration. Activity was highest at 4-5 h into the dark period. There was no demonstratable circadian rhythm in contractures (Figueroa et al. 1985).

Effect of Inhibition of Prostaglandin Synthetase. We have demonstrated that 4-aminoantipyrine and antipyrine are both inhibitors of prostaglandin synthetase. Contractures are inhibited by 4-aminoantipyrine administered via either the fetal or maternal route intravascularly at 125-143 days' gestation (Figs. 5, 6). Within 60 min of infusion of 4-aminoantipyrine, maternal uterine vein plasma PGFM concentration was reduced to 14% of resting level (Fig. 7). At the same time contracture frequency decreased to 30% of preinfusion values (El Badry et al. 1984). In a subsequent study

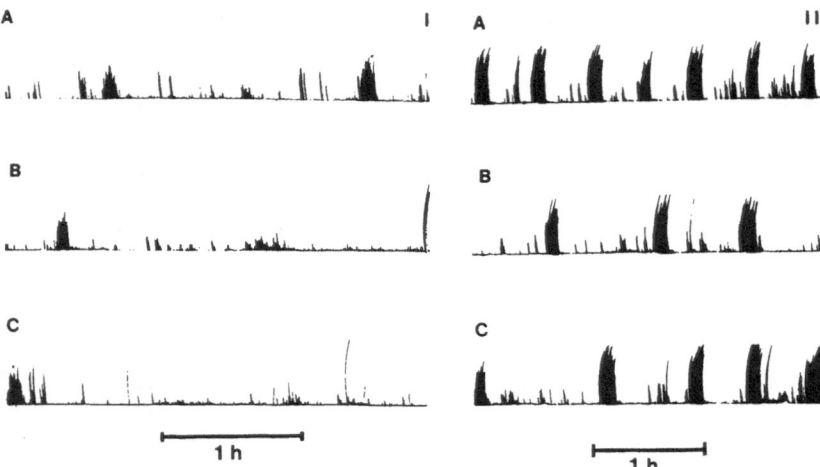

Fig.5A–C. Myometrial electromyographic trace in two pregnant ewes *(I, II)* before **(A)**, during **(B)**, and after **(C)** the administration of a bolus of 300 mg 4-aminoantipyrine to the fetus followed by infusion of 4-aminoantipyrine at 15 mg min^{-1} for 3 h (reproduced with permission from El Badry et al. 1984)

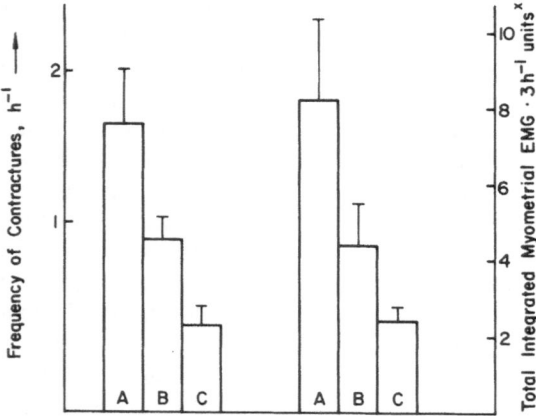

Fig.6. Frequency of contractures (h^{-1}) and total integrated myometrial electromyographic activity (1 unit = 2.5 µV s) in the pregnant sheep in the 3-h preinfusion of 4-aminoantipyrine *(A)*, during the 3-h infusion period *(B)*, and 3 h after infusion *(C)*. Infusion of 4-aminoantipyrine to the fetus was made at 125–143 days' gestation (*n* = 10, mean ± SEM) (reproduced with permission from El Badry et al. 1984)

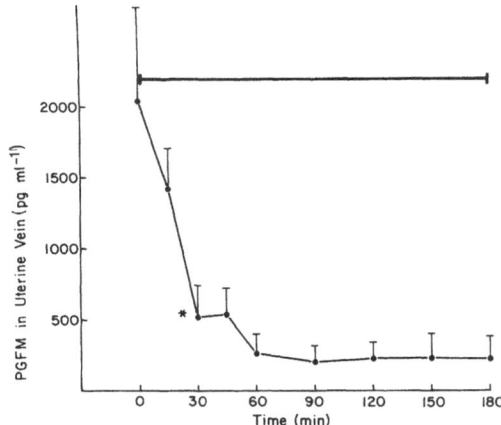

Fig. 7. Maternal uterine vein plasma concentrations of PGFM (mean ± SEM) immediately before and during the infusion of 4-aminoantipyrine into the fetal jugular vein at 15 mg min^{-1} *(bar).* *By 30 min the maternal uterine venous concentration of PGFM had fallen significantly ($p < 0.05$) (reproduced with permission from El Badry et al. 1984)

we demonstrated that the effect on prostaglandin metabolism was dose dependent. Antipyrine infused into the fetal jugular vein at rates of 4 and 15 mg min^{-1} induced a significant decrease in maternal uterine vein PGFM, whilst infusion at 1 mg min^{-1} was without effect (Pimentel et al. 1986). We have also demonstrated using a microsome – enriched preparation of bovine seminal vesicles that antipyrine and 4-aminoantipyrine will inhibit in vitro prostaglandin synthetase activity (Cohen et al. 1985). In a recent unpublished study we have shown that arachidonic acid administration to the pregnant ewe increases the frequency of contractures; this is followed by a later decrease. Interestingly, control animals receiving bovine serum albumin (BSA) vehicle alone underwent the late decrease in frequency. This raises interesting questions as to whether BSA contains an endogenous inhibitor of prostaglandin synthesis.

Acknowledgement. I would like to thank Karen Moore for her help with the manuscript.

References

Cohen D, Corbin J, Figueroa JP, Nathanielsz PW, Mitchell MD (1985) Inhibition of arachidonic acid metabolism by antipyrine and 4-aminoantipyrine. Am J Obstet Gynecol 154: 420–423

El Badry A, Figueroa JP, Poore ER, Sunderji S, Levine S, Mitchell MD, Nathanielsz PW (1984) Effect of fetal intravascular 4-aminoantipyrine infusions of myometrial activity (contractures at 125-143 days gestation in the pregnant sheep. Am J Obstet Gynecol 150: 474-481

Figueroa JP, McDonald TJ, Nathanielsz PW, Poore ER, Wentworth RA (1985) Circadian variation in myometrial activity in the chronically instrumented pregnant sheep at 120-130 days gestation. J Physiol (Lond) 369: 116P

Fowden AL, Silver M (1983) The effect of the nutritional state on uterine prostaglandin F metabolite concentrations in the pregnant ewe during late gestation. Q J Exp Physiol 68: 337-349

Fowden AL, Silver (1985) The effects of food withdrawal on uterine contractile activity and on plasma cortisol concentrations in ewes and their fetuses during late gestation. In: Jones CT, Nathanielsz PW (eds) The physiological development of the fetus and newborn. Academic, London, pp 157-161

Harbert GM (1977) Biorhythms of the pregnant uterus (*Macaca mulatta*). Am J Obstet Gynecol 129: 401-408

Jansen CAM, Krane EJ, Thomas AL, Beck NFG, Lowe KC, Joyce P, Parr M, Nathanielsz PW (1979) Continuous variability of fetal PO_2 in the chronically catheterized fetal sheep. Am J Obstet Gynecol 134: 776-783

Milvae R, Mitchell MD, Nathanielsz PW, Pimentel G, Rosen ED (1986) The effect of food withdrawal on myometrial electromyographic (EMG) activity and maternal plasma concentrations of 13,14-dihydro-15-keto prostaglandin $F_{2\alpha}$ (PGFM) and oestrone sulphate in the pregnant ewe at 122-127 days gestation (d G.A.). J Physiol (Lond): 41P

Nathanielsz PW, Ratter S, Thomas AL, Rees L, Jack PMB (1976) The role and regulation of corticotropin in the fetal sheep. CIBA Found Symp 47: 73-91

Nathanielsz PW, Frank D, Gleed R, Dillingham L, Poore ER, Figueroa JP (1984a) Methods of investigation of the chronically instrumented pregnant rhesus monkey preparation maintained on a tether and swivel system. In: Nathanielsz PW (ed) Animal models in fetal medicine, vol 3. Perinatology Press, Ithaca NY, pp 110-160

Nathanielsz PW, Jansen CAM, Yu HK, Cabalum T (1984b) Regulation of myometrial function throughout gestation and labor: effect on fetal development. In: Beard RW, Nathanielsz PW (eds) Fetal physiology and medicine: the basis of perinatology. Perinatology Press, Ithaca NY, pp 629-653

Nathanielsz PW, Poore ER, Brodie A, Taylor NF, Pimentel G, Figueroa JP, Frank D (1984c) Update on molecular events of myometrial activity during pregnancy. In: Nathanielsz PW, Parer JT (eds) Research in perinatal medicine. Perinatology Press, Ithaca NY, pp 87-111

Novy MJ, Walsh SW, Cook MJ (1980) Chronic implantation of catheters and electrodes in pregnant nonhuman primates. In: Nathanielsz PW (ed) Animal models in fetal medicine, vol I. Perinatology Press, Ithaca NY, pp 133-168

Pimentel G, Figueroa JP, Mitchell MD, Massmann A, Nathanielsz PW (1986) Effect of fetal and maternal intravascular antipyrine infusion on maternal plasma prostaglandin concentrations in the pregnant sheep at 104-127 days gestation. Am J Obstet Gynecol 155: 1181-1185

Taylor NF, Martin MC, Nathanielsz PW, Seron-Ferre M (1983) The fetus determines the circadian oscillation of myometrial electromyographic activity in the pregnant rhesus monkey. Am J Obstet Gynecol 146: 557-567

Regulation of Electrical Activity in the Myometrium of the Pregnant Ewe

G. D. Thorburn[1], H. C. Parkington, G. Rice, G. Jenkin, R. Harding, J. Sigger, M. Ralph, V. Shepherd and K. Myles

Introduction

Periodic contractions have been recorded from the uterus of many species during pregnancy (Taverne et al. 1979; Germain et al. 1982; Bell 1983; Demianczuk et al. 1984). In ewes, two contractions per hour occur after day 60 of gestation, and when pairs of electromyographical (EMG) electrodes are placed at several locations on the uterus, bursts of EMG activity can be recorded that occur approximately synchronously at the various sites (Harding et al. 1982). Since each burst of EMG activity across the uterus is associated with a small rise in intrauterine pressure (IUP), measured via an intra-amniotic catheter, it seems reasonable to assume that activity is generalized throughout the uterus during a burst. In this paper we will consider the generation of the bursts of EMG activity rather than their propagation throughout the organ. We will address the question as to whether the bursts are an intrinsic property of the muscle or whether activity is initiated by the periodic release of spasmogen.

Uterine Activity In Vitro

To explore more fully the possibility that the pattern of activity observed in the pregnant uterus might be exclusively myogenic in origin, we have studied activity in myometrium isolated in vitro (Shepherd et al. 1986). Strips of uterine muscle (15 mm long and 3 mm wide) were obtained from sheep from day 50 of pregnancy up to, and including, labour. Longitudinal and circular muscle components were investigated separately and also strips of 'whole' uterus that consisted of both muscle layers, in their normal association, with endometrium intact. In strips of longitudinal or circular muscle, spontaneous contractions appeared only after they had been placed in the organ bath for 30–60 min. After the onset of spontaneous activity, the frequency of contractions was greater in circular than in longitudinal myometrium at all stages of gestation studied. The mean frequency in strips of circular muscle was 0.57 ± 0.09 contractions per minute whereas that in longitudinal myometrium

[1] Department of Physiology, Monash University, Clayton, Victoria, Australia 3168

The Endocrine Control of the Fetus
Ed. by W. Künzel and A. Jensen
© Springer-Verlag Berlin Heidelberg 1988

was 0.22 ± 0.02 contractions per minute. The frequency of contractions in strips of whole uterus was intermediate between that observed in circular and longitudinal myometrium. Addition of indomethacin (10^{-5} μM) to the incubation medium caused a significant decrease in the frequency of contractions in all preparations.

The pattern of activity observed in myometrial strips in vitro (Parkington 1985) differed markedly from that occurring in the uterus of the intact animal. The delay in onset of spontaneous activity in tissue excised from the animal, together with the higher frequency of contractions in isolated strips of myometrium, suggest that uterine smooth muscle may be under tonic inhibition in vivo. Upon removal of the inhibitory agent, presumably by washout in vitro, the frequency of contractions greatly exceeds that observed in the uterus of the intact animal. It is possible that prostaglandins are involved in this intrinsic activity, since both the duration and the frequency of contractions were altered by indomethacin in vitro, the greatest reduction in frequency being observed in strips of whole uterus with endometrium intact. The potential for prostaglandins of endometrial origin to influence spontaneous activity was clearly demonstrated in these experiments.

Circulating Factors

Since the behaviour of segments of uterus removed from animals is quite different from that of the uterus in intact animals, it was of interest to study the behaviour of segments isolated from the uterus and translocated to another part of the animal. Pieces of myometrium (25 mm × 35 mm) were surgically removed from the body of the uterus and sutured to steel frames to ensure maintenance of their original dimensions. Following attachment of pairs of EMG electrodes, each isolated segment was fixed in place in a fold of the omentum. A regular pattern of EMG activity became established approximately 10 days after the isolation procedure. The capacity of the isolated segment to respond to a circulating spasmogen was confirmed by the injection of 100 mU oxytocin or 2 mg $PGF_{2\alpha}$ via the jugular vein of each ewe. Myometrial activity started in both isolated and intact uterus within 3 min of injection.

Analysis of the frequency and duration of EMG bursts and the total amount of EMG activity per hour recorded by all electrodes revealed no significant difference in these parameters between intact and isolated myometrium. However, when the precise occurrence of each burst of spontaneous activity was examined it was found that in 13 out of 15 animals there was no temporal association of EMG bursts either between intact and isolated segments or between different isolated segments, at any stage of gestation.

Two conclusions can be drawn from these results. Firstly, the failure to demonstrate a difference in the frequency of bursts between intact and isolated myometrium suggests that a circulating inhibitory factor may be responsible for the low frequency of bursts in the intact animal, compared with in vitro. Secondly, since there was no temporal relationship between bursts, it is unlikely that a circulating factor is responsible for initiating bursts of activity in the uterus of pregnant ewes. Furthermore, the results obtained in isolated segments are not consistent with the notion that local factors, produced by the feto-placental unit, are essential for the initiation of activity.

The Prostaglandins and Their Inhibition

In view of the inhibitory effects of indomethacin on spontaneous activity in strips of whole uterus in vitro, and since the capacity of the endometrium to synthesize prostaglandins has been demonstrated in vitro (Evans et al. 1981), it seemed important to investigate the effects of prostaglandins and their inhibitors on EMG activity in the intact ewe. Bolus doses of 0.2–2.0 mg $PGF_{2\alpha}$, administered via the jugular or uterine arteries, were followed by a burst of EMG activity and an increase in IUP. The duration of the prostaglandin-induced burst increased with increasing concentrations of the prostaglandin. Both electrical and mechanical responses were similar in amplitude to those occurring spontaneously. Indomethacin (60 mg/h) was administered intravenously for 30 min to four ewes at 64–66 and six ewes at 133–142 days of pregnancy. The interval between contractions increased significantly during indomethacin infusion from a control value of 42 ± 10 min to 96 ± 22 min (Fig. 1). The inhibition was long-lasting.

These results demonstrate that PGF can induce activity in the uterus of the intact ewe during pregnancy. Furthermore, the involvement of prostaglandins in generating spontaneous activity is suggested by its inhibition during and following infusion of indomethacin.

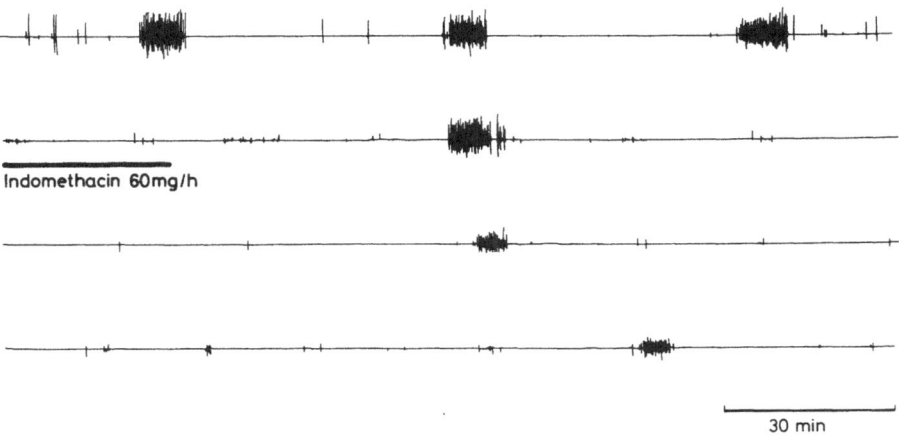

Fig. 1. Indomethacin (60 mg/h) infusion for 30 min prolongs the interval between bursts of EMG activity for many hours

The production of PGE_2 by the ovine placenta increases markedly during the last third of gestation and results in increasing arterial concentrations of PGE_2 in the ewe during this time (Fowden et al. 1987). Thus the pregnant uterus is exposed to increasing concentrations of PGE in its arterial supply. Assuming that PGE stimulates the uterus of the ewe, the placental PGE may be an important factor in determining myometrial activity in pregnant sheep. Recent studies (Andrianakis et al. 1987) have demonstrated that PGE production by the placenta is inhibited by the infusion of a PG synthetase inhibitor, 4-aminoantipyrine, and that myometrial activity is also inhibited. These results suggest that placental PGE may be an important stimulant for uterine activity.

The Nature of Tonic Inhibition

Several factors associated with pregnancy are known to suppress contraction in uterine smooth muscle. These include progesterone, prostacyclin, relaxin and β-adrenoceptor activation. The possible involvement of these agents in the tonic inhibition of uterine activity in vivo, as suggested by our earlier findings, will now be explored.

Prostacyclin

We showed previously that indomethacin prolongs the interval between bursts of spontaneous activity in vivo. This result does not support a role of prostacyclin in tonic inhibition of uterine activity, since an increase in the frequency of contractions following blockade by indomethacin would be expected.

Relaxin

Injection of porcine relaxin into the non-pregnant ewe inhibits spontaneous activity (Porter et al. 1981) but, despite efforts to isolate ovine relaxin in this and other laboratories, there has been no demonstration of the existence of the peptide in ewes to date.

Innervation

The degeneration of adrenergic nerves within the uterus during pregnancy has been well documented in several species (Thorbert et al. 1978, 1979) and we have confirmed that a similar phenomenon takes place in the uterus of the ewe (Sigger and Parkington 1983). However, the palsma membrane of the smooth muscle cells still contain adrenoceptors and therefore remain subject to the influences of circulating catecholamines. We investigated the effects on spontaneous EMG activity of infusing a variety of adrenoceptor agonists and antagonists in intact ewes. The effects of adrenaline, administered at 0.95 mg/h, changed progressively during pregnancy from excitatory in the early stages (greatest prior to day 60) to inhibitory at term, while noradrenaline infusion caused a reduction in integrated EMG activity in all cases, an effect which became progressively greater towards term. The α-adrenoceptor antagonist phenoxybenzamine, infused concurrently with or 1 h after the start of a 3-h infusion of adrenaline, caused EMG activity to return close to the control value. Bolus doses of the β-adrenoceptor antagonist propranolol (12 mg) administered during adrenaline infusion were each followed by a contraction, but when administered alone propranolol was without observable effect on integrated EMG activity or on the pattern of uterine contractions. Thus the data do not support the hypothesis that catecholamines, released either from nerve endings or as circulating hormones, have an important role in regulating myometrial activity during pregnancy in this species.

Vasointestinal polypeptide (VIP) is present in nerves within many smooth muscle organs, including the uterus (Fahrenkrug 1979; Alm et al. 1980). While it has been shown that VIP has a profound inhibitory effect on contractions when applied exogenously to isolated preparations of uterine and other smooth muscle, its release from nerves in response to perivascular or field stimulation has never been demonstrated in the uterus. Furthermore, all catecholamine fluorescence, indicative of noradrenergic nerves, disappears from segments of uterus that have been translocated to the omentum and it is assumed that no nerves of any kind survive the procedure. Since those segments display a *pattern* of activity that is not different from that observed in the intact organ (although they do not contract in synchrony with the uterus), an involvement of nerves of any kind in modulating spontaneous activity seems unlikely.

Progesterone

The concept of progesterone as the major inhibitor of uterine motility was first mooted by Csapo in 1956. In the sheep, as in many other species, there is a clear relationship between the decline in circulating progesterone levels and the increase in frequency of uterine contraction as the animal goes into labour. The precise mechanism of progesterone inhibition is incompletely understood. While it is clear that progesterone does not interfere with the ability of uterine smooth muscle cells to sustain an action potential and contract (Parkington 1983), the available evidence suggests that the steroid may be responsible for the limited ability of electrical activity to spread from cell to cell during pregnancy (Sims et al. 1982; Parkington 1983). Thus, by restricting spread of activity, progesterone would limit the effectiveness of tension development. As the ewe goes into labour electrical coupling between myometrial cells is enhanced several-fold (Parkington 1985) as is the apparent conduction velocity (Parkington et al. 1987). Thus, coordinated contraction of the entire uterus and greater development of tension is possible.

Progesterone may also influence uterine motility via an effect on the biosynthesis of prostaglandins (Wilson et al. 1985; Jeremy and Dandona 1986; Rice et al. 1987). Recently the effects of progesterone on PGE_2 and $PGF_{2\alpha}$ biosynthesis by rat myometrium in vitro was examined during mid-pregnancy and at term. Following incubation of strips of myometrium for 18 h, prostaglandin synthesis was measured in the presence or absence of arachidonic acid. Prostaglandin synthesis was low in tissue obtained from mid-pregnant rats, with no increase observed with the addition of arachidonic acid, indicating that cyclo-oxygenase activity was limiting prostaglandin production in this tissue.

In myometrium obtained from parturient rats, there was a marked increase in the capacity of myometrium to synthesize prostaglandins both basal and arachidonic acid-stimulated, that is, the activity of cyclo-oxygenase was increased markedly. Whether this represents an increase in the amount of cyclo-oxygenase present or the removal of tonic inhibition remains to be established. The inclusion of progesterone (10^{-9} to 10^{-6} M) in incubations did not affect prostaglandin synthesis by tissue obtained from mid-pregnant rats, presumably because it is already under the influence of high concentrations of progesterone. In tissue obtained from parturient rats, progesterone suppressed both basal and arachidonic acid-stimulated prostaglandin synthesis. These data indicate that progesterone suppresses myometrial PGE and PGF synthesis, at least, at the level of cyclo-oxygenase, but do not exclude the

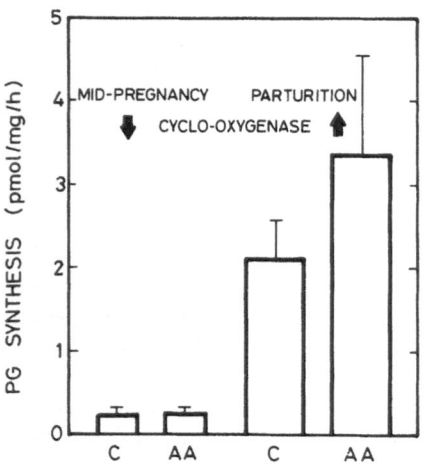

Fig. 2. Prostaglandin E_2 biosynthesis by rat myometrium in vitro. Myometrium from three mid-pregnant (day 15) or parturient rats was incubated in medium 199 for 60 min at 37 °C under an atmosphere of carbogen, in the absence *(C)* or presence *(AA)* of arachidonic acid (30 μM). Prostaglandin E_2 released into the incubation medium was quantified by radioimmunoassay. Data are expressed as mean ± SE

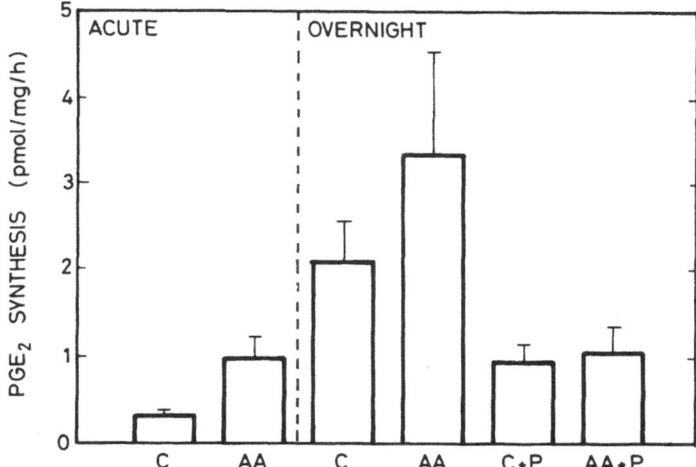

Fig. 3. Effect of progesterone on the biosynthesis of PGE_2 by rat myometrium in vitro. Basal *(C)* and arachidonic acid-stimulated *(AA)* PGE_2 biosynthesis by rat myometrium after 2 h *(acute)* or 18 h incubation in vitro is presented. The effect of the inclusion of progesterone (100 μM) on basal *(C + P)* and arachidonic acid-stimulated *(AA + P)* PGE_2 biosynthesis is also shown. Incubation conditions were as described for Fig. 2

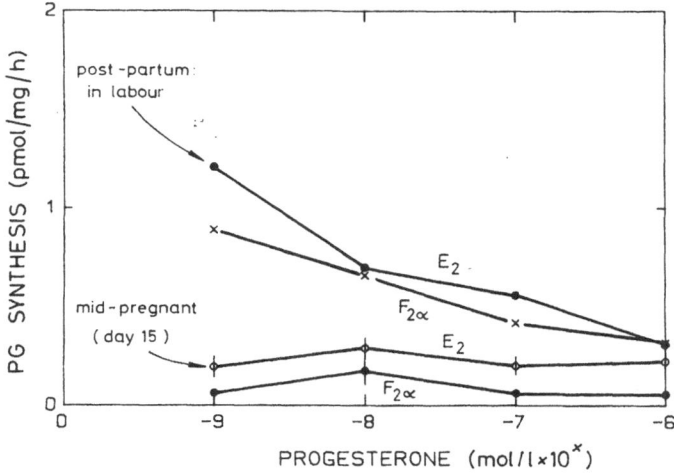

Fig. 4. Progesterone dose response curve. The effect of including increasing concentrations of progesterone in the incubation medium for 18 h on PGE_2 and $PGF_{2\alpha}$ biosynthesis by rat myometrium is presented. Conditions of incubation were as described in Fig. 2

possibility that progesterone may also have an effect on the level of phospholipase A_2 (PLA_2) activity.

Progesterone-regulated PLA_2 inhibitory proteins have been identified previously in human amniotic fluid (Wilson et al. 1985). These proteins suppress prostaglandin synthesis by human endometrial cells in vitro and may be involved in the tonic suppression of uterine prostaglandin synthesis during gestation. A decrease in the synthesis of these proteins may result in enhanced uterine prostaglandin synthesis and myometrial contractility (Figs. 2–4).

In sheep, myometrial cytosol also shows PLA_2-inhibitory activity. Cytosolic preparations from myometrium obtained from early pregnant ewes inhibit PLA_2-stimulated hydrolysis of phosphatidylcholine in vitro (Rice et al. 1987). Inhibition of PLA_2 activity in myometrium may result in a decreased ability for arachidonic acid liberation and its subsequent metabolism to uterotonic prostaglandins. The effects of myometrial PLA_2 inhibitory activity in vivo remain to be established.

Oestrogen

In early pregnancy electrical activity recorded intracellularly in strips of circular myometrium (Parkington 1985) consists of continuous action potentials; that recorded extracellularly from the uterus of intact ewes (Sigger et al. 1984) consists of individual spikes. The grouping of electrical activity into bursts occurs around days 50–60 of pregnancy, both in vivo and in vitro, and this coincides with the time at which oestrogen levels in the circulation increase (Challis and Patrick 1981). Thus the appearance of quiescent periods is associated with a rise in plasma oestrogen.

The effects of exogenously administered oestrogen on uterine motility vary with the muscle layer and concentration of steroid. High concentrations of steroid inhibit smooth muscle and the ratio of oestrogen in myometrium to oestrogen in the plasma increases in guinea-pigs and rabbits during pregnancy (Batra et al. 1980). In rats, oestrogen induces rhythmic contractions in circular myometrium while rendering the longitudinal muscle component completely quiescent (Chamley and Parkington 1984). Clearly, additional work is required in order to elucidate the role of oestrogen in modulating uterine activity.

Conclusion

Uterine smooth muscle in pregnant ewes undergoes prolonged periods of quiescence, with brief episodes of electrical and mechanical activity occurring approximately twice per hour. We have presented evidence which suggests that tonic inhibition may be imposed by circulating factor(s). Furthermore, our results implicate prostaglandins in the generation of the bursts of activity.

References

Adrianakis P, Walker DW, Ralph MM, Thorburn GD (1987) Effect of 4-aminoantipyrine on prostaglandin (PG) concentrations in uterine and umbilical circulations of pregnant sheep. Proc Endocrinol Soc Aust 30: 115

Alm P, Alumets J, Håkansson R, Owman C, Sjöberg NO, Sundler F, Walles B (1980) Origin and distribution of VIP (vasoactive intestinal polypeptide)-nerves in the genito-urinary tract. Cell Tissue Res 205: 337

Batra S, Sjöberg NO, Thorbert G (1980) Sex steroids in plasma and reproductive tissues of the female guinea-pig. Biol Reprod 22: 430-437

Bell R (1983) The prediction of preterm labour by recording spontaneous antenatal uterine activity. Br J Obstet Gynaecol 90:884-887

Challis JRG, Patrick JE (1981) Fetal and maternal oestrogen concentrations throughout pregnancy in the sheep. Can J Physiol Pharmacol 59: 970-978

Chamley WA, Parkington HC (1984) Relaxin inhibits the plateau component of the action potential in the circular myometrium of the rat. J Physiol (Lond) 353: 51-65

Csapo AI (1956) Progesterone block. Am J Anat 98: 273-291

Demianczuk N, Towell ME, Garfield RE (1984) Myometrial electrophysiologic activity and gap junctions in the pregnant rabbit. Am J Obstet Gynecol 149: 485-491

Evans CA, Kennedy TG, Patrick JE, Challis JRG (1981) Uterine prostaglandin concentrations in sheep during late pregnancy and adrenocorticotropin-induced labor. Endocrinology 109: 1533-1538

Fahrenkrug J (1979) Vasoactive intestinal polypeptide: measurement, distribution and putative transmitter function. Digestion 19: 149

Fowden AL, Harding R, Ralph MM, Thorburn GDT (1987) The nutritional regulation of plasma prostaglandin E concentrations in the fetus and pregnant ewe during late gestation. J Physiol (Lond) 394: 1-12

Germain G, Cabrol D, Visser A, Sureau C (1982) Electrical activity of the pregnant uterus in the cynomolgus monkey. Am J Obstet Gynecol 142: 513-519

Harding R, Poore K, Bailey A, Thorburn GDT, Jansen CAM, Nathanielsz PW (1982) Electromyographic activity of the nonpregnant and pregnant sheep uterus. Am J Obstet Gynecol 142: 448

Jeremy JY, Dandona P (1986) RU486 antagonizes the inhibitory action of progesterone on prostacyclin and thromboxane A$_2$ synthesis in cultured rat myometrial explants. Endocrinology 119: 655-660

Parkington HC (1983) Electrical properties of the costo-uterine muscle of the guinea-pig. J Physiol (Lond) 335: 15-27

Parkington HC (1985) Some properties of the circular myometrium of the sheep throughout pregnancy and during labour. J Physiol (Lond) 359: 1-15

Parkington HC, Sigger JN, Harding R (1987) Coordination and regulation of electrical activity in the myometrium of the pregnant ewe. Proc Aust Physiol Pharmacol Soc 18 (1): 32P

Porter DG, Lye SJ, Bradshaw JMC, Kendall JZ (1981) Relaxin inhibits myometrial activity in the ovariectomized non-pregnant ewe. J Reprod Fertil 61: 409-414

Rice GE, Wong MH, Thorburn GDT (1987) Ovine myometrial cytosol inhibits phospholipase A$_2$ activity in vitro. Prostaglandins 34: 563-578

Shepherd V, Parkington H, Jenkin G, Ralph MM, Thorburn GDT (1986) Activity of ovine circular and longitudinal myometrium during pregnancy and parturition. Proc Aust Soc Reprod Biol 18: 107 (abstract)

Sigger JN, Parkington H (1983) Changes in uterine innervation during pregnancy. Aust Perinatal Soc Inaugural Congress. Excerpta Medica Asia Pacific Congress Series 14: 23-24

Sigger JN, Harding R, Bailey A (1984) Development of myometrial electrical activity during the first half of pregnancy in the sheep. Aust J Biol Sci 37: 153-162

Sims SM, Daniel EE, Garfield RE (1982) Improved electrical coupling in uterine smooth muscle associated with increased numbers of gap junctions at parturition. J Gen Physiol 80: 353-375

Taverne MAM, Naaktegeboren C, van der Weyden GC (1979) Myometrial activity and expulsion of fetuses. Anim Reprod Sci 2: 117-131

Thorbert G, Alm P, Owman C, Sjöberg NO, Sporrong B (1978) Regional changes in structural and functional integrity of myometrial adrenergic nerves in pregnant guinea-pig, and their relationship to the localization of the conceptus. Acta Physiol Scand 103: 120

Thorbert G, Alm P, Björklund A, Owman C, Sjöberg NO (1979) Adrenergic innervation of the human uterus. Disappearance of the transmitter and transmitter-forming enzymes during pregnancy. Am J Obstet Gynecol 135: 223

Wilson T, Liggins GC, Aimer GP, Skinner SJM (1985) Partial purificaiton and characterization of two compounds from amniotic fluid which inhibit phospholipase activity in human endometrial cells. Biochem Biophys Res Commun 131: 22-29

Induction of Labour in Domestic Animals: Endocrine Changes and Neonatal Viability

M. Silver[1] and A. L. Fowden

Introduction

Since nature has provided mechanisms which normally ensure safe, reliable delivery at term, reasons for advocating the artificial induction of labour require some justification. Apart from purely scientific considerations, the induction of labour in domestic animals has both economic and clinical relevance. Thus, if gestation can be shortened significantly without jeopardizing neonatal viability, this will have economic advantages; alternatively, if the value of the dam and/or her offspring is sufficiently great, a predictably timed delivery would be useful. Clinically, there are occasions when termination of pregnancy is necessary, and induction, if predictable and safe, has advantages over surgery.

The present review attempts to answer two specific questions on the induction of labour. First, when should it be performed in relation to normal term, so that viable young are produced with minimal maternal stress and postnatal complications? Second, how is the procedure best carried out in the light of present knowledge and given the range of prepartum endocrine changes which occur in different species? The discussion has been confined largely to pig, sheep and mare, which differ widely in their reproductive physiology.

Fetal Maturation and Readiness for Birth

An essential prerequisite for the satisfactory induction of preterm labour is that the fetus be in a state of readiness for delivery. Thus, its energy reserves must be adequate, its lungs mature, its gut functional, and it must be able to suck, swallow and maintain its body temperature. Many maturational processes are associated with the prenatal increase in adrenocortical activity which has been observed in virtually all mammals (Liggins 1976). The precise timing of this rise in circulating fetal corticosteroids varies widely: in the sheep fetus, the changes begin about 15 days before term (147 days) and escalate in the last 4–6 days (Fig. 1). The changes in the fetal pig are somewhat similar in time course, although the final surge is much shorter in

---

[1] The Physiological Laboratory, Downing Street, Cambridge, CB2 3EG, UK

The Endocrine Control of the Fetus
Ed. by W. Künzel and A. Jensen
© Springer-Verlag Berlin Heidelberg 1988

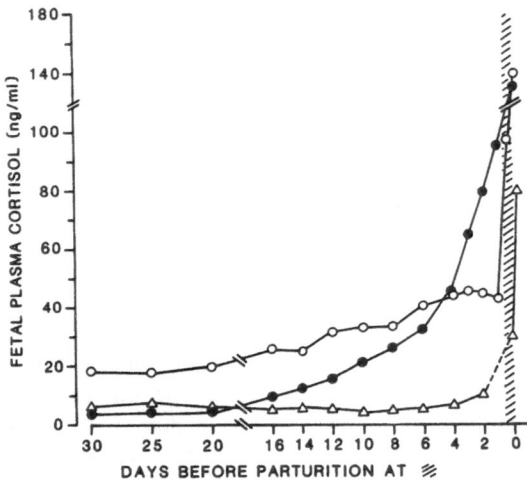

Fig. 1. Typical prenatal changes in plasma cortisol in fetal lamb (●), piglet (○) and foal (△), catheterized in late gestation and sampled until delivery at term

duration. By contrast, little or no change in adrenocortical activity appears to occur in the fetal foal until just before birth, and certainly the absence of a clearly defined cortisol rise in this species poses a major threat to its postnatal survival if delivered early (Rossdale and Silver 1982).

A further indication of fetal adrenocortical competence is shown by the short-term responsiveness of the gland to exogenous adrenocorticotrophic hormone (ACTH). This increases markedly near term in the fetal lamb (Challis et al. 1984), and even at 110 days (i.e. well before any prepartum changes) a rapid two-fold rise in fetal plasma cortisol is evoked by ACTH (M. Silver, unpublished observation). The fetal piglet adrenal is also responsive to ACTH from about 0.75 of gestation both in vitro (Lohse and First 1981) and in vivo (Fowden and Silver 1988), whereas the fetal foal presents a complete contrast, in that no rise in cortisol after ACTH can be detected in the catheterized fetus in late gestation (see Fig. 2, test 1). Tests on the neonatal adrenocortical response to ACTH have shown that in premature foals the adrenals are incapable of more than a 20%–50% increase in secretion compared with a 300% rise in fully mature neonates (Rossdale et al. 1982).

Maternal Signs and Signals

Maternal Endocrine Changes and Their Relationship to the Fetus

The chain of events which eventually results in parturition may involve a whole endocrine cascade, the sequence of which differs in different species (Thorburn et al. 1977). The gradual prepartum fall in progesterone, one of the first maternal changes, coincides with the increase in fetal plasma cortisol in the sheep. Indeed, there is now much evidence for fetal corticosteroid activation of placental 17 β-hydroxylase which leads to oestrogen synthesis from progesterone, which in turn affects the formation of prostaglandins (PGs) and receptors for oxytocin as labour begins (Thorburn et al. 1977). Only in the ruminants (sheep, cow, goat) does the fetus seem to play an unequivocal role in triggering the full prepartum cascade of endocrine changes. In these species a deficient pituitary-adrenal axis prolongs gestation, while ACTH or cortisol administration to the fetus initiates premature parturition (Thorburn et al. 1977). The subsequent maternal changes differ in detail; thus in the sheep the prepartum rise in oestrogen occurs in the last 24 h, followed closely by rises in PGF, whereas in the cow and goat the changes begin somewhat earlier.

In the pig, in which the progesterone source is luteal and the overall maternal changes are somewhat similar to those in the ruminant, there appears to be some fetal involvement in the chain of prepartum endocrine events. Thus, hypophysectomy or decapitation of the whole litter prolongs gestation (Stryker and Djuik 1975). However, intrafetal ACTH does not induce a clear-clut shortening of gestation (Bosc 1973; Randall et al. 1984). Final luteolysis in the pregnant sow may be provoked by exogenous PGs or their analogues (First and Bosc 1979; Silver et al. 1983) but no prepartum rise in uterine venous PG is detectable; in fact PGs only increase during the first stage of labour (Silver et al. 1979).

Prepartum endocrine changes in the mare are quite unlike those seen in either the ruminants or the pig. Not only is there a slight rise rather than a fall in total circulating progestogens, but oestrogen levels (e.g. equilin and equilenin) decline over the the last third of gestation, in association with a fall in fetal gonadal weight, although a small rise in oestradiol may occur just before term (Barnes et al. 1975; Pashen 1984). Surprisingly, equine pregnancy is not terminated after fetal gonadectomy, although maternal circulating oestrogens fall to undetectable levels due to lack of precursors. Maternal plasma PGs show no significant changes in the last few weeks of pregnancy and only rise during late first or early second stage of labour (Pashen 1984). There are thus no clear hormonal signals of approaching parturition in the mare, and, unlike the pig or sheep, term cannot be predicted to within a day or two since the gestation period ranges from 320 to 360 days.

Prepartum Mammary Changes

In species bred for milk there is no difficulty in obtaining samples of preterm milk but in others (e.g. pig) it is almost impossible to extract any secretion until labour is already in progress. The major changes in the composition of mammary secretion occur just before parturition in sheep and goat (Fleet et al. 1975). In the mare, mammary secretions can be obtained starting a few weeks before delivery and the dramatic fall in Na^+ and rises in K^+ and Ca^{++} which occur in the last 72 h provide a much better indication of delivery than any hormonal events (Ousey et al. 1984).

Prepartum Myometrial Activity

In the sheep the duration and frequency of myometrial contractures increase towards term but few 'proper', labour-type contractions occur in this species before labour begins (Silver 1988). The pregnant sow also shows random episodes of myometrial activity near term (Taverne et al. 1979). Recently Haluska et al. (1987) reported that considerable electromyographic activity was present in equine myometrium during late gestation, which suggests that the uterus of the mare may be more active towards the end of gestation than that of the other two species.

Induction of Labour

General Methods of Induction

The foregoing summary of fetal and maternal prepartum changes in the three species gives some indication of when induction of labour may be safely carried out. If the hormonal cascade is stimulated from the beginning, the likelihood of neonatal survival will be greater than if the process is initiated near the end of the chain. In ruminants labour can be induced reliably by stimulating the fetal adrenal cortex with exogenous ACTH to trigger the whole sequence of events ending with the delivery of viable young much earlier than normal (Thorburn et al. 1977). By contrast, in the sow, Bosc (1973) reported that a single dose of ACTH to each fetus induced premature labour 10 days later, whereas Randall et al. (1984) found ACTH infusion to have no such effect. The first comparable experiment on the mare has shown that ACTH, infused into the fetus, increases adrenocortical activity, with the birth of a viable foal after 4.5 days of treatment (Fig. 2). However, induction of labour by fetal manipulation is clearly impracticable both economically and clinically in any of these species, even when such methods are reliable. Hence, the effectiveness of maternally administered induction agents has received much wider attention.

Maternally administered dexamethazone and other glucocorticoids in ruminants and pig will eventually induce labour provided large amounts are given. However, this may well depress endogenous fetal adrenal activity and can also lead to retained placenta (Thorburn et al. 1977).

Since progesterone is the hormone which maintains pregnancy, any drugs that will reduce or inhibit its production are potential interceptive agents. Thus, in species where the corpus luteum is essential throughout, PGs are generally luteolytic and can be used to terminate gestation, while their effectiveness in other species will depend largely upon the sensitivity of the myometrium to their action. More recently, drugs which prevent the synthesis of progesterone from pregnenolone [Trilostane, Epostane (Sterling Winthrop)] have been used to induce parturition in a number of species (Creange et al. 1981; Taylor et al. 1982). They are potentially active at luteal and placental sites, but can inhibit the adrenal steroidogenesis as well.

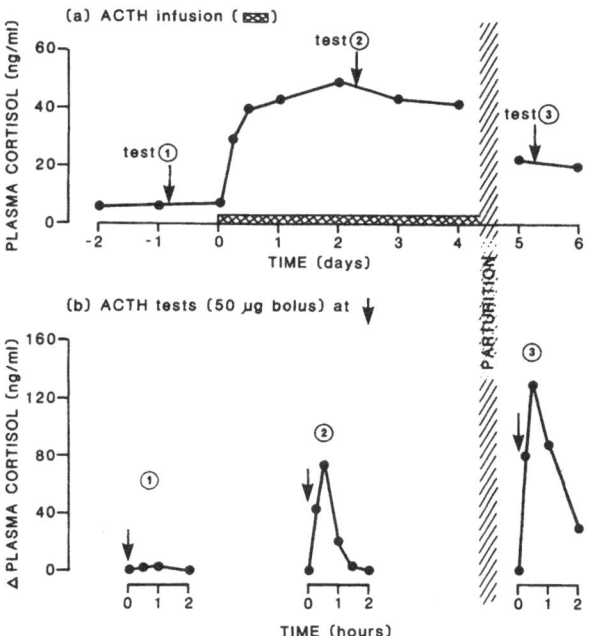

Fig. 2. a Effect of ACTH infusion (2 μg kg^{-1} h^{-1}) on plasma cortisol in a catheterized fetal foal (310 days' getation) and **b** effect of single ACTH bolus test *1* before and *2* during infusion and *3* about 6 h after birth

Maternal oestrogen administration is relatively ineffective in almost all species (Thorburn et al. 1977). Similarly, oxytocin has little effect when given before term, since the induction of oxytocin receptors is usually dependent on the preceding endocrine changes; only in the mare and the human does oxytocin seem to be an efficient inducing agent in the absence of any other known endocrine changes (Rossdale et al. 1982; Thorburn et al. 1977).

Specific Techniques for the Sow, Ewe and Mare

In the pig high neonatal mortality rates are still encountered even at term, and therefore any procedures which will induce farrowing at a specific time - so that attendance at the delivery can be ensured - must be potentially useful. PGs and their analogues, which are luteolytic in the sow, have been extensively studied and are now widely used to induce labour within the last 2-3 days of term (115 days); farrowing usually occurs within 30 h and neonatal viability and survival rates are equivalent to those at normal delivery at term (First and Bosc 1979).

Silver et al. (1983) examined the effects of induction with the PG analogue cloprostenol earlier in gestation (105-110 days). Delivery of the first piglet occurred 25 ± 2 h after injection of 200 μg of the drug with 97% live births. The sequence of endocrine changes in both fetuses and sow following cloprostenol showed that even as early as 106 days a rapid surge in fetal plasma cortisol occurred once first stage labour had begun. This rise coincided with the escalation in maternal PG production which did not start until long after the maximum drop in progesterone levels. All neonates were active and had no respiratory problems, but those born before 108-109 days had poor suck reflexes and appeared unable to find the teats.

More recently, Epostane has been tested as a potential inducing agent (Silver and Fowden, 1988). Of 11 sows infused with the drug between 106 and 111 days, 7 delivered 27 ± 2 h later with the birth of 95% live piglets. In the remaining 4 sows progesterone levels were reduced temporarily by the Epostane and then returned to normal after 6-12 h. Apart from lowering fetal and maternal plasma progesterone levels (Fig. 3), Epostane also had an inhibitory effect on adrenocortical activity in both sow and fetuses, but fetal cortisol concentrations were eventually restored after 6-12 h by enhanced ACTH secretion. There was a further escalation during labour, so that by the time of birth high cortisol levels were present in all the neonates. Like those induced with cloprostenol, the newborn piglets were viable, and those born after 108-110 days were able to suckle normally.

The reasons for the capricious nature of Epostane induction in the sow are not immediately apparent; oestrone levels were reduced in addition to progesterone in all sows (Fig. 3), but in those which failed to farrow the oestrone/progesterone ratio remained much lower than that in the successfully induced animals.

In the ewe, until recently, maternally administered inducing agents have either been ineffective (oestrogens, PGs, oxytocin) or imprecise (dexamethazone). Initial tests with Trilostane, which drastically reduced maternal progesterone levels (Taylor et al. 1982; Jenkin and Thorburn 1985) did not always induce labour when given between 115-130 days gestation, i.e. before the prepartum fetal cortisol rise. More recently, the effects of Epostane were examined in ewes between 137 and 141 days (Silver 1988), by which time increased fetal adrenocortical activity had begun (see Fig. 1). Parturition occurred 33 h later in all 20 animals tested; the lambs were healthy and their subsequent growth was normal. The fall in maternal progesterone was as rapid and prolonged as that reported previously (Taylor et al. 1982) and was accompanied by significant rises in uterine venous PGFM (Fig. 4). However, no changes in myometrial activity were detectable until 3-6 h after the Epostane, when the number but not the duration of contractures increased. Figure 4 shows that Epostane had the same inhibitory effect on maternal and fetal adrenocortical activity as in the sow; fetal cortisol levels were later restored to normal by very large increases in ACTH (Silver 1988). There was a further rapid surge in cortisol before delivery (Fig. 4) such that the final concentration attained at birth was within the range seen in lambs born naturally. Epostane appeared to be effective whether administered intravenously, intramusculary or orally, although the induction time was slightly longer when given by mouth (Silver 1988).

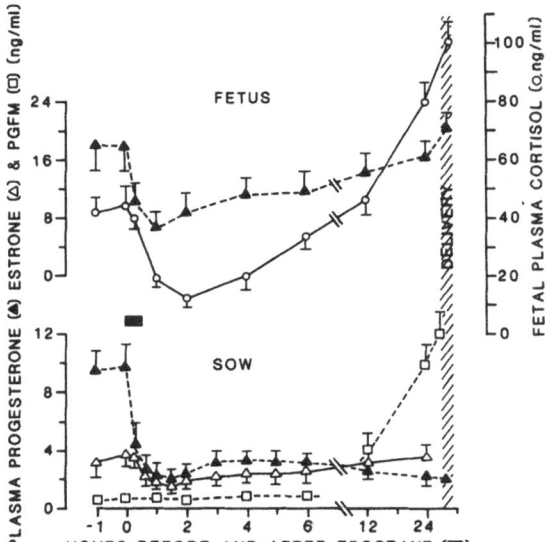

Fig. 3. Effect of Epostane (1-2 mg/kg body wt.) administered intravenously to the sow on plasma hormone levels in mother and fetus (mean ± SE, n=7). First piglet delivered at

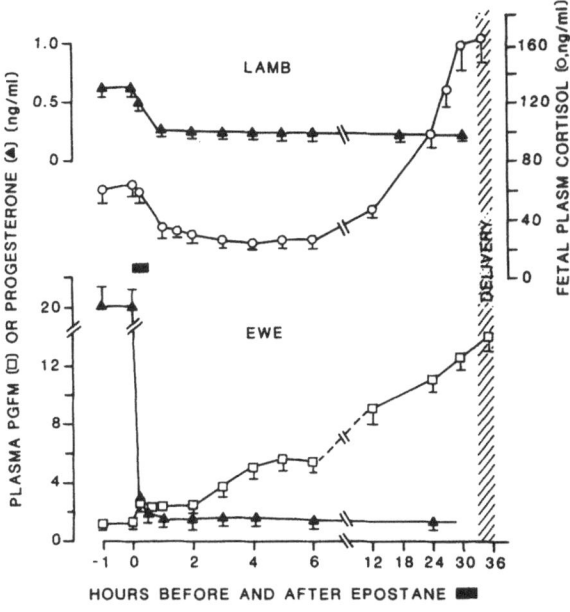

Fig. 4. Effect of Epostane (1-2 mg/kg body wt.) administered intravenously to the ewe on plasma hormone levels in mother and fetus (mean ± SE, n=10). Time of delivery

In the mare there is no clearly defined stage in late gestation when induction of labour can be carried out safely with the certainty that a viable foal will be delivered, partly because full term in the mare is so variable but also because the period of fetal maturation appears to be so short (Rossdale and Silver 1982). Even the changes in milk composition are not entirely reliable, as there is no clear correlation between these and fetal maturity (Ousey et al. 1984).

Oxytocin is the most effective inducing agent in the mare, irrespective of the stage of gestation (Leadon et al. 1982). PGs and their analogues have also been used, with more variable results both in terms of success rate and interval between administration and delivery (Leadon et al. 1982). However, because PGs and oxytocin stimulate myometrial activity, any preceding endocrine changes are circumvented, and these agents thus terminate gestation rather than stimulate parturition. Indeed, the rapidity with which oxytocin may act in the mare is surprising; birth may occur between 15 min and 2 h after intravenous injection of 5 IU oxytocin, during which time the foal may turn, and the cervix becomes relaxed and dilated before the membranes break and the foal is expelled (Leadon et al. 1982). During this truncated first and second stage of labour endogenous PG production escalates quickly, but there is little or no change in maternal plasma progesterone or estrogen (Pashen 1984). Figure 5 shows some of the changes in two mares with catheterized fetuses which were induced on the basis of 'near-term' milk changes; one fetus was clearly premature since its plasma cortisol rise was minimal even after birth, while the other was more mature, showing a slight predelivery rise and a near-normal postnatal surge in cortisol. Maternal plasma progesterone concentrations were higher in the mare which delivered the viable foal. In fact, Ousey et al. (1987) have sug-

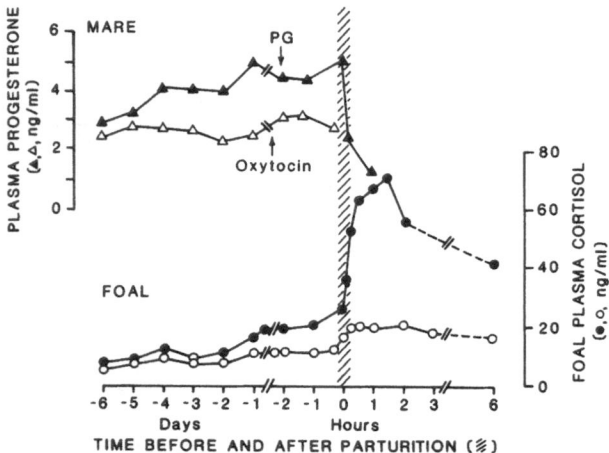

Fig. 5. Maternal progesterone and fetal cortisol changes in two mares given either 250 µg fluprostenol i. v. (PG, closed symbols) or 5 IU oxytocin i. v. (open symbols) at ↓ to induce labour

gested that the rise in total progestagen may be a further pointer to the proximity of 'full term' in the mare.

Attempts at reducing equine placental progesterone production with Epostane have failed (Fowden and Silver 1987); in this study maternal progesterone was lowered for only 2–3 h following Epostane in late gestation. Even when the Epostane infusion was accompanied by the injection of a PG analogue, labour was not induced. During Epostane treatment no detectable rise in uterine venous PGFM occurred even at the time of minimal progesterone levels, while repeated tests with different doses of the drug had no effect on the outcome of pregnancy (Fowden and Silver 1987). There is no doubt that increased uterine PG production can occur under other circumstances and may even lead to premature delivery, as, for example, during food deprivation, particularly if the mares are lipaemic (Silver and Fowden 1982). However, neither temporary reduction in progesterone levels nor permanent removal of the source of oestrogen precursors results in parturition in the mare.

Conclusions and Prospects

If the aim is to interrupt pregnancy rather than produce a viable neonate, then high doses of PGs are effective in most species with a luteal-based pregnancy, while oxytocin or large amounts of dexamethazone can be used in other species, and in either case assessment of fetal maturity is irrelevant. On the other hand, if a live healthy neonate is required, then the stage at which induction is performed and the method used become of prime importance. In species like the ruminants and the pig, where the gestation period is relatively precise, induction can be started within the last 15th of pregnancy, when fetal maturation is nearing completion. Methods which reduce maternal progesterone in these species evoke the remaining prepartum endocrine changes, including a shortened, though sharp, fetal cortisol surge, and result in live births and good neonatal survival rates. The exact way in which the final escalation in fetal adrenocortical activity is stimulated varies; Epostane inhibits cortisol production and thereby provokes a feedback response from the fetal pituitary. PGs or their analogues may well have an effect on the fetal adrenal itself or on ACTH production. Whatever the mechanism, the resultant increase in fetal adrenocortical activity is dependent upon adequate responsiveness of its pituitary adrenal axis: it is this which appears to be absent in the fetal foal until very close to spontaneous parturition. Int he mare, therefore, there is a very great need for an induction technique which incorporates some form of fetal pituitary adrenal stimulation to ensure a degree of prenatal maturation before delivery. Meanwhile, the search continues for other indicators of fetal 'readiness for birth' and the approach of full term in the mare. Signs and signals which can be detected reliably either from the mare or the fetus would be of considerable clinical use. Otherwise, the danger exists that induction with the usual agents (oxytocin, PGs) will result in delivery of a premature, nonviable foal.

Acknowledgements. We thank the many members of the Laboratory who have helped in this work, which was supported by the Agriculture and Food Research Council and the Horserace Betting Levy Board.

References

Barnes RJ, Nathanielsz PW, Rossdale PD, Comline RS, Silver M (1975) Plasma progestagens and oestrogens in fetus and mother in late pregnancy. J Reprod Fertil [Suppl] 23: 617-623

Bosc MJ (1973) Modification de la durée de gestation de la truie apres administration d'ACTH aux fétus. C R Acad Sci 276: 3183-3186

Challis JRG, Mitchell BF, Lye SJ (1984) Activation of fetal adrenal function. J Dev Physiol 6: 93-105

Creange JE, Anzalone AJ, Potts GO, Schane HP (1981) WIN 32, 729, A new, potent interceptive agent in rats and rhesus monkeys. Contraception 24: 289-299

First NL, Bosc MJ (1979) Proposed mechanisms controlling parturition and the induction of parturition in swine. J Anim Sci 48: 1407-1421

Fleet IR, Goode JA, Hamon MA, Laurie MS, Linzell JL, Peaker M (1975) Secretory activity of goat mammary glands during pregnancy and the onset of lactation. J Physiol 251: 763-773

Fowden AL, Silver M (1987) Effects of inhibiting 3β-hydroxysteroid dehydrogenase on plasma progesterone and other steroids in the pregnant mare near term. J Reprod Fertil [Suppl] 35: 539-545

Fowden AL, Silver M (1988) Adrenocortical activity in the fetal pig. J Physiol (in press)

Haluska GJ, Lowe JE, Currie BW (1987) Electromyographic properties of the myometrium correlated with the endocrinology of the prepartum and postpartum periods and parturition in the mare. J Reprod Fertil [Suppl] 35: 553-564

Jenkin G, Thorburn GD (1985) Inhibition of progesterone secretion by a 3β-hydroxysteroid dehydrogenase inhibitor in late pregnant sheep. Can J Physiol Pharmacol 63: 136-142

Leadon DP, Rossdale PD, Jeffcott LB, Allen WE (1982) A comparison of agents for inducing parturition in mares in the pre-viable and premature periods of gestation. J Reprod Fertil [Suppl] 32: 597-602

Liggins GC (1976) Adrenocortical-related maturational events in the fetus. Am J Obstet Gynecol 126: 931-939

Lohse JK, First NL (1981) Development of the porcine fetal adrenal in late gestation. Biol Reprod 25: 181-190

Ousey JC, Dudan F, Rossdale PD (1984) Preliminary studies of mammary secretions in the mare to assess fetal readiness for birth. Equine Vet J 16: 259-263

Ousey JC, Rossdale PD, Cash RSG, Worthy K (1987) Plasma concentrations of progestagen, estrone sulphate and prolactin levels in pregnant mares subject to natural challenge with EHVI. J Reprod Fertil [Suppl] 35: 519-528

Pashen RL (1984) Maternal and fetal endocrinology during late pregnancy and parturition in the mare. Equine Vet J 16: 233-238

Randall GCB, Kendall JZ, Tsang BK, Taverne MAM (1984) Role of the fetal adrenal in the timing of parturition in the pig. Proc Soc 10th Internat Congr Anim Reprod and Artific Insem 2: 108-110. Cited in: Tumbleson ME (ed) (1986) Swine in biomedical research. Plenum, New York, pp 1179-1185

Rossdale PD, Silver M (1982) The concept of readiness for birth. J Reprod Fertil [Suppl] 32: 507-510

Rossdale PD, Silver M, Ellis L, Frauenfelder H (1982) Response of the adrenal cortex to tetracosactrin (ACTH) in premature and full term foals. J Reprod Fertil [Suppl] 32: 545-553

Silver M (1988) Effects on maternal and fetal steroid concentrations of induction of parturition in the sheep by inhibition of 3β-hydroxysteroid dehydrogenase. J Reprod Fertil 82: 457–465

Silver M, Fowden AL (1982) Uterine prostaglandin F metabolite production in relation to glucose availability in late pregnancy: possible influence of diet on time of delivery in the mare. J Reprod Fertil [Suppl] 32: 511–519

Silver M, Fowden AL (1987) Inhibition of 3β-hydroxysteroid dehydrogenase in the sow near term: effects on fetal and maternal steroids and on delivery. Q J Exp Physiol (in press)

Silver M, Barnes RJ, Comline RS, Fowden AL, Clover L, Mitchell MD (1979) Prostaglandins in the foetal pig and prepartum endocrine changes in mother and foetus. Anim Reprod Sci 2: 305–322

Silver M, Comline RS, Fowden AL (1983) Fetal and maternal endocrine changes during the induction of parturition with the PGF analogue, cloprostenol, in chronically catheterized sows and fetuses. J Dev Physiol 5: 307–321

Stryker JL, Dzuik PJ (1975) Effects of fetal decapitation on fetal development, parturition and lactation in pigs. J Anim Sci 40: 282–287

Taverne M, Naaktgeboren C, Elsaesser F, Forsling M, Weyden WGL, Ellendorff F, Smidt D (1979) Myometrial electrical activity and plasma concentrations of progesterone estrogens and oxytocin during late pregnancy and parturition in the pig. Biol Reprod 21: 1125–1134

Taylor MJ, Webb R, Mitchell MD, Robinson JS (1982) Effect of progesterone withdrawal in sheep during late pregnancy. J Endocrinol 92: 85–93

Thorburn GD, Challis JRC, Currie WB (1977) Control of Parturition in domestic animals. Biol Reprod 16: 18–27

Subject Index

W. Künzel, H. Gips, University Gießen
(Hrsg.)

Gießener Gynäkologische Fortbildung 1987

XV. Fortbildungskurs für Fachärzte der Frauenheil-
kunde und Geburtshilfe

1988. 79 Abbildungen, 146 Tabellen.
Etwa 280 Seiten.
ISBN 3-540-17851-1

Das Buch enthält die Vorträge des XV. Gießener
Fortbildungskurses für Fachärzte der Frauenheil-
kunde und Geburtshilfe. Das *erste Thema* ist der
Kinder- und Jugendgynäkologie gewidmet, ein
Gebiet, dem in Zukunft mehr Aufmerksamkeit als
bisher geschenkt werden sollte.
Das *zweite Thema* beschäftigt sich mit der Diagnostik
und Therapie der Sterilität.
Das *dritte Theama* behandelt gynäkologisch-opera-
tive und diagnostische Fragen. Ist die Hysterektomie
eine Methode, von der großzügig Gebrauch
gemacht werden kann oder gibt es Einschränkun-
gen? Was bietet die Ultraschalluntersuchung für
Möglichkeiten in der Erkennung gynäkologischer
Tumoren? Gibt es neue Wege in der Behandlung
des Korpuskarzinoms und Ovarialkarzinoms?
Der *vierte Themenbereich* beschäftigt sich mit Fragen
aus der täglichen Praxis, wie perioperative Antibioti-
kaprophylaxe, Nebenwirkungen von Kontrazeptiva
und das Osteoporoserisiko in der Postmenopause.
Im *fünften Themenkreis* schließlich werden geburts-
hilfliche Fragen besprochen, so z.B. die Übertragung
der Schwangerschaft, die Einleitung der Geburt mit
Prostaglandinen, die Infektionen während der
Schwangerschaft und deren Prophylaxe und Therapie sowie diagnostische und therapeutische Fragen
zur Frühschwangerschaft.

Springer-Verlag
Berlin Heidelberg New York
London Paris Tokyo

W. Künzel (Ed.)

Fetal Heart Rate Monitoring

Clinical Practice and Pathophysiology

1985. 157 figures. XII, 244 pages.
ISBN 3-540-15313-6

In this book, physiologists and obstetricians active in experimental studies of fetal heart performance address themselves to current problems in cardiac monitoring. The book opens with an historical survey, followed by descriptions of the techiques of cardiac monitoring as practiced in the USA, Great Britain and Germany. The physiologic regulatory mechanisms underlying fetal heart rate are used as the basis for the interpretation of pathologic patterns.

Springer-Verlag
Berlin Heidelberg New York
London Paris Tokyo

Springer